MW01625930

Mineral Property Evaluation

Handbook for Feasibility Studies and Due Diligence

Senior Editor: Richard L. Bullock
Associate Editor: Scott Mernitz

PUBLISHED BY THE
SOCIETY FOR MINING, METALLURGY & EXPLORATION

Society for Mining, Metallurgy & Exploration (SME)
12999 E. Adam Aircraft Circle
Englewood, Colorado, USA 80112
(303) 948-4200 / (800) 763-3132
www.smenet.org

The Society for Mining, Metallurgy & Exploration (SME) is a professional society whose more than 15,000 members represents all professionals serving the minerals industry in more than 100 countries. SME members include engineers, geologists, metallurgists, educators, students and researchers. SME advances the worldwide mining and underground construction community through information exchange and professional development.

ISBN 978-0-87335-445-5
eBook 978-0-87335-447-9

Library of Congress Cataloging-in-Publication Data

Names: Bullock, Richard Lee, 1929- editor. | Mernitz, Scott, editor.
Title: Mineral property evaluation : handbook for feasibility studies and due diligence / senior editor, Richard L. Bullock ; associate editor, Scott Mernitz.
Description: Englewood, Colorado, USA : Society for Mining, Metallurgy & Exploration (SME), [2018] | Includes bibliographical references and index. Identifiers: LCCN 2017037073 (print) | LCCN 2017045780 (ebook) | ISBN 9780873354479 | ISBN 9780873354455 | ISBN 9780873354479 (ebook)
Subjects: LCSH: Mine valuation.
Classification: LCC TN272 (ebook) | LCC TN272 .M655 2018 (print) | DDC 333.33/9--dc23
LC record available at https://lccn.loc.gov/2017037073

Contents

Preface

Having prepared numerous mineral property feasibility and evaluation studies as well as due diligence reports over many years, it is quite obvious to the authors of this handbook, and to many others to whom we have spoken, that there is a need for more guidance in the mining industry regarding these types of studies and reports. This handbook provides such tools in the form of general guidelines to follow in (a) performing these studies and (b) preparing proper documents for "bankable" presentations, that is, those that will be accepted by financial institutions as having adequate detail for fine-point evaluation of the financial risks to a project.

As is industry practice, many of the smaller companies try to perform some of the early stages of feasibility and evaluation of their mineral properties in-house and then turn to consulting firms to produce the additional pieces for what they expect to be a bankable study. Then these companies are often very surprised at the amount of work that still needs to be accomplished—beyond what they have already done—before a report can be used as bankable. Many of the larger companies, performing these studies themselves or contracting the study, find that (after the fact) many of their projects that become operating mines simply do not yield the return on the investment that was projected at the time of the evaluation study. In fact, it is estimated that less than 30% of the projects that are developed in the mineral industry yield the return on investment that was projected from the project feasibility studies. Some of the tools described in this handbook will greatly improve the probability of meeting those projections and minimizing project execution capital cost "blowout" that have become so prevalent in this industry in recent years.

The great work of various organizations to produce the Australasian VALMIN Code, the South African JORC Code, the Canadian National Instrument 43-101 requirements, and the U.S. Securities and Exchange Commission's modification of Industry Guide 7 have all contributed to the much-improved standards of defining the resources and the reserves and to minimize fraud in these areas. *But why do they offer only a few sentences or paragraphs of guidance* when it comes to the rest of what goes into the engineering of the feasibility studies and offer *no minimum standards*?

Finally, to its credit, the Society for Mining, Metallurgy & Exploration has set some minimum standards with its new recommendations in *The SME Guide for Reporting Exploration Information, Mineral Resources, and Mineral Reserves* (2017 edition) as to the amount of engineering that should go into engineering feasibility studies at the three levels as well as the expected accuracy of the cost estimates and the amount of contingency that should be used. This is a great step in the right direction that hopefully will start a trend in the codification of these standards in the codes of the various countries.

In the past, the primary reasons for the poor performance of mineral properties compared to projections were

- Improper identification of problems related to the minable reserve;
- Errors of projection of the recoveries, either in mining or metallurgy;
- Underestimating dilution (loss of recovered value) during the mining process;
- Insufficient engineering, geologic studies, and/or design in the feasibility studies;
- Omitting or underestimating large items of operating and/or capital cost;
- Failing to completely define the real problems of constructing and operating the facilities to produce the products of the property; and
- Lack of understanding of the property rank competitively with the rest of the world.

Several books have been written in recent years on performing the financial analysis of mineral property evaluation, and most concentrate on the economic techniques of evaluation. None of these books, however, cover more than about 10% to 20% of the work that must to be completed prior to performing the proper financial analysis of a mineral property. Nor do any of them describe what is done or what to expect when a due diligence study is performed in checking many of the findings of a feasibility study, completed either in-house or by a consultant.

To change the way in which feasibility studies are performed, this handbook will serve a real purpose in helping to train students, young professionals, and even those in their mid-careers within the mining industry in all that must be considered for a full-blown mineral property evaluation and feasibility study. In recent years, there have been hundreds of major mineral project cost overruns, some amounting to several billion dollars. It is hoped that if the methodology is followed that is presented in this handbook, these overruns will be mitigated, or at least minimized.

Just as a mineral property evaluation project team may be comprised of several disciplines, so are the writers of this text. The intent is to give each discipline the opportunity to describe those issues that must be considered and understood about that discipline that will affect the eventual success or failure of the mineral development should it continue into the exploitation phase of property development. The information is presented so that it can be understood by the average mining professional, no matter what his or her specialty may be. Likewise, most of the text as it is presented will be fully understood by those interested parties in the financial institutions dealing with mineral property financing. Certainly, it will give the interested person a guideline, or checklist, to compare against what is or should be in a report, and at least recognize what was omitted, so that all the pieces can be gathered to present a complete picture.

Most of this handbook was written by Behre Dolbear associates and is based on the years of experience of the authors, each of which has at least 15 years' experience performing this type of work with the company. But that does not mean they are necessarily expressing a policy of Behre Dolbear; rather it is the opinion of the writer based on his or her experience.

This handbook should be used by those involved in many areas of the mining industry:

- Professionals in the minerals industry that perform mineral property evaluations, not only within the United States and Canada but overseas as well. (The only countries in which these principles may not apply are those where the governments get involved and mines are developed that do not necessarily require economic viability for that particular operation.)

- Companies that have mineral properties and perform mineral property feasibility studies and evaluations or are buying properties based on property evaluation.
- Financial institutions, both domestic and overseas, that finance or raise capital for the minerals industry.
- Consulting firms and A/E (architectural and engineering) contractors that get involved with mineral property feasibility studies and need standards to follow (though they might be the last to admit that they need such a book).
- Probably the most important are the mining and geological engineering students and the geology and economic geology students that need course textbooks for topics that are included within the handbook. This is the opportunity to teach them the standards of the industry that they should follow throughout their careers. A textbook of this kind has been lacking over the years and is greatly needed.

Acknowledgments

This handbook was conceived more than a decade ago and had the early endorsement and encouragement of the management of Behre Dolbear. Without the dedication of its organization of associates and staff, this volume would have never been started, let alone completed. The experience of the authors on hundreds of projects, many of which were under the direction of Behre Dolbear's policies and guidelines, forms the foundation for this project and, consequently, the knowledge and experience that is presented in many of its chapters. I acknowledge with gratitude the many Behre Dolbear associates who peer reviewed various chapters: David M. Abbott Jr., Mark A. Anderson, Amy E. Jacobsen, Bernard Guarnera, Mike Martin, Scott Mernitz, and Nina Rice.

Thanks are conveyed to Roderick G. Eggert, professor and Viola Vestal Coulter Foundation chair in Mineral Economics at the Colorado School of Mines, for his marketing contribution.

Also expressed is sincere appreciation to the management of InfoMine USA for their generous contribution of examples of cost models referred to or used in the three cost estimating chapters.

The senior editor also acknowledges the unique experience that he was afforded in developing methods and techniques of mine project evaluation through his work assignments in St. Joe Minerals Corporation and Exxon Minerals Company. Of the more than 250 properties that he has been involved, approximately 90% of them were with these two mining companies.

About the Editors

Richard L. Bullock, D.Eng. (now retired), was an independent mining engineer consultant and also a senior associate serving with Behre Dolbear for 18 years. Bullock also served as the Robert Quenon Endowed Chair and professor of mining engineering, teaching numerous courses for his alma mater, Missouri University of Science and Technology (Missouri S&T) for six years and, as professor emeritus, teaching online courses for 10 years. He has more than 50 years of experience in the mining industry, having managed mineral property feasibility evaluations, mine developments and projects, ongoing mining operations, mining research, and multidisciplinary engineering design groups. Recognized for his strong technical, economic, and operating background, Bullock has been called upon during his career to perform various engineering studies, mineral property feasibility evaluations, and due diligence studies for more than 250 prospective or operating properties. These facilities included underground, open pit, and placer properties in the United States, Canada, Mexico, Panama, Honduras, Brazil, Chile, Peru, Bolivia, Argentina, Germany, Ireland, Great Britain, Australia, and China, and included commodities of lead, zinc, iron, copper, tungsten, uranium, gold, silver, platinum, palladium, molybdenum, coal, ilmenite/rutile, chromite sands, fluorspar, trona, and borates.

During his professional career, Bullock gained hands-on experience in every operating mining position, from mine shift foreman to division manager and corporate administrative positions of director of mine development and research, vice president of engineering and research, and project executive of an 80,000-tpd copper project in Chile. He has served on the corporate operating committees of St. Joe Minerals Corporation, A.T. Massey Coal, Tennessee Consolidated Coal, and Compañía Minera Disputada de Las Condes in Chile. Bullock currently serves on the board of directors for Purcell Tire and Rubber Company.

Bullock earned bachelor of science, master of science, and doctorate of engineering degrees. He is a registered professional engineer in Missouri, Nevada, New York, and Tennessee and is an internationally certified qualified professional in mining and ore reserves evaluations from the Mining and Metallurgical Society of America.

Bullock is a distinguished member of SME and has achieved Legion of Honor status. He is an elected member of the Missouri S&T Mines and Metallurgy Academy. Bullock has been honored with many awards and has been recognized for notable industry achievements. He received SME's prestigious Daniel C. Jackling Award in 2011, the SME–Pittsburgh Section and Pittsburgh Coal Mining Institute of America's Stephen McCann Memorial Educational Excellence Award in 2002, and the Distinguished Service Award from SME's Mining & Exploration Division in 1999. He received the Alumni Merit Award, eight individual teaching awards, and a professional engineer of mines degree from Missouri S&T.

Importantly for the industry, Bullock was co-editor of two SME underground mining books, *Techniques in Underground Mining* (1998) and *Underground Mining Methods* (2001). He was also a valuable contributor to the third edition of the *SME Mining Engineering Handbook* (2011), organizing and soliciting chapters for four sections and writing six chapters. Collectively, these books as well this current handbook on feasibility studies have captured much of his lifelong experience and knowledge for future mineral industry engineers.

Scott Mernitz, Ph.D., is an independent environmental due diligence consultant for the minerals industry, governments, and financial institutions. He evaluates the risks to investment from environmental, social, and community perspectives for mining projects at various stages of development. Following initial work for state environmental agencies in Wisconsin and Colorado, he has consulted in affiliation with engineering and environmental consulting firms for more than 30 years on the impacts of mining and natural resources development projects to the human, biological, and physical environment. His experiences over the past 20 years have been primarily with Behre Dolbear, headquartered in Denver, Colorado, as a senior associate, where he was involved with numerous and varied precious and base metals, industrial minerals, and energy projects in locations spanning North and South America, Central America, the Caribbean, Africa, the Middle East, and Australia.

Regarding notable achievements, Mernitz is a qualified professional of the Mining and Metallurgical Society of America and a published author in his field. He has authored journal articles, conference presentations, and a 200-page textbook titled *Mediation of Environmental Disputes: A Sourcebook* (New York: Praeger, 1980). In addition, he wrote two chapters in this current book and also assisted with author relationships and editing of other chapters.

Mernitz earned a B.A. in geography and history from Elmhurst College, Illinois; an M.A. in physical geography (earth sciences), cultural and historical geography from the University of Colorado–Boulder with a thesis on the impacts of coal mining near Boulder; and a Ph.D. from the University of Wisconsin–Madison in the interdisciplinary Land Resources Degree Program at the Gaylord Nelson Institute for Environmental Studies, with a dissertation on the potential for mediation of environmental disputes in 1978.

Contributing Authors

David M. Abbott Jr.

Principal and Senior Associate for Behre Dolbear and Former U.S. Securities and Exchange Commission Geologist

Mark A. Anderson

Senior Associate and Former Member of the Board of Directors of Behre Dolbear

Richard L. Bullock

Professor Emeritus, Department of Mining and Nuclear Engineering at Missouri University of Science and Technology, and Former Senior Associate of Behre Dolbear

John T. Crawford

Senior Associate of Behre Dolbear

Qingping Deng

Former Vice President and Senior Associate of Behre Dolbear

Roderick G. Eggert

Professor and Viola Vestal Coulter Foundation Chair in Mineral Economics, Division of Economics and Business at Colorado School of Mines, and Deputy Director, Critical Materials Institute, an Energy Innovation Hub (research consortium) funded by the U.S. Department of Energy

Amy E. Jacobsen

Senior Associate and Chairwoman, Behre Dolbear Board of Directors

Rachal H. Lewis Jr.

Senior Associate of Behre Dolbear

Scott Mernitz

Senior Associate–Environmental of Behre Dolbear

CHAPTER 1

Introduction to Mineral Property Feasibility Reporting

Richard L. Bullock

Believe it or not, mine feasibility studies are as old as the industry itself. In the first recorded writing on mining by Georgius Agricola (1556), he gave many clues about what to look for in evaluating a mine. For example, he said (with a bit of paraphrasing in braces):

> *Now a miner, before he begins to mine the veins, must consider seven things, namely: the situation* [the geologic/mineralogical setting], the conditions [the geotechnical mining conditions], *the water* [the area hydrology], *the roads,* [the infrastructure], *the climate* [the weather and the environment], *the right of ownership* [the land and legal situation of the property], *and the neighbors* [the socioeconomic conditions of the area].

Agricola then proceeds to elaborate on what *he* means by "situations," "conditions," and so on, which the authors of this text liberally translated to modern-day meaning. Putting it in this context, one begins to realize that the miner has always had to evaluate many things to determine if a mine was really feasible. Today, we study all of the same things, plus a few more, but we do it in a much more systematic method.

Skipping ahead some 350 years, another famous miner, Herbert C. Hoover, spoke out on mine valuation (Hoover 1909):

> *It is a knotty problem to value the extension of a deposit beyond a short distance from the last opening. For a short distance it is proved ore and for a further short distance is probable ore. Mines are very seldom priced at a sum so moderate as that represented by the profit to be won from the ore insight, and what value should be assigned to this unknown portion of the deposit admits no certainty. No engineer can approach the prospective value of a mine with optimism, yet the mining industry would be non-existent today were it approached with pessimism.*

Hoover's place, and his early influence on our industry and mine valuation, is elaborated upon in Chapters 2 and 3 when ore reserves and resource classifications are discussed.

This introductory chapter introduces a few of the main topics that are discussed throughout this handbook. It identifies the issues to be considered and how they are interrelated, and it should indicate why the successful conclusion of any property evaluation is dependent upon the development, work, and conclusions of the project team.

Most mineral engineers, geologists, mineral company executives, and mineral development lending agencies think of the feasibility study concept as the formal methodology that brings the necessary information about the raw mineralogical data from a property, through the feasibility and preliminary design process, all the way to the point where a comparable economic analysis of the envisioned project can demonstrate financial viability. But the feasibility studies have to be completed by

- Many different people;
- On different commodities;
- For many different types of mines and process plants;
- In many different climatic, political, and social environments that take many different forms;
- And yet, in the end, all of the feasibility studies must accomplish the same thing: demonstrate comparable financial opportunity of investment potential.

The forgoing discussions focus our attention to the fact that there does need to be a consistent, systematic methodology in performing evaluation and feasibility work, which the authors of this handbook believe is possible.

EVALUATION VERSUS VALUATION: WHAT IS THE DIFFERENCE?

This is probably a good place to acquaint the reader with the differences between the main subject of this handbook, *mineral property feasibility and evaluation*, and that of a *mineral property valuation*. Both types of studies examine the property variables: geologic-, mining-, processing-, marketing-, social-, and environmental-related considerations and how they affect profitability and worth of the property.

The objective of a feasibility and evaluation study should be to develop the value of the undeveloped or developed mineral property to the company that is considering applying technical and physical changes to the property to bring it into production of a mineral product. The analysis needs to determine the net present worth returned to the company for investing in these changes and to reach that decision point as early as possible, and with the least amount of money spent on the evaluation study. The results of determining this value can take several forms, which could indicate

- Moving to the next phase of the study,
- Going forward with the development of the property,
- Putting the project on hold until more information (geologic, metallurgical, etc.) is obtained or there is an economic market change,
- Beginning to look for a joint venture partner,
- Trying to sell the property, or
- Cutting losses and walking away from the property.

The objective of a valuation study determines the worth of the property, considering all of the property variables listed above, and then determines the worth of the asset to the company, considering two major factors (Guarnera and Martin 2011):

1. **The highest and best use of the mineral property as it exists.** The property may or may not have value just because it has mineral; or even if it has mineral of value, there may be a use for the property that has a higher value then the extraction of the mineral. Developing the property for real estate is an example.
2. **The fair market value (FMV) of the asset.** The FMV is "the price an asset would be exchanged for with the parties being a willing buyer and seller, with both parties having access to the same information about the asset, and with neither party being under compulsion to buy or sell the asset" (Guarnera and Martin 2011)."

Both valuations and evaluations are extremely important in the industry, but this handbook only covers the recommended procedures for evaluations. Readers who wish to learn more about the valuation of mineral properties are referred to Chapter 4.6 of the *SME Mining Engineering Handbook* (Guarnera and Martin 2011).

WHO SHOULD PERFORM THE MINERAL PROPERTY FEASIBILITY AND EVALUATION STUDY?

The first matter to focus on is who should do the feasibility study. Some might argue that the very people who found the resource obviously know the most about it; therefore, those within the exploration group should perform the early feasibility study and even the early bulk sampling and/or test mine. These individuals argue that they already have an organization in place within that area or country, so it makes sense to continue their work in the feasibility mode instead of exploration elsewhere. Furthermore, these explorationists have demonstrated their ability as "ore finders," and they should be rewarded the next opportunity to demonstrate their ability as "mine builders." To admit that they have a vested and biased interest in the success of the property becoming a "world class mine" would be putting it mildly. There is absolutely no way that the discoverer of a new mineral resource can look on the outcome of that resource during a feasibility study with totally unbiased feelings any more than a mother could sit on the jury trial of her own child. This is not to say that the exploration group should not have a very large part of the early input. But from that point on, the project team must be organized with persons of unbiased, multidisciplinary thinking. The subjects of building this multidisciplinary project team are discussed later, but for now, consider the organization that should have the responsibility of performing all of the project feasibility/evaluation studies. Sometimes this group is referred to as project development or mine evaluation and development.

This single multidisciplinary organization should be assembled within a medium- to large-size mineral company that has the responsibility for mineral property feasibility. It should perform or supervise consulting organizations performing all evaluation studies of mineral deposits and mineral processing facilities for projects discovered or acquired, whether the projects are located in the country of the home office or any other nation. Project acquisitions, joint ventures, or project expansions may need special treatment, but this is also discussed later.

Assignment of evaluation studies of all types to a centrally headquartered organization has the following advantages:

- It ensures that all of the projects are studied and evaluated in exactly the same manner.
- It makes sure that all projects will have people of specific disciplines available to work on every aspect of the projects to be studied; thus, each phase will be technically evaluated properly for that depth of study.

- It ensures centralized project planning and scheduling.
- It provides an experience-based group to consider the results of all candidate projects.

More on how the project teams should be organized is discussed in Chapter 11.

WHEN MIGHT AN EVALUATION STUDY NEED TO BE PERFORMED?

There are several reasons why an evaluation may need to be performed on a mineral property. Let us assume that you, the readers of this handbook, are employees of a large, multinational minerals corporation and work within its project evaluation and development group. With that in mind, you might be called upon to do your job when one of the following situations exist:

- The company is looking at the potential purchase of a new mineral property where someone claims to have an ore-grade mineral reserve in place and now must know the price that the company can bid and still meet the desired return on its investment.
- Your exploration group brings you a property that they believe has ore-grade mineral that could be developed and you must determine the optimum value to the company.
- Your company is looking at buying a property that already has a mine/mill in place and it needs to know its value to the company.
- Your company is considering a change to the mining or milling method at an existing operating property and needs to determine whether it will yield a greater return.
- Your company is looking to expand one of its existing mines (or mills) and needs to determine which of the mining or milling systems proposed will yield the greatest return.
- Your company has found more mineral resource within the existing property boundaries and needs to know if it can be mined profitably.
- The company is looking toward starting a new mine and wants to know where the mine's productions costs will rank in comparison to the other major mines in the world of the same commodity.
- Your company is considering the purchase of an entire company that has many mines and needs to determine whether this acquisition meets the company's economic objectives.
- Another company has all of its mines for sale, and your company is trying to select the best one to buy and determine the price to bid for the property while maintaining the required economic return.
- Your company wants to evaluate a joint venture with another company and needs to determine how that will affect the return not only to the project, but to your company.
- The owner of an ore reserve wants to borrow money to develop the property and must demonstrate to the lender that the project will remain economically viable throughout the loan repayment period.

Obviously, the evaluations that need to be done under each of the preceding circumstances is not exactly the same type, but they have many of the same work procedures in common. For example, being able to accurately determine the mining cost is common to all of the evaluations.

STAGES OF A PROJECT

As typified by the hundreds of thousands of mineral projects that have been observed over the years, the following five stages of a project have been humorously identified (Laird 1997):

- Stage 1. Excitement, euphoria
- Stage 2. Disenchantment
- Stage 3. Search for the guilty
- Stage 4. Punishment of the innocent
- Stage 5. Distinction for the uninvolved

The purpose for studying this subject is to ensure that, throughout your careers, all of you will remain in the first stage and that distinction is appropriately granted to all of those who were involved. The stages listed above imply that if the results of a feasibility study show less than euphoria, then it is a failure, yet nothing could be further from the truth. Successful mining companies rely on the project team to develop the true potential for any mineral property. However, if that property is not one that is economically viable, the project team deserves credit for learning the truth, even when they may be under considerable duress from those who are promoting the project. The evaluation team should be acknowledged for making recommendations for other dispositions for the project at that point in time, before any more company money is spent on the property.

Because mining is a business that is constantly depleting assets, mining companies must likewise be constantly increasing their mineral reserve assets, either through exploration or acquisition, or both. This generally means that most successful companies will have numerous potential prospects that they are considering, either from a raw exploration point of view or through acquisitions. Thus, a growing mining company might have as many as six to twelve active projects in its portfolio at various stages of exploration, evaluation, and development.

Given that the company may have several projects that must be evaluated, the prospects

- May be for different commodities,
- Will likely have different individuals performing the evaluations,
- Will probably be starting at different points in time,
- Will no doubt have unequal lives, and
- May be located in different countries.

Does this list alert you to a problem? It should, since the most important element in performing complete property evaluations for a company is that each evaluation should be done exactly like every other evaluation within that company. Therefore, there must be procedures in place that will require different evaluators to follow the same procedures on different properties, and develop equivalent feasibility studies that can be compared. *This is the single most important principle that must be faithfully followed by any company doing property evaluations.* Likewise, it would be helpful to investment houses if all of their potential clients had projects with equivalent feasibility studies that were somewhat comparable, at least in respect to completeness.

This is one of the primary purposes of this handbook: to instill in each reader the concept that there must be a very regimented method in performing complete property evaluations that lead to feasibility reports. When dealing with a new property, the following steps are the customary and proper approach to correctly bring the property from a prospect to a money-making operation or to reject the property early on from further company expenditures:

1. Exploration and land/water position control
2. Feasibility with evaluation studies
3. Appropriation and financing
4. Design basis established, followed by design
5. Construction of facilities
6. Development of the mine and process plant
7. Startup of operations
8. Production buildup to full production
9. Closure and reclamation

These steps may tend to overlap somewhat, with some activities proceeding simultaneously. Each step is important to the success of a project, but the front-end feasibility/evaluation step is particularly critical because it will establish the bases for the design, construction, and production steps.

Most of the steps listed previously are topics for which at least one book could be developed (i.e., they are not simple subjects that can be learned overnight). But the focus of this handbook—consideration and methods of mine feasibility studies—is covered within the following chapters.

WHAT MUST BE CONSIDERED FOR A PROPER FEASIBILITY STUDY?

Everything sums up what must be considered for a properly documented property evaluation. But just saying "everything" doesn't really help you know how to start and what to look for. Hustrulid and Kuchta (1995) try to get the evaluation process started by providing a list of everything that the people of Kennecott could think of that should be studied for a potential open pit mine. That list is not repeated here, though it is a good checklist for an open pit mine evaluation. There are a few problems, however, with the inexperienced in working with such a list. In the first place, it is not necessarily presented in a working order that things should be done. The other problem with such a list is that the beginner has no clue as to how much work should be done on each subject at the beginning of the project, during the project, or at the end of it. Consequently, a detailed list is given in the appendixes to Chapter 11 (Appendixes 11A, 11B, 11C, and 11D) to give the correct sequence of study and what should be included.

WHAT SHOULD BE THE OBJECTIVES OF A COMPLETE MINERAL PROPERTY FEASIBILITY STUDY?

One of the first concepts is to learn the purpose of the mineral property feasibility study. The following statement, or something very similar, is often quoted as the purpose: "The feasibility study has one primary goal: to demonstrate that the project is economically viable if it is designed, constructed and operated in accordance with the concepts set forth in the study" (Laird 1997). But is that the true purpose of the feasibility study? What if the project is not viable under any design conditions?

It would be absolutely fantastic if all the projects that were turned over to a project team were worthy of being designed, constructed, and operated. Unfortunately, many feasibility projects get evaluated at an early stage and are often found to be not worthy of any further study, let alone full development. On the other hand, the evaluator must always take the approach that the project is economically viable when he or she first starts. It is only after a systematic, engineered approach to evaluation of the property that the evaluator can determine whether or not the project is economically and environmentally feasible, and if it is not, what it would take to make it so. Some mineral projects cannot be made to be economically viable at that point in time, no matter how brilliant the engineers and geologists are, and in spite of the amount of money that is applied to it. It is as simple as: all resources are not reserves, nor are all minerals an ore. Yet the above-stated objective is often the premise that gets projects into trouble before they start.

So what should be the objective of a mineral property feasibility study? Obviously, the objective should be to maximize the value of the property to the company either by exploiting it, selling it, or doing nothing. But the objective should also be to reach that decision point as early as possible, with the least amount of money spent on the evaluation.

But how can you do this? How do you know when you have studied each of the hundreds of items of information enough so that you have confidence in the feasibility study, and the economic analysis based on that study, that have been assembled?

You learn to do it by a phased approach to mine evaluation. Several books, and in fact most mineral companies, take a similar approach to mineral property evaluation.

THE PHASED APPROACH TO MINERAL PROPERTY FEASIBILITY

Several individuals who have written about performing proper feasibility studies have treated the activities as only one continuous process, from the time that the resource was identified until a decision could be made to develop the property. This one-step approach to feasibility leading directly to development may sometimes be the correct approach with extremely high-grade ore bodies or if the company requires mandatory development for some reason. But the one-step approach is very risky from a technical point of view. It is the opinion of the authors that such methods might very well develop a suboptimal operation, even though it could still meet the company's needs. Furthermore, it may cost the company far too much money to find out that the project economics prove inadequate if this approach is taken.

Most companies, and books on the subject, recommend a phased approach to mineral property evaluation. The most commonly used terms to describe the phases of feasibility in recent years are those adopted by the Canadian National Instrument (NI) 43-101 regulation, which uses *preliminary*, *prefeasibility*, and *feasibility* to define the three phases. In many places in this handbook, we have added these terms in parentheses to our preferred usage of terms. To cite a few other terminology examples, Table 1.1 was compiled to illustrate these various names used since the mid-1970s. The bottom line is that someone could call his study almost anything, but the investor did not know what was really implied by the name.

Likewise, Hustrulid and Kuchta (1995) describe a three-phase system as *conceptual study*, *preliminary study*, and *feasibility study*. They later referred to the second study as an "intermediate valuation report" and defined some of the content details of the intermediate and feasibility reports as originally described by Taylor (1977). The authors of this handbook agree with the concepts as laid out by Taylor. Totally independent of his work, one of this handbook's

TABLE 1.1 Equivalence of feasibility study terminology

Reference	Date	Level 1	Level 2	Level 3	Level 4
Taylor	1977	Preliminary	Intermediate	Feasibility	
Hustrulid and Kuchta	1995	Conceptual	Preliminary	Feasibility	
Barnes (AusIMM)	1997	Resource Calculations	Preliminary Evaluation	The Feasibility	
Barnes (AusIMM)	1997	Preliminary	Indicative	Definitive	
Barnes (AusIMM)	1997	Scoping	Prefeasibility	Bankable	Basic Engineering
Noort and Adam (AusIMM)	2006	Scoping	Prefeasibility	Definitive	
Bullock (SME)	2011	**Preliminary**	**Intermediate**	**Final**	Basic Engineering
NI 43-101	2000	Preliminary	Prefeasibility	Feasibility	Basic Engineering
NI 43-101	2011	Preliminary Economic Assessment	Prefeasibility	Feasibility	Basic Engineering
Hickson and Owen (SME)	2015	Scoping	Prefeasibility	Feasibility	

authors developed a very similar approach to mine evaluation, which is detailed at length in Chapter 11. The authors of the terms given in Table 1.1 still use them because they indicate to the investor what phase of the feasibility study has been completed, which relates to the amount of engineering that has been accomplished and therefore represents the accuracy of the forecast return. However, the terms of NI 43-101 have become the most commonly used in North and South America.

When writing about the methods of Cyprus Australia Coal Company, Stone (1997) describes a five-phase study that he classifies as *preliminary screening*, *preliminary and site visit*, *secondary assessment*, *prefeasibility study or detailed assessment*, and *feasibility study or final acquisition study*. The accuracy of the study at each phase improves, of course, as more engineering effort goes into the study. It moves from ±33% in the second lowest phase to ±5% in the final phase. This increase of estimate accuracy is certainly common to all of the systems.

What you will learn in this handbook is a three-phased approach to conducting a mine feasibility study. Obviously, this is not the only way to do it, but the authors believe that it is the safest and most prudent method. As different situations arise on various commodities, you and your company may believe that steps can and should be skipped. This may be true, but be aware of the potential consequences where you are taking shortcuts, particularly if your company's experience is weak in that area of the new project.

The three steps for conducting feasibility studies recommended in this handbook are simply

1. Preliminary feasibility,
2. Intermediate feasibility (or prefeasibility), and
3. Final feasibility.

Although these may seem a lot like some of the systems referred to earlier in this discussion, they are not the same. Learning what is in these three studies and how to apply the work from one level of effort to the next are important parts of this handbook.

Another important part of the concept is to learn how to apply controls to portions of the study so you can prepare cost estimates and schedules to the planning of your work and follow the cost of doing your work completely throughout the project. To do this, you must first organize a list of subjects to be studied into work categories and assign numbers to them. This is known as a work breakdown structure, or WBS. No two people will develop a WBS exactly alike, and that is okay. The important point is to get the work organized so that you can track it, both from an accounting and schedule point of view, and do it on a computer.

Chapters 2 and 3 deal with all aspects of examining the discovered geologic resource. Chapter 2 starts from what information should be transferred from the exploration group to the project team and then describes the work of examining the initial sampling that was done and the additional sampling that needs to be done, the resource and reserve calculations and classifications, the geology and geostatistics that need to be applied to those reserve calculations, tonnage factors, cut-off grades, and dilution and deletion of ore. All of this finally leads to producing an acceptable geologic models of the resource or reserve in Chapter 3.

How to approach the mine planning of the project, whether it is a surface or underground operation, and how to choose the mining method, size the mine, select the equipment, and then develop the personnel levels are discussed in Chapters 4, 5, and 6. Also covered is a discussion on the importance of data gathering and geotechnical testing for the mine design.

Likewise, selecting the mineral processing system for the project—whether it is simply crushing and screening or a full-blown flotation or oxidation/leaching process that must be used to extract the valuable mineral—and valuing the importance of metallurgical testing are discussed in Chapter 7.

Chapter 8 covers what is probably the most overlooked and understudied item of all the early phases of the feasibility study: marketing. The methodologies and strategies used for the different commodities of the mineral industry are covered. Also, the importance of building a cost seriatim of competitive producers and placing your project in that seriatim are discussed. It has become obvious to some that this recent extended downturn in the minerals commodity market proves this point; you must be in the lower quartile of product cost seriatim to survive profitability for such an extended period. This view was expressed recently in a presentation titled Staying Competitive in a Low Price Global Market by W.P. Goranson (2016).

Environmental and sustainability considerations that must be addressed during all phases of the feasibility study as well as how they will be carried through the operating years and on into closure and reclamation are discussed in Chapters 9 and 10, respectively. These chapters cover how it all starts with what you put into your feasibility reports, which then feeds into the environmental permit applications. Very early in the project, you must consider how and when to involve the stakeholders in the sharing of information and getting feedback to the project. The early phase involves looking for fatal environmental flaws that could stop the project, scoping out the agencies, scoping out the regulations, and understanding the social and environmental climate of the area in question. The mid-phase considerations entail organizing the baseline studies, developing environmental plans, developing early plans to mitigate any identified problems, filing for long lead time permits, and developing preliminary mine/plant closure and rehabilitation plans. The final phase considers the filing for short lead time permits, initiating any studies required to mitigate environmental problems that may have been uncovered, continued stakeholder involvement, preparing final mine/plant closure and reclamation plans for permit application, obtaining permits, preparing bonding, and implementing the environmental plans. In recent years, the social issues involved with starting a mining and metallurgical operation have come under severe pressure to assure the communities in the surrounding area that the operation will indeed improve the lives of those people already in the area. These issues involve sustainability and the social license to operate.

Chapter 11 describes all the activities at each level of feasibility study and explains how to move a project from an exploration project, through the feasibility phase, and then to engineering design, or whether to recommend the project to the "back burner."

Out of the final feasibility study will develop the design criteria that have been decided upon by the project team, the value of which was demonstrated by the economic analysis and approved by company executives. This design criteria will be turned into a design basis report (DBR). How to prepare such a document and what it must include is the subject of Chapter 12. It is this document that is used as the basis for the subsequent engineering design. Not only does the DBR contain the technical data and information decided on by the company during the final feasibility study, but it also encompasses the project execution plan for the contracting, building, and constructing of the project and also the operating plan, which will guide the engineers and builder to construct the mine/plant to be built so that the operating philosophy of the company can be quickly achieved and maintained. One of the key items in avoiding cost overruns is the development of the DBR. The DBR will identify many of the detailed activities of project execution and exploitation. Once it is completed, the contractor will bid or submit exceptions and by doing so will minimize project change orders (Bullock 2013).

How to build a project schedule and perform cost estimates for all of the functions that will be performed in the final constructed project is the subject of Chapters 13 through 16. To laypeople who are not familiar with building mines and metallurgical plants, it may seem that it always takes much too long and that cost overruns are always present in the building of mines and plants, but this should not be true. What is true is that the expectations based on most final feasibility studies are overly optimistic, and consequently, the project gets off to a bad start. Benchmark data from case histories are presented in Chapter 14 showing what should be expected. Different cost estimating methods and tools are discussed that are appropriate for the various levels of feasibility, which include the appropriate accuracies and contingencies.

Chapter 17 covers investment vulnerability and risk that must be considered for every project. With today's economic, social, political, and environmental conditions, which are in such a state of change, this aspect of mineral property assessment must be heavily considered and an investment vulnerability allowance added to the potential cost of building and operating the property in question. Specific investment risk issues and their impacts are included. The magnitude and frequency of recent massive blowout cost overruns illustrates how important it is to consider risk during the execution phase of the project, which is also covered in this chapter.

Covering the professional ethics involved and the liabilities that might occur while doing mineral property reserve and feasibility work is covered in Chapter 18. It is critical that everyone doing any aspect of this type of work understands the principles covered here.

The work that must be done to verify the authenticity of all the work covered in the preceding chapters (and more) is covered in Chapter 19 and is known as due diligence studies and reporting. In today's financial world, every company that seeks an outside source of financing to acquire or develop the mineral property must have a due diligence study on the findings of its evaluation as well as feasibility reporting, or what is being offered to that company to buy. These documents can take several different forms and can cover individual aspects to the property or everything that was done on it. The examinations can last a few days or take months to perform, but all approaches are discussed in the chapter.

Although there are many good books that have been written recently describing all of the methods of actually performing the financial analysis, no book of mineral property feasibility would be complete without covering the subject in enough detail to illustrate current good practice of the mineral industry. This is covered in Chapter 20 and illustrates the use of discounted cash-flow analysis, net present value, and internal rate of return; describes how to determine the cost of capital; and discusses the sensitivities that should be considered with a Monte Carlo simulation. Each of us who perform mineral property studies and appraisals and write reserve and feasibility reports subject ourselves to considerable liability. Understanding all aspects of this detail is very important to each of us as individuals and to every company performing any part of the work.

REFERENCES

Agricola, G. 1556. *De Re Metallica* (Translation by H.C. and L.H. Hoover, 1912), 1950 edition. New York: Dover.

Barnes, E., ed. 1997. Editor's note on feasibility. In *MINDEV 97*. Carlton, VIC: Australasian Institute of Mining and Metallurgy. p. 1.

Bullock, R.L. 2011. Mineral property feasibility studies. In *SME Mining Engineering Handbook*, 3rd ed. Edited by P. Darling. Littleton, CO: SME. pp. 227–261.

Bullock, R.L. 2013. Mine feasibility studies still need improving. Presented at Innovations in Mining Engineering Conference. Sept. 2013. Rolla, MO: Missouri University of Science and Technology.

Goranson, W.P. 2016. Staying competitive in a low price global market. Presented at MINExpo 2016, U.S. Mine Projects Session, Sept. 27, 2016. www.minexpo.com/education-sessions.

Guarnera, B.J., and Martin, M.D. 2011. Chapter 4.6, Valuation of mineral properties. In *SME Mining Engineering Handbook*, 3rd ed. Edited by P. Darling. Littleton, CO: SME.

Hickson, R.J., and Owen, T.L. 2015. *Project Management for Mining: Handbook for Delivering Project Success.* Englewood, CO: SME.

Hoover, H.C. 1909. *Principles of Mining, Valuation, Organization and Administration*, 1st ed. New York: McGraw-Hill.

Hustrulid, W., and Kuchta, M. 1995. *Open Pit Mine Planning and Design*, Vol. 1. Rotterdam/Brookfield: A.A. Balkema.

Laird, A.M. 1997. How to develop a project. In *MINDEV 97*. Carlton, VIC: Australasian Institute of Mining and Metallurgy. pp. 3–8.

NI 43-101. *Standards of Disclosure for Mineral Projects.* Canadian National Instrument 43-101. Ontario Securities Commission. www.osc.gov.on.ca/en/15019.htm.

Noort, D.J., and Adam, C. 2006. Effective mining project management systems. In *International Mine Management Conference: Proceedings.* Melbourne: The Australian Institute of Mining and Metallurgy. pp. 87–96.

Stone, I. 1997. Orebody definition and optimisation (coal). In *MINDEV 97*, Carlton, VIC: Australasian Institute of Mining and Metallurgy. pp. 39–45.

Taylor, H.K. 1977. Mine valuation and feasibility studies. In *Mineral Industry Costs.* Compiled by J.R. Hoskins and W.R. Green. Spokane, WA: Northwest Mining Association. pp. 1–17.

CHAPTER 2

Quality Assurance and Quality Control in Sampling and Sample Analysis

David M. Abbott Jr.

This chapter examines several topics related to the determination that a mineral resource exists, as defined by the internationally accepted Committee for Mineral Reserves International Reporting Standards (CRIRSCO) mineral resource and mineral reserve classification systems.* The following were CRIRSCO member organizations as of June 2017:

- Joint Ore Reserves Committee (JORC), Australasia
- Brazilian Commission for Resources and Reserves (CBRR), Brazil
- Canadian Institute of Mining, Metallurgy and Petroleum (CIM), Canada
- Comisión Minera, Chile
- Pan-European Regional Council (PERC), Europe
- KAZRC Association, Kazakhstan
- Mongolian Professional Institute of Geosciences and Mining (MPIGM), Mongolia
- National Association for Mineral Resources (NAEN), Russia
- SAMCODES Standards Committee (SSC), South Africa
- Society for Mining, Metallurgy & Exploration (SME), United States

Table 2.1 lists the more common code documents for mineral resource and mineral reserve classification systems. Although there are other code documents, most countries follow one of the codes listed in this table.

Other countries are considering joining this committee. The CRIRSCO members' reporting codes are used for international financing of mining ventures. For example, most North American mining firms are asked to submit National Instrument 43-101–type reports, which include the CIM definitions, even though the firms are not registered on a Canadian stock exchange.

Other classification systems, such as those issued by bureaus of mines or geological surveys, exist but are not applicable for mine financing. For example, the U.S. Bureau of Mines and U.S. Geological Survey (USGS) issued their *Principles of a Resource/Reserve Classification System for Minerals* in USGS Circular 831 in 1980. The first two sentences of Circular 831 state,

* The acronym CRIRSCO comes from the original name for the committee, the Combined Reserves International Reporting Standards Committee.

TABLE 2.1 Common code documents

CRIRSCO Member Organization	Country	Code Document	Abbreviated Title	Reference
Canadian Institute of Mining, Metallurgy and Petroleum (CIM)	Canada	*National Instrument 43-101: Standards of Disclosure for Mineral Projects*	NI 43-101	NI 43-101
Joint Ore Reserves Committee (JORC)	Australasia	*Australasian Code for Reporting of Exploration Results, Minerals Resources and Ore Reserves*	JORC Code	JORC 2012
Society for Mining, Metallurgy & Exploration (SME)	United States	*The SME Guide for Reporting Exploration Information, Mineral Resources, and Mineral Resources*	SME Guide	SME 2017
South African Mineral Asset Valuation (SAMVAL) Working Group	South Africa	*The South African Code for the Reporting of Mineral Asset Valuation*	SAMVAL Code	SAMVAL Working Group 2016
South African Mineral Resource Committee (SAMREC)	South Africa	*The South African Code for the Reporting of Exploration Results, Mineral Resources and Mineral Reserves*	SAMREC Code	SAMREC 2016

Through the years, geologists, mining engineers, and others operating in the minerals field have used various terms to describe and classify mineral resources, which as defined herein include energy materials. Some of these terms have gained wide use and acceptance, although they are not always used with precisely the same meaning.

The USGS and similar government agencies around the world are charged with identifying potential sources of mineral commodities for use in the future. This is a legitimate task. But it is not the same as identifying the mineral deposits that can be economically (profitably) exploited today, which is the focus of the CRIRSCO members' reporting codes. When such government and other classification systems use the same terms but with different meanings, this results in unwarranted public confusion that should be avoided (Abbott 2001).

This chapter begins with a discussion of the quality assurance/quality control procedures for samples. These procedures are required to demonstrate that the basic data used in mineral resource or mineral reserve estimation are reliable and repeatable. The chapter then moves on to details of defining a mineral resource or mineral reserve in specific deposits, that is, applying the CRIRSCO definitions to actual deposits. This discussion includes a subsection that specifically addresses industrial minerals.

QUALITY ASSURANCE AND QUALITY CONTROL*

The purpose of sampling is to obtain representative portions of a mineral deposit that can be analyzed for a variety of purposes, particularly mineral content (quantity) and quality. Additional properties, such as geotechnical information, density, amenability to various types of processing, and so on, can be collected from at least some types of samples. A wide variety of sample types (rock, soil, water, and air) are collected in the process of exploring for and delineating the details of a mineral deposit and determining the potential environmental impacts

* Much of the material in this section is adapted from Abbott 2007a.

resulting from exploitation of the deposit. This is not the place for a detailed discussion of the types of samples that can be collected (see, for example, Scott and Whateley 1995). Rather, the focus in this section is on the procedures employed to ensure that the samples collected and the analytical results obtained from the samples provide reliable and repeatable data that can be used to model the deposit and estimate mineral resources and mineral reserves within acceptable degrees of assurance. This is the function of quality assurance and quality control programs, henceforth referred to as QA/QC programs. QA/QC programs should commence with the first samples taken at the beginning of exploration and continue throughout the life of the property, including the production stage, to ensure that the sampling data continue to be reliable and repeatable. Failure to include appropriate QA/QC procedures renders the sampling results at least suspect and potentially as totally unreliable.

Prior to the Bre-X gold mining scandal that was uncovered in 1997, little formal attention was paid to QA/QC programs. For example, in his generally excellent book, *Exploration and Mining Geology*, Peters (1987) covers the topic in two short paragraphs. Scott and Whateley (1995) provide an excellent summary discussion of various sampling and drilling methods and statistics for determining sample sizes, and so forth, but do not mention QA/QC programs. Even Pitard's *Pierre Gy's Sampling Theory and Sampling Practice* (1993) does not really address QA/QC procedures of the type discussed here. Following the Bre-X fraud, more detailed attention has been paid to the subject, for example, by Bloom and Titaro (1997), Bloom (2000), Sinclair and Blackwell (2002; Section 5.7), and Roden and Smith (2014).

Roden and Smith (2014) point out that

> *the key message that needs to be remembered in the area of field sampling is that errors introduced at this stage of the data generation process are, in most instances, the largest errors introduced into a program and that these errors cannot be rectified in the subsequent processing of the sample. Errors created in the field can only be rectified in the field.*

Roden and Smith (2014) note that the two most common problems in field sampling are sample losses and poor sample splitting. Losses depend on the sampling method involved but can involve dust, extensive water flows during drilling, poor field handling, and inadequate strength or seals on sample bags or containers. Losses of fines or "heavies" are the common outcome and result in biased samples. Riffle splitters are the preferred type of splitter but they can be labor intensive. Grab sampling presents well-known problems, which can be reduced by combining several grabs that incorporate all relevant areas. Eyde and Eyde (1992) address sampling problems for industrial minerals, which include identifying the presence and effects of even small amounts of contaminants and the need to preserve the character of the in-situ deposit in sampling clays.

Implementing a QA/QC Program

Roden and Smith (2014) and, particularly, Bloom (2000) set out several mechanisms that can be used to monitor sample data, which include the following methods:

- Routine insertion of unprepared, barren (blank) samples
- Routine submission of duplicate field samples
- Resubmission of 5%–10% of sample preparation duplicates (sample pulps)
- Insertion of control samples

- Insertion of reference (standard) samples
- Randomization of sample numbers prior to submission to the laboratory
- Comparison of multi-element trends for elements determined by different laboratory procedures
- Comparison of the results for the same element determined by different methods
- Analysis of 5%–10% of the sample pulps at different analytical labs or an umpire assay

Blank samples are sample materials that are similar to the mineralized field samples and that are known to contain no or negligible amounts of the minerals or elements of interest. The blank samples are submitted to check on sample preparation procedures as well as the analytical procedures. If the mineral or element of interest is reported in the analysis, attention must then be directed at why the anomalous result occurred; it could come from any step in the process of sample collection through analysis or from mislabeling of the samples during the process. Regardless of the source of the error, a problem within the sampling and analytical procedures has been identified and its source can be identified and corrected through additional testing. This is particularly true when blank samples routinely have anomalous results. Bloom (2000) recommends that a blank be inserted every 20 to 50 samples.

Duplicate field samples are collected at the same time, from the same place, and in the same manner as the other field samples. Duplicate field samples, and other types of duplicate samples—for example, resubmitted sample pulps—provide information on the repeatability of the sampling and analytical procedures (depending on which steps are duplicated). The analytical results from duplicate samples should be within accepted analytical limits. If they are not, this may indicate a problem with the collection, preparation, and analytical procedures, or they may indicate that there is a significant nugget effect in the deposit. Bloom (2000) recommends that the numbers for duplicate samples should be at least 20 numbers apart so that the duplicate samples are analyzed in different batches in the lab. As with blank samples, the number of duplicate samples required varies with the confidence in the sampling and analytical processes. Early in a project, a higher percentage is needed, say 20%; once the reliability of the sampling and analytical processes is established, the number of duplicates can drop to one in every 20 to 50 samples. Abbott (2014a) points out that core splitting to obtain duplicate field samples assumes the uniform mineralization throughout the core. If this assumption is not met, the core splitting is not a viable method of obtaining duplicate field samples and alternative methods must be used.

The *nugget effect*—first named in studies of gold deposits—results from inhomogeneities within the sample. A gold nugget or other large particle may represent the total gold content of a large volume, for example, a cubic meter. But being a single particle, it will only be present in one sample of that volume even though duplicate samples from that volume were collected. Although gold nuggets provide dramatic examples of the nugget effect (although less of an effect than gem-quality diamonds in a diamond deposit), nugget effects occur in a variety of deposits (Abbott 2014a). Where a nugget effect is known or suspected, alternative sampling methods, and perhaps analytical methods, must be employed to obtain the repeatable analytical results required for mineral resource and reserve estimation.

Randomized sample numbers are numbers assigned to the samples in a different sequence than the samples were collected. Randomization is used before shipping the samples for the preparation and/or analytical steps and allows for identification of drift or bias in the sampling results. The drawback of randomization is the increased potential for introducing transcription errors and some increased handling procedures. Randomization is also best performed in large sample lots and where computer-generated sample number tags using bar codes are applied to identify samples throughout the analytical process, which reduces transcription errors.

Reference or standard samples are samples with known quantities of the elements or minerals of interest. Reference or standard samples have been carefully prepared in large, thoroughly homogenized batches, and the analytically repeatability and analytical error limits have been determined by repeated analysis performed by several laboratories. *Control samples* are similar in that they are homogenized samples with known quantities of the minerals or elements of interest, but they have been prepared internally to the company or project. Bloom and Titaro (1997) recommend inserting a control sample in every 20-field-sample group.

The particular QA/QC program adopted will depend on several factors:

- The deposit's delineation stage (preliminary exploration, advanced exploration, production)—during the initial and preliminary stage of exploration, grab samples may be collected to get an idea of the range of mineralization even though grab samples are often not representative.
- Type of minerals occurring in the deposit and their abundance (precious and base metals versus coal versus an industrial mineral).
- Whether a significant nugget effect exists (large variance between samples taken at the same location).
- The degree to which contamination between successive samples is likely to occur and will materially affect the analytical results—this successive sample contamination can occur at the collection stage (e.g., successive composite lengths in a rotary drill hole [particularly a problem with rotary drilling in gold deposits located below the water table]) or during the sample preparation process.
- The ability to submit duplicate, blank, and reference (standard and/or control) samples in a form that conceals their identity during the sample preparation and analysis steps that follow the submission of the duplicate, blank, and reference samples; difficulties include
  - Duplicate samples may not be available in cases where whole cores are submitted as part of the routine sampling program; and
  - Reference (standard and/or control) samples are already in pulp (finely ground) and form to ensure the homogenization required to create useful reference standards and when the color of the reference sample may not be close to the pulps of the field samples.

The overall QA/QC program should include the use of different laboratories to test the accuracy and repeatability of the analytical results. For example, many labs claim to be able to test for platinum group metals (PGMs), but testing has shown that very few labs can accurately analyze for PGMs (BLM 2002; Whyte 2000). In 2000, thirteen people associated with Intertek

Testing Services Environmental Laboratories, Inc., of Richardson, Texas, were indicted for failure to comply with standard and accepted laboratory procedures designed to prevent cross-sample contamination and to ensure accurate results. The indictment charged that the soil, water, and air samples submitted came from more than 59,000 separate projects and involved as many as 250,000 separate analyses (*U.S. v. Jeffus, et al.* 2000).* This example demonstrates that not only must one use reputable laboratories known to be able to provide reliable results for the types of analyses being run, but that these results must be independently checked and verified as a part of a thorough QA/QC program. Bloom and Titaro (1997) recommend that one in every 10 sample pulps should be sent to a second lab for reanalysis.

Bloom and Titaro (1997) describe some of the difficulties encountered in finding good labs outside Australia and North America. These include lack of infrastructure, poor communications, bureaucracy, outdated lab equipment with high detection limits, poor lab practices, limited lab capacity, lack of ready access to high-quality consumables, lack of computers that lead to transcription errors, and political risks.

An important part of any QA/QC program is the regular monitoring of the results. X-Y correlation plots are a common and easily prepared safeguard, although flagging of significant variances in spreadsheets is another means of checking results. While anomalous values may indicate that a problem has cropped up, investigation of the source of the problem may isolate it to a particular sample. Was a piece of core submitted as a blank sample taken from an interval sufficiently close to the ore zone that a stray anomalous value was indeed present? Was there a transcription error in sample numbering? Anomalies should not be ignored; they should be explained. Repeated anomalous values provide the justification for detailed testing of the procedures in order to identify and correct the problem producing the anomalous results.

Once the anomalous results have been resolved, Roden and Smith (2014) point out that the precision of the sampling program is easily determined from the duplicate samples with the mean percent difference (MPD) approach, which is calculated by

$$\text{absolute } (x_1 - x_2) \,/\, ((x_1 + x_2)/2) \times 100 \qquad \textbf{(EQ 2.1)}$$

where x_1 and x_2 are duplicates of the same sample. The individual MPD results are then averaged over the range of similar samples, which provides the expected variability in analytical value for any similar sample in the database. Roden and Smith (2014) state, "Statistical analysis of this MPD measurement has shown it to be an extremely robust measure that closely approximates the relative standard deviation. Doubling this number will therefore provide a 95 per cent confidence interval around the assay value."

Figure 2.1 shows a correlation plot of 455 duplicate sample pairs. The R^2 correlation coefficient for these pairs is 0.9685 and the MPD is 16.2%. Because the data are from a "nuggety" deposit, the MPD value is not considered to be a problem. The nuggety character of the deposit used in this example is exacerbated by the nature of the MPD calculation. When one of the duplicate pairs reports 0.0 grade units (ppm, in this case) and the other reports some detectable quantity, the MPD between the two samples can be large. By eliminating those

* Following a trial that concluded in November 2001, 8 of the 13 defendants were acquitted of the charges against them. Five pleaded guilty prior to the trial. The not-guilty verdicts were reached in part because the sample log-in and other procedures were so lax that the government was unable to prove who was operating which piece of analytical equipment at a particular time (Abbott 2002).

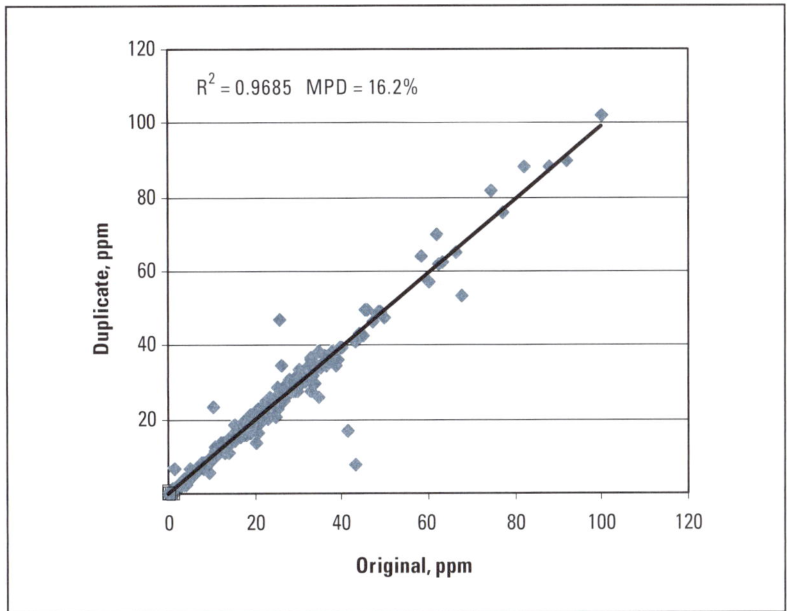

FIGURE 2.1 Correlation plot of 455 duplicate sample pairs

pairs reporting less than 1 ppm (well below the cut-off grade), the MPD of the remaining 379 sample pairs dropped to 6.8%.

Bloom and Titaro (1997) estimate that a QA/QC program would add about 15% to direct assay costs and 1% to overall exploration program costs. The value received from this extra expenditure is assurance that the analytical results from the sampling, on which all mineral resource and reserve estimates depend, are reliable.

Other Sampling Issues

When reviewing a sampling program, the following questions should be asked:

- Who did the work? Were the appropriate procedures followed?
- Was there an unbroken "chain of custody"? That is, were the samples collected by someone with responsible charge for the sampling, were the samples securely stored, were they transported to the analytical lab by a responsible person, and was a written record kept of each transfer of the samples from one person to the next? Does this record list names, dates, signatures, and a list of sample descriptions and numbers?
- How were the samples taken? Were they
  - Grab samples?
  - Chip samples?
  - Channel samples?
  - Core samples?
  - Drill cuttings?
  - Some other type?

- How was the location of the samples determined? If GPS (global positioning system) was used, which reference geoid was used and what are the distance errors in location?
- Drilling:
  - Were drill holes surveyed (both the location of the top of the hole and downhole surveys that determine hole deviation)?
  - Was the drill-hole spacing adequate? How was this determined? Has the spacing been checked?
- How and where were the assays or other analyses and tests done?
- Why were the analytical methods and test runs selected?
- Did the laboratory perform and report on its internal QA/QC program?

The appropriate answers to these and other questions will depend on the type of deposit being examined and the purpose for which the samples were taken. Reconnaissance geochemical samples need not be as carefully located or analyzed to the same precision as later deposit delineation samples. Rowe and Hite (1992) describe the sampling and drilling done to delineate the Crandon, Wisconsin, volcanogenic massive sulfide deposit. They note that the global resource estimate did not significantly change after 40 holes were drilled. But they also observed that the additional drilling done (more than 180 holes) considerably improved the confidence in the knowledge of the deposit's continuity, distribution, and variability. The appropriate analyses and tests for an industrial mineral deposit vary widely depending on the industrial mineral being examined and the potential market(s) for that mineral (Eyde and Eyde 1992).

RESERVE OR RESOURCE?

Georgius Agricola (1556) observed that "the first and principal cause [that mines fail] is that they do not yield metal, or if, for some fathoms, they do bear metal they become barren with depth." For this reason, estimation of the amount of metal, or other mineral products, present in a mineral deposit has always been a matter of considerable interest to mining investors. This section addresses several topics related to the estimation of mineral resources and reserves:

- Fundamental concepts in mineral reserve and mineral resource classification systems and definitions, including their biases toward precious and base metals.
- The differences between the mining industry's preferred classification and the U.S. Securities and Exchange Commission's (SEC's) Industry Guide 7 (1992).*
- The application of the principles developed to the estimation of industrial mineral resource and reserve estimates.

Before proceeding, we should remember H.C. Hoover's (1909) admonition regarding the business of estimation and valuation:

* On June 16, 2016, the SEC announced proposed mining disclosure rules that would delete Industry Guide 7 and replace it with definitions and rules purportedly closely aligned with international standards such as NI 43-101 and the JORC Code. The process of accepting and reviewing comments on the proposed rules, the issuance of proposed rule revisions and following comment period, and so on, can be expected to take some time (SEC 2016).

Any value assessed must be a matter of judgment and this judgment based on geological evidence. Geology is not a mathematical science, and to attach money equivalence to forecasts based on such evidence is the most difficult task set for the mining engineer. It is here that his view of geology must differ from that of his financially more irresponsible brother in the science.

Hoover (1909) also noted:

It is hardly necessary to argue the relative importance of the determination of the cost of production and the determination of the recoverable contents of the ore. Obviously, the aim of mine valuation is to know the profits to be won, and the profit is the value of the metal won, less the cost of production.

The cost of production embraces development, mining, treatment, management. Further than this, it is often contended that, as the capital expended in purchase and equipment must be redeemed within the life of the mine, this item should also be included in production costs.

Models of geologic domains and their boundaries underlie all of our statistical analyses and fancy computer models. If these models are wrong, our estimates will be wrong.

Definitions of "Reserves" and Classification Systems

The SEC's Industry Guide 7 states that a *reserve* is "that part of a mineral deposit which could be economically and legally extracted or produced at the time of the reserve determination" (SEC 1981, 1992). The SEC's definition of *mineral reserve* is the most concise. The CRIRSCO standard definition (2012) essentially arrives at the equivalent of the SEC's definitions through a circuitous route that first defines *mineral resources* and then *mineral reserves.*

Using either the SEC's or the mining industry's definition of *mineral reserve*, the key elements are

- The part of a deposit that can be recovered and sold *(geology, mining, and process engineering)*,
- Economically (profitably) extractable *(market)*, and
- Legally extractable *(title and permits)*
- At the time of determination (Abbott 1997, 1999).

For industrial minerals, economic and legal (permitting) issues are usually much more important than the geologic aspects of a deposit. This contrasts with precious and base metals, for which geology is usually the greatest risk.

Figure 2.2 reproduces "Figure 1" in the SME Guide (2017) that is essentially the same as "Figure 1" in the CRIRSCO (2012) template and other guides used by CRIRSCO members.* Figure 2.2 illustrates the standard definitions of a mineral resource and mineral reserve classification system. As shown in the figure, mineral or ore reserves are distinguished from mineral

* The minor differences between these two figures include whether *proven* (U.S. usage) or *proved* (British usage) and *geological* or *geoscientific* are used; these terms are functionally identical. The SME (U.S.) terminology is used in this chapter.

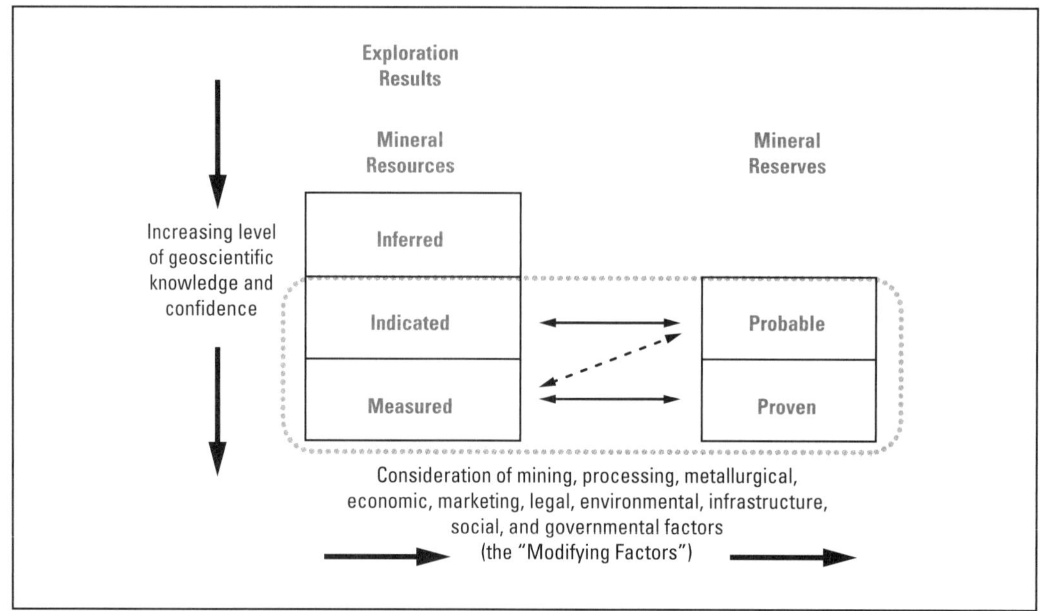

Source: SME 2017

FIGURE 2.2 General relationship between exploration information, mineral resources, and mineral reserves

resources by the requirements of the "modifying factors," the factors requiring the profitable and legal requirements of the foregoing key elements of the definition of mineral reserves. Figure 2.2 also includes the subdivision of mineral resources into the inferred, indicated, and measured categories and of mineral or ore reserves into the probable and proven categories based on the degree of geoscientific knowledge and confidence in the estimated quantities. The term *ore reserve* is often used, for example, in the JORC Code (2012). However, *ore* is mostly synonymous with *reserves*, but *reserves* is a broader term that includes mineral commodities such as coal and some industrial minerals whose economically minable portions are not generally referred to as "ore." This distinction is pointed out in the note to the definition of *reserve* in Industry Guide 7.

The CRIRSCO standard definitions (2012) document defines *mineral resource* as

> *a concentration or occurrence of solid material of economic interest in or on the Earth's crust in such form, grade or quality and quantity that there are reasonable prospects for eventual economic extraction. The location, quantity, grade or quality, continuity and other geological characteristics of a Mineral Resource are known, estimated or interpreted from specific geological evidence and knowledge, including sampling.*

The phrase "reasonable prospects for eventual economic extraction" is not currently defined by CRIRSCO. The SME Guide (2017) includes the CRIRSCO definition in paragraph 33. The notes to paragraph 35 state,

> *The term "reasonable prospects for eventual economic extraction" implies a judgment (albeit preliminary) by the Competent Person with respect to the technical and economic factors likely to influence the prospect of economic extraction, including the approximate mining*

parameters, such as dilution, mining recovery, and minimum mining thickness. In other words, a Mineral Resource is not an inventory of all mineralization drilled or sampled, regardless of cut-off grade, likely mining dimensions, location, or continuity; rather it is a realistic estimate of mineralization which, under assumed and justifiable technical and economic conditions, might become economically extractable. Portions of a deposit that do not have potential for eventual economic extraction, or which contain significant amounts of deleterious elements/minerals for which adequate test work has not been carried out, cannot be included.

Additional notes provide further explanation and guidance on the definition of mineral resources. These notes should be carefully considered.

Biases in the CRIRSCO and Other Mineral Resource/Reserve Classification Systems

Two biases exist in the CRIRSCO and other mineral resource and mineral reserve classification systems. The first is a bias toward precious and base metals reflected by the assumption that all troy ounces, pounds, and so forth, of the metal can be sold. The more recent editions of the major world mining organization classification systems alleviate this bias by requiring discussion about the marketing of the metal and/or mineral products produced. The importance of marketing for industrial minerals is discussed later in this chapter.

The second bias in the classification system is the assumption that the mineral deposit being classified is newly explored and delineated. Exploration will come first, followed by further geological delineation that will allow the increasingly detailed classification of mineral resources. There is an inherent assumption that the deposit will be fully delineated by surface drilling, which allows an appropriate computer program to design an open pit shape. This in turn will allow detailed estimation of the mining costs as the number and size of shovels, haul trucks, blasthole drills, support vehicles, explosive consumption, and so on, are optimized. Mineral processing studies will occur as the process continues, as will the various environmental studies and social licensing agreements. In short, the geological delineation will occur prior to the final estimation of the modifying factors that must be satisfied to convert mineral resources to mineral reserves. As a result, some classification systems require that mineral resource estimates be made prior to estimating mineral reserves.

This second bias is reflected when preparing mineral reserve reports of operating mines. For example, Section 5.3, Metallurgical and Testwork, of SAMREC Table 1 (SAMREC 2016) focuses on the testing and development of the processing system. For a mine that has been operating for 20 years, such test work is not occurring (except for the minor tweaking that continually occurs). For operating mines, the relevant technical reports describe the processing flow sheet that is being used.

The second bias also occurs when the mineral deposit is a steeply dipping tabular body that extends to great depth (e.g., the Silver Valley deposits in Idaho, Stillwater Mining Company's J-M Reef (in Montana) or when the deposit involves the many hydrothermal vein systems or other deposits that are sufficiently deep (e.g., the Mississippi Valley Pb-Zn deposits) that detailed drilling from the surface is prohibitively expensive. In such deposits, only limited reserve quantities are known, and further reserves are identified as the underground workings are extended to greater depths and laterally—the sinking and drifting form of exploration commonly employed prior to World War II. A classic example is the Homestake mine in South Dakota that produced for more than 100 years with no more than 3 years of reserves delineated during any one year. For these deposits, the modifying factors (permits in place, the mining methods and costs, the processing costs, etc.) are well known. What is not known is the detailed geology. As underground workings are extended, the sampling required to delineate reserves is conducted, and, as the results come in, a reserve stope can be designed and waste areas delineated in little time. In such cases, preparing a mineral resource estimate first is unwarranted and essentially useless.

The Associated Adjectives: *Proven, Probable, Measured, Indicated,* and *Inferred*

The adjectives associated with mineral reserves, *proven* and *probable*, and those associated with mineral resources, *measured*, *indicated*, and *inferred*, are defined in terms of the degree of geologic assurance, as illustrated in Figure 2.2. The terms *proved* and *probable* were originally defined by Hoover (1909):

- ***Proved Ore:*** *Ore where there is practically no risk in failure of continuity.*
- ***Probable Ore:*** *Ore where there is some risk, yet warrantable justification for assumption of continuity.*
- ***Prospective Ore:*** *Ore which cannot be included in the above classes, nor definitely known or stated in any terms of tonnage.*

These were the definitions used by the SEC prior to the adoption of the revised definitions now in Industry Guide 7 (Abbott 1985, 1999).

As indicated by Hoover's definitions, the basic difference between the adjectives is the degree to which sampling and other data support a degree of geologic assurance with which the mineral deposit being considered is known. The mining industry codes (CRIRSCO, SME Guide, JORC Code, etc.) modify this degree of geologic assurance approach, which is the basic discriminator, by allowing measured mineral resources to be equivalent to probable mineral reserves if the modifying factors introduce a sufficient level of uncertainty (see Figure 2.2). For example, if there is uncertainty about how well a proposed processing method will scale up from bench to actual operational size, this may warrant converting a measured mineral resource to a probable rather than proven mineral reserve.

The definitions used in the guide or code specified for the jurisdiction for which a report is being written (or that selected by the client if a public filing is not contemplated) are the specific definitions that should be used. The amount of sampling and other geologic information required between the various categories (e.g., proven and probable) depends on the deposit and the analysis and interpretation of that data. Although at first glance it would appear that debates about the appropriate category would be common, in practice they are not. This issue is more fully addressed in the discussion of "Boundary Issues" later in this chapter.

Common Assumptions in the Mineral Classification Systems

Three fundamental assumptions are inherent in both the CRIRSCO and related industry classification systems:

1. A "blue sky" boundary exists in each system.
2. Geology represents the greatest risk of failure in the system.
3. If you produce a quantity of mineral product (e.g., ounces of precious metals or pounds of base metals), the quantity can be sold.

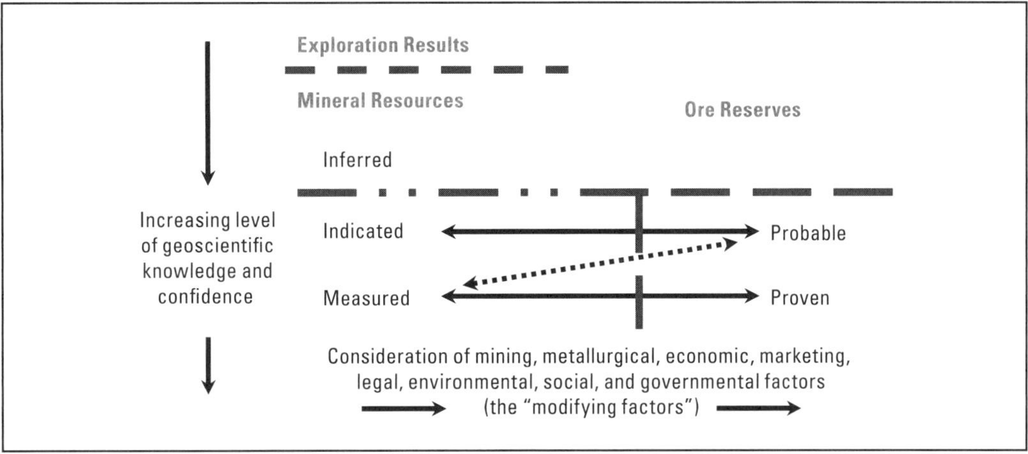

Adapted from SME 2017

FIGURE 2.3 Blue sky boundaries within the mineral resource and mineral reserve classification system

The last two assumptions are not true for industrial minerals.

The blue sky boundary in each system represents the boundary beyond which estimates of amount and quality are too speculative to justify quantification, as stated by Hoover (1909) in his definition of *prospective ore*, "ore which cannot be included in the above classes, nor definitely known or stated in any terms of tonnage." Figure 2.3 illustrates the blue sky boundaries.

The boundaries in Figure 2.3 are between "exploration results" and "inferred mineral resources," between "inferred mineral resources" and "indicated and measured mineral resources," and between "indicated and measured mineral resources" and "probable and proven ore reserves." The dashed boundary between "exploration results" and "inferred mineral resources" is the mining industry's boundary between the nonquantifiable and quantifiable estimates; "nonquantifiable" being, in Hoover's sense, that estimates based on this information are too speculative to reasonably or justifiably warrant quantification. The long, dashed boundary around "probable and proven ore reserves" and separating these categories from the mineral reserve categories is the boundary set in Industry Guide 7 between "mineral reserve estimates" and "mineralized material" estimates. The extension of this boundary separating the "inferred mineral resources" and the "indicated and measured mineral resources" separates the SEC's "mineralized material" category from the more speculative "inferred mineral resources" category.

The second common assumption in mineral resource and mineral reserve classifications is that geoscientific factors present the greatest risk to the accuracy of the mineral resource and mineral reserve estimates. This assumption is true for most metallic deposits but is not true for coal and most industrial minerals—the precious and semiprecious gems and minerals in pegmatites, hydrothermal veins, and diatreme deposits are the notable exceptions. Most industrial mineral deposits, such as limestones, marbles, dimension stones, clays, aggregates, and many others, are geologically relatively simple and exhibit much more consistency of quality over both thickness and along strike than is true of almost all metallic deposits. Indeed, the existence of a very large quantity of material that meets minimum specifications is required for most industrial mineral deposits to be considered for economic extraction.

The third basic assumption in mineral resource and mineral reserve classifications is that if you produce a quantity of mineral product, for example, ounces of precious metals or pounds of base metals, the quantity can be sold. Although it is true that the price of a metal can vary, sometimes by significant amounts, over a period of days or months, the fact remains that the quantity of metal can be sold at some price whenever it is offered for sale and in the amount offered by the mining firm. Nor does the location of the deposit affect the marketability of metal. Gold or copper produced anywhere in the world can be sold in the appropriate metals market. Again, for most industrial minerals, this assumption does not hold. Delineation of an industrial mineral deposit does not mean that the product produced from that deposit can be sold. Even where permitting is assumed, the ability to successfully market the product remains a major stumbling block. The demand for the product is not infinite (the assumption for precious and base metals). The industrial minerals mantra, "Geology is important but marketing is paramount," is repeated in this chapter.

Finally, those claiming that mineral resources and/or mineral reserves exist in a particular deposit have the burden of proof for demonstrating the validity of their claims. The mining industry classification systems include a table (usually identified as "Table 1") that outlines the types of information required to claim that a particular category of mineral resources or mineral reserves exists.

Are Mineral Reserves Part of or Additional to Mineral Resources?

The answer to this question is not as simple as it seems. Yet knowing the answer is very important in mineral resource and mineral reserve classification (Abbott 1999, 2001). Figure 2.4 illustrates the issues involved.

From one viewpoint, mineral reserves should be part of the total mineral inventory, which includes the associated mineral resources. Paragraph 36 in the 2012 JORC Code states, "In situations where figures for both Mineral Resources and Ore Reserves are reported, a statement must be included in the report which clearly indicates whether the Mineral Resources are inclusive of, or additional to the Ore Reserves." The problem with this viewpoint is its failure to recognize that mineral resources are estimates of in-situ material while mineral reserve estimates, ultimately, are estimates of the amount of valuable mineral (or its constituents, such as metals) that can be sold. Thus, mineral reserve estimates must be net of all mining and processing losses. They are not in-situ estimates. The mining and processing losses that must be included in a mineral reserve estimate include waste material: the barren, low-quality, and weathered zones; physical losses from belts, trucks, loading, pillars, and so forth; contaminated material; and processing losses. North American practice has always separated mineral reserve and mineral resource estimates, recognizing that mixing the two is less like mixing apples and oranges and more like mixing apples and chickens; the two categories are distinctly different.

Because the JORC Code allows the combination of mineral resources and mineral reserves, one must pay attention to reports to make sure whether the Australasian or North American viewpoint has been followed.

Figure 2.5 illustrates another significant difference between mineral resource estimates and mineral reserves pointed out by Noble (1993). Initial estimates of deposit size increase as drilling continues, reaching a maximum as the limits of the deposit are delineated. This becomes the in-situ mineral resource estimate of the deposit's size, the left-hand column in

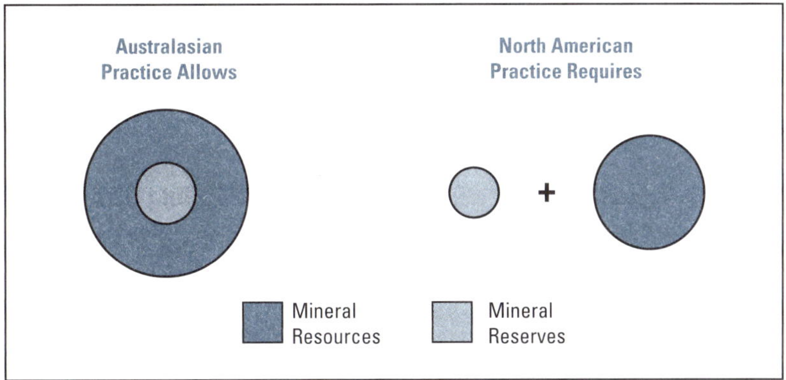

FIGURE 2.4 Australasian and North American answers to the question, "Are mineral reserves part of or additional to mineral resources?"

Figure 2.5. As mining engineering, processing planning, and testing proceed, particularly as economic constraints are placed on the planning, the size of the deposit decreases. The in-situ minable estimate, shown in the middle column in Figure 2.5, is the estimated amount of the total in-situ deposit that can be economically extracted. The estimate of greater interest is that of the recoverable amount of valuable mineral(s), shown in the right-hand column. This column is shorter than the in-situ minable amount because of mining and processing losses. The relative sizes of the three columns in Figure 2.5 will vary from deposit to deposit. But the size of the three columns will always decrease to the right. The point is that estimates of in-situ deposit size, the mineral resource estimate, are very different from and larger than the estimated amount of recoverable mineral reserves.

Boundary Issues Between Inferred, Indicated, Measured, Probable, and Proven

What is the boundary between one mineral resource and/or mineral reserve category and another? What characteristics result in a sufficient change in the confidence between one set of estimates and another that results in the first set being classified as an indicated mineral resource and the other being a measured mineral resource? These and similar questions are considered boundary issues.

When contemplating diagrams of mineral resource and mineral reserve classifications like those illustrated in Figures 2.2 and 2.3, one would expect that boundary issues should present real problems. For example, deciding which category a particular drill spacing represents should be difficult.

In practice, boundary issues do not present the difficulties one would expect. Although there are tests that can be made to determine the adequacy of a particular drill spacing, experienced exploration geologists should have a pretty good feel for the degree of assurance associated by a particular drill spacing simply by determining whether a spacing decrease provides significant amounts of additional information. The degree of assurance with a particular drill spacing may vary across the deposit. There may be geologic boundaries, such as structural discontinuities, requiring more careful delineation than required for confirming general deposit characteristics away from the discontinuity. While there may be legitimate differences of professional opinion

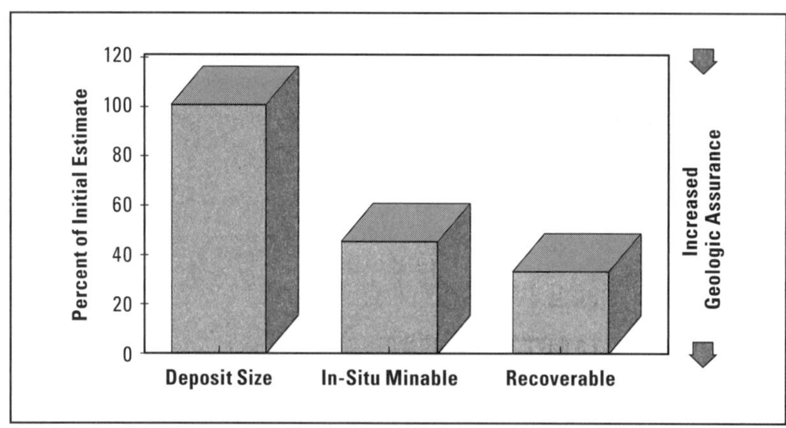

Source: Noble 1993

FIGURE 2.5 Changes in estimated deposit size as more information is gathered

regarding the risk associated with a particular geological or other modifying factor within a classification scheme, they can be resolved with various methods of risk testing.*

Reconciliation of estimated versus actually produced quantities provides the best test of the estimation method used. Although reconciliation cannot begin until after mining commences, collecting appropriate reconciliation data should be a priority for both confirming estimation methods and determining whether some problem may be cropping up. Table 2.2 presents reconciliation data from mining three pits within an industrial mineral property.

The data in Table 2.2 came from a mining operation that delineated its ore bodies from the surface using a 100-ft × 100-ft grid. After a substantial thickness of overburden was stripped from the ore, a 25-ft × 25-ft grid was drilled for ore control. The "100-ft/actual product" row of Table 2.2 compares the estimated with the actual amount of product produced based solely on the initial 100-ft × 100-ft grid. Although the calculation method used for the foregoing estimates was simple and done by hand calculator rather than computer modeling, the estimated-to-actual amounts varied between pits, and the accuracy of the estimates was within generally accepted limits. The variation in estimation between pits suggests that the estimates were getting better as more familiarity with the ore bodies was acquired through mining experience.

Confusion About the Meaning of *Resources* and *Reserves*

The confusion between *resources* and *reserves* is twofold. First, the two words are very similar, and most native speakers of English do not readily distinguish between them without reflection. But even when the difference between the words is recognized, the mining industry uses the reverse of the common English usage, as shown in Table 2.3.

* There are those who would like mathematically specified risk limits, for example, ±10% at a 95% confidence level. The problem is deciding how to accurately measure the specified risk and which statistical test to use. Differing opinions on this issue lead to differing results and interpretations, requiring a fallback to the qualitative assessment of risk.

TABLE 2.2 Percentage of estimate divided by production

Drill Spacing	Pit 1	Pit 2	Pit 3
100-ft/25-ft tons	92.1%	107.5%	100.5%
100-ft/25-ft product	86.3%	102.2%	96.3%
100-ft/actual product	87.7%	—	—

Note: The 100-ft/25-ft spacing indicates that a 100-ft × 100-ft surface drill spacing was followed by 25-ft × 25-ft ore control drilling after stripping.

TABLE 2.3 Confusion between the words *resources* and *reserves*

	Resources	Reserves
Common usage	Available for current use	Saved for the future
Mining usage	May be available in the future	Currently available for mining

When defining "proven ore" and "probable ore," Hoover (1909) observed the following:

> *The old terms "ore in sight" and "profit in sight" have been of late years subject to much malediction on the part of engineers because these expressions have been so badly abused by the charlatans of mining in attempts to cover the flights of their imaginations.... In fact, the substitutes are becoming abused as much as the originals ever were. All convincing expressions will be misused by somebody.*

Disclosures About Exploration Information

The case of *Securities and Exchange Commission v. Texas Gulf Sulphur Co.* (1966, 1968) grew out of the discovery of the Kidd Creek copper/zinc massive sulfide deposit near Timmins, Ontario, and originally defined illegal insider trading.* In this case, the initial information that was found to be material, undisclosed, or inside information was exploration information—specifically, the results of the first drill hole into a geophysical anomaly. The first hole drilled into Kidd Creek was very good, but one drill hole, however good, is not sufficient to delineate a mineral resource or a mineral reserve, even when supported by a geophysical anomaly. One drill hole and an anomaly may suggest that a mineral reserve may exist; nevertheless, further exploration and drilling are required (Peters 1987).

The Kidd Creek discovery was unusual. Most drilling programs proceed with more fits and starts. Yet the property undergoing initial exploration drilling may be the most significant asset of a junior mining company and the exploration of the deposit its only significant business activity. As such, exploration information must be reported. A great deal of exploration information can actually be qualitatively disclosed without reaching the issue of whether mineral resources or mineral reserves exist. What type of deposit is being explored? What deposit model is being used? Are the results encouraging or discouraging? Are the results prompting a change in plans?

The discovery of the Kidd Creek deposit provides an excellent example of the significance of qualitative information. The results of the initial drill hole at Kidd Creek were known

* Other aspects of the Kidd Creek discovery relating to insider trading and the confidentiality of client information are discussed in Chapter 18.

to very few individuals and prompted an immediate and quiet program of additional land acquisition. Following completion of the land acquisition, the drilling program resumed with additional rigs being added at frequent intervals. Unlike the land acquisition program, the drilling campaign could not be kept from the local Timmins mining community. An aggressive drilling program is not undertaken unless the results are very good and the project has been moved onto a faster track. Without any hard information on drilling results, rumors of a discovery started simply by observation of drilling activity. This qualitative information led to an increase in the share price of Texas Gulf Sulphur's stock even though no quantitative data had been released by the company.

The insider trading was committed by the few people who knew something about the actual drilling results. One of the interesting findings in the legal decisions is the discovery that there were legitimate corporate reasons for not making the initial drilling results public and for limiting the number of people who knew the details. What was found to be illegal were the trades made by those who had the information prior to its public release and dissemination.

The types of qualitative disclosures about exploration information that can be made include the location of properties being explored, the types of deposit being sought, the plan of exploration, updates on the execution of those plans, whether the results warrant changes in the plans (e.g., the rapid increase in the number of drill rigs at Kidd Creek), and the general views of those in charge of the exploration program on the results of the program. None of this information includes the quantitative results of the exploration program. But it does let everyone know what those who are running the program are planning to do, have done, and what they think about the results to date.

Selected disclosure of quantitative exploration information is a potential invitation to a securities fraud lawsuit. The SEC's antifraud rule 10b-5 states,

> *It shall be unlawful for any person, directly or indirectly, by the use of any means or instrumentality of interstate commerce, or of the mails or of any facility of any national securities exchange,*
>
> *(a) to employ any device, scheme, or artifice to defraud,*
> *(b) to make any untrue statement of a material fact or to omit to state a material fact necessary in order to make the statements made, in the light of the circumstances under which they were made, not misleading, or*
> *(c) to engage in any act, practice, or course of business which operates or would operate as a fraud or deceit upon any person, in connection with the purchase or sale of any security.* (17 CFR Part 240.10b-5; SEC 2014)

Mining industry press releases including phrases such as

- "...samples assay as high a ...,"
- "...selected samples...,"
- "...ore-grade mineralization...,"
- "...preliminary estimates...,"
- "...near the well-known ___ Mine...,"
- "...along the prolific ___ trend...,"

which all suggest that material information has been omitted and therefore that the disclosures are potentially fraudulent. The reasons for this are as follows:

- Some assay information or sample results are clearly omitted from the disclosure; what about these undisclosed results?
- How can one know what the "ore grade" is for an incompletely explored deposit?
- Proximity to well-known mines or locations along a trend, particularly when illustrated by a small-scale map, generally suggests unwarranted potential.

What is really needed is the unbiased opinions of those who have examined and analyzed all the data. What do they think, even in qualitative terms? These opinions are what really matters, particularly if rendered by independent observers.

Legally Extractable—Mineral Tenement

The "legally extractable" part of the definition of mineral reserves consists of two major parts: first, owning, leasing, or otherwise having the right to extract the minerals; and second, obtaining the permits required to build and operate a mine or quarry and the associated processing facilities.

The first issue in determining the ability to legally exploit a mineral deposit is having the appropriate mineral tenement or title. Mineral tenement can take a variety of forms, such as direct ownership of the land overlying the deposit along with ownership of the right to extract the minerals. Mineral title can be separated from surface ownership, which is something that has been a problem for the oil and gas industry in some areas and could also apply to deposits of coal or uranium. The mining laws of different countries provide for different forms of mineral title. In some countries, like the United States and Canada, the location of mining claims is allowed in some cases and for some minerals. Leases of mineral rights may be granted by private owners or the government. Other countries allow mining firms to obtain exploration permits that can be converted to exploitation permits. Mineral tenement issues are legal questions requiring the assistance of legal counsel. Mining firms should be familiar with and comply with the mineral tenement provisions in the area(s) of their current and planned operations.

Legally Extractable—Permitting and Social Licensing

This second issue in determining the ability to legally exploit a mineral deposit, obtaining the required building and operating permits, has become the more significant part of legal extractability. Opposition to mine and quarry permits comes from neighbors, environmental groups, and other entities for a variety of reasons. The NIMBYs (not in my backyard) and the BANANAs (build absolutely nothing anywhere near anything) are just two common acronyms signifying opposition. Even those mines, quarries, and processing facilities that use no chemicals face opposition and increasingly lengthy and costly permit application processes. Dealing with such opposition has come to be known as the social license to operate in the 21st century.

Social license to operate refers to the concept of informal or unwritten "permission" from immediate neighbors, local communities, indigenous peoples, and nongovernmental organizations (NGOs) to conduct operations. The operators can be natural resource companies or operations performed by an increasingly broad range of geoscience endeavors including geohazards, waste disposal, highway development or redevelopment, and so on. The general public is generally uninformed or misinformed about the complex scientific and technical challenges accompanying these issues and is concerned about risks of a proposed project, whether real or imagined, that may be unknown to them. Social license to operate programs seek to educate the public and concerned groups about the real impacts of a project, the measures that will be taken to minimize adverse impacts, and the benefits that will be derived from the project. The public's concerns should be listened to and considered in determining how the project will proceed. Geoscientists are likely to be the first "boots on the ground" examining proposed projects, and their public interactions can have a significant impact on the public's perception of, and concerns about, the project (Swarthout 2014). Geoscientists must be viewed as honest and trustworthy and work toward building trust in the information about the proposed project. Abbott (2015), Vecchia (2015), and Hulse (2016) provide reviews and insights into evolving development of the social license to operate. Consideration of the social license to operate is included in Table 1 of the SME Guide (2017).

The following are common objections to mines and quarries:

- Visual impacts of the operations
- Disturbance of river systems, which particularly affects sand and gravel operations
- Dust and noise, which are part of mining, comminution, and sizing operations
- Truck traffic, particularly for high-volume producers
- Actual or alleged effects on groundwater such as a lowered water table and/or contamination
- Vibration from blasting inducing damage to neighboring structures

As Sandman (2012) points out, the actual, perceived risk of a project to a community equals the actual hazard plus the public's outrage at the perceived character of the risk. Perceptions may not be realistic, but they drive the outrage and the vehemence driving opposition. Sandman presents strategies for dealing with these problems that a project may ignore at its peril.

To have a mineral reserve, the required permits must be in place or there must be a firm or reasonably assured basis for believing that the required permits will be issued in due course. The reasonably assured types of unissued permits are things like certificates of occupancy for buildings under construction. Operating and environmental permits seldom fall into this category. (For more information on environmental permitting and sustainability, including the social license to operate, see Chapters 9 and 10.)

COMMON PROBLEMS WITH MINERAL RESERVE ESTIMATES

Prenn (1992) identified several common problems with reserve estimates, which include the following:

- **Sampling errors.** Are the samples representative of the part of the deposit being sampled? Has appropriate compositing been done? Have the appropriate analytical tests of physical and chemical properties been conducted by qualified laboratories? Does an appropriate QA/QC program exist to ensure that analytical results are accurate?
- **Geological errors.** Has the deposit been sufficiently delineated so that no structural or stratigraphic surprises will occur? Have appropriate specific gravity or density tests been

conducted (this can be a problem with swelling clays, vuggy host rocks, and porous materials like diatomite)?

- **Modeling errors.** Has a correct geologic model been chosen for the deposit? Have the relevant geologic boundaries (formation boundaries, facies changes, faults, folds, property boundaries including setbacks, etc.) been incorporated in the model? Have any anisotropies in projection distances been identified? Has the accuracy of the data used in the modeling been verified? Have detailed geologic maps and sections been constructed? Has an appropriate algorithm been used in setting areas of influence? Has the model been tested? How?
- **Engineering errors.** Is the size of the mining equipment appropriate for the amount of material being moved? Have appropriate rock mechanics studies been performed? Can processing bench tests be scaled up to production streams without problems? Why? Is a new process being used? What is the basis for believing it will work? Are there bottlenecks in the mining or processing streams? What is being done to correct them?
- **Market evaluation errors (industrial minerals).** How was market supply and demand estimated? Why will the produced product be able to penetrate the market? Have potential customers used test batches of product to determine whether the product will meet their needs? Does that entity have the working capital required to get the project constructed and running and producing during the market penetration period?

The Tonnage Factor

The *tonnage factor*, the densities of the ore rocks and the gauge rocks, is a vital number in mineral resource and reserve calculations. This is the value (in grams per cubic centimeter [which equals metric tons per cubic meter] or cubic feet per short ton), rarely containing more than three significant digits, that converts volume measurements to weight measurements. It is the value whose significant digits, more than any other, that limits the significant digits that can be used in the estimates (despite the ability of modern computers to calculate to far greater, but misleading, precision). Errors in the tonnage factor determination are multiplied through the whole estimate.

Determining the tonnage factor can be very difficult where the rocks being tested are vuggy and/or contain swelling clays, a common occurrence for many hydrothermal deposits, or for porous rocks like diatomite or perlite deposits. While there are various methods of calculating the tonnage value, one marble quarry, whose royalty owners didn't trust the math, simply had a local dimension stone operation carefully shape and polish a cubic foot of marble, which then could be placed on a scale. Unfortunately, this solution is not always practicable, though perhaps it should be tried more frequently than it is. Variations in the tonnage factor throughout the deposit complicate the calculations. Therefore, it is important that an appropriate number of samples to test the tonnage factor are collected and properly tested.

Dilution and Deletion

Dilution results from the addition of waste rock to the ore stream, and *deletion* is either the inclusion of ore in the waste stream or the non-extraction of the estimated ore. Both dilution and deletion have a variety of causes.

TABLE 2.4 Ore zone thickness and minimum mining-width grade*

Ore zone thickness	1 ft	2 ft	3 ft	4 ft
Grade of 5-ft minimum mining width	2 GU	4 GU	6 GU	8 GU

*Assumes a constant 10 grade units (GU) ore grade. Grade units are units of measure such as ounces per ton or parts per million. The specific unit is not relevant to the example.

A primary cause of dilution and deletion is the character of the ore–waste contact. Is it regularly shaped (planar is optimal) or not (replacement deposits can have very irregular contacts). The mining method employed can also have a significant impact. The minimum effective mining unit is the volume of material for which the ore–waste decision is made. This can be by the shovel or scoop load, but frequently is larger—by the blast round or at least partial bench blast round.

In underground mines, the minimum mining width has a significant dilution effect. Both the thickness of the ore zone and its dip combine with the size of the selected mining equipment to determine the amount of dilution. For example, a vertical ore zone with variable thickness must be diluted to a 5-ft minimum mining width for a particular mining method and machine size. The width of the ore zone thus affects the effective cut-off grade, as shown in Table 2.4.

The following calculation is used in Table 2.4:

$$\text{(ore thickness} \times \text{grade) / total thickness = minimum mining width grade} \quad \textbf{(EQ 2.2)}$$

More generally, the calculation is a length-weighted average of all ore and waste zones, divided by the total width of the ore zones, plus the remaining amount of minimum mining width or the required added thickness of waste that must be mined in addition to the ore zone. Table 2.4 demonstrates the situation where the 5-ft minimum mining width exceeds the width of the ore zone (maximum 4 ft). Assume that the ore zone used in Table 2.4 increases in width to 6 ft and that a 1-ft waste zone is required on each side to achieve the minimum mining width. The total width is then 8 ft, and the minimum mining width grade then becomes

$$(6\text{ ft} \times 10\text{ GU}) / 8\text{ ft} = 7.5\text{ GU} \quad \textbf{(EQ 2.3)}$$

Dilution also results from such things as overbreak and internal dilution. *Overbreak* results from the blasting of unplanned waste into a stope or on an ore bench because of nearby structural weaknesses or the use of too much explosive. *Internal dilution* results from the need to mine at minimum widths, as discussed previously, or from the need to mine through low-grade zones to reach and exploit high-grade pockets. While commonly thought of in underground mining, internal dilution also can occur in open pits characterized by irregular and pod-like grade distributions or irregular (nonplanar) contacts between ore and waste. Stone and Dunn (2012) provide a good discussion of dilution and some exercises in their Chapter 7.

Grade Cutting and Capping

Deciding whether, when, and how to cut or cap grades is an important issue, one that has been a matter of increased debate since statistical techniques for limiting the influence of "outlier" grade values were developed. On the one hand, capping or cutting high-grade analyses is a method of artificially adjusting data (Pocock 2001). On the other hand, failing to cap or cut anomalously high grades can result in overestimation of the grade and contained ounces in a precious metal deposit.

The events of November 2012 through November 2013 at Pretium Resources' Brucejack deposit in northwest British Columbia provides an especially interesting example of whether and how to cap or cut grades. On November 22, 2012, Pretium released an NI 43-101 Mineral Resources Update Technical Report by Snowden Mining Industry Consultants on the Brucejack gold deposit in British Columbia (Snowden 2012). This report contained an estimate of indicated mineral resources in the Valley of the Kings (VOK) zone containing 16.1 million t (metric tons) grading 16.4 g of gold per metric ton (g Au/t) and 14.1 g Ag/t with a 5-g/t cut-off. Open pit extraction was contemplated. While most of the 57,895 assays for the VOK zone were low grade, extremely high-grade samples (9,383 g Au/t maximum assay) were encountered. The mean assay was 8.85 g Au/t and the variance was 3,172 g Au/t. The skewness was 93.4. Clearly, this is an unusual deposit that is very heterolithic and has extreme grade variations, making mineral resource estimation difficult. The first recommendation for mineral resource estimation made in the November 2012 technical report was to undertake the mining and processing of a 10,000-t representative bulk sample to test the validity of the deposit model and mineral resource estimation approach used by Snowden.*

In late 2012, Pretium hired Strathcona Mineral Services to help select the area to be mined for the bulk sample and to use a sample tower† as part of the validation process. Strathcona personnel have used a sample tower on many remote mining projects to obtain grade verification of predicted grades from bulk samples of deposits with a high "nugget" effect (very high grade samples) without having to ship the entire bulk sample from the remote location (Thalenhorst and Dumka 2010). The VOK zone is an example of an extremely high nugget deposit. Some electrum particles (the gold–silver proportion overall is 54% to 46% silver) are reported as being centimeters in size. Photos are posted on Pretium's website (Pretium 2017).

Work on the bulk sampling program began in the spring of 2013. On September 23, 2013, Pretium issued a press release stating that the entire 10,000-t bulk sample had been shipped to the processing mill in Montana and that 6,000 t had arrived at the mill (Pretium 2013a). This press release also reported on seven assays taken from drilling done in connection with the bulk sampling program that exceeded 1,000 g Au/t, including one from the Cleopatra vein, which was discovered in the bulk sample area, assaying 2,140 g Au/t over 0.5 m (1.6 ft) and 195 g Au/t uncut over 9.29 m (30.48 ft).

On October 9, 2013, Pretium issued a press release stating that Strathcona had resigned from the project on October 8 (Pretium 2013b). No details about the reason for Strathcona's resignation were included in this press release. On October 10, Pretium released details of drilling results from 16 of 17 drill fans from the bulk sample area, including 3½ pages of detailed assay results, and also posted a series of maps and cross sections from this drilling on its website (Pretium 2013c). Again, there was no mention of the Strathcona resignation, and Strathcona did not publicly release any information about its resignation. The price of Pretium's stock fell 50% in October 2013.

---

* A maximum 10,000-t bulk sample size is specified by British Columbian regulations.

† A sample tower is a structure containing a series of riffle splitters that reduce the size of the sample by half at each splitter. Because the splitters are in sequence, a large reduction in volume that is hoped to be unbiased is achieved. The lack of bias needs to be verified. Sample towers may not work in high nugget deposits such as precious metals and diamond.

In an October 22, 2013, press release, Pretium announced preliminary results from the processing of the bulk sample by the mill (Pretium 2013d). It also noted that prior to resigning, Strathcona provided Pretium with preliminary assay results from the sample tower, including the 426585E crosscut, which averaged 2.08 g/t. Based on the 2,167 t extracted from the 426585E crosscut, an estimated 145 ozt (troy ounces) would be recovered based on the sample tower results. Preliminary mill processing of the 2,167 t extracted from the 426585E crosscut recovered 281 ozt of gold. In the opinion of both Pretium and Snowden, there is a significant difference in the contained gold estimated by the selective sampling of the 426585E crosscut by the sample tower and the actual contained gold determined by milling. Both Strathcona and Snowden are highly regarded mining consulting firms. This October 22nd press release contained the first specific statements that there was an expert opinion dispute between Strathcona and Snowden, Pretium's independent expert consultants.

On November 25, 2013, the *Northern Miner* published an interview with Graham Farquharson, Strathcona's founder and president, responding to Pretium's announcements of its bulk sampling results (Preston 2013). Strathcona's assessment contrasts starkly with Pretium and Snowden's mineralization model and mineral resource estimate. In its October 8th resignation letter to Pretium, Strathcona stated (Preston 2013),

> *There are no valid gold mineral resources for the VOK zone, and without mineral resources there can be no mineral reserves, and without mineral reserves there can be no basis for a feasibility study.... Statements included in all recent press releases about probable mineral reserves and future gold production* [from the VOK zone] *over a 22-year mine life are erroneous and misleading.*

Strathcona believed that the discovery of the high-grade Cleopatra vein skewed the results.

The expert-opinion dispute between Strathcona and Snowden centers on the estimation methodology and the interpolation method used—how far can the latent values of very high-grade assays scattered throughout the deposit be projected? Farquharson also stated that there is gold in the VOK, that it is not a Bre-X–type fraud, but that Strathcona believes the project needs a very different geological model and a change in the mining approach to small-tonnage, high-grade vein mining from bulk mining.* This expert-opinion dispute also highlights an interesting professional ethics issue discussed in Abbott 2014b.

Further discussion of this subject is addressed in the "Determining the Capping Grades" section of Chapter 3.

Has Geostatistics Been Properly Applied?

Geostatistical methods are increasingly being applied to all types of mineral deposits. But are the proper geostatistical methods being used? Should ordinary, log-normal, indicator, or some other type of kriging be used? If an inverse-distance algorithm is used to limit the influence of samples at greater distances from the block whose value is being calculated, what power (squared, cubed, etc.) is being used? These issues are addressed in greater detail in subsequent sections of this chapter and in Chapter 3.

---

* The 1997 exposure of the fraudulent salting of samples from Bre-X's Busang deposit in Indonesia, uncovered by Strathcona, led to the adoption of Canadian NI 43-101 as a means of preventing such fraud in the future.

Calculating the Cut-off Grade

Cut-off grade is defined as "the lowest grade of mineralization that qualifies as ore in a given deposit; rock of the lowest assay included in an ore estimate" (AGI 1997). Stone and Dunn (2012) provide a fuller definition:

> *The grade that will just cover all the costs incurred by (or charged to) the operation is usually referred to as the cut-off grade. In some instances, this figure may include a minimum profit, but for most evaluations it is better to look first at the break-even grade (if for no other reason than to avoid considerations of income tax). Likewise, it is sometimes argued that the developed ore should be treated if it will pay for all subsequent treatment costs, since the sunk development costs are no longer relevant.*

As Stone and Dunn (2012) note at the end of their definition, arguably, material that will pay for the cost of milling should be included in the cut-off grade. They conclude, "**Unless absolutely unavoidable, ore that will not pay for all of the costs with which it should be charged should never be mined or sent to the mill**" (emphasis in the original). Running substitute cut-off-grade material through a mill can only be justified when there is greater mill throughput capacity than delivered ore amounts, and then only when the subgrade material would have to be mined to expose ore for extraction. The problem is that running such material can lull an operation into thinking that processing such material is always okay even when it is obvious that it adversely affects profitability. Another case is when normally subgrade material is run through the mill and is at the peak of a price cycle when the normally subgrade material can be mined and processed at a profit. But even in this case, care should be taken that the overall mine plan and its economic basis are not being unduly compromised.

The aforementioned definitions are clear, but figuring out what an appropriate cut-off grade should be is frequently a more difficult task. Sometimes analysis of the geologic model of the deposit and examination of the analytical data will reveal a natural cut-off grade. Figure 2.6 is a histogram of assay data from the J-M Reef at the Stillwater palladium–platinum mine in south-central Montana. It shows a natural cut-off grade of about 0.3 ozt of palladium plus platinum per short ton.

Geotechnical characteristics of a deposit may also directly affect the cut-off grade. For example, where the boundary of a geologic domain containing the ore, or located in close proximity to the ore, is a structurally weak zone to which blasting will break during mining, the resulting dilution will impact the cut-off grade that should be used.

Wooller (2001) points out that processing criteria may also affect the cut-off grade. Plant capacity, comminution characteristics (both physical and costs), consumable consumption, and other processing factors can be optimized just like the mine plan.

King (2001) notes that there can be a close correlation between mine production scheduling and the cut-off grade policy. He notes that

> *as more of a reserve is mined in a period, the period cut-off grade must increase until the processing and market limits are met. With low* [reserve] *production rates relative to the mill and market capacities, the cut-off grade would stay on the economic break even grade, referred to as "mine constrained."*

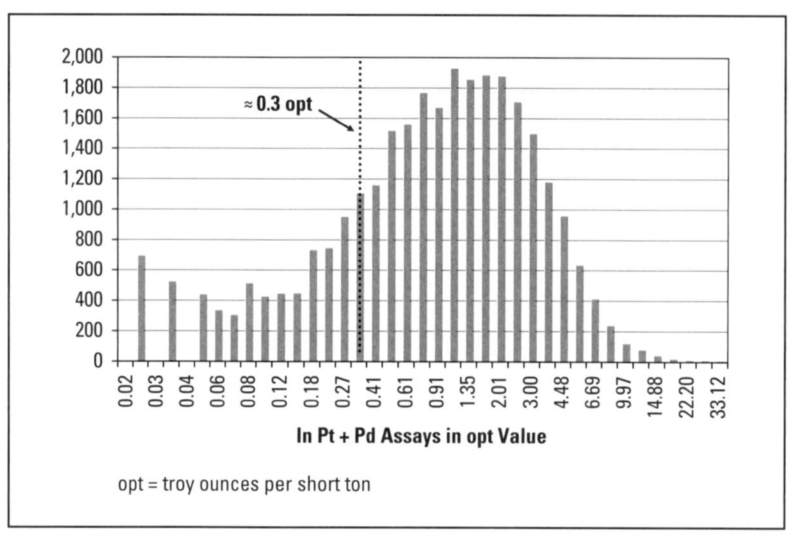

Adapted from Abbott 2003

FIGURE 2.6 Histogram of the log of palladium plus platinum assays from the J-M Reef at the Stillwater mine showing a natural cut-off at about 0.3 opt

Nevertheless, the cut-off grade is ultimately determined by subtracting the costs of mining, processing, and product sales from the income derived from the product sales. The project economics are tested in a cash-flow analysis, with and without assumed discount rates. Prices and costs are normally assumed to be constant unless there are contractual or other very good reasons for adjusting these parameters. The cut-off grade is the minimum grade required to produce a positive cash flow. But as noted above, the cut-off grade is really dependent on geological, mining, and processing characteristics and so the economic analysis is the last, but necessary, step in the analysis. Iterative or optimization analysis is frequently required to select between alternatives.

Summary

Arseneau and Roscoe (2000) provide an interesting summary for this section:

> *It has been* [Roscoe Postle Associates'] *experience that the main problem in resource estimation does not necessarily lie with the estimation method itself but with the basic application or misapplication of basic geological principles. The main issue revolves around establishing continuity of mineralization and grade within a mineral deposit prior to estimating the resource.*

Stone and Dunn (2012) echo this conclusion, quoting Harry M. Parker (1994, personal communication):

> *Although there are occasional, and in our opinion "lucky" exceptions, most successful mining ventures are proven by accumulation of representative short-range data at the feasibility stage by drilling close-spaced holes or from bulk sample pits or underground workings. Conversely, most mines which have been disappointing or have failed because of reserve problems have skipped this step in their development.*

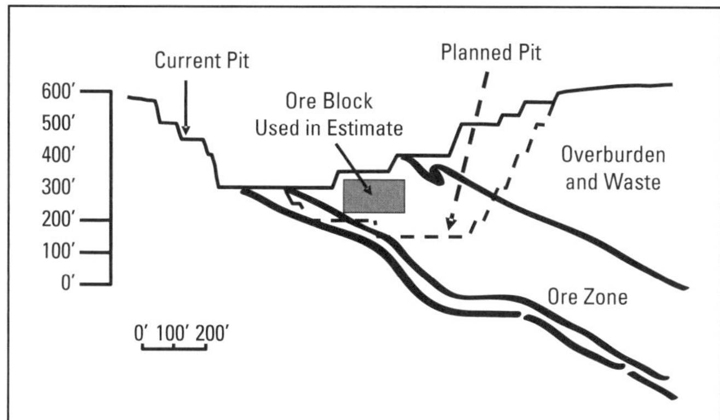

FIGURE 2.7 Example of poor geologic delineation of an industrial mineral reserve

SPECIFIC CONSIDERATIONS FOR INDUSTRIAL MINERALS

The term *industrial minerals* covers a very diverse group of minerals used for an extremely wide variety of processes. On the bases of product volume, weight, and value, industrial minerals outstrip the products of base and precious metals. SME's *Industrial Minerals and Rocks: Commodities, Markets, and Uses* (Kogel et al. 2006) provides a reasonably comprehensive overview of the industrial minerals world. The proceedings volumes of the annual Forum on the Geology of Industrial Minerals contain detailed descriptions of the types of industrial minerals deposits that are the focus of each year's forum along with more general papers.* Delineating the mineral resources and mineral reserves of industrial minerals is generally fairly simple, geologically—there are exceptions such as diamonds and minerals recovered from pegmatites—while determining potential products and markets is very complex and challenging (Holmes and Abbott 1995; Abbott 2007b). These characteristics contrast with precious and base metal deposits for which geology usually presents the greatest uncertainties and risks. Newly developed industrial minerals properties have generally been known about for years; they are not normally identified by grassroots exploration programs like those used for precious and base metal exploration.

The realities of many industrial minerals projects are that the deposit's geology is frequently very simple—for example, bedded deposits. As a result, the detailed geologic studies common at precious and base metal mines may not exist. Given the extremely simple calculation methods used to calculate mineral resource and mineral reserve quantities, reconciliation of estimated quantities and grades versus actual production is frequently the only way of demonstrating the accuracy of the estimates. If the reconciliation data exist and if the reconciliation demonstrates that the estimates are accurate within acceptable limits, the reconciliation is the best demonstration that the method in use works, as shown in Table 2.2. Unfortunately, in far too many cases, accurate reconciliation data do not exist.

Figure 2.7 shows an example of poor geologic delineation of an industrial mineral reserve. The current and planned pit outlines are shown. The geologic boundaries extend to the lower

* The annual Forum on the Geology of Industrial Minerals is hosted by a different state geological survey or foreign country's geological survey, and the host organization publishes the proceedings of each conference. Searching for "forum on the geology of industrial minerals" in a web browser will find most of the proceedings volumes. For example, Abbott 2007b and Reed 2007 are from the 43rd forum sponsored by the Colorado Geological Survey.

right below the current pit and show the hanging and footwalls of the ore zone along with a footwall border zone. As can be seen by the hanging and footwall boundaries, the deposit has been subjected to some folding. While not shown, there is some drill control on the illustrated section, although the drilling is not that closely spaced. The reserve estimate for the ore zone was based on using a block, illustrated with the gray block in the figure. The height of the block was the bench height, and the width of the block perpendicular to the strike was a set amount. The reserve estimate was based on the number of bench levels remaining to be mined times the strike length over which the levels would be mined.

The estimated mineral reserves based on the foregoing methodology clearly existed, but the estimate drastically miscalculated the amount of ore present. Because the crude estimation method used was sufficient for several decades of future mining, doing additional drilling to more accurately estimate the size of the reserve was not deemed a priority. It was a question of "why do more?" The answer, at least in part, is that with the increasing consolidation of the industrial minerals business into a few large, public companies, securities regulations are requiring stricter compliance with the applicable mineral classification scheme. Reed (2007) describes modern methods of the volumetric analysis and three-dimensional visualization of industrial minerals deposits.

Processing—Where the Money Is

Processing is, with rare exceptions, where the money is in industrial minerals projects. The capital and operating costs of the processing plant are frequently much higher than those for mining equipment and extraction. In addition, the value of a ton of material delivered to the processing plant is much lower, often an order of magnitude or more, than the value of a ton of product. However, the minerals in the deposit must be suitable for the processing methods employed. The failure of process flow sheets, failure to identify critical contaminants, failure to test via pilot plant, and so forth, have led to failed projects. Additionally, does one mineral interfere with the processing of another? In some cases, the economic processing of one mineral and the economic processing of a second mineral are less economic than the joint processing of normally subeconomic-grade minerals.

These situations are in part responsible for the complex, branch testing of physical and chemical characteristics of an industrial minerals deposit to determine what potential products can be made from the deposit and at what cost. Is a clay suitable for paper or for ceramics? Having made this branch decision, the question becomes: which ceramics or which papers? Can a "waste" stream become a salable product? The dust recovered from crushing and sizing some limestone deposits can be used as a soil amendment. This testing requires larger samples than those normally collected for base and precious metals deposits. The required industrial minerals tests are usually far more complicated than the assaying methods used for metals. The time and costs required for industrial minerals testing is not appreciated by those more familiar with metallic deposits and can lead to significant underfunding of exploration and evaluation programs.

The presence of even small amounts of contaminates can kill an industrial minerals project. For example, fibrous zeolites in an otherwise zeolite-rich deposit can destroy the deposit's economic potential because the fibrous zeolites are deemed asbestiform. Likewise, the presence of arsenic or mercury in volcanic sulfur deposits can eliminate their cost advantages. The presence of thin seams of halite in a nahcolite deposit can result in failure to meet specification and the loss of large quantities of what were believed to be reserves. Contamination can also occur because of poor mining practices.

Figures 2.8 and 2.9 illustrate two different contamination problems. In Figure 2.8, the dike in a clay deposit near Bovill, Idaho, has contaminated the white clay with yellow and red iron oxides, forming a waste volume within the deposit. In Figure 2.9, the failure to cleanly strip the overlying red Amsden Formation from underlying Madison Limestone causes the Amsden material to contaminate the Madison Limestone product.

In summary, it must be remembered that a mineral deposit is not an ore body. An ore body is a mineral reserve. A mineral deposit may or may not contain mineral reserves or ore; it will contain material that cannot be mined for various engineering reasons—like pillars in an underground mine—and material that cannot be economically mined, processed, and sold. In the grand scope of industrial mineral project evaluation, geology is important, but permitting and, especially, marketing are more important.

FIGURE 2.8 The yellow dike in this white clay deposit near Bovill, Idaho, contaminates the clay with iron oxides, creating a waste volume within the clay deposit

FIGURE 2.9 Failure to cleanly strip the overlying red Amsden Formation from underlying Madison Limestone results in Amsden material contaminating the Madison Limestone product

FIGURE 2.10 The "industrial minerals Catch-22"

Economically Extractable—Markets

As noted under earlier in this chapter under the "Common Assumptions" section, mineral reserve classification schemes assume that if a quantity of metal is produced, it can be sold. Therefore, the economically extractable part of the definition of mineral reserves focuses on the cash-flow analysis that demonstrates profitability. While such cash-flow analyses remain a significant part of the economic analysis for industrial minerals, the more important question is, can a quantity of product be sold? Is there one or more markets for the product produced that will purchase product from the proposed operation? Such sales cannot be assumed for most industrial minerals.

The SEC's engineering staff has taken the position in comment letters that in order for an industrial mineral operation to claim that reserves exist, the operation must either have a sustained history of profitable production or have firm sales contracts to purchase the product to be produced. Banks adopt a similar position when deciding whether to finance the construction of an industrial mineral operation. This has led to what this author calls the "industrial minerals Catch-22," which is illustrated in Figure 2.10 (Abbott 1997, 1999).

The Catch-22 works like this: To have industrial reserves, one has to be able to demonstrate economic viability by having a sales contract. Getting a sales contract requires that one can demonstrate to customers that the needed quantities of mineral can be produced and the required specifications can be met. This requires an operating quarry and processing plant. But the loan to develop the quarry and build the plant cannot be obtained without the sales contract. And so the circle, or Catch-22, goes around. Again, the mantra: "Geology is important, but marketing is paramount."

Unlike the markets for precious and base metals, consumers of industrial minerals tend to represent final- or near-final-stage markets. In many cases, the consumer is the end user, or the final processor selling to the end user. For example, the bentonite plants in northeastern Wyoming and adjacent South Dakota sell drill mud and low-end, non-clumping kitty litter as final products to the end user. They also sell products by railcar to companies that will upgrade the kitty litter by adding other components such as sodium bicarbonate.

Market structure for any particular industrial mineral is based on the competitive market for the end-use product. For example, the market structure comparative area for natural trona is global, while the market structure comparative area for road-building aggregates has been typically viewed as 40–64 km (25–40 mi) in North America (although, for various reasons, this radius may expand significantly in particular markets). Common characteristics of industrial mineral product markets include the following:

- **High producer concentration.** Generally only a few well-financed firms are able to compete in the market for a particular product. Even the aggregate and sand and gravel businesses—which have long been the prime example of small, family-owned enterprises—are becoming increasingly dominated by a few internationally operating firms. Hence, each market is dominated by a monopoly or oligopoly.
- **No monopsonies or oligopsonies.** There are only one or a few consumers of the industrial mineral product, at least in terms of major industrial mineral markets.
- **Brand name recognition and strength.** Industrial mineral producers strive to achieve these characteristics to establish and maintain particular markets. Castle & Cooke's Arm & Hammer brand of sodium bicarbonate is a particularly good example. Another example is World Minerals' (now part of Imerys) diatomite that it sells under the Celite brand name.
- **Product development according to customer needs.** The technical assistance, sales service, and company research and development (R&D) departments can define market share by identifying particular customer needs and developing a product that meets the required specifications. The differences between many products may be very slight, although these differences may be critical to the customer.
- **Combined markets.** Given that certain products can be combined, many successful companies pursue mineral projects servicing similar markets (e.g., construction minerals).
- **Company secrets.** Because the same parent company can produce different industrial minerals products that compete with each other, intra-company secrecy can be as tight as extra-company secrecy. Consultants to such firms must be sensitive in this area.

Having the ability to produce high-quality commodities does not necessarily result in sales. Sales efforts may fail because of

- Lack of customers,
- Total delivered price (cost product plus transportation to the user),
- Lack of company history (the brand name issue),
- Lack of technical assistance provided to the customer, and
- Lack of understanding product specifications.

Marketing and sales strategies for industrial minerals, like those for marketing any product, have short-, medium-, and long-term aspects. Short-term strategies address immediate customer needs (e.g., single orders and focus on aspects such as product quality, product technical specifications, product deliverability, delivered product price, and beating the competition from similar products). Medium-term strategies are based on competitive bids and historical performance (e.g., aggregate bids for road building). Long-term strategies are established by becoming quasi partners with the consumer by not only supplying product with reliable quality meeting the customer's technical specifications, quantity deliverability requirements, and price, but also but working on combined R&D efforts focused on improving existing products or developing new ones. These improved or new products, whether developed in customer partnerships or not, can provide improved or new product competition. Corporate consistency and longevity that provide needed technical support and service, consistently avoid impurity or contamination problems, and possibly providing product substitution also foster long-term relationships.

Industrial minerals producers are increasingly obtaining ISO (International Organization for Standardization) certification of their plants and products to improve or maintain their competitive position. Market dominance can be achieved through being the primary supplier in a particular market, by concentrating on the production of those minerals needed by a selected market (e.g., construction materials or paper products). Alternatively, a different type of market dominance can be achieved by being the dominant supplier of a selected industrial mineral with the ability to meet the needs of all customers for that mineral by supplying the mineral products, R&D, technical assistance, dependable service, and reliable deliverability. Either market dominance strategy has the goal of making "price" not an issue for customer orders.

Industrial Mineral Pricing

Because of the chemical- and/or physical-specification nature of industrial mineral products, prices are significantly less volatile than those for precious and base metals. For many industrial mineral commodities, prices have had a near constant growth, commonly at 1% to 3% per year for several years. For many industrial mineral products, when managed correctly, the returns are similar to that of an annuity. The ability to generate higher-than-normal profits tends to be project and/or product specific and not industry-price related. Public company multiples tend to range between 10 and 15 times the earnings.

Prices are typically quoted on a FOB (free on board) plant or CIF (cost, insurance, and freight) shipping location basis. However, the negotiated price is almost always on a delivered cost basis. For many industrial minerals, the cost of transportation exceeds the value of the commodity. This is principally why the market for aggregates has traditionally been limited to a 40-to-64-km (25-to-40-mi) radius of the processing plant. However, limited deposit availability due to both geologic and permitting constraints combined with existence of "super" quarries located at tidewater or adjacent to main rail lines has significantly altered delivered-product costs. Thus, tidewater quarries in the Maritime provinces of Canada or Scotland are successfully delivering aggregate to Florida. Likewise, tidewater quarries in British Columbia are increasingly supplying aggregate to the West Coast of the United States. Tidewater quarries in Alaska, Maine, and other states cannot compete in this market because of Jones Act requirements for shipping in U.S.-registered vessels. Greek perlite can be delivered to the East Coast of the United States at prices that are competitive with domestic production.

Determining the Profitability of an Industrial Mineral Operation

Determining the profitability of an industrial mineral operation, a requirement for determining that reserves exist, appears to be a very complicated exercise for two principal reasons. First, one or a few crude types are blended in differing proportions and are routed through a variety of processing steps resulting in the production of many products. These products can differ in chemical and/or physical specifications. In addition, a single-specification product can be made from different mixes of the available crude types (with differing processing costs and

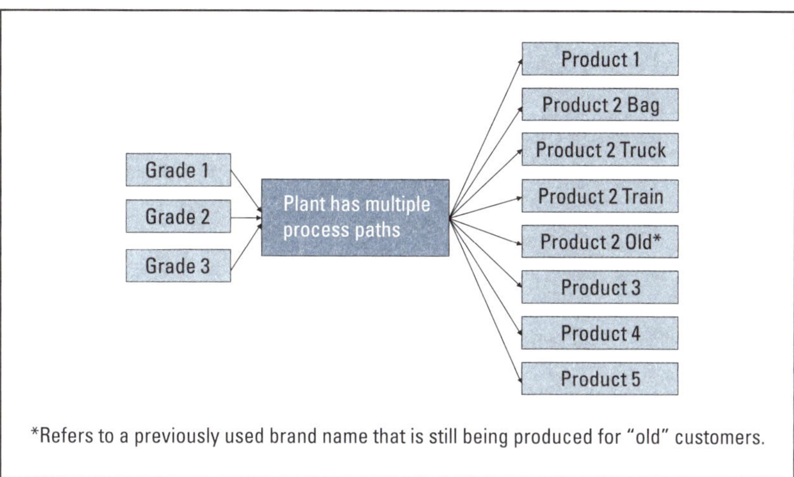

FIGURE 2.11 Schematic showing an industrial minerals facility in which a limited number of crude ore types or grades are processed into a variety of products

losses). A single-specification product can be shipped in consumer packages, 40- or 50-pound bags, tote bags of various sizes, by bulk truck, and by bulk railcar. Sometimes a single-specification product is sold under different brand names to different customers. Each of these many product and shipping options is sold for a different price and in variable quantities. These variables are schematically illustrated in Figure 2.11.

A second complexity occurs where two or more of the products can compete with each other or with competing products sold by other parts of the same parent company. For example, some perlite and diatomite products can compete in the same filtration or filler markets. Likewise, high-quality and whiteness ground calcium carbonates and kaolins compete in some paper filler and coatings markets. This leads to strict corporate secrecy policies that restrict access to market pricing, quantity, and market projection information within parts of the same parent corporation.

In addition, a single-specification product can be made from different mixes of the available crude types with differing processing costs and losses. Some industrial mineral operations have what is known as a "driver" crude. A driver crude is the crude type whose quantity and quality are such that its availability for processing makes a significant or critical difference in a project's economic viability. Ensuring availability of the driver crude in the quantities required for the planned product mix is a critical part of mine planning and sequencing.

Constructing a matrix accounting for all the variables in crude types, processing stream alternatives, product specification types, packaging and shipping types, and product pricing and quantities can easily become an extremely complex exercise. The easiest way of cutting this Gordian knot is by examining the operation as whole. All the costs of mining the crude grades and all the processing and packaging costs are summed and compared with the revenue from all product sales. Even when the prices and volumes of particular products vary over time, if the combined operation is profitable over a sustained period, the profitability of the operation is demonstrated and the appropriately delineated crude ore types and quantities can be classified as mineral reserves. The occurrence of an occasional unprofitable year is acceptable if the cumulative sum of profit and losses over a reasonable period of years is profitable. (For more information on industrial minerals marketing, see Chapter 7.)

SUMMARY

It must be remembered that a mineral deposit is not an ore body. An ore body is a mineral reserve. A mineral deposit may or may not contain mineral reserves or ore; it will contain material that cannot be mined for various engineering reasons, such as pillars in an underground mine, and material that cannot be economically mined, processed, and sold. In the grand scope of industrial mineral project evaluation, geology is important, but permitting and, especially, marketing are more important.

REFERENCES

Abbott, D.M., Jr. 1985. SEC reserve definitions—Principles and practice. *Colorado Professional Geologist* 15(1):9–12.

Abbott, D.M., Jr. 1997. Reporting requirements of the SEC's Industry Guide No. 7. In *Prospectors and Developers Association of Canada (PDAC) Proceedings of a Short Course: Securities Law Reporting Requirements for Ore Reserves.* Toronto: PDAC. March 12, 1997.

Abbott, D.M., Jr. 1999. Mineral resource and reserve definitions: Their evolution, meaning, and current status, and regulation of mineral resource and reserve estimators. In *1999 AIPG Annual Meeting Short Course, Slides and Related Papers.* CD-ROM.

Abbott, D.M., Jr. 2001. "There's glory for you": Definition-related problems between the CMMI, SEC, and USGS mineral reserve and resource classification systems. Abstract. *Geological Society of America Abstracts with Programs* 33(7):184.

Abbott, D.M., Jr. 2002. Testing lab case defendants acquitted. *The Professional Geologist* (March): 24.

Abbott, D.M., Jr. 2003. Mineral reserve audit as of December 31, 2001. Unpublished report. Denver, CO: Behre Dolbear.

Abbott, D.M., Jr. 2007a. Assuring the reliability of your sampling results. *The Professional Geologist* (Nov./Dec.): 33–38.

Abbott, D.M., Jr. 2007b. Industrial minerals reserves and resources classification and evaluation. In *Proceedings of the 43rd Forum on the Geology of Industrial Minerals.* CD-ROM. Denver, CO: Colorado Geological Survey.

Abbott, D.M., Jr. 2014a. Use professional judgment: Know the assumptions underlying a best practice. *Trans. SME* 336:421–425.

Abbott, D.M., Jr. 2014b. Tough ethical decision case—The Brucejack deposit example. *The Professional Geologist* (Oct/Nov/Dec): 46–47.

Abbott, D.M., Jr. 2015. Social licensing—A new responsibility for exploration geoscientists. *The Professional Geologist* (Jan/Feb/Mar): 32–33.

AGI (American Geological Institute). 1997. *Dictionary of Mining, Mineral, and Related Terms,* 2nd ed. Alexandria, VA: American Geological Institute.

Agricola, G. 1556. *De Re Metallica* (Translation by H.C. and L.H. Hoover, 1912), 1950 edition. New York: Dover.

Arseneau, G.J., and Roscoe, W.E. 2000. Practical applications of geology to mineral resource/reserve estimation. In *Mining Millennium 2000: Prospectors and Developers Association of Canada Conference.* CD-ROM.

BLM (U.S. Bureau of Land Management). 2002. *Results of Analyses of Standard and Blank Samples Tested at Selected Assay Laboratories in North America.* BLM Mineral Report. Available from BLM National Training Center in Phoenix, AZ.

Bloom, L. 2000. Implementing quality assurance programs. In *Mining Millennium 2000: Prospectors and Developers Association of Canada Conference.* CD-ROM.

Bloom, L., and Titaro, D. 1997. Building confidence in assays. *Mining Engineering* 49(7):46–48.

CRIRSCO (Committee for Mineral Reserves International Reporting Standards). 2012. Standard definitions. www.crirsco.com/news_items/CRIRSCO_standard_definitions_oct2012.pdf.

Eyde, T.H., and Eyde, D.T. 1992. Application of geology to problems of sampling and grade control unique to industrial mineral deposits. Applied Mining Geology: Problems of Grade Control (1985 symposia). Edited by R.E. Metz. pp. 85–89. In *Applied Mining Geology: General Studies, Problems of Sampling and Grade Control, and Ore Reserve Estimation.* Edited by A.J. Erickson Jr., R.E. Metz, and D.E. Ranta. Littleton, CO: SME.

Holmes, D.A., and Abbott, D.M., Jr. 1995. Defining industrial mineral reserves: Common and subtle problems. In *Proceedings of the 28th Annual Forum on the Geology of Industrial Minerals.* Edited by B.J. Walker. West Virginia Geological and Economic Survey Circular C-46. pp. 116–123.

Hoover, H.C. 1909. *Principles of Mining, Valuation, Organization and Administration*, 1st ed. New York: McGraw-Hill.

Hulse, D. 2016. International social license in mining projects. *Mining Engineering* 68(2):24–29.

JORC (Joint Ore Reserves Committee). 2012. *The Australasian Code for Reporting of Exploration Results, Mineral Resources and Ore Reserves* (The JORC Code). Gosford, New South Wales: JORC. http://jorc.org/docs/jorc_code2012.pdf

King, B. 2001. Optimal mine scheduling. In *Mineral Resource and Ore Reserve Estimation—The AusIMM Guide to Good Practice.* Edited by A.C. Edwards. Melbourne: Australasian Institute of Mining and Metallurgy. pp. 451–458.

Kogel, J.E., Trivedi, N.C., Barker, J.M., and Krukowski, S.T. (eds.) 2006. *Industrial Minerals and Rocks: Commodities, Markets, and Uses*, 7th ed. Littleton, CO: SME.

NI 43-101. *Standards of Disclosure for Mineral Projects.* Canadian National Instrument 43-101. Ontario Securities Commission. www.osc.gov.on.ca/en/15019.htm.

Noble, A.C. 1993. Geologic resources vs. ore reserves. *Mining Engineering* 45(2):173–176.

Peters, W.C. 1987. *Exploration and Mining Geology*, 2nd ed. New York: John Wiley and Sons.

Pitard, F.F. 1993. *Pierre Gy's Sampling Theory and Sampling Practice*, 2nd ed. Boca Raton, FL: CRC Press.

Pocock, J.A. 2001. Why feasibility resource estimates under-valued the Peak orebody. In *Mineral Resource and Ore Reserve Estimation—The AusIMM Guide to Good Practice.* Edited by A.C. Edwards. Melbourne: Australasian Institute of Mining and Metallurgy. pp. 299–314.

Prenn, N.B. 1992. Reserve calculations: An adventure in geo-fantasy? SME Preprint No. 92-196. Littleton, CO: SME.

Preston, G. 2013. Strathcona's Farquharson on the problem with Brucejack. *Northern Miner*, Nov. 27. www.northernminer.com/news/strathconas-farquharson-responds-to-pretiums-bulk-sample-result/1002752061/ (accessed February 2017).

Pretium Resources. 2013a. Pretium Resources Inc.: Bulk sample and exploration drilling update. Press release, Sept. 23. www.pretivm.com/news/news-details/2013/Pretium-Resources-Inc-Bulk-Sample-and-Exploration-Drilling-Update/default.aspx (accessed January 2017).

Pretium Resources. 2013b. Pretium Resources Inc.: Bulk sample update. Press release, Oct. 9. www.pretivm.com/news/news-details/2013/Pretium-Resources-Inc-Bulk-Sample-Update/default.aspx (accessed January 2017).

Pretium Resources. 2013c. Pretium Resources Inc.: Underground bulk sample drilling continues to intersect high-grade gold. Press release, Oct. 10. www.pretivm.com/news/news-details/2013/Pretium-Resources-Inc-Underground-Bulk-Sample-Drilling-Continues-to-Intersect-High-Grade-Gold/default.aspx (accessed January 2017).

Pretium Resources. 2013d. Pretium Resources Inc.: First bulk sample cross-cut processing results. Press release, Oct. 22. www.pretivm.com/news/news-details/2013/Pretium-Resources-Inc-First-Bulk-Sample-Cross-Cut-Processing-Results/default.aspx (accessed January 2017).

Pretium Resources. 2017. High-grade gold. www.pretivm.com/projects/photo-gallery/high-grade-gold/ (accessed Jan. 20, 2017).

Reed, J.P. 2007. Volumetric analysis and three-dimensional visualization of industrial mineral deposits. In *Proceedings of the 43rd Forum on the Geology of Industrial Minerals.* CD-ROM. Denver, CO: Colorado Geological Survey.

Roden, S., and Smith, T. 2014. Sampling and analysis protocols and their role in mineral exploration and new resource development. In *Mineral Resource and Ore Reserve Estimation—The AusIMM Guide to Good Practice.* Edited by A.C. Edwards. Melbourne: Australasian Institute of Mining and Metallurgy. pp. 73–78.

Rowe, R.G., and Hite, R.G. 1992. Applied geology: The foundation for mine design at Exxon Minerals Company's Crandon deposit. Applied Mining Geology (1984 symposia). Edited by A.J. Erickson Jr. pp. 9–27. In *Applied Mining Geology: General Studies, Problems of Sampling and Grade Control, and Ore Reserve Estimation.* Edited by A.J. Erickson Jr., R.E. Metz, and D.E. Ranta. Littleton, CO: SME.

SAMREC (South African Mineral Resource Committee). 2016. *The South African Code for the Reporting of Exploration Results, Mineral Resources and Mineral Reserves (The SAMREC Code).* www.samcode.co.za.

SAMVAL (South African Mineral Asset Valuation) Working Group. 2016. *The South African Code for the Reporting of Mineral Asset Valuation (The SAMVAL Code).* www.samcode.co.za/samcode-ssc/samval.

Sandman, P.M. 2012. *Responding to Community Outrage: Strategies for Effective Risk Communication.* Fairfax, VA: American Industrial Hygiene Association. http://psandman.com/media/RespondingtoCommunityOutrage.pdf.

Scott, B.C., and Whateley, M.K.G. 1995. Evaluation techniques. In *Introduction to Mineral Exploration.* Edited by A.M. Evans. Oxford; Cambridge, MA: Blackwell Science. pp. 161–202.

SEC (U.S. Securities and Exchange Commission). 1981. Adoption of amendments to Form S-18. Securities Act Release 33-6299.

SEC (U.S. Securities and Exchange Commission). 1992. *Industry Guide 7: Description of Property by Issuers Engaged or to Be Engaged in Significant Mining Operations.* sec.gov/about/forms/industryguides.pdf (accessed Jan. 17, 2017). Originally issued as part of Form S-18 in 1981 in Securities Act Release 33-6299; Industry Guide 7 was issued July 30, 1992.

SEC (U.S. Securities and Exchange Commission). 2014. Rule 10b-5. Employment of manipulative and deceptive devices. 17 CFR Part 240.10b-5. https://www.gpo.gov/fdsys/pkg/CFR-2014-title17-vol4/xml/CFR-2014-title17-vol4-sec240-10b-5.xml (accessed Jan. 18, 2017).

SEC (U.S. Securities and Exchange Commission). 2016. 17 CFR Parts 229, 239 and 249 [Release Nos. 33-10098; 34-78086; File No. S7-10-16], RIN 3235-AL81. Modernization of property disclosures for mining registrants; proposed rule. www.sec.gov/rules/proposed/2016/33-10098.pdf.

Securities and Exchange Commission v. Texas Gulf Sulphur Co., 258 F. Supp. 262 (1966) (the District Court case) and the Court of Appeals case, *Securities and Exchange Commission v. Texas Gulf Sulphur Co.*, 401 F.2d. 833 (1968).

Sinclair, A.J., and Blackwell, G.H. 2002. *Applied Mineral Inventory Estimation.* New York: Cambridge University Press.

SME (Society for Mining, Metallurgy & Exploration). 2017. *The SME Guide for Reporting Exploration Information, Mineral Resources, and Mineral Reserves.* Englewood, CO: SME.

Snowden Mining Industry Consultants. 2012. *Pretium Resources Inc. Brucejack Project: Mineral Resources Update Technical Report* (accessed February 2017 on Sedar.com).

Stone, J.G., and Dunn, P.G. 2012. *Ore Reserve Estimates in the Real World*, 4th ed. Special Publication No. 3. Littleton, CO: Society of Economic Geologists.

Swarthout, A.T. 2014. Another important role for an economic geologist. *SEG Newsletter*, no. 98 (July): 8.

Thalenhorst, H., and Dumka, D. 2010. Bulk sampling of mineral projects using a sample tower: Lessons from the field. *CIM Journal* 1(1):44–54.

U.S. v. Jeffus, et al. 2000. Indictment, U.S. District Court for the Northern District of Texas, case 300-CR-375-D. www.txnd.uscourts.gov/pdf/Notablecases/300cr375_1.pdf (accessed March 12, 2007).

USBM and USGS (U.S. Bureau of Mines and U.S. Geological Survey). 1980. *Principles of a Resource/Reserve Classification for Minerals.* USGS Circular 831. Reston, VA: USGS. https://pubs.usgs.gov/circ/1980/0831/report.pdf.

Vecchia, P. 2015. The oil and gas industry: the challenge for proper dissemination of knowledge. In *Geoethics: The Role and Responsibility of Geoscientists.* Special Publication 419. Edited by S. Peppoloni and G. DeCapua. London: Geological Society of London. pp. 133–139.

Whyte, J.B. 2000. Black magic and white lab coats: Pseudoscience in precious metals exploration. In *Mining Millennium 2000: Prospectors and Developers Association of Canada Conference.* CD-ROM.

Wooller, R. 2001. Cut-off grades beyond the mine—Optimising mill throughput. In *Mineral Resource and Ore Reserve Estimation—The AusIMM Guide to Good Practice.* Edited by A.C. Edwards. Melbourne: Australasian Institute of Mining and Metallurgy. pp. 459–468.

CHAPTER 3

Geological Resource Modeling

Qingping Deng

The development of computer technologies and geostatistics in the last 30 years has made it possible for most modern mining projects to use computerized methods for resource\reserve estimation and mine planning. Compared with traditional manual methods, computerized methods are fast and editable. The assay database, no matter how big it is, can be administrated, presented, and used in grade estimation with ease. Advanced geostatistical methods can be applied to grade estimation, and multiple estimation methods can be applied to the same database to validate each other. And grade estimation can be easily modified. Another big advantage is the block model, normally the result of computerized grade estimation, which can be effortlessly used in open pit optimization and mine planning.

Computerized methods themselves, however, do not guarantee the quality of the results, and recently, numerous mining projects that were modeled and designed using computerized methods have failed. Furthermore, grade estimation using advanced geostatistical methods is difficult for many geologists and mining engineers to understand, and the estimation results are not readily verified by a manual method. This leads to the distrust of geostatistics by some tradition-oriented mining professionals.

Actually, geostatistics itself should not be blamed for those failed mining projects with ore reserves estimated by geostatistical methods. First, geostatistics is only a tool for grade estimation. It cannot resolve existing problems in an assay database or the geological interpretation. A biased assay database and/or a biased geological interpretation will generally result in a biased resource model, no matter which advanced geostatistical method is used in grade estimation. Second, some empirical factors, such as mining dilution and mining losses, need to be applied to the in-situ geological resources estimated by any geostatistical method to produce the minable reserves. Generally, appropriate factors are not completely accounted for by a geostatistical method. Third, kriging, the most popular geostatistical method, is supposed to produce the best linear unbiased estimate under specified conditions. If those conditions are not met or approached by a mineral deposit, the grade estimate produced by kriging could be biased. This requires that the practitioners have sufficient understanding of their tools; in many instances, common sense is very important to avoid biased grade estimation.

In a computerized ore reserve estimation, there are three major steps. The first step is to generate an assay and geological database, which is the foundation for the ore reserve estimation. Generating and verification of the assay and geological database is discussed in detail in Chapter 2. The second step is to estimate what is in the ground, that is, the geological resources. This chapter will only cover what should be done when an electronic database is received. The third major step for a computerized ore estimation is to estimate which part of the resources can be mined at a profit at current economic and legal conditions—the minable reserves—which is the topic related to most of the remaining chapters of this handbook.

ELECTRONIC DATABASE VERIFICATION

An assay and geological database in electronic form is the starting point of geological resource modeling for a mining project. A database generally consists of a number of drill holes and sometimes additional surface and/or underground channel samples. Those channel samples are normally incorporated in the database as pseudo–drill holes. The information for each drill hole and/or pseudo–drill hole includes collar location survey, downhole deviation survey, assays or check assays, and geology. The topographic information is part of the electronic database, although it is generally presented as a separate file from the database. Digital models of any existing underground workings should also be included or generated for those historical underground mining projects.

The database needs to be verified before it can be used for geological resource modeling. Data entry errors are often present in a database, especially for assays entered by hand. Database verification is generally part of a reserve audit. Assays in at least 5% of randomly selected drill holes should be checked with the original assay certificates or other data sources. Data entry errors found in database verification should be carefully analyzed. In general, the assay intervals with data entry errors should be significantly less than 1% of the total assay intervals in a database, and the errors should be random in nature. If the error percentage is getting uncomfortably high or if there is any indication of the presence of a systematic error, the whole database should be checked against the original certificates or other data sources, and the errors found should be corrected in the database before any further work is done.

The quality of the drill-hole and topographic surveys can be checked against each other by plotting both on cross sections and plans. The surface drill-hole collars generally should not be floating above or buried beneath the topography. However, there are exceptions. For example, if the topography has been modified by mining activities after drilling, some of the drill-hole collars may not conform to the current topography. Inconsistent drill-hole collar location and topography indicates poor database quality, and the geological resource model generated from this database will be less reliable.

GEOLOGICAL RESOURCE MODELING PROCEDURES

When the database verification indicates that the assay and geological database is in good-quality condition, quality, a geological resource model can be generated from the database. The following are the general procedures to generate a computerized geological resource model:

1. Construct geological models, including digital models for any existing underground workings for a historical underground mining district.
2. Define the three-dimensional (3-D) computerized block model.
3. Code geological domains to assays and model blocks.
4. Statistically analyze assay data.
5. Determine the capping grades.

6. Generate composites from individual assays.
7. Model variograms from composite data.
8. Estimate model block grade and verify grade estimation.
9. Classify geological resources into categories with different confidence levels.

The execution order of the preceding steps can be changed, and some of the steps can even be omitted based on the specific condition and resource modeling requirements for each deposit. How to exploit each geological resource modeled is discussed in detail in the following chapters of this handbook.

CONSTRUCTING AN UNBIASED GEOLOGICAL MODEL

A satisfactory and unbiased geological model is just as important as a good assay database for a well-founded geological resource estimate. Sometimes, especially at the early stage of a project when the drill-hole density is not sufficiently high, a biased geological model can make the resource estimation very different from the ground truth. When performing reserve auditing, the author has experienced biased geological modeling that resulted in 20% to more than 50% resource overestimation. Hence, a biased geological model could have a much larger impact to a resource estimate than a biased estimation method.

Depending on the complexity of the deposit and the grade-controlling factors, a geological model can consist of one or more of the following:

- Lithological model, which separates the deposit into different lithology or rock domains
- Alteration model, which separates the deposit into different alteration domains
- Structural model, which separates the deposit into different structural domains
- Weathering model, which separates the deposit into different weathering domains
- Mineralization model, which separates the deposit into different mineralization domains
- Grade model, which is predominantly defined by one controlling grade or a combination grade with the consideration of geological constraints

The most important role of a geological model for the computerized ore reserve estimation is to control the grade projection in grade estimation. Therefore, the geological factors that control the grade distribution in a deposit should be modeled.

For example, if the mineralization is mostly controlled by one or few rock or alteration types in a deposit, the rock or alteration domain model should be constructed. However, if there is not an apparent relationship between rock or alteration types and grade distribution, the rock or alteration domain model may not be important for grade estimation, and it can be omitted for the deposit. Furthermore, if the quality of the drill-hole geological logging is poor, the rock or alteration domain model that is constructed may even be misleading. A structural model is generally needed if the mineralization is offset by some post-mineral faults and the mineralization is not continuous across the structures. Some of the metals, such as copper, are not stable in the supergene process and will be leached in some parts of the supergene zones and enriched in other parts of the supergene zones; therefore a weathering domain model will be very important for these deposits. It is quite common that the only major grade-controlling factor that can be modeled in a deposit is the grade itself. Therefore, the grade model is the most common geological model constructed for reserve estimation in the mining industry.

Another role of a geological model is to define the bulk density distribution in a deposit. This is very important if the bulk density is significantly different in different rock or alteration types in a deposit. However, it is not uncommon for a deposit using an overall average bulk density when the bulk density is not significantly different in different rock or alteration types.

Sometimes, a geological model is used to control the metallurgical behavior of different ore types. For example, the gold distribution may not change much over the oxide–sulfide boundary in a deposit, but the metallurgical behavior of ore is generally very different. The gold grade classified as ore in the oxide zone could only be classified as waste in the sulfide zone because of the lower metallurgical recovery and/or high processing costs.

The 3-D computerized modeling technique developed in recent years for many of the mining software packages has significantly improved the geological modeling process. In the early years of computerized geological modeling, cross sections or plans with assays and geological information were plotted on paper. Geological interpretation was done on paper and then digitized into the computer software system. Current 3-D graphic systems in many mining software packages allow generating section and/or plan geological interpretations directly on the computer screen. Furthermore, the data points of the geological interpretation can be snapped on the actual drill-hole data points, which makes the geological interpretation more accurate. The geological modeling process generally consists of the following steps:

1. Project cross-sectional views of assays and geological information on computer screens.
2. Produce sectional geological interpretation on computer screens.
3. Build a 3-D solid geology model from a sectional geological interpretation.
4. Check for consistency.

Some simple examples are used here to illustrate the importance of unbiased geological interpretation for grade estimation. Figure 3.1 shows four drill holes and a surface outcrop with blue mineralized intercepts for a shear-zone-controlled or stratabound deposit. By simply

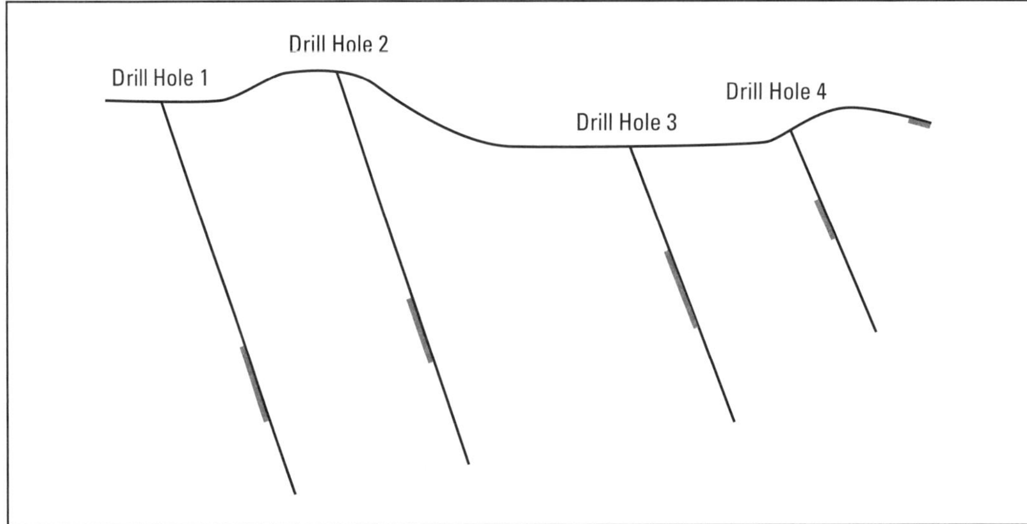

FIGURE 3.1 Four drill holes with mineralized intercepts (heavy black lines). This deposit could be a shear-zone-controlled or stratabound deposit.

connecting the mineralized intervals in the drill holes, the outcrop will produce an interpretation as shown in Figure 3.2.

However, if the surface geological mapping indicates the presence of a post-mineral normal fault between drill holes 2 and 3, Figure 3.3 will be a better geological interpretation. The major difference between interpretations in Figures 3.2 and 3.3 is the location of the ore body, but the volume of the ore body has not been changed. Therefore, either interpretation should not produce a tonnage-biased resource estimate.

The mapped fault between drill holes 2 and 3 could be interpreted as a feeder structure for the deposit, and it could also host the root of the ore body. The geological interpretation

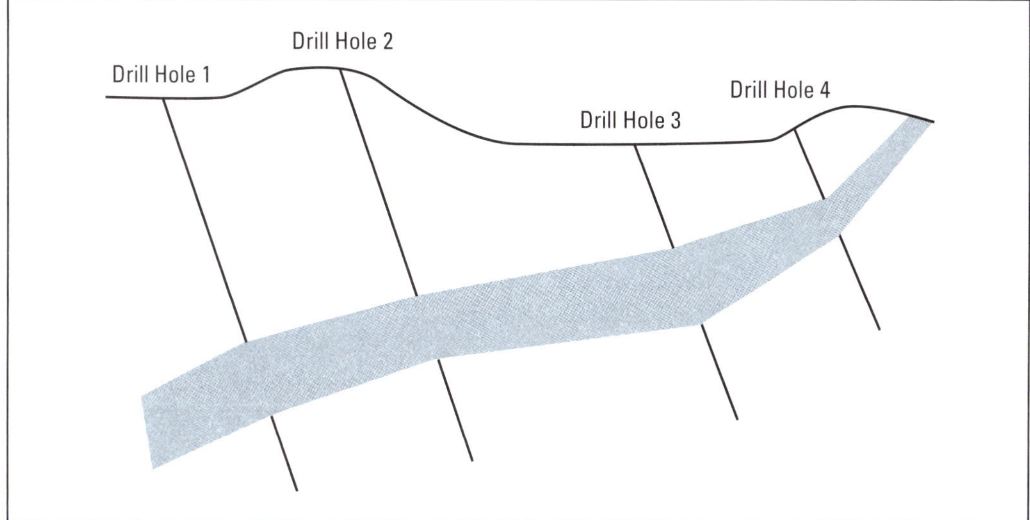

FIGURE 3.2 Interpretation produced by connecting the mineralized intervals in the drill holes

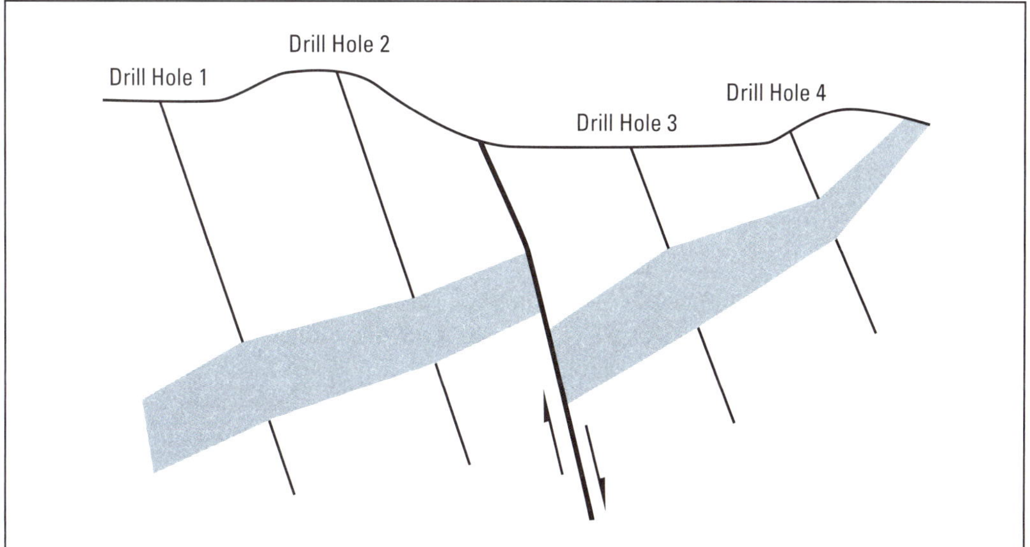

FIGURE 3.3 Geological model interpreted with a post-mineral, normal fault between drill holes 2 and 3

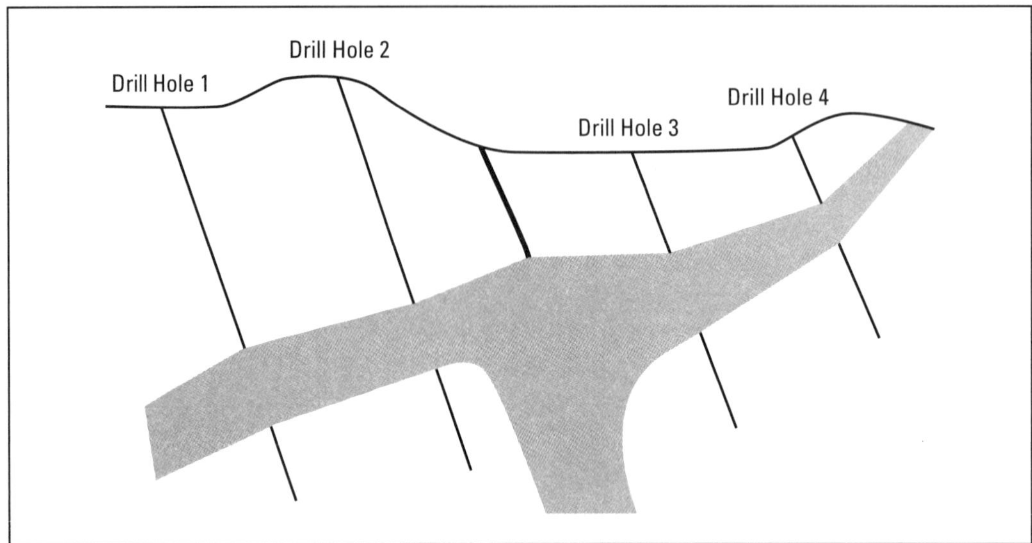

FIGURE 3.4 Possibly biased feed structure interpretation

is shown in Figure 3.4. For the purpose of exploration, this interpretation is probably not too bad, and the geologist should then design some follow-up drill holes to prove the existence of the root structure. However, for the purpose of resource modeling, this represents a biased geological interpretation, as it is based on the geologist's imagination, not on hard geological evidence. Similarly biased geological interpretations have been seen by this author in reserve audits.

If one of the drill holes did not intercept the mineralized interval, the geologist will need to interpret the ore bodies around the barren hole. The ore body could terminate at any point between the mineralized hole and the barren hole, but there is no way to know it until additional holes are drilled or the ore body is mined. For resource modeling purposes, an unbiased interpretation will draw the ore body boundary near the middle of these two holes. However, if the boundary was drawn closer to the waste hole (Figure 3.5), the ore tonnage could be overestimated. The opposite could also happen if the boundary was drawn closer to the mineralized hole, resulting in underestimation of the ore tonnage (Figure 3.6).

In addition to traditional manual geological modeling, the mining industry also uses an indicator kriging method to define the grade mineral envelope (Pan 1994). This method is especially useful for deposits with complicated ore body boundaries but a constant attitude, such as a shear-zone-controlled precious metal deposit with a constant strike and dip. Care should be taken in defining the geostatistical mineral envelope around the edges and the bottom of the drilling database, as the technique will tend to produce some abnormal blocks inside and/or outside the mineral envelope.

The importance of a well-founded and unbiased geological model cannot be overemphasized. A good, unbiased geological interpretation is the basis for a good resource/reserve estimate. It guides the computer programs in grade estimation and limits the extrapolation to the reasonable. Bad and/or biased geological models could produce extremely biased resource/reserve estimations, which is one of the leading factors in mine project failure.

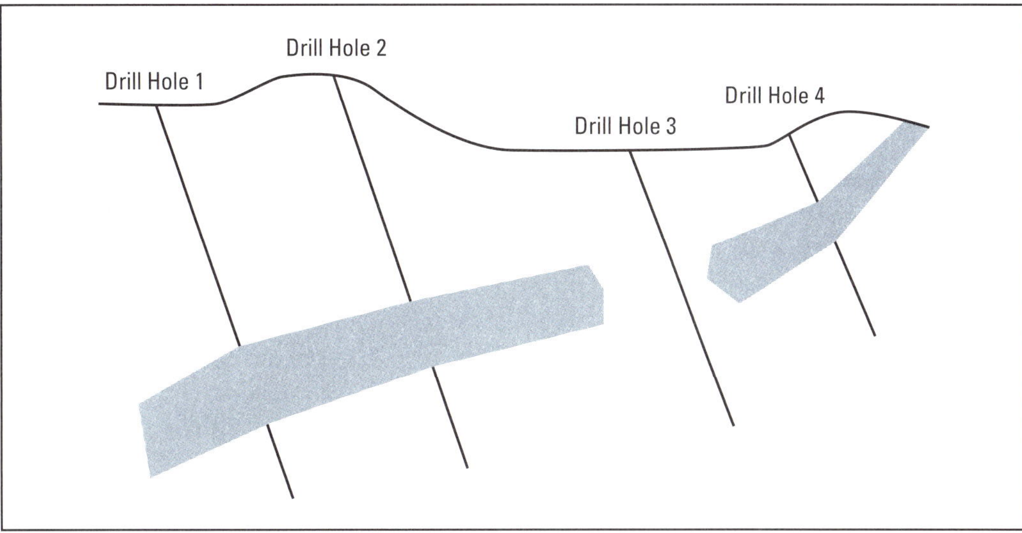

FIGURE 3.5 This geological model may overestimate the ore tonnage

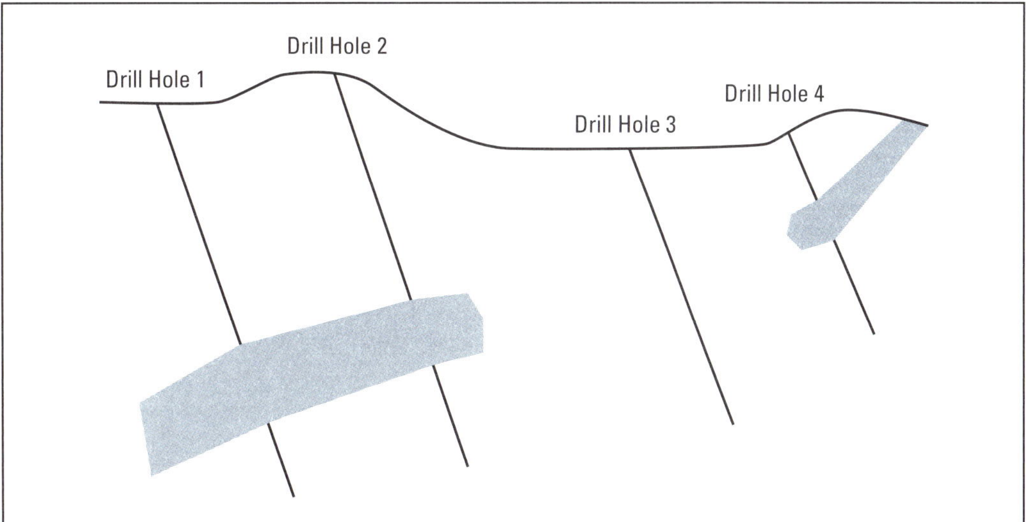

FIGURE 3.6 This geological model may underestimate the ore tonnage

DEFINING A 3-D COMPUTERIZED BLOCK MODEL

The computerized resource estimation is generally achieved through a 3-D block model. Two important considerations are needed in defining a block model: the orientation of the block model and the block size.

For a deposit with no apparent anisotropy, such as a porphyry copper deposit, the block model orientation is not critical for resource modeling, and the model axes are generally oriented with the geographic directions. For an anisotropic deposit, such as a vein-type deposit, however, the block model orientation becomes an issue, and the axes of the block model should follow the direction of mineralization. Different mining software packages have different ways to orient the block model with the mineralization. Some of the software packages rotate the axes of the block model to the direction of the mineralization, and the drill-hole database maintains the true geographic orientation. Other software packages rotate the drill-hole database and the topography to make the major mineralization direction follow the major block model axes. It is very important to understand how the rotation is achieved when transferring a block model from one mining software package to another.

The vertical block model axis can also be rotated in some mining software packages. This is probably good for an underground mining project, but it will cause problems for open pit projects as the computerized pit optimization and planning will be difficult to perform on a block model with a rotated vertical axis.

An appropriate block size should be used for a block model. Large blocks are easier and quicker to model, but the grade distribution could be overly smoothed and the ore–waste boundary cannot be correctly determined. Small blocks give a better representation of the grade distribution, but the selection built in the resource model may not be able to be achieved in actual mining operation. Selection of the model block size should be based on the size of the anticipated project selective mining unit (SMU), which is generally determined from the deposit grade distribution and the size of the mining equipment to be used in a mining operation. In the early years of computerized resource modeling, it was impractical to use an SMU as the model block size for a large mining project because of computer processing limitations. Because computer processing capacities are now significantly improved, it is possible to model almost any projects using an SMU block size.

CODE GEOLOGICAL DOMAINS TO ASSAY SAMPLES AND MODEL BLOCKS

The geological domains defined in geological modeling need to be coded to the assay samples in the database and the model blocks.

Different methods are used to code the geological domains to assay samples. It is generally not too difficult to code the broad geological domains to the assay samples. However, when the geological domains are narrow, coding becomes a challenging task. A commonly employed method is to use the 3-D solid geology models of the geological domains to code the assay samples. Because the 3-D solid models were generated from 3-D section interpretations snapped on actual data points, the coding results should match the original interpretation exactly. It is important to carefully check that the coding has been done correctly. One of the worst coding methods uncovered by this author from ore reserve auditing is using a cut-off grade to code the assay samples, resulting in an extremely positive bias in grade estimation.

Coding the geological domains to the block model is relatively easy and makes it more difficult to produce a bias than coding the assay samples. Depending on how much detail in grade distribution is needed, model blocks can be coded as whole blocks or block partials. The whole block coding generally uses the centroid rule or the majority rule, in which a model block is coded based on the location of its block centroid or the majority of the block. The partial block coding may code one mode block with more than one geological domain, and the domain codes and the proportions of the domains in each block are recoded in the model file. The partial block coding is important for some narrow structure-controlled deposits, especially for underground mining projects. The sub-block coding method is also commonly used in many mining software packages.

The domain boundaries can be treated as "hard" or "soft" in grade estimation. A hard boundary means that the assay samples and model blocks are totally isolated by domain, and only samples located within a domain will be used to estimate the grade for blocks inside the domain. A soft boundary means that the assay samples and model blocks are not totally isolated by domain. Some of the samples outside but near a domain boundary will be used together with the assay samples inside the domain to estimate the block grades inside the domain. This will produce a smoother transition at the domain boundary in grade estimation. When a boundary is utilized as soft, it should be soft in both directions; otherwise, some kind of grade bias may be introduced in grade estimation.

STATISTICAL ANALYSIS OF ASSAY DATA

Classic statistics is one of the most vital parts of a good ore reserve estimate. It should be used to guide the selection of the appropriate modeling procedures and the determination of the capping and/or cutting grades and to evaluate the final results.

Classic statistics should be performed on original assays before the geological model definition. For example, classic statistics of original assays by rock type or alteration type can be used to determine if the rock type and/or the alteration type is the primary grade-controlling factor in a deposit. If it is determined that the rock type or alteration type is the most important grade-controlling factor, a rock or alteration domain model should be constructed to separate the original assays into different populations than to estimate block grade in each population separately.

Classic statistics of assays of different sample types or different sampling campaigns can be used to determine if there is any systematic grade bias for different sample types or drilling campaigns. Care should be taken in sample type or sampling campaign comparison, and the samples selected should be from comparable locations. If the sample locations are not comparable, the comparison result could be misleading.

The probability distribution sometimes will provide important sample population distribution information. If it is determined that more than one assay population is present in a deposit, then these different assay populations may need to be separated in grade estimation. Generally, the different assay populations can be related to one or more grade-controlling geological factors in a deposit.

Classic statistics should be performed on original assays coded by the geological domains. Assay statistics by domain can be used to check the quality of the geological domain model. If one or more geological domains are supposed to control the mineralized assay samples but the relationship is not clear for assay statistics by domain, something could have gone wrong in the geological modeling or assay domain coding. Typically, capping grade analysis should be conducted on original assays by domain. Classic statistics of capped original assay samples by domain can be used to check the compositing procedures.

Classic statistics should be also performed on assay composites in different geological domains. The composite grade distribution can be used to select the grade estimation algorithm. The composite grade variance should be used to guide the selection of the total sill in variogram modeling.

The declustered composite grade mean and the declustered grade distribution generated by the nearest-neighbor declustering method can be used to check the global grade bias in block grade estimation. In general, the block grades should have a similar distribution as the declustered composite grades; the average grade of the estimated blocks should not be higher than the average declustered composite grade at zero cut-off grade. And the variance of the block grades should be smaller than the composite grade variance in a geological domain because a model block generally has a much larger volume than an assay composite.

DETERMINING THE CAPPING GRADES

To prevent grade overestimation, outlier samples that do not fit an assay distribution should be properly capped before grade estimation. Grade capping is more critical for precious metal deposits as precious metals are generally distributed as trace elements in a deposit, and their distribution is generally highly skewed. The top 1% of the high-grade samples can carry as much as 20% or more of the total metal in a drill-hole database. Grade capping is less critical for base metal deposits as base metals are generally distributed as minor or sometimes even major elements in a deposit, and their grade distribution is not as skewed as precious metals.

Capping grade determination is more of an art than a science. The highest grade samples should be sufficiently capped to prevent grade overestimation, but they should not be over-conservatively capped to kill a good project. Different grade capping methods are preferred by different people. Some arbitrary capping grades, such as 30 g/t (grams per metric ton) (approximately 0.88 opt [troy ounce per short ton]) gold, have been used in the mining industry. These arbitrary capping grades might be appropriate in some instances, but their use is not recommended for most of the mining projects. A capping grade of 30 g/t (0.88 opt) gold might be still too high for a disseminated gold deposit, but it could be too low for a structure-controlled high-grade deposit.

Capping grade analysis is preferably carried out on the original assay samples, as the sample compositing process tends to smooth the grade distribution and protect some of the outlier samples in the original assays. However, capping on composites is also an acceptable practice in the mining industry. In general, the capping grade determined on composites should be lower than the capping grade determined on original assays because the grade distribution has been smoothed by the compositing process.

When analyzing the capping grade for the original assays, special care should be paid to sample populations with different length supports. For example, the database for one gold project contains two types of assay samples: One of the samples is the diamond-core drill hole with a nominal sample length of 1 m (3 ft). Another of the samples is the surface channel with a nominal sample length of 5 m (16 ft). The capping grade determined for the 1-m (3-ft) drill-core samples is 70 g/t (2.04 opt) gold, while the capping grade determined for the 5-m (16-ft) surface channel samples is 30 g/t (0.88 opt) gold. The capping effect of these two very different capping grades on sample populations with different length supports is very similar, as indicated by the percentage of metals capped in the database. When the uncapped 1-m (3-ft) drill-hole assays were composited to 5-m (16-ft) composites, it was found that the appropriate capping grade determined for the 5-m (16-ft) drill-hole composites is also 30 g/t (0.88 opt) gold.

Capping grade for a deposit should be determined from its sample distribution. One of the commonly used methods for capping grade determination is using the break point on a grade probability plot in a log-log scale. If the assay samples follow a log-normal distribution, the data points in the probability plot should follow a straight line. For example, if the assay silver grade probability distribution for a shear-zone-controlled silver deposit has a break point at around 10,000 g/t, it is appropriate to cap the silver assay grade at the break point of 10,000 g/t. The probability plot could also show that there are two or more major assay grade populations in the database, as indicated by the two or more straight lines in the diagram. The lower straight line could represent the silver distribution of premineralization background, and the upper straight lines could represent the mineralized silver assay distribution. These populations of two or more should be separated in grade estimation if possible and appropriate.

A different approach to limit the influence of the outlier samples in the mining industry is to reduce the search distance of outlier samples in grade estimation. This method, when used correctly, can also have a similar effect to the grade estimation as capping the outlier samples. The advantage of this method over grade capping is that it keeps the assay upper tail grade distribution in the estimated block grades, but how much of the search distance should be reduced for the outlier samples is generally arbitrary.

The ultimate check for a correct capping grade is the production reconciliation. If the production reconciliation indicates that the capping grade used in grade estimation is too high or too low, the capping grade should be adjusted accordingly. One danger with production reconciliation is to use a small portion of the production for comparison, as the comparison could be very different in different parts of the deposit. The volume used for comparison should be as large as possible. And also, different styles of mineralization or different geological domains should be separately studied if possible.

COMPOSITING

The original assay samples are generally composited before variogram modeling and grade estimation. The objectives of compositing are to

- Provide equal or similar length support for samples used in variogram modeling and grade estimation,
- Ensure that the grade estimation samples have similar resolutions as the ore control samples, and
- Reduce the total number of samples used in variogram modeling and grade estimation.

In most mining software systems, there are two compositing algorithms: bench compositing and length compositing. Bench crest and toe are used as composite boundaries in bench compositing, and all the samples inside the bench limit are combined into one composite without considering the actual sampling length for the composite. The midpoint location of the composite is assigned as are the composite coordinates. Some mining software systems store the composite elevation at the bench toe level. Some problems exist with bench compositing. For example, if the holes are drilled at different dip angles, the length support of bench compositing could be different for different composites, especially for the low-angle drill holes. And bench compositing sometimes causes excessive smoothing in grade estimation, as the composite interval is not related to the original assay interval, thus one 1-m (3-ft) high-grade assay could produce two 5-m (16-ft) moderate-grade composites.

Length compositing, also referred to as downhole compositing, overcomes those bench compositing weaknesses. The compositing starts at the collar location of a drill hole, and composite intervals are based on drill-hole depth. Although length compositing is preferred in grade estimation, bench compositing is still an acceptable industrial practice, especially for those projects with mostly vertical drill holes.

Composite length is generally selected based on the bench height for an open pit mining project, as the bench height will be the mining resolution. If the composite length is much shorter than the bench height, the dilution built into the grade model may not be sufficient.

However, if the composite length is longer than the bench height, the dilution built into the grade model might be excessive.

VARIOGRAPHY

Variography, or spatial statistics, is the determination of the spatial correlation of sample values, that is, how sample values vary as a function of distance. Variography study is very important for computerized ore-reserve estimation in the mining industry. This is because sample variogram (also referred to as semi-variogram) models are generally used to guide the selection of search parameters in grade estimation and to guide resource/reserve confidence level classification. If kriging is used for grade estimation, the sample variogram models are part of the kriging formula, which assigns the kriging weights to each sample point. Numerous geostatistical textbooks thoroughly cover variography theory (Journel and Huijbregts 1978; Clark 1979; Rendu 1981; Isaaks and Srivastava 1989; Rendu and Mathieson 1990); however, this handbook concentrates on the practical applications of variography study.

The major components of an ideal spherical variogram model are the nugget, sill, and range. The nugget is the variogram value at zero distance, and it represents the pure randomness of the sample values, as samples taken from locations next to each other may have different grades for a nuggety deposit. Higher nugget value indicates poorer sample value continuity, and the grade estimation in deposits with a higher variogram nugget tends to be more difficult. The sill is the change in correlation of the sample values over distance. The range is the maximum distance of sample value spatial correlation. Sample values are no longer correlated beyond the variogram range. A variogram model could have more than one sill and range, which are referred to as nested structures. Theoretically, the sum of the nugget and sills, referred to as the total sill, should be equal to the sample variance, if there is not a trend in the spatial distribution of sample values.

Following are the three basic steps in variogram modeling:

1. Select the variogram modeling parameters, including variogram directions, angle tolerance, lag distance, and lag tolerance.
2. Calculate a sample experimental variogram for each lag distance.
5. Fit a variogram model to the calculated experimental variogram points.

Variogram modeling parameter selection should be based on the geology and sample configuration of the deposit. For an isotropic deposit, such as a large porphyry copper deposit, an omnidirectional variogram model could be all that is needed for grade estimation. However, in most cases, the sample value correlation in different directions is not the same, and variograms are generally modeled for the three axis directions normal to one another to define the spatial distribution of the sample value correlation. The direction for the best sample value continuity is referred to as the major axis, and the variogram range along this direction should be the longest. The direction for the least sample value continuity within the plan perpendicular to the major axis is referred to as the minor axis, and the variogram range along that direction should be the shortest. The direction perpendicular to the plan determined by the major and minor axes is referred to as the semi-major axis, and the variogram range along this direction should be between the variogram ranges for the major and minor axes. In some of the mining software systems, the semi-major axis is referred to as minor axis, and the minor axis is referred to as vertical axis.

Determination of the three axis directions should be based on the geological control of the deposit. If the grade distribution is strongly directional, such as that in a vein-type or structure-controlled deposit, the major and semi-major axes should be located within the structure plane, and the minor axis should be perpendicular to the structure plane. For a practical purpose, the strike direction and the downdip direction of the controlling structure can be selected as the first two axis variogram directions. Which of them is the major axis will depend on the variogram range. For some other deposits, the axis mineralization directions are not obvious, and variograms in many directions should be modeled to find out the primary axes. There is some sophisticated 3-D variogram modeling software available in the mining industry, that makes finding the primary variogram directions much easier. Care should be taken in using this type of 3-D variogram modeling software, and it should always be kept in mind that the geological controls and the data configuration should be carefully considered.

Selection of the angle of tolerance and lag distance is generally a trial-and-error process. Variograms modeled using a small angle of tolerance will better reflect the sample value continuity in the variogram direction. But sometimes, the sample pairs found within a small angle of tolerance are not sufficient to produce a meaningful variogram model. Variograms modeled using a larger angle of tolerance will reflect the average sample value continuity within the sample search cone. The extreme is using an angle of tolerance of 90 degrees, which will produce an omnidirectional variogram.

A good starting point for selecting the lag distance is the average sample distance along the variogram direction, and it can be adjusted longer or shorter to produce the best variogram model. The lag tolerance is generally selected at half of the lag distance. As the drilling data are generally directional, the lag distance for different directions should be adjusted to reflect the sample configuration. For example, the lag distance along the drill-hole direction can be selected at the composite length, whereas lag distance perpendicular to the drill-hole direction can be selected at the average drill-hole spacing.

Once the variogram modeling parameters are selected, most variogram modeling software will automatically calculate the variogram value at different lag distances along one or more directions using the following formula:

$$\gamma(h) = \frac{1}{2n}\sum(S_i - S_j)^2 \qquad \textbf{(EQ 3.1)}$$

where $\gamma(h)$ is the variogram value, n is the total number of selected sample pairs, and S_i and S_j are the sample grades for each selected sample pair.

There are some basic rules in fitting a variogram model to the calculated experimental variogram points on a $\gamma(h)$–distance scatter plot. If these rules are not followed, very different variogram models could be fit into the same experimental variogram calculation.

The first rule is that the total sill (= nugget + sills) of the variogram model should be equal to the sample variance if there is not a trend in sample value spatial distribution. A slight change of the total sill value could significantly increase or decrease the variogram range. Two very different variogram models could be fitted into the same experimental variogram calculation. The experimental variogram calculation could be normalized by dividing the variogram value by the sample variance; therefore, the total sill of the variogram model should be equal to 1. The variogram model with a correct total normalized sill (1.0) has a range of 33 m (108 ft); however, if the total sill is selected at 1.12, the variogram range would be increased to 100 m (328 ft), which is three times the correct variogram range. This long variogram range will give a false picture of much better continuity for the mineralization and mislead the grade estimation and reserve classification. Resource/reserve classification based on this kind of variogram model has been seen in due diligence reserve audits.

The second rule is that the nugget and sill should be consistent in different directions. Because of the directional nature of the drill-hole assay database for most mining projects, the sample density is generally very different in different directions. The direction along the drill hole generally has the highest sample density, which is generally much higher than the directions perpendicular to the drill-hole direction. Therefore, the nugget and sills of the experimental variogram calculation in different directions could appear quite different. Generally speaking, the variogram nugget and sills should be determined from the direction with the highest sample density, and the same nugget and sills should then be applied to variogram models in other directions. Only one set of nugget and sills can be input into the kriging formula in kriging grade estimation. Even if there are different nuggets and sills in different directions, only one set of them will be used. One danger with using different nuggets and sills for different directions is that the anisotropy ratio could be obscured by the incorrect nuggets and sills.

The third rule for variogram modeling is that the three axis directions should be normal to each other. If a good variogram model cannot be produced in one of the axis directions, a direction close to the axis direction should be selected to produce an approximate variogram model for that axis direction. Using variogram models in three random directions in kriging grade estimation is not an acceptable method in the mining industry.

It should be kept in mind that the samples used in variogram modeling are only the *samples* of the whole population. The spatial correlation of the samples is just a statistic of the spatial correlation of the whole population. When the quality of the samples is poor because of incorrect measured sample values or insufficient sample density, the spatial correlation obtained from the samples could be significantly different from the true spatial correlation of the whole population. This means that the variogram model is data dependent. Generally, both the variogram model nugget and variogram range will decrease with the increase of the sampling density until a certain limit is reached. Variogram models generated using drill-hole databases in the early stage of an exploration project tend to have higher nuggets and longer ranges than the true variogram. Using this type of variogram model as the only guide in selecting search parameters for grade estimation and classifying resources and reserves should be avoided.

One of the common problems in variogram modeling is that the variograms are not calculated for the mineralized samples. If assay samples with mixed mineralized/waste populations are not separated by geological domains, and all the samples are used in variogram calculation, then the variogram models that are produced could be overwhelmed by the waste samples, resulting in nice, smooth, well-behaved variogram models with long variogram ranges. This will usually lead to the incorrect conclusion that the mineralized samples correlate at unreasonably long distances. One simple method to eliminate the influence of the waste population in this instance is to use a cut-off grade in variogram calculation, and only samples with a grade equal to or higher than the cut-off grade will be used in variogram calculation. One example of the influence of waste samples on variogram calculation shows that a variogram calculated using all the samples has a range of 64 m (210 ft), which more than doubled the variogram range of 31 m (102 ft), calculated using a cut-off grade of 0.4 g/t (0.01 opt) gold.

Another common problem in variogram modeling is caused by the directional nature of the drill-hole database. From the exploration point of view, the holes should be drilled as close as possible in the direction perpendicular to the mineralized plane. Therefore, it is common to see that the majority of the drill holes in a database were drilled toward one direction. For example, for a near-vertical structure-controlled mineralized system, the holes should be drilled in a dip angle as low as possible. However, because of the limitation of the current drilling technology, surface drill holes are rarely drilled in a dip angle less than 45 degrees. If the drill-hole spacing is larger than the average mineralized zone thickness, which is the case in most instances, it is generally difficult to model the variogram range directly for the direction perpendicular to the mineralized zone. However, a reasonable variogram model can normally be produced along the drill-hole direction, and the variogram range in the direction perpendicular to the mineralized zone can be indirectly obtained from the variogram model along the drill holes. It should be kept in mind that the variogram range in the direction perpendicular to the mineralized zone should be shorter than the variogram range along the drill-hole direction, and the precise variogram range in the direction perpendicular to the mineralized zone can be calculated from the variogram ranges along the drill-hole direction and inside the mineralized plane using the ellipse equation.

Variogram models should make sense when comparing with the geology. The major variogram axis should be within the mineralized plane for a structure-controlled deposit; otherwise, something could be wrong with the variogram model.

In conclusion, the variogram model is both data dependent and user dependent. It is critical to use a good variogram model in grade estimation and resource/reserve classification in an ore reserve study.

GRADE ESTIMATION AND VERIFICATION

Modern reserve grade estimation mostly uses computerized methods, as traditional manual methods are generally more time-consuming, tedious, and the result is difficult to modify and use in open pit mine planning. However, manual methods are still used for ore reserve estimation, especially for underground operations, in some parts of the world. Furthermore, manual methods are often used as a tool to verify computerized grade estimation when there is a disagreement on the ore reserves estimated by the computerized methods.

The traditional manual ore reserve estimation methods include the sectional methods and polygonal and triangular methods (Stone and Dunn 1996). The sectional methods can be further separated into cross-sectional methods and longitudinal sectional methods. The manual methods are not discussed in detail, as the computerized methods are emphasized in this handbook.

The most commonly used computerized grade estimation methods include the nearest-neighbor method, inverse-distance method, and kriging method.

The *nearest-neighbor method* is the block model equivalent of the manual polygonal method. It uses the closest sample to determine the grade for a model block. This method actually produces a declustered sample data set within the study space with no variance reduction. It should give similar grade and tonnage figures as the manual sectional or polygonal method. As there is no variance reduction from the sample grades to model blocks grades, ore reserves estimated by the nearest-neighbor method are generally too optimistic, which generally have higher grade and lower tonnage at a given cut-off grade than the actual production. Therefore,

ore reserves estimated by the nearest-neighbor method are generally not acceptable in the mining industry. However, the nearest-neighbor method is generally considered as a globally unbiased grade estimation method. The average grade for a nearest-neighbor model at the zero cut-off grade is generally unbiased, which is commonly used to check the global grade bias for block model estimated by other methods.

The inverse-distance method assigns grades to model blocks from one or more samples within a specified search neighborhood using the linear weighted average method. The sample weights are based on the inverse of the distance, or the anisotropy distance, from the composite to the block centroid raised to a user-defined power. The sum of all the sample weights is normalized to one. The formula for inverse-distance grade estimation is as follows:

$$\text{grade} = \frac{\sum x_i \times \left(\frac{1}{d_i}\right)^n}{\sum \left(\frac{1}{d_i}\right)^n} \tag{EQ 3.2}$$

where x_i is the sample grade, d_i is the distance or anisotropy distance from sample grade x_i to the block centroid, and n is the user-defined inverse-distance power ranging from 1 to ∞.

The *inverse-distance method* has a long history of utilization and is still a commonly used grade estimation method in the mining industry. Different inverse-distance power can be used to adjust the amount of smoothing built into the grade estimation model. Generally speaking, the higher the inverse-distance power, the less smoothing will be built into the grade estimation. When the inverse-distance power is sufficiently high, such as 4 or 5, the result of the grade estimation will be similar to that of the nearest-neighbor method. The most commonly used inverse-distance power in the mining industry is between 2 and 3.

The primary criticism of the inverse-distance method is that a sample declustering function is not built into the formula. Therefore, the inverse-distance method is more commonly used for deposits drilled on a uniform grid, such as a large porphyry copper deposit. To compensate for the lack of a sample declustering function, quadrant or octant search is commonly used to limit the sample influence from a particular direction in inverse-distance grade estimation.

Kriging is considered a "best linear unbiased estimator" under specified conditions, which assigns a grade to a block using one or more samples within a specified search neighborhood:

$$\text{grade} = \sum b_i \times x_i \tag{EQ 3.3}$$

where b_i is the kriging weight of sample grade x_i.

The sample kriging weight, b_i, is a function of the sample variogram model and sample configuration, which is determined by a set of simultaneous linear equations. Numerous geostatistical textbooks cover the theory of kriging in depth (Journel and Huijbregts 1978; Clark 1979; Rendu 1981; Isaaks and Srivastava 1989; Rendu and Mathieson 1990).

Similar to the inverse-distance method, the *kriging method* is a linear weighted average method using one or more samples. However, the key difference from the inverse-distance method for kriging is that the sample kriging weight, b_i, is determined not only by the anisotropy distance from the sample to the block centroid, but also by the anisotropy distance between samples. That means a declustering function has been built into the kriging method. Therefore, the quadrant or octant search is not as critical for kriging as it is for the inverse-distance method.

Following are the three key assumptions for the kriging method:

1. The sample grade distribution is Gaussian (normal).
2. The sample grade distribution is stationary; that is, the mean and variance of the sample grades do not change with distance.
3. The sample variogram models are positive definite, which guarantees that the kriging equations have only one set of solutions.

These assumptions are rarely completely met for a real deposit. For example, gold and silver grades in most precious metal deposits generally have a near lognormal distribution, and more than one sample population is often present in a deposit. Therefore, the first assumption is no longer valid for those deposits. In some other deposits, the sample grades decrease with depth, and the second assumption is not valid for them. (It should be noted here that stationary grade distribution in samples is also a prerequisite for the inverse-distance method.) As previously mentioned, the sample variogram model is data dependent and user dependent; therefore, the sample variogram model used in kriging may not approach the true data spatial correlation. Because of these problems, grades estimated by kriging for a deposit are not always the best grade estimation, and special care should be taken to ensure that the grade estimation environment approaches the key kriging assumptions.

A variety of kriging techniques have been developed for grade estimation in the mining industry. The most commonly used kriging techniques are ordinary kriging and indicator kriging.

Ordinary kriging is the classic kriging method for grade estimation and its sum of the sample weights is normalized to 1. Ordinary kriging has been proven as a reliable grade estimation technique and it is the most widely used kriging technique in the mining industry.

Indicator kriging is a much more complicated geostatistical technique than ordinary kriging. It is actually the ordinary kriging technique performed on indicators. The sample grades are transformed to indicators of either a zero or 1 based on a cut-off grade before variogram modeling and grade estimation in indicator kriging. This technique does not estimate the grade of a block but the probability or proportion of a block above the selected cut-off grade. Indicator kriging is generally performed on a set of user-defined cut-off grades, normally ranging from 8 to 15, in grade estimation, which produces a grade probability distribution for a model block. The estimated block grade probabilities are adjusted to ensure statistical consistency before summarizing the tonnage and grade for the resource model.

Theoretically, the estimated block grade probability distribution represents the grade distribution of a selective size of the samples, which is generally smaller than the SMU, which in turn is generally smaller than the indicator kriging model block size for a mining project. Therefore, the estimated block probabilities need to be adjusted again to approach the grade probability distribution of the SMU. This procedure is generally referred to as variance reduction, as the variance of the grade distribution of the selective size of the sample is generally higher than that of the SMU. However, how much of the variance should be reduced is generally subjective. For a producing mining operation, the mine production data can be used to determine the amount of variance reduction. An empirical variance reduction factor has to be used for a mining project in the feasibility stage.

After the variance reduction, the block model tonnage and the average grade above a selected cut-off grade based on each partial block tonnage and grade above the cut-off grade can be summarized as the geological resources for the project.

The ordinary kriging model is often an overly smoothed grade model, and the advantage of indicator kriging over ordinary kriging is that an indicator kriging model can reduce the grade smoothing based on a user-defined factor. Indicator kriging is generally considered as a better grade estimation tool than ordinary kriging for deposits with high nugget effect and high variance for the sample grades because the sample indicator variogram models are generally much more robust than the sample grade variogram models.

Because different variogram models can be used for different cut-off grades, the grade distribution variation at different grade ranges can be handled easily in indicator kriging. For example, in a shear-zone-controlled gold deposit, the high-grade ore shoots within a mineralized zone may have a different dip angle than the entire mineralized zone, and the sample indicator variogram models at the higher cut-off grades can have a different dip angle than those at the lower cut-off grades. The resource model generated using these variogram models will maintain the sample grade distribution at different grade ranges.

Besides the grade probability distribution, an average block grade is also generated from all block partials in indicator kriging. Some of the indicator kriging users use a model block size similar to the SMU, and the average block grades are used as the estimated block grades for the resource model. Block models estimated by this type of indicator kriging will generally be very similar to the ordinary kriging model.

The disadvantage of indicator kriging over ordinary kriging is its complexity and numerous empirical adjustments, which prohibit its wide use in the mining industry.

Other varieties of kriging techniques derive from ordinary kriging and/or indicator kriging and are used in the mining industry. For example, in restricted kriging, the higher-grade samples above a user-defined cut-off grade are projected within a shorter distance than the lower-grade samples below the cut-off grade. This technique is especially good for deposits with a two-population grade distribution, that is, a higher-grade population and a lower-grade population. The resource model estimated by restricted kriging will preserve the grade distribution in the sample data better than that estimated by ordinary kriging. A restricted kriging resource model is also generally more conservative than an ordinary kriging model using the same capping strategy because the influence range of the higher-grade samples is reduced in restricted kriging.

Both inverse distance and kriging should produce acceptable resource block models if properly used. Other than the grade estimation technique, the search strategy also plays an important role in grade estimation. The search strategy consists of search orientation, anisotropy ratio, and sample selection.

The search orientation and anisotropy ratio should be based on the sample variogram model and the geological constraints. As the drill-hole data are generally directional, that is, the sample density is much higher in the drill-hole direction than that in the directions perpendicular to the drill-hole direction, the actual search distance along the drill-hole direction used is often shorter than the distance defined by the anisotropy ratio. Reducing the search distance, especially along the drill-hole direction, can reduce the amount of smoothing built into a resource model.

Sample selection includes the number of samples used for estimating a block and the spatial restriction on sample selection. The minimum number of samples used generally ranges from one to three, but the minimum number of samples used to define a measured/indicated resource block is generally at least two. The maximum number of samples for estimation is widely ranged, from as low as four to as high as unlimited within the search neighborhood. Generally, the maximum number of samples for indicator kriging is quite high in order to produce a probability grade distribution for the estimated blocks, but it can be much lower for inverse distance and ordinary kriging. The maximum number of samples used in grade estimation is commonly used to adjust the amount of smoothing built into the resource model. The lower the maximum number of samples used, the less smoothing will be built into the block model. The spatial restriction on sample selection is mostly used for declustering the samples selected for grade estimation. This restriction includes limiting the number of samples from a single drill hole, quadrant, or octant.

It is very important to avoid bias in selecting an estimation method and search strategy. To compensate for the smoothing nature of the inverse-distance squared and/or ordinary kriging methods for deposits with more than one sample population, a process that uses more than one pass for grade estimation is sometimes used in the mining industry, and this could introduce bias. One example of the grade bias introduced by a two-pass inverse-distance procedure was found in a reserve auditing on a shear-zone-controlled silver deposit. An inverse-distance squared pass using a longer search distance and all the samples in the database was first applied to produce estimation for the background silver distribution, and a second inverse-distance cubed pass using a shorter search distance and only the higher-grade samples above a selected cut-off grade was then applied to produce estimation for the high-grade zones, as these high-grade zones were not apparent in the first inverse-distance squared pass because of over-smoothing. The final block grade for blocks that estimated a grade in the two passes was the average of the two inverse-distance passes. As the second pass used a biased sample selection strategy, that is, only the higher-grade samples above a cut-off grade, the resulting grade estimation is extremely biased.

The grade bias can be classified into global grade bias and local grade bias. The global grade bias refers to the bias in total contained metal in a deposit. The local grade bias refers to the bias in grade distribution within the deposit. A globally biased estimation is generally biased locally, but a globally unbiased estimation could also be locally biased. One commonly used method to check the global grade bias, as discussed previously, is comparing the average grade of the estimation with that of a nearest-neighbor grade model at the zero cut-off grade.

The local grade bias can be checked on cross sections or plans with both block grades and the samples used for grade estimation plotted. The block grade distribution, although smoother, should be similar to the sample grade distribution. For a producing mine, the local grade bias can also be checked by comparing the production grade distribution with the block model grade distribution. A theoretical block grade distribution can be calculated from the nearest-neighbor model grade estimation, the sample variogram model, and the block size, which is also used to check the local grade bias for some projects.

A minable reserve generally refers to the material that will be mined from a deposit and delivered to the mill or heap for processing. It should include the mining dilution and exclude the mining losses that occurred in the blasting, mining, and transporting process.

The mining dilution is the waste material below the cut-off grade that is mined and delivered to the mill and heap for processing together with the ore-grade material, and the mining losses are the ore-grade material lost in the blasting, mining, and transportation process.

There are two different ways to deal with mining dilution and mining losses in ore reserve estimation. One is to construct a resource model with appropriate mining dilution and mining loss built in, and the amount of smoothing in the resource model will account for not only the change of support, but also the mining dilution and mining loss. The second way is to construct a resource model similar to the in-situ grade distribution, then apply a dilution factor and a mining loss factor on the resource model. The dilution built into the resource model plus the additional dilution factor applied on the resource model should be equal to the actual mining dilution for a project.

It is very important to understand that the amount of dilution built into a resource model can be very different because of the different modeling techniques used. For some of the resource models, very little or no mining dilution has been built into the resource model because the amount of smoothing in grade estimation is kept to a minimum. For some other deposits, more than sufficient mining dilution has been built into the resource model because of excessive smoothing in grade estimation, as indicated by the higher tonnage and lower average grade above the cut-off grade in the resource model than that of the actual production. No additional mining dilution factor should be applied to the latter type of resource model in producing the minable reserves, and the resource model may have to be modified to reduce the amount of smoothing.

The grade capping, the mining dilution, and mining losses are interrelated adjustments for a resource model. The objective for these adjustments is to produce a minable reserve similar to the actual production. Therefore, the best check for these adjustments is the production reconciliation.

Production reconciliation should be carried out using a volume as big as possible. A production reconciliation using a small mined volume is less useful, and sometimes it is even misleading. One project that this author audited used the production reconciliation to adjust the minable reserve estimation. The production grade for the first half of the first year was considerably lower than the model-predicted grade, and a very conservative resource model was generated based on the production reconciliation. However, the production grade of the first half of the second year turned out to be quite higher than the grade predicted by not only the adjusted resource model, but also the original model. A much more optimistic resource model then was constructed to guide future production. It is believed that the second model adjustment likely overestimated the minable reserve, as only a small good portion of the actual production was used for production reconciliation.

RESOURCE CLASSIFICATION

The resource/reserve classification regulations are discussed in detail in Chapter 2. These regulations, however, only give qualitative definitions of different resource/reserve categories, and the quantitative parameters for resource/reserve classification have to be defined for each mining project.

For computerized block model resource estimation, the blocks with grade estimation are generally classified into measured, indicated, and inferred categories based on user-defined parameters. *Only the measured and indicated resource blocks will be used to generate minable reserve estimation, as the confidence level for the inferred material is generally not sufficiently high for reserve estimation.* Methods commonly used to categorize block confidence level in the mining industry include the following:

- Anisotropic distance to the closet sample or average anisotropic distance to all samples used to produce the block grade estimate
- Anisotropic distance plus a minimum number of samples or drill holes
- Kriging variance or kriging error

It is preferable to use the anisotropic distance plus the number of samples or drill holes for resource classification, as these parameters are straightforward and easily understandable. Kriging variance or kriging error is also a good resource classification tool if used properly. However, kriging variance is not a straightforward parameter, and it changes if different variogram models and search strategies are used in grade estimation.

Less emphasis has been put on separating the measured and indicated resources than on separating the indicated and inferred resources, as the measured and indicated resources are generally used together to produce the minable reserve in the mining industry. The sample variogram range plus a minimum number of samples or drill holes is commonly used as the distance parameter to classify the indicated resources. However, the sample variogram range should not be used as an absolute guide to separate the indicated and the inferred resources, as the sample variogram model is data dependent and user dependent. For a project at the early stage of exploration, the sample variogram range is often much longer than the true variogram range. Therefore, the sample variogram range is not a good measure of grade spatial continuity, and the maximum distance used to define the indicated resource blocks should be shorter than the sample variogram range. For a sample variogram model with a high nugget value, the maximum distance for defining the indicated resource block should also be less than the sample variogram range.

REFERENCES

Clark, I. 1979. *Practical Geostatistics*. London; New York: Elsevier Applied Science.

Isaaks, E.H., and Srivastava, R.M. 1989. *An Introduction to Applied Geostatistics*. New York; Oxford: Oxford University Press.

Journel, A.G., and Huijbregts, C.J. 1978. *Mining Geostatistics*. London: Academic Press.

Pan, G. 1994. A geostatistical procedure for defining mineralization envelopes and modeling ore reserves. SME Preprint No. 94-35. Littleton, CO: SME.

Rendu, J.M. 1981. *An Introduction to Geostatistical Methods of Mineral Evaluation*. Monograph series. Johannesburg: South African Institute of Mining and Metallurgy.

Rendu, J.M., and Mathieson, G. 1990. Statistical and geostatistical methods. In *Surface Mining*, 2nd ed. Edited by B.A. Kennedy. Littleton, CO: SME. pp. 301–374.

Stone, J.G., and Dunn, P.G. 1996. *Ore Reserve Estimates in the Real World*. Special Publication No. 3. Littleton, CO: Society of Economic Geologists.

CHAPTER 4

Introduction to General Mine Planning

Richard L. Bullock

Many details must go into the planning of a mine. This detailed information must come from several sources. First is the geological, structural, and mineralogical information combined with the resource/reserve data. This information leads to the preliminary selection of potential mining method and sizing the mine production. From this, the development planning is done, the equipment selection is made, and the mine staffing projections are completed, all leading to the economic analysis of the foregoing scenario of mine planning.

However, one cannot assume that the planning just described will guarantee the best possible mine operation unless it is the best possible mine planning and it has been done correctly. Any sacrifice in the best possible mine planning introduces the risk that the end results may not reach the optimum mine operation desired. Planning is an iterative process that requires looking at many options and determining which, in the long run, yields the optimum solution.

This chapter addresses many of the factors to be considered in the initial phase of all mine planning. These factors have the determining influence on the mining method, operation size, pit slope (if an open pit), size of the mine openings (if underground), mine productivity, mine cost, and, eventually, economic parameters used to determine whether the mineral reserve even should be developed.

INFORMATION NEEDED FOR PRELIMINARY MINE PLANNING*

Technical Information

Assuming that the resource to be mined has been delineated with prospect drill holes, the items listed in Chapter 11 should be available for mine planning for the mineralized material. If this is an exploration project that has been drilled out by the company exploration team, this information should have been gathered during the exploration phase and turned over to the mine evaluation team or the mine development group.

More information on each of these subjects might have to be gathered, but if it can be started during the exploration phase of the project, much time will be saved during the feasibility/evaluation and development phases of the project.

* Most of the text in this section is taken from Bullock 2001.

Geologic and Mineralogic Information

General knowledge of similar rock types or structures in existing mining districts is always helpful. In developing the first mine in a new district, there is far more risk of making costly errors than in the other mines that may follow.

The geologic and mineralogic information needed includes the following:

- The size (length, width, and thickness) of the areas to be mined within the overall area to be considered, including multiple areas, zones, or seams
- The dip or plunge of each mineralized zone, area, or seam, noting the maximum depth to which the mineralization is known
- The continuity or discontinuity noted within each of the mineralized zones
- Any swelling or narrowing of each mineralized zone
- The sharpness between the grades of mineralized zones within the material considered economically minable
- The sharpness between the ore and waste cut-off, including whether
  - This cut-off can be determined by observation,
  - This cut-off must be determined by assay or some special tool,
  - This cut-off also serves as a natural parting, resulting in little or no dilution,
  - The break between ore and waste must be induced entirely by the mining method, or
  - The mineralized zone beyond (above or below) the existing cut-off represents submarginal economic value that may become economical at a later time.
- The distribution of various valuable minerals making up each of the potentially minable areas
- The distribution of the various deleterious minerals that may be harmful in processing the valuable mineral
- Whether the identified valuable minerals are interlocked with other fine-grained mineral or waste material
- The presence of alteration zones in both the mineralized and the waste zones
- The tendency for the ore to oxidize once broken
- The quantity and quality of the ore reserves and resource with detailed cross-sections showing mineral distribution and zones of faulting or any other geologic structure related to the mineralization

Physical and Chemical Information

The needed structural information includes the following:

- A detailed description of the cover including
  - Depth of cover;
  - Type of cover;
  - Structural features in relation to the mineralized zone;
  - Structural features in relation to the proposed mine development; and
  - Presence of and information about water, gas, or oil that may be encountered.

- The quality and structure of the host rock (back, floor, hanging wall, footwall) including
  - Type of rock,
  - Approximate strength or range of strengths,
  - Any noted weakening structures,
  - Any noted zones of inherent high stress,
  - Noted zones of alteration,
  - Porosity and permeability,
  - The presence of any swelling-clay or shale interbedding,
  - Rock quality designation throughout the various zones in and around all of the mineralized area to be mined out,
  - Rock mass classification of the host rock (rock mass rating, Q-system, or modified rock mass rating),
  - Temperature of the zones proposed for mining, and
  - Acid-generating nature of the host rock.
- The structure of the mineralized material, including all of the factors in the preceding list, plus
  - The tendency of the mineral to change character after being broken (e.g., oxidizing, degenerating to all fines, recompacting into a solid mass, becoming fluid),
  - The siliceous content of the ore,
  - The fibrous content of the ore, and
  - The acid-generating nature of the ore.

The Need for a Test Mine

From this long list of information that is badly needed to do a proper job of mine planning, it becomes evident that all of this information cannot be developed just from the exploration data acquired during that phase of the operation. Nor is it likely that it can all be obtained accurately from the surface. If this is the first mine in this mining area or district, then what is probably needed during the middle phase of the mine feasibility study is a test mine development. While this may be an expense that the ownership was hoping that they would not have to endure in advance, the reasons for a test mine are quite compelling. Listed in Chapter 11 are several dozen valid reasons why a test mine should be developed prior to completing the mineral property evaluation and feasibility study.

Property Information

The needed property information includes the following:

- The details on the land ownership and/or lease holdings, including royalties to be paid or collected identified by mineral zones or areas
- The availability of water and its ownership on or near the property
- Details of the surface ownership and surface structures that might be affected by subsidence, or mining, of the surface

- The location of the mining area in relation to any existing roads, railroads, navigable rivers, available power, the community infrastructure, and available commercial supplies
- The local, regional, and national political situations that have been observed with regard to the deposit

OTHER FACTORS OF EARLY MINE PLANNING

Mine Production Sizing

There is a considerable amount of available literature on the selection of a production rate to yield the greatest value to the owners (Lama 1964; Tessaro 1960; McCarthy 1993; Christie 1997; Smith 1997). Basic to all modern mine evaluations and design concepts is the desire to optimize the net present value or to operate the property in such a way that the maximum internal rate of return is generated from the discounted cash flows. Anyone involved in the planning of a new operation must be thoroughly familiar with these concepts. Equally important is the fact that any entrepreneur planning a mining operation *solely from the financial aspects of optimization* and not familiar with today's operating problems of maintaining high levels of concentrated production at a low operating cost per ton over a prolonged period is likely to experience unexpected disappointments in some years with low (or no) returns.

Other aspects of the problem of optimizing mine production relate to the potential effect of net present value. Viewed from the purely financial side, producing the product from the mineral deposit at the maximum rate yields the greatest return. This is because of the fixed cost involved in mining, as well as the present-value concepts of any investment. Still, there are "...practical limitations to the maximum intensity of production, arising out of many other considerations to which weight must be given" (Hoover 1909). There can be many factors limiting mine size, some of which are listed here:

- Market conditions, including current price of the product(s) versus the trend price
- Mineral grade and the corresponding reserve tonnage
- Effect of the time required before the property can start producing
- Attitude and policies of the local and national government and the degree of stability of existing governments and their policies, taxes, and laws that affect mining
- Availability of a source of energy and its cost
- Availability of usable water and its cost
- Cost and method of bringing in supplies and shipping production
- Physical properties of the rock and minerals to be developed and mined
- Amount of development required to achieve the desired production related to the shape of the mineral reserve
- Size and availability of the workforce that must be obtained, trained, and maintained
- Future potential instability of the government causing a company to develop a smaller, high-grade mine in the beginning until receiving its objective return, then using the income from the existing property to expand it to mine out the lower-grade ores with much less return

While all of the preceding factors must be taken into consideration, another approach to sizing the mine is to use the Taylor formula (Taylor 1977). Taylor studied more than

200 mining properties and used regression analysis to determine the formula for sizing a mine. Taylor implies that the formula was not applicable to steeply dipping mineral reserves or deep shafts. Actually for most other types of deposits, this author has found it to be a fairly good place to start, but it must be tested against all of the other physical variables listed earlier. The formula is

$$\begin{aligned}\text{life of the mine} &= 0.20 \times (\text{minable reserve ore tons})^{0.25} \\ &= \text{life of mine (years)} \pm 1.2 \text{ years}\end{aligned} \quad \textbf{(EQ 4.1)}$$

For mineral resources that are steeply dipping, and for deeper mining, one must consider the shape of the resource and how much development it will take to sustain the desired production. According to McCarthy (1993), for Australian underground narrow-vein mines approximately 50 vertical meters (165 vertical feet) per annum is currently economically appropriate for modern mechanized mines. Thus, for example, if you can block off reserves for mining 10,000 metric tons per vertical foot, then the production would be 500,000 metric tons per annum. Properties above McCarthy's "best fit" trend line are usually overly capitalized or have higher than average operating cost.

Not only does the resource's tonnage affect the mine size, but the distribution of grade can certainly affect the mine planning. Unless a totally homogeneous mass is being mined, it may make a considerable economic difference as to which portion is mined first and which is mined last. Furthermore, no ore reserve has an absolute fixed grade-to-tonnage relationship; trade-offs must always be considered. In most mineral deposits, lowering the mining cut-off grade means that there will be more tons available to mine. But the mine cut-off must balance the value of each particular block of resource against every type of cash cost that is supported by the operation, including all downstream processing cost, plus, in this authors opinion, the amortization of the capital that was used in constructing the new property.

Even in bedded deposits, such as potash or trona, the ability or willingness to mine a lower seam height may mean that more tons can eventually be produced from the reserve. In such cases, the cost per unit of value of the product generally increases. Also narrow-seam mining, just as narrow-vein mining, greatly reduces the productivity of the operation compared to high-seam and wide-vein or massive mining systems.

In considering the economic model of a new property, after all of the physical and financial limits are considered, all of the variables of grade and tonnage, with the related mining costs, must be tried at various levels of mine production that, in the engineer's judgment, are reasonable for that particular mineral resource. At this point in the analysis, the various restraints of production are introduced to develop an array of data that illustrate the return from various rates of production at various grades corresponding to particular tonnages of the resource. At a later time, probability factors can be applied as the model is expanded to include other restraining items.

Mine Production Timing

For any given ore body, the development required before production startup is generally related to the size of the production and dependent on the mining method. Obviously, the stripping time for most large porphyry copper deposits is very large compared to the stripping time for even a large quarry. For an underground example, a very large production may require a larger shaft or multiple hoisting shafts, more and larger development drifts, simultaneously opening more minable reserves, and a greater lead time for planning and engineering all aspects of the mine and plant. The amount of development, on multiple levels for a sublevel caving operation or a block caving operation, will be extensive compared to the simple development for a room-and-pillar mining operation. In combination, all of these factors could amount to a considerable difference in the development time of a property. In the past, this time for mine development has varied from two to eight years. In turn, this would have an indirect economic effect: The capital would be invested over a longer period of time before a positive cash flow is achieved. To aid the engineer in making rough approximations of the time parameters related to the size and depth of the mine shafts, the reader is referred to Bullock (2001).

For mines that must have extensive development in depth compared to those that are primarily developed one or two levels and have extensive lateral development, the intensity of development can be much different. The lateral development on each level of a room-and-pillar mine opens up new working places, and the mine-development rate can be accelerated each time a turnoff is passed, provided that there is enough mining equipment and hoisting capacity available. That is in contrast to mines that have a very limited number of development faces per level but more levels where it is shown that a vertical development of about 50 m (165 ft) per annum is about normal for a modern mechanized vertical mine, as discussed earlier.

The timing of a cost is sometimes more important than the amount of the cost. Timing is an item that must be studied in a sensitivity analysis of the financial model for the mine being planned. In this respect, any development that can be postponed until after a positive cash flow is achieved without increasing other mine costs certainly should be postponed.

Government Attitudes, Policies, and Taxes

Government attitudes, policies, and taxes generally affect all mineral extraction systems and should be considered as they relate to the mining method and the mine size. Assume that a mine is being developed in a foreign country and that the political scene is currently stable but impossible to predict beyond five to eight years. In such a case, it would be desirable to keep the maximum amount of development within the mineral zones, avoiding development in waste rock as much as possible. That would maximize the return in the short period of political stability. Also, it might be desirable to use a method that mines the better ore at an accelerated rate to get an early payback on the investment; if the investment remains secure at a later date, the lower-grade margins of the reserve then might be exploited. However, one must be careful that the remaining resource still contains enough good ore so as not to ruin the potential for mining the remaining resource, providing that the government's stability does continue. This author is not advocating rampant high grading, which leaves the bulk resource as worthless after only a few years of mining. But there is merit for carefully planned optimization of mining some of the higher-grade portions of the reserve while not jeopardizing the remaining reserve.

Some mining methods, such as room-and-pillar mining, allow the flexibility of delaying development, which does not jeopardize the recovery of the mineral remaining in the mine. In contrast, a mining system such as block caving or longwall mining might be jeopardized by such delays.

Similar situations might arise as a result of a country's tax or royalty policies, sometimes established to favor mine development and provide good benefits during the early years of production; in later years, the policies change. That would have the same effect as the preceding case; again, the flexibility of the mining rate and system must be considered, but not to the extent of jeopardizing the remaining resource.

SPECIFIC PLANNING RELATED TO PHYSICAL PROPERTIES

The physical nature of the extracted mass and the mass left behind are very important in planning many of the characteristics of the operating mine. Four aspects of any mining system are particularly sensitive to rock properties:

1. The competency of the rock mass in relation to the in-situ stress existing in the rock determines the unsupported open dimensions unless specified by government regulations. It also determines whether additional support is needed.
2. When small openings are required, they have a great effect on the productivity, especially in harder materials for which drill-and-blast cycles must be used.
3. The hardness, toughness, and abrasiveness of the material determine the type and class of equipment that can extract the material efficiently.
4. If the mineral contains or has entrapped toxic or explosive gases, the mining operation will be controlled by special provisions in the government regulations.

Preplanning from Geologic Data

Preplanning from geologic data is imperative for all mining operations. The details of open pit design are explained in Chapter 5. The reader can find additional information on the needed geologic data and its implication for open pit planning and design in a text by Hustrulid and Kuchta (1995).

Chapter 6 explains the details of underground mine planning for various methods. But for all underground mining, using geologic and rock-property information obtained during preliminary investigations of sedimentary deposits, isopach maps should be constructed to show the horizons to be mined and those that are to be left as the roof and floor. Such maps show variances in the seam or vein thickness and identify geologic structures such as channels, washouts (wants), and deltas. Where differential compaction is indicated, associated fractures in areas of transition should be examined. Areas where structural changes occur might be the most favored mineral traps, but they usually are areas of potentially weakened structures. Where possible, locating major haulage drifts or main entries in such areas should be avoided; if intersections are planned in these areas, they should be reinforced as soon as they are opened to an extent greater than that ordinarily necessary elsewhere in the opening.

Again referring to flat-lying type deposits, extra reinforcing (or decreasing the extraction ratio) may also be necessary in a metal mine where the ore becomes much higher in grade than is normal. Thus the rock mass usually has much less strength. Where the pillars already have been formed prior to discovering the structural weakness, it probably will be necessary to reinforce the pillars with fully anchored reinforcing rock bolts or cables. It is advisable to map all joint and fracture information obtained from diamond-drill holes and from mine development, attempting to correlate structural features with any roof falls that might occur.

Practical Considerations of Layout Design

The following subsections and figures are taken from the works of Spearing (1995), which is taken from and represents years of operating experience in very strong rock of the South African deep mines. While it may be true that all of this information applies to strong rock under considerable stress, it would apply equally to weaker ground under low to moderate stress.

Spacing of excavations. The following rules are based on the theory of stress concentrations around underground openings and the interaction of those stress concentrations. The usefulness of these guidelines has been borne out by experience obtained underground. It should be noted that the accompanying sketches are not necessarily to scale. Stress interaction between excavations can obviously be controlled by an increase in the installed support, but costs will also increase significantly. If there is adequate available space, it is generally more cost-effective to limit stress interaction between excavations.

Flat development:

1. Square cross section (Figure 4.1)
   a. Spaced horizontally at three times the combined width of the excavations.
   b. Spaced vertically at three times the width of the smaller excavation, provided that the area of the larger excavation is less than four times the area of the smaller.
2. Rectangular cross section (Figure 4.2)

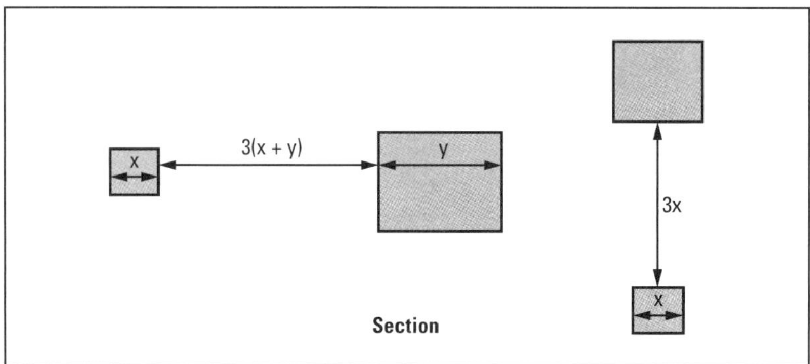

Source: Spearing 1995

FIGURE 4.1 Square cross section

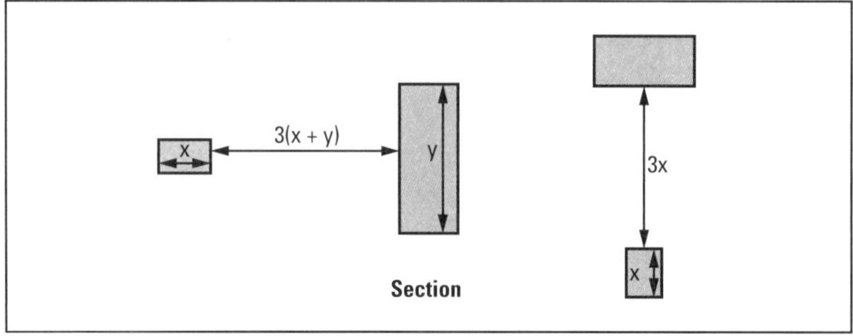

Source: Spearing 1995

FIGURE 4.2 Rectangular cross section

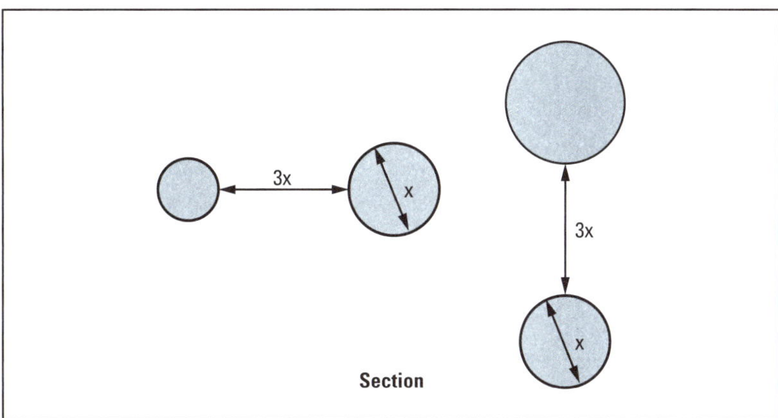

Source: Spearing 1995

FIGURE 4.3 Circular cross section

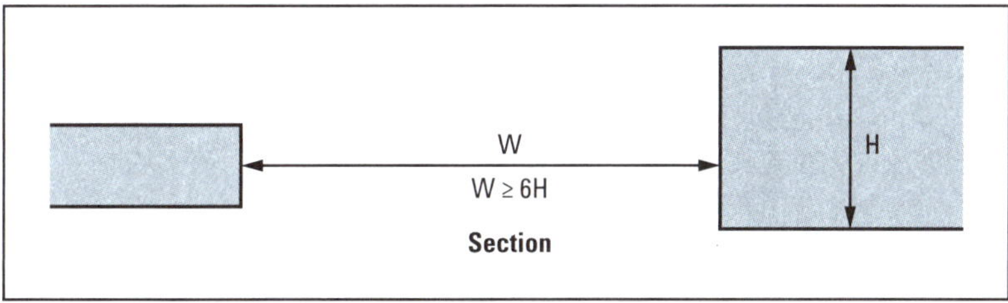

Source: Spearing 1995

FIGURE 4.4 Pillar height-to-width relationship

a. Spaced horizontally at three times the combined maximum dimensions of the excavations.

b. Spaced vertically at three times the maximum dimension of the smaller excavation, provided that the height-to-width ratio of either excavation does not exceed 2:1 or 1:2.

3. Circular cross section (Figure 4.3)

   a. Spaced horizontally at three times the diameter of the larger excavation.

   b. Spaced vertically at three times the diameter of the smaller excavation provided that the area of the excavation is less than four times the area of the smaller.

Vertical development (e.g., shafts):

1. Square cross section at three times the combined widths of the excavation.
2. Rectangular cross section at three times the combined diagonal dimensions of the excavations.
3. Circular cross section at three times the diameter of the larger excavation.

Pillar sizes. The pillar between irregularly shaped excavations should maintain a height-to-width ratio of a least 1:6 (i.e., pillar width must exceed six times the maximum pillar height [Figure 4.4]). For a pillar design under conditions of high stress (e.g., at depth), the height of the excavation should include an approximation of the fractured rock in the immediate vicinity of the excavation. (It is the author's opinion that the "pillar sizes" criterion applies, as

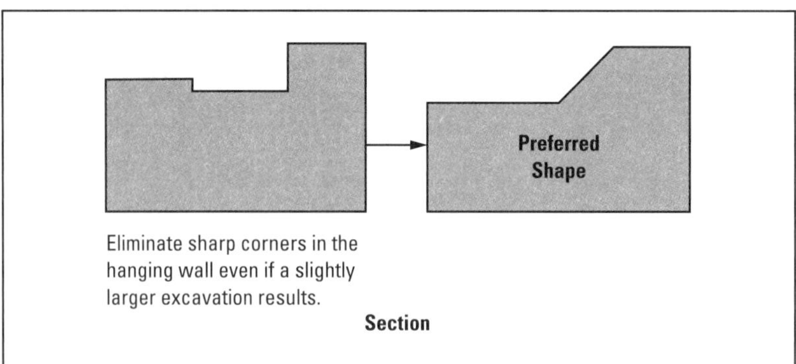

Source: Spearing 1995

FIGURE 4.5 Preferred opening shapes

Spearing states, to areas of high stress and thus does not apply to the many room-and-pillar operations in rather shallow environments, such as the many limestone properties and lead-zinc mines in the mid-continental United States.)

Shape of excavations. *Cross-sectional shape.* To achieve a given cross-sectional area for an excavation, it is often better to utilize a square shape. High, narrow excavations lead to excessive sidewall slabbing, which requires more intensive support, and low, wide excavations lead to large, unsupported hanging wall spans that generally require very long supports. Therefore, as a guideline, the width and height of excavations should be kept to the absolute minimum, and width and height should be as equal as possible. However, under conditions of very high stress, a low, wide excavation is generally the best (i.e., the shape of a horizontal ellipse).

Uniform shape. Sharp corners in excavations lead to unnecessarily high concentrations of stress, with a likelihood of excessive fracturing and premature failure. Therefore, the shape of openings should be kept as regular as possible, and any changes in shape should be "contoured" (Figure 4.5).

Breakaways. Multiple breakaways should be avoided to reduce dangerously large hanging wall spans. The rule is that breakaways should be spaced at six times the width of the excavation between successive tangent points (Figure 4.6A).

Acute breakaways (i.e., less than 45°) should be avoided since these result in "pointed" bullnoses, which are unstable (Figure 4.6B). Fracture and failure of a bullnose results in large, unsupported hanging wall spans.

The breakaways for inclines are often brought too close to the connecting crosscut. The length of the connection between an incline and the flat should be three times the diagonal dimension of the flat end (Figure 4.6C).

Orientation of adjacent excavations. The most highly stressed part of an excavation is the corner. Therefore, positioning of development in unfavorable orientations, as shown in Figure 4.7, should be avoided. In plan also, similar precautions should be taken to avoid this type of unfavorable orientation, especially where existing development is slipped out for excavations such as substations and battery bays.

Geology. In all cases, the geology of the area should be taken into consideration. Known weak geological horizons (e.g., the Upper Shale Marker and the Khaki beds associated with

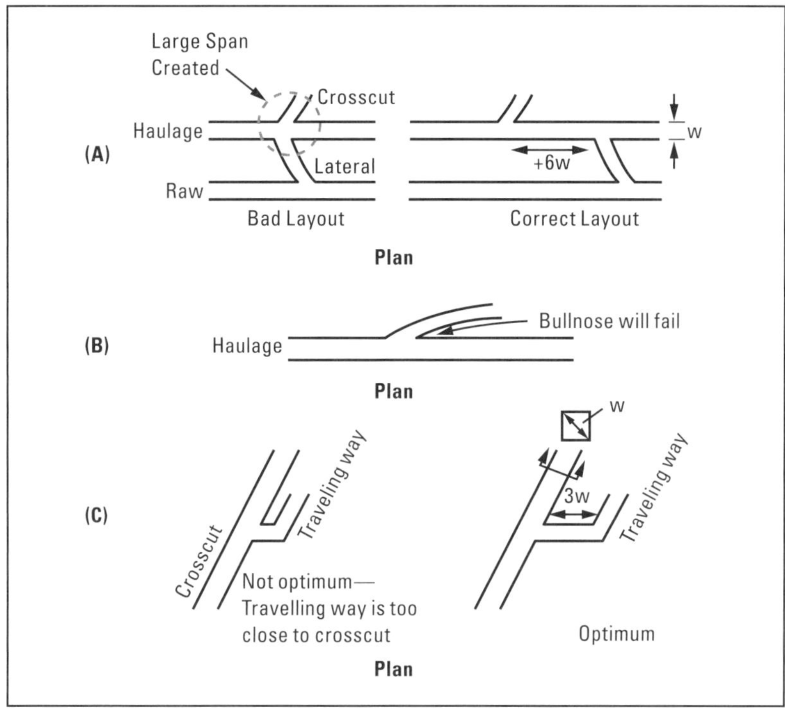

Source: Spearing 1995

FIGURE 4.6 Stable and unstable breakaways: (A) Offsetting breakaways; (B) angle of breakaways; (C) distance between breakaway and larger crosscut

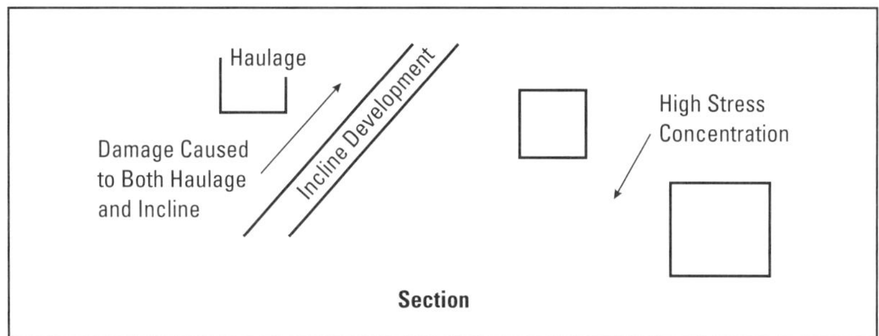

Source: Spearing 1995

FIGURE 4.7 Preferred orientation of development openings

the Orange State gold-bearing reef structures) should be avoided even at the expense of longer crosscuts.

In permanent excavations (sumps, settlers, hoist chambers, etc.), the position of faults and the orientation of joints sets are critical to the stability of the development. Layouts must therefore cater to such geological features.

All excavations should be kept away from dikes where possible (Figure 4.8A). When a dike is traversed, this should be done by the most direct route. Breakaways should not be sited in dikes, even at the expense of extra development.

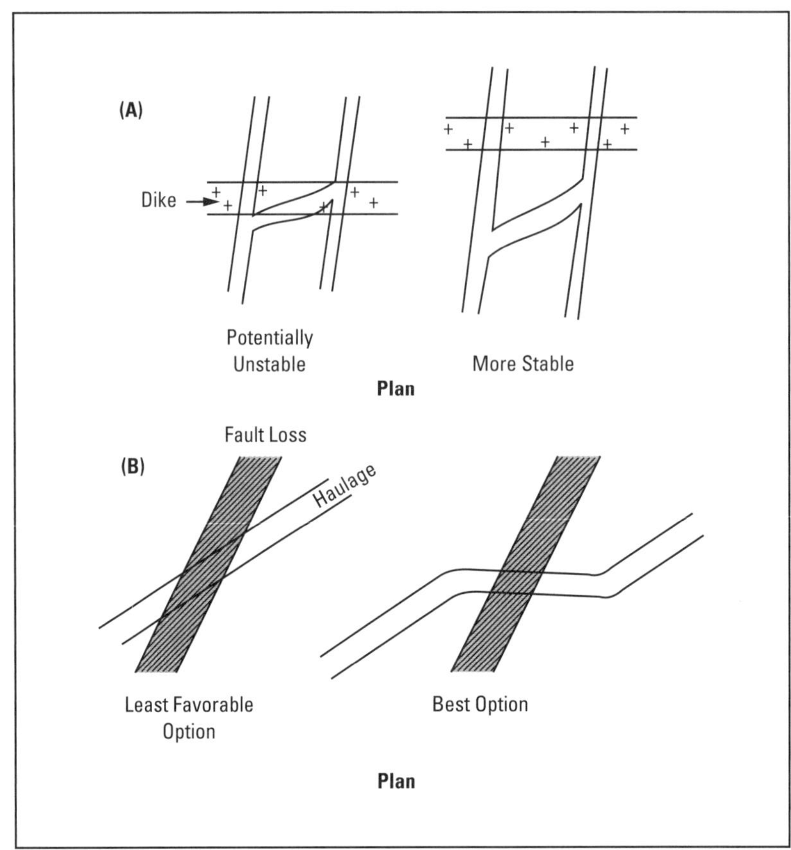

Source: Spearing 1995

FIGURE 4.8 Preferred approach to and intersections with (A) dikes and (B) faults

Haulages should not be positioned in fault losses, and development should not occur alongside a fault. A fault should always be intersected at an angle as near to normal as possible (Figure 4.8B).

Overstoping. When haulages are positioned beneath mined-out areas, consideration should be given to the 45° destressing guideline (Figure 4.9). An overstoping angle of 45° is generally required to destress a haulage. At an angle greater than 45°, stress concentrations are higher. This rule applies to reefs (ore seams) of 0° to 20°. Dips greater than this often call for computer modeling to show the extent of overstoping needed. Haulages should not be laid out too close to remnant pillars, which are very highly stressed.

Stoping. Stoping as applicable to advancing longwall panels and room-and-pillar stoping should not be carried out toward adverse geological features with the overall face shape parallel, or near parallel, to the feature. Mining should take place toward (and through) the geologic feature at as large an acute angle as possible (Figure 4.10).

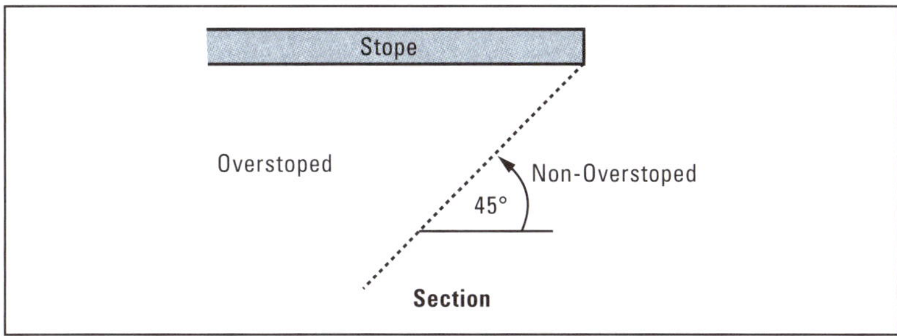

Source: Spearing 1995

FIGURE 4.9 Diagram of 45° overstoping angle for destressing haulageway

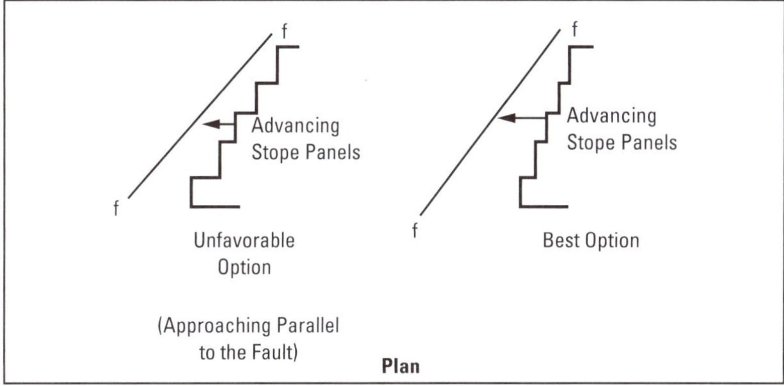

Source: Spearing 1995

FIGURE 4.10 Preferred approach to stoping through geologic feature

Hardness, Toughness, and Abrasiveness of Extracted Material

The hardness, toughness, and abrasiveness of the material determine whether the material can be extracted by some form of mechanical cutting action, by drilling and blasting, or by a combination of both methods.

The mechanical excavation by Tenneco of the borate minerals from the Death Valley open pit mine in California using roadheaders and later using the same type of roadheader at the nearby underground Billie mine is a case in point. Bucket-wheel excavators have been used extensively in Germany and Australia for stripping the overburden from the brown coal fields.

Technological advances in hard-metal cutting surfaces, steel strengths, and available thrust forces allow increasingly harder and tougher materials to be extracted by continuous mining machines. The economics of continuous cutting or fracturing as compared to drilling and blasting are gradually being changed for some of the materials that are not so tough or abrasive. However, for continuous mining (other than tunnel boring machines) to be competitive with modern high-speed drills and relatively inexpensive explosives, it appears that the rock strengths must be less than 103,400–124,020 kPa (15,000–18,000 psi) and have a low abrasivity. However, if the rock is full of fractures, then this also is a great aid to mechanical excavation. In one case, this author knows where a roadheader is being used in a welded volcanic tuff even though the rock strength is well over 137,800 kPa (20,000 psi) but contains many close spaced fractures. This entire subject is covered in an article on the gradual trend toward mechanical excavation in underground mining (Bullock 1994).

At times, reasons other than the first cost of extraction favor the use of one mining system over another. Using a mechanical excavation machine is nearly always advantageous in protecting the remaining rock where blasting might be prohibited. Likewise, the continuous nature of mechanical excavation can be used to speed mine production. This was seen in the development openings driven by the Magma Copper Company in developing the Kalamazoo ore body (Chadwick 1994; Snyder 1994) and Stillwater Mining Company in developing their original ore body, as well as their East Boulder ore body (Tilley 1989; Alexander 1999). A continuous boring tool may also be desirable for totally extracting an ore body without personnel having to enter the stoping area. Certainly, continuous mining machines, where they are applicable, are much easier to automate then cyclic drilling and blasting equipment. The automation of the Robbins mobile miner at the Broken Hill mine in Australia is one example (Willoughby and Dahmen 1995). Another more dynamic case is that of the complete automation of the potash mines of the Potash Corporation of Saskatchewan (Fortney 2001).

PLANNING THE ORGANIZATION AND REQUIRED EQUIPMENT

The amount of equipment or personnel for the needs of all mines cannot be specified in general terms. Chapters 5 and 6 deal with the specifics of staffing and equipment of surface and underground mines, respectively. The purpose of this discussion is to mention some of the general problems that may be encountered and actions to mitigate those problems.

Workforce and Production Design

It is necessary to consider several factors concerning planning the workforce to operate the mine. Many questions needing investigation will be difficult to answer, but they have profound effects on the financial success of any mining project and eventually must be faced:

- Is the supply of labor adequate to sustain the production level dictated by other economic factors? If not, can the needed labor be brought in, and at what cost?
- What is the past history of labor relations in the area? Are the workers accustomed to a five-day work schedule and, if so, how will they react to a staggered six- or seven-day schedule?
- Are the local people trained in similar production operations, or must everyone be trained before production can achieve full capacity?
- Will a camp have to be built and the workers transported in on a weekly schedule?
- Can people with maintenance skills be attracted to the property, or will the maintenance crew have to be built up through an apprenticeship program?

Apprenticeship programs are very slow processes. Accordingly, some state laws restrict the number of people who can be trained each year in such programs. That one item could cause a mine designed and equipped for a very large daily production to fall far short of its goals.

It is important that the "people problems" be investigated at the same time that the property is being evaluated and designed. This will provide adequate time for specialized training, minimizes unexpected costs, and prevents basing economic projections on policies which, if implemented, could destroy employee morale or community confidence. The productivity and profitability difference between an operation with good morale and good labor relations and an operation with poor morale and poor labor relations (with many work stoppages) can make the difference between profit or loss of the mining operation. Of all the items involved in mine design, this one is the most neglected and can be the most disastrous.

Field-Tested Equipment

The selected equipment should be produced by manufacturers that field-test their equipment for long periods of time before marketing it to the mining industry. Too many manufacturers build a prototype machine and install it in a customer's mine on the contingency that they will stand behind it and make it work properly. Eventually, after both user and manufacturer redesigns, rebuilds, reinforces, and retrofits, a workable machine is finally obtained. However, the cost in lost production is imposed on the mine operator, not on the manufacturer. The manufacturer then can proceed to sell the "field-tested" retrofitted model to the entire industry, including competitive mines. In the case of a load-haul-dump (LHD) unit introduced several years ago in an underground mine, there were 53 design changes between the prototype sold to the mine and the final production model, all made by the mining company and all adapted by the manufacturer. In another case, a prototype drill jumbo was modified so extensively over a two-year period that nearly every auxiliary component was retrofitted in the user's operation; even after two years, the drill jumbo still was not performing up to the specifications at the time of purchase. Even though manufacturers do need the help and cooperation from the mining industry to develop equipment, the manufacturer must pursue longer periods of testing in the industrial environment, rather than selling the units to the industry and then pursing continuous research on the prototype models.

Equipment Versatility

The equipment selected for the mining operation should be as versatile as possible. Normally for large surface mines, this is not so much of a problem. But for the smaller quarries and for most underground mines, it can be a problem if the equipment cannot be adapted to several types of mining operations. For example, in one room-and-pillar mining operation (Bullock 1973), the same high-performance rotary percussion drilling machines were used for drilling the bluff or brow headings and were mounted on standard drill jumbos for drilling holes for burn-cut drifting and stoping rounds and slabbing rounds in the breast headings. Because the drills on these jumbos penetrate extremely fast, they also were used to drill the holes for roof bolts and, in some cases, to drill holes for reinforcing pillars. The same front-end loader was used to load trucks in one stope and to perform as an LHD unit in another stope. By switching working platforms, the same forklift tractors served as explosive-charging vehicles and as utility service units for handling air, water, and power lines. They also served as standard forklifts for handling mine supplies. This equipment philosophy results in the following advantages:

- There is less equipment to purchase and maintain.
- Less training is required for operators and maintenance personnel. At the same time, all personnel have a better chance of becoming more efficient at their jobs.
- Having fewer types of machinery means having less inventory to obtain and maintain.

There are some possible disadvantages of this philosophy:

- A more efficient machine may be available to do the job being done by the versatile machine. Therefore, if a great amount of that type of work is to be done, it may be advisable to use the specialized machine.
- The mine may become too dependent on a single source to supply their equipment.

As an example of the first disadvantage, the mine previously cited was drilling a great number of high bluffs, some brows, and breakthrough pillars. As a result, the mine eventually switched over to air-track drills because those machines were more efficient in the particular applications.

Likewise, a high-speed roof bolting jumbo eventually was acquired because it was much more efficient for installing roof bolts.

Equipment Acceptance

The equipment selected should have a very broad acceptance and is desirable if it is in common use throughout both the mining and construction industries. Because underground mines impose a headroom restriction not encountered on the surface, that is not always possible. However, where headroom is not a problem, selecting a standard piece of equipment means that the components will have endured the rigorous use of the construction industry. Furthermore, parts for such equipment normally are off-the-shelf items in the distributors' warehouses across the country.

Application Flexibility

The selected equipment should be flexible in application. That is, the equipment should be able to accelerate and move rapidly, have good balance and control at high speeds, be very maneuverable, and have plenty of reserve power for severe applications. Both trucks and loaders should have ample power to climb all grades in the mine and to accelerate quickly to top speed on long, straight hauls.

ORE STORAGE POCKET

Some surge capacity must be provided between the normal mine production gathered from various parts of the mine and the conveyance that transports the material to the processing plant. In an open pit operation or quarry, where trucks are used to haul from the faces to the initial crusher, this is not a problem. But it is a problem in a surface operation where a single conveyor transports the material from a central location to the processing plant or in an underground mine where the need arises because the material usually has many production paths to reach the shaft, slope, or adit, but usually only one path out of the mine. Therefore, the single-path conveyance is vulnerable to downtime for maintenance and repair, but such downtime cannot be allowed to disturb the rest of the mining cycle. Similarly, the multipath production can operate intermittently (one or two shifts per day), while the material can flow from the mine continuously.

The correct size of the surge pocket depends on what it is intended to accomplish. For example, if management decided to try to hoist 20 shifts per week but run the stoping operations only 2 shifts per day in a mine producing 4,535 t/d, or metric tons per day (5,000 stpd [short tons per day]) but stopped mine production between midnight Friday and 7:00 a.m. Monday, the minimum pocket size would be found from

$$C = [Sm/Sh] \times Smd \times Tms \quad \textbf{(EQ 4.2)}$$

where

C = ore pocket capacity needed in short tons or metric tons
Sm = number of shifts per week the mine operates
Sh = number of shifts per week the hoist operates
Smd = number of shifts the mine is down
Tms = number of metric tons or short tons produced per mine shift

Continuing on with the preceding example:

Sm = 10 shifts
Sh = 20 hours
Smd = 6 shifts
Tms = 2,268 t/shift (2,500 st/shift)

For this example, the capacity would be calculated as

$$C = [10/20] \times 6 \times 2{,}268 = 6{,}804 \text{ t}$$
$$C = [10/20] \times 6 \times 2{,}500 = 7{,}500 \text{ st}$$

In this example, the storage was calculated to be 150% greater than the daily mine capacity. Although this is typical for some underground operations, it is much larger than the requirement for many others. Obviously, there is no general agreement on the optimum size of an ore pocket; both the nature of the operations and the management priorities differ. However, the two major considerations are (1) the size of this storage and whether interruptions before the next step are critical and (2) whether there is additional storage at the discharge of the initial crushing and before the ore proceeds to the next step in the flow.

REFERENCES

Alexander, C. 1999. Tunnel boring at Stillwater's East Boulder project. *Mining Engineering,* 51(9):15–24.

Bullock, R.L. 1973. Mine-plant design philosophy evolves from St. Joe's New Lead Belt operations. *Mining Congress Journal,* 59(5):20–29.

Bullock, R.L. 1994. Underground hard rock mechanical mining. *Mining Engineering,* 46(11):1254–1258.

Bullock, R.L. 2001. General planning of the noncoal underground mine. In *Underground Mining Methods: Engineering Fundamentals and International Case Studies.* Edited by W. Hustrulid and R. Bullock. Littleton, CO: SME. pp. 15–28.

Chadwick, J. 1994. Boring into the lower Kalamazoo. *World Tunnelling, North American Tunneling Supplement* (May).

Christie, M.A. 1997. Ore body definition and optimisation. In *MINDEV 97: The International Conference on Mine Project Development*, Sydney, Australia, November 24–26. Victoria, Australia: Australasian Institute of Mining and Metallurgy. pp. 47–56.

Fortney, S.J. 2001. Advanced mine-wide automation in potash. In *Underground Mining Methods: Engineering Fundamentals and International Case Studies.* Edited by W. Hustrulid and R. Bullock. Littleton, CO: SME.

Hoover, H.C. 1909. *Principles of Mining.* New York: McGraw-Hill.

Hustrulid, W., and Kuchta, M. 1995. *Open Pit Mine Planning and Design,* Vol. 1. Rotterdam: A.A. Balkema.

Lama, R.D. 1964. Some aspects of planning of deep mines. *Colliery Engineering,* (June): 242–245.

McCarthy, P.L. 1993. Economics of narrow vein mining. In *Proceedings, Narrow Vein Mining Seminar*, Bendigo, Victoria, June. Victoria, Australia: Australasian Institute of Mining and Metallurgy. pp. 89–97.

Smith, L.D. 1997. A critical examination of the methods and factors affecting the selection of an optimum production rate. *CIM Bulletin,* (Feb.): 48–54.

Snyder, M.T. 1994. Boring for the lower K. *Engineering and Mining Journal,* (April): 20ww–24ww.

Spearing, A.J.S. 1995. *Handbook on Hard-Rock Strata Control.* South African Institute of Mining and Metallurgy Special Publication Series SP6. Cape Town, South Africa: CPT. pp. 89–93.

Taylor, H.K. 1977. Mine valuation and feasibility studies. In *Mineral Industry Costs.* Spokane, WA: Northwest Mining Association. pp. 1–17.

Tessaro, D.J. 1960. Factors affecting the choice of a rate of production in mining. *Canadian Mining and Metallurgical Bulletin,* (Nov.): 848–856.

Tilley, C.M. 1989. Tunnel boring at the Stillwater mine. In *Proceedings, Rapid Excavation and Tunneling Conference (RETC).* Edited by R. Pond and P. Kenny. Littleton, CO: SME. pp. 449–460.

Willoughby, R., and Dahmen, N. 1995. Automated mining with the mobile miner. In *Mechanical Mining Technology for Hard Rock Short Course*, June, Colorado School of Mines, Golden, CO.

CHAPTER 5

Surface Mine Planning

John T. Crawford

OPEN PIT AND STRIP MINING

Mine planning is a continuous, iterative, multidisciplinary process. It is a key part of all phases of the mine evaluations, feasibility studies, and due diligence examinations forging links between and integrating geology, engineering, metallurgy, operations, and economics, bringing them together in the design, evaluation, development, and operation of an ore body throughout its life. In staged feasibility studies, the differences in the mine planning process for each stage are more a function of the amount of data available than differences in the steps and their sequence in preparing a mining plan. This is largely because of the data and format requirements of mine planning computer software.

Mine planning plays a powerful role in transforming an ore body from a mineral deposit into an economically viable business entity. Aspects from collection and analysis of raw geologic and drilling data to production and sale of salable product are involved. It has both technical and business analysis attributes. In addition to determination of mining limits and mine design, other important parts of mine planning are operational equipment planning, mining sequencing, production scheduling, ore stockpiling, waste rock disposal, reclamation, and water management.

It is important to have people with field and operations experience on the mine planning team and to avoid overreliance on people with primarily computer expertise. Operating input is critical for the mine planning steps that have such important impacts on operating flexibility and field execution of the plan and operations responsibility and accountability. The mine planning steps primarily involved here are pushback design (including roads, ramps, waste disposal, and ore stockpiles), mining sequencing of pushbacks, and production scheduling. A balanced team is needed for proper setup of planning methods and parameters and interpretation of results.

The ore mined and processed in an optimized life-of-mine design and production schedule is the minable reserve. The mine design, plan, and production schedule for an ore body must be operationally realistic.

In this chapter, *open pit mining* generally refers to base and precious metals, such as copper and gold deposits, while *strip mining* applies primarily to coal and Florida phosphate. Some bedded or seam deposits, such as Mesabi Range and Western Australia iron ore and deeper strip mining ore bodies, have attributes of both open pit and strip mining.

Choice of Mining Method

Whether an ore deposit will be planned and evaluated as a surface or underground mine depends on several factors:

- Depth of ore body or high stripping requirements, which could favor underground mining
- Severity of surface terrain, which could hamper surface mining access as well as production rates and efficiencies
- Regulatory and environmental requirements, which could favor underground mining
- Economics of surface mining compared to underground mining
- Mining losses, dilution, and cut-off criteria, which could favor surface mining
- In the later stages of a deep open pit, determination of when to convert to an underground mine.

In some instances, the configuration of an ore body will indicate that it has both surface and underground minable zones. It is likely that the location of the transition between the two mining methods will be based on relative economics or possibly pit slope stability circumstances that would impose an engineered limit on the pit depth. The most common occurrence for an ore body being developed using both surface and underground mining methods is large porphyry copper and disseminated gold deposits.

Surface Mine Planning

Surface mine planning includes both long- and short-range plans covering the life of mine, from inception or current status to production exhaustion. The results must be operationally realistic and executable in the field. In this regard, it is necessary to analyze and interpret a mining plan and financial results for each year during the life of an ore body in addition to the overall picture. This is because operating personnel will ultimately be responsible and accountable for achieving the plan for a given year. Operating managers do not want their plan years to have unfavorable or unrealistic goals and expectations and unrealized performance.

Long-range planning generally considers the requirements to cover the period after the third to fifth year of production to the life of mine. Short-range planning, which is more detailed than long-range planning, comprises the first three to five years of development and operation. It is a key interface between planning and operations and involves budgeting and production accountability. The emphasis of many mining company business plans is on the next three to five years of operation. Very often, the first few years of a production schedule in a feasibility study for a new mine may serve as the budget for initial operations. Short-range planning also includes attributes of ore control.

It is necessary that all surface mining plans meet federal, state, and local safety, environmental, and other regulatory requirements throughout the life of the property (construction, operation, and closing).

At operating properties, achieving and sustaining operationally realistic and field-executable mining plans, from design and evaluation phases to development and production, depend on monitoring and feedback from existing operations. Because of this function, mine planning is continuous during the life of an operating property because of the need to adapt to unforeseen and changing conditions, and because of the unceasing competitive demand to identify and implement innovative methods and techniques. When practical, it is most desirable to use benchmarked data from comparable mining and processing operations for study, design, cost analysis, and evaluation of greenfield projects.

The selection of open pit or strip mining design for an ore body is based on the geology and geometry of the deposit and the mining equipment to be used. Usually truck and shovel operations are used in open pits, whereas draglines are associated with strip mining. In coal mining, however, both methods are often used to mine the same ore body, such as in the Powder River Basin of Wyoming.

To set up a process that will lead to a set of high-order decisions and alternatives to be tested as part of the study, the results of the analysis process is a strategy table, an example of which is shown in Table 5.1.

TABLE 5.1 Example of a strategy table showing key decisions (column headings) and alternatives*

Mining	Processing	Water	Power	Transportation
Current plan (sublevel stope mining)	4.5 Mt† (current)	3,000 ML† artesian (current)	Gas-fired (current)	0.75 Mt truck and rail (current)
Open pit in northern zone; continue sublevel stop mining in southern zone	6.0 Mt (brownfield)	6,400 ML artesian	Expand gas-fired to pipe capacity	**Concentrate slurry pipeline**
Open pit in northern zone with new decline from base of pit	10 Mt (brownfield)	**3,000 ML artesian plus desalination to meet demand**	**Grid connection**	1.5 Mt truck and rail
Open pit everything	**11.5 Mt (7-Mt greenfield)**			

Source: Whittle 2011.
*A hypothesis is represented by the choice of one alternative from each column. A "big capital" hypothesis is indicated in bold type.
†Mt = megatons; ML = megaliters

Following are typical steps recommended for creating the technical design of an open pit mine. Many of the steps are iterative and must be repeated when new data are acquired:

1. Choose a hypothesis from the strategy table (Table 5.1) to test. Prepare an assumption set and a resource model for analysis.
2. Apply a pit optimization tool to generate a set of pit shells.
3. Determine a theoretical best-case plan and choose the final pit.
4. Choose the production rate.
5. Choose a plan for a final pit.
6. Choose the number of pushbacks and their basic design.
7. Optimize the schedule and processing cut-offs (Whittle 2011).

Source Data Inputs to the Process

Surface mine planning makes extensive use of specialized computer software, both commercial packages and systems developed in-house. The two most common computerized methods for generating open pit mine designs are the floating cone (Carlson et al. 1966) and the

Lerchs–Grossmann (Lerchs and Grossmann 1965). Both methods have been well accepted in the mining industry for several decades. Additional capabilities for support of analysis and convenience are provided by "office suite" software packages (spreadsheets, presentation preparation). The main reasons a company develops mine planning software in-house is because it needs capabilities that are not available in commercial packages. Examples of these capabilities are the modeling of complex mine and processing production flow sheets and detailed deposit-specific financial analysis requirements. It is important that the use of computer applications is not by rote and that the results are not treated as flawless, especially in the early stages of mine planning.

All the mine planning software and subsequent analysis, evaluation, and interpretation are built around the following major categories of open pit criteria, parameters, and source or input data:

- Ore-body gridded block model
- Demonstrated resource base
- Surface topography and surface control boundaries
- Pit slope angles
- In-situ and broken densities (for ore and waste materials)
- Metallurgy as applicable to surface mine planning
- Adjustments to ore-body block model input data
- Mining losses and dilution
- Block economic value modeling (downstream processing and sales economics, property mining and processing, operating performance and cost data)

It is important to remember that the accuracies of the computer output are only as accurate as the input data from these sources. Errors in basic data tend to be magnified and become more apparent with computer modeling. These basic data or equivalents are used in mine planning for both open pit and strip mining. They would also be used for mine planning employing manual methods.

Outputs from the Process

The principal outputs from the mine pit design and planning process for a surface mine are as follows:

- Ultimate pit limit design (generate location and design of the pit or mining area)
- Pushback design
  - Pushback bench widths (parameters and designs of pushbacks, mining zones, segments, or cuts)
  - Nest of pits (provide economic guidance for design, positioning, and mining sequence of pushbacks, mining zones, segments, or cuts)
  - Roads and ramps
  - Waste rock disposal areas

  - Ore stockpiles
  - Surface facilities (requirements and design for maintenance, utilities, and ancillary facilities)
- Production scheduling
  - Production rates
  - Operating schedule
  - Cut-off criteria
  - Operating flexibility
  - Preproduction development
  - Blending and stockpiling strategies
  - Mining losses and dilution
- Financial analysis
  - Equipment performance and costs (for production, service, and support equipment)
  - Actual operating performance and cost data
- Optimization analyses
- Sensitivity analyses
- Operations monitoring and reconciliation

These outputs are similar for both open pit and strip mining.

OPEN PIT CRITERIA, PARAMETERS, AND DATA INPUT NEEDED

Ore-Body Gridded Block Model

The ore-body gridded block model is the cornerstone of modern mine planning practices used in developing a mining plan for an ore body. It is the most common method of preparing a three-dimensional quantitative model of the ore-body characteristics needed to develop a mining plan. The gridded block model is the basis around which effective mine planning computer applications have been developed and are used efficiently. It is a key link between geology and engineering in the design, evaluation, and development of an ore body.

Data are typically assigned or related to individual blocks by several industry-recognized methods of assignment, use of location references, or by regression equations related to assigned data. Data that are assigned or related to individual or groups of blocks include

- Location coordinates and elevation,
- Ore grades,
- Rock type and geologic structure,
- Geologic and metallurgical domain boundaries, and
- Bed and seam thicknesses.

In-situ density and metallurgical data may be assigned to blocks or related to them by equations as functions of assigned data variables. Economic values of blocks are calculated using economic models and assigned directly.

The ore-body block model is designed to cover the area and depth of the ore body being modeled. The model should also encompass enough area and depth surrounding the ore body to include the waste areas that would be included in the mining plan for an open pit or strip mine and to accommodate additional data in the future that could enlarge the ore body and the resulting mining plan.

Block Size

Typically, the size of the blocks is in the range of 15 to 30 m square (50 to 100 ft square). An exception to this is the 100-m-square (330-ft-square) block used in Florida phosphate mining where that is the typical exploration and development drilling pattern. There are some occasions when the use of rectangular-shaped blocks is appropriate. One-quarter size subblocks may be used in some ore-body block models for improved resolution along domain boundaries. The height of the block is normally the bench height planned for mining operations. The bench height for mining is generally based on the size of loading equipment planned and can be influenced by blasting design, geology, and ore grade patterns, which may suggest a bench height that would improve mining selectivity and grade control and reduce mining losses and dilution. The bench height should permit single-pass drilling for blastholes and be lower than the height of the bucket reach of a rope shovel, hydraulic shovel, or front-end loader so the crest of the bench will be actively cut by the bucket teeth to avoid unsafe overhangs. This is particularly important for operations with frozen ground conditions. In some instances, different bench heights may be used in different zones of the mine, such as waste benches being higher than for ore. Bench heights generally range between 9 m and 18 m (30 ft and 60 ft). The selection of bench height can have implications for mining loss and dilution adjustments, which is discussed later in this chapter. There are federal and state safety regulations on bench heights.

In seam or bedded ore bodies, the block height may be the thickness or interval between marker beds or seams. In addition, the blocks may be of uniform height but contain data for more than one bed or seam. In rare circumstances where mining operations parallel the dip of the seam or bed, the top and bottom surface of a block may be sloped.

Domains

Assigning data values to blocks involves defining geologic and metallurgical domain boundaries, compositing drill sample data and interpolation, and extrapolation of composite value data. Domain boundary codes are assigned to blocks from geologic and metallurgical data bench maps and cross sections. Domains are generally used to define the zones of different material and ore types within the ore body based on physical, chemical, and metallurgical characteristic differences. In seam or bedded ore bodies, some domain boundaries may correspond to key marker seams or beds. A key part in setting up domain boundaries is determining if data from one domain can be used in another. Spacing or intervals between drilling and other sampling should be adjusted as needed to provide sufficient data for each domain, with special attention being given to obtaining data along the margins of a domain boundary.

Composite Values

Sample data from drilling, underground workings, and so forth, are usually composited before interpolation and extrapolation are done. There should be down-the-hole surveys for vertical holes when geology and structure may cause significant deviation from the vertical and for angle-drilled holes. Where the sample data are aligned near vertical, such as samples from drill holes, the compositing is done by block height interval or by seam or bed thicknesses within the block height. For samples aligned more horizontally, the composite interval is based on geologist and mining engineer interpretation of the sample data. Statistical analysis of sample data may provide guidance here. The selection of composite value intervals can have implications for mining loss and dilution adjustments. Implicit in compositing values are blending and dilution within benches of ore-body block model block-height intervals.

Interpolation and Extrapolation Methods

Interpolation and extrapolation of composite data to assign values to blocks can be done by a variety of methods:

- **Direct assignment.** Direct assignment assigns a code or composite value to blocks within a zone of influence or fixed distance of a single code or composite value. The zone of influence is generally three-dimensional. The fixed distance may be uniform or variable in different directions. If there are overlapping areas of influences for assigning a code or composite value to a block, the closest of the allowable codes or composite values governs and is assigned. An example of this method is some Florida phosphate deposits.
- **Polygons.** Using polygons is a common method of determining the area of influence for direct assignment of values. Because of the methodology, it is necessary that the area of influence be horizontal or in a plane. It is necessary to determine if the area of influence should reach beyond the limits of the composite value data. If drilling is too widely or irregularly spaced, it may be necessary to establish the maximum allowable dimensions of a polygon. Blocks not included within a polygon would not be assigned values.
- **Contouring.** Topographic contouring methods may be used for assigning seam and bed thicknesses and other values to blocks for some seam and bedded ore bodies.
- **Inverse-distance weighting and geostatistical methods.** Both inverse-distance weighting and geostatistical methods use interpolation and extrapolation procedures for assigning composite values to blocks by a weighted average using a distance measure as the weighting variable. For inverse-distance weighting, usually squared or cubed, the measure is the distance between a composite value and the center of a block. Using geostatistical methods, the physical distance is modified by statistically derived adjusting variables. Both methods can be applied selecting composites from within benches or planes, or three-dimensionally.

Key parts in the use of these methods are determining the number of composite values to use to assign a value to a block and the maximum allowable distance of influence for a composite value. These parameters need to be determined for both interpolation and extrapolation. It is important that the results of the interpolation and extrapolation provide a reliable representation of the ore body and related waste areas. Localized variability needs to be reliably portrayed. Although it may not necessarily have a significant impact on the overall minable reserve, it may be important in determining production schedules and cut-off strategies.

The number of composite values used to assign a value to a block influences the resolution of localized variability in the deposit. Using an excess number of composites may result in too much smoothing of the values over an area or zone of the ore body, thereby losing the texture of localized variability. Usually there are fewer composite values used with extrapolation than with interpolation because it is used primarily in the fringe areas of the ore body where there are usually fewer eligible composite values available.

It is also important to consider the angular distribution of the composite values being used for a given block. If the overall angular distribution of the eligible composite values is less than 180 degrees, then the interpolation process becomes extrapolation. It may also be important to consider the angle, sometimes referred to as a *shadow angle*, between individual composite values. If there is a cluster of composite values within a narrow angle, their contribution to the average may be unrepresentative, and it may be desirable to use only the closest composite value within the angle. This is a matter of geological and statistical interpretation of angle and distance. Over the years, the increasing use of geostatistical methods has resulted in less attention being given to the distinction between interpolation and extrapolation of composite values and to shadow angles.

The maximum distance that a composite value can influence is determined by technical interpretation and judgment of geology and composite value data. The use of variograms is an industry-recognized and effective analysis tool in this process. In many situations, the maximum allowable distance for extrapolation is less than for interpolation. Variograms provide a statistical basis for determining the maximum distance. The distances determined from variogram analyses can be applied with both inverse-distance weighting and geostatistical methods for interpolation and extrapolation.

The matter of the maximum distance used for extrapolation is a subject of considerable discussion among geologists and mining engineers. Many mine planning experts believe that even with geostatistical analysis, data assigned to blocks by extrapolation are inherently less reliable than those assigned by interpolation. It is the difference between projecting beyond and working within the data field of composite values. This difference in reliability can be managed by limiting the maximum allowable distance of influence for extrapolation or by carefully adding artificial composite value data. These additional composited values will provide additional control needed outside the field of composite values and along domain boundaries to enable the use of interpolation. Examples exist of both approaches being used to control or substitute for extrapolation. It is important that the use of artificial composite values be considered very carefully so that they are used only to control interpolation and not as substitutes for necessary additional drilling or other sampling.

The following four approaches are generally acknowledged to establish maximum distances of influence for extrapolation compared to those used for interpolation:

1. The variogram range(s), same as interpolation
2. One-half the variogram range(s)
3. One-half the drill-hole spacing in the geologic ore mineralization domain(s)
4. The width or height of two blocks

The choice of the distance parameters for interpolation and extrapolation can have an important bearing on defining the demonstrated resource base.

Some mines have developed block models with blocks about 6 m square (20 ft square) for blasthole drilling data. These models are used for short-term and day-to-day operations planning, budgeting, scheduling, and ore control.

An important difference in ore-body models for strip mines from open pit mines is that the block height may be more oriented to the thicknesses of key seams or beds and that the block data may contain data for multiple seams or beds. The methods used for compositing and assigning composite values to the blocks are the same as for open pit mines.

Demonstrated Resource Base

Numerous regulatory, government, and professional bodies have defined various classifications regarding the reliability and accuracy of mineral resources and ore reserves. One set of common terms used is *measured*, *indicated*, and *inferred* ranging from best to worst. It has been determined that only the measured and indicated portions of a resource should be used as the base for mine planning to determine the minable ore reserves. This base for mine planning is referred to as the demonstrated resource base.

The *demonstrated resource base* comprises those parts or zones of an ore body where the reliability and accuracy of the geological, drilling, metallurgical, and other sampling data meet the standards to be classified as measured and indicated. This applies to geologic zones, domains and boundaries, and particularly quantitative values. It is important to evaluate them because they affect the resulting mining plan and production schedule being operationally realistic and executable in the field.

All inferred material must be treated as barren waste according to U.S. Securities and Exchange Commission regulations. Careful attention must be observed regarding its inclusion in other parts of the world. Characteristics of inferred material may be used in estimating the effects of mine dilution on mining and processing performance.

The demonstrated resource base should not be influenced by economics or cut-off criteria, as these are determined during development of the mining plan. Thus, it is important to remember that the demonstrated resource does not define what ore is. The unminable portions or zones of the demonstrated resource base are determined during the mine planning process.

The definition of measured and indicated material is quite subjective among geologists and mining engineers. Setting forth recommended criteria herein is impractical because the specifics are unique to any given deposit. Typical criteria could be similar to the following:

- **Measured or proven**—values assigned by interpolation, four or more composites used, and maximum distance of influence limited to variogram range values
- **Indicated or probable**—values assigned by interpolation or extrapolation, less than four composites used for interpolation, maximum distance of influence for interpolation limited to variogram range values, maximum distance of influence for extrapolation limited to 50% of the variogram range values or two blocks outside the field of composite values

It is possible that an adverse ore or metallurgical characteristic associated with a block may eliminate it from the demonstrated resource base because of the excessive risk associated with handling it in production, such as inability to blend it because of inhibiting processing performance or unacceptable salable product contamination.

A specific criterion for determining the demonstrated resource base needs to be established and followed for each ore body being evaluated, even if it indicates that there are interstitial blank or waste zones created within geological mineralized domains. Such blank or waste zones provide strong support for drilling along domain boundaries and for fill-in. Complex domain configurations will require more closely spaced drilling and sampling to provide sufficient data to fill in interstitial blank or waste zones.

Some geologists believe that the demonstrated resource base should be made up of all of the geologically defined ore mineralization domains without regard for the density of drilling and other sampling data. This can be a major contention between geologists and mine planners. There needs to be a distinction made between the amount of data required for reasonable definition of domain and other boundaries and the amount of data needed to assign reliably specific quantitative values to individual blocks. The data required for the latter are generally regarded as being greater. The quantitative values are generally more important factors in planning and evaluations than domain or other boundaries. Being operationally realistic should be the goal in defining the demonstrated resource base and over-optimism should be avoided.

Interstitial blank or waste zones should be expected in the demonstrated resource base and included in mining plans as waste and dilution until the fill-in data from drilling and sampling are obtained and completed.

One must keep in mind the important relationship between interstitial or blank areas within ore domains in an ore-body block model and subsequent fill-in data obtained from drilling or other sampling. Under normal ore-body modeling conditions and parameters, fill-in data will result in a reduction in the demonstrated resource base. The only conditions under which the demonstrated resource base will be enlarged by fill-in data are when there are interstitial waste or blank zones in ore domains and they are partially or entirely shown to be ore by the fill-in data. There is the potential for fill-in data to alter domain boundaries.

In an optimized pit design and mining plan, fill-in data may affect the optimized minable ore reserve, depending on the fill-in data values being better or worse than the optimization threshold values.

Surface Topography and Surface Control Boundaries

Surface topographic information is obtained through aerial and ground surveys from government and private sources. If the data are obtained from outside sources, they need to be validated by in-house-generated information to the extent practical. It will be necessary to establish a survey net for the ore body and surrounding area for use during exploration, development, and operations. Establishing it from a reliable base reference will normally serve to validate the topographic data. More than one mine has experienced severe problems during development or operation as the result of topographic or surveying errors.

In remote areas, however, outside information and established reference survey grids may have significant errors. A practical way to remedy such situations now is the use of satellite-based surveying methods to establish a reliable survey net and then adjust outside data, if feasible, or run new aerial or ground topographic surveys linked to the satellite-based survey net.

Topographic information for mine planning is normally digitized for use in the computer from aerial photos or contour maps. The digitization is usually done on a grid that corresponds to the grid for the ore-body block model or a smaller interval, typically one quarter of the block spacing.

Property boundaries or other surface control boundaries—such as un-relocatable, permanent roads and utilities and environmentally sensitive areas or zones excluded from mining that may influence the areas or limits of mining, waste disposal, or other surface plant and ancillary facilities—should be generated, validated, and, if necessary for mine planning, digitized for the computer.

Pit Slope Angles

Operating and final pit slopes are designed from analysis and evaluation of geotechnical and hydrologic data. These data are collected from exploration or special drilling and other sampling in and surrounding the ore body being evaluated. In addition, the geologic studies of the ore body and surrounding zones are important. Within a pit design there are likely to be several different pit slopes with zones of influence defined by vertical sector planes and groups of levels.

The geotechnical data, analysis, and evaluation are directed at assessing the structure and strength of the rock that will influence and be affected by the slope passing through it. Structural considerations include faults, joints, bedding planes, gouge zones, and orientation to the proposed pit wall. The expected effects of operating blasting designs need be included.

Hydrologic considerations include aquifers, water tables, and flows of surface and subsurface water. In some instances, drainage of water bodies or relocation of streams may be involved. Understanding of hydrologic characteristics are critical for such engineering and operating considerations, such as pit wall stability, support and safety factors, pit slope dewatering assessment, and pit drainage and dewatering for production equipment. It is important to understand the trade-offs between the costs of pit slope dewatering and support and the slope angle, and their impacts on the pit limit location and stripping requirements.

An overall pit slope is made of four components:

1. Pit sectors and zones
2. Bench face height and angle
3. Safety bench interval and width
4. Road and ramp width allowances

A schematic cross section of the components in an overall pit slope is shown in Figure 5.1.

Inter-Ramp Slope Angles

The geotechnical and hydrologic data are used to determine pit sectors and inter-ramp slope angles. Selection of pit sectors and zones are based on various rock type, geologic, structure, and hydrologic characteristics that could result in sufficient differences in inter-ramp slope angles to need to have separate values for different parts of the pit ultimate limit surface. Typically for a large open pit, there are several sectors and zones that are described by vertical planes and bench elevations.

The inter-ramp slope angle is an engineered value and includes a suitable safety factor. Technically it is the maximum allowable slope for the vertical interval between roads and ramps. A given inter-ramp slope angle is valid for a specified maximum vertical interval. If as mine design proceeds it becomes apparent that the allowable maximum vertical interval is being exceeded, the affected inter-ramp slope angle and portion of the pit design need to be reevaluated and modified.

Geometrically, an inter-ramp slope angle includes the bench face height and slope angle and the safety bench interval and width. Normally the bench face height and slope angle result from the interactions between geology, rock, and structure characteristics; blasting design; and loading equipment size. The safety bench is frequently located on alternate bench elevations and is wide enough to catch material raveling from benches above. Safety for operations below

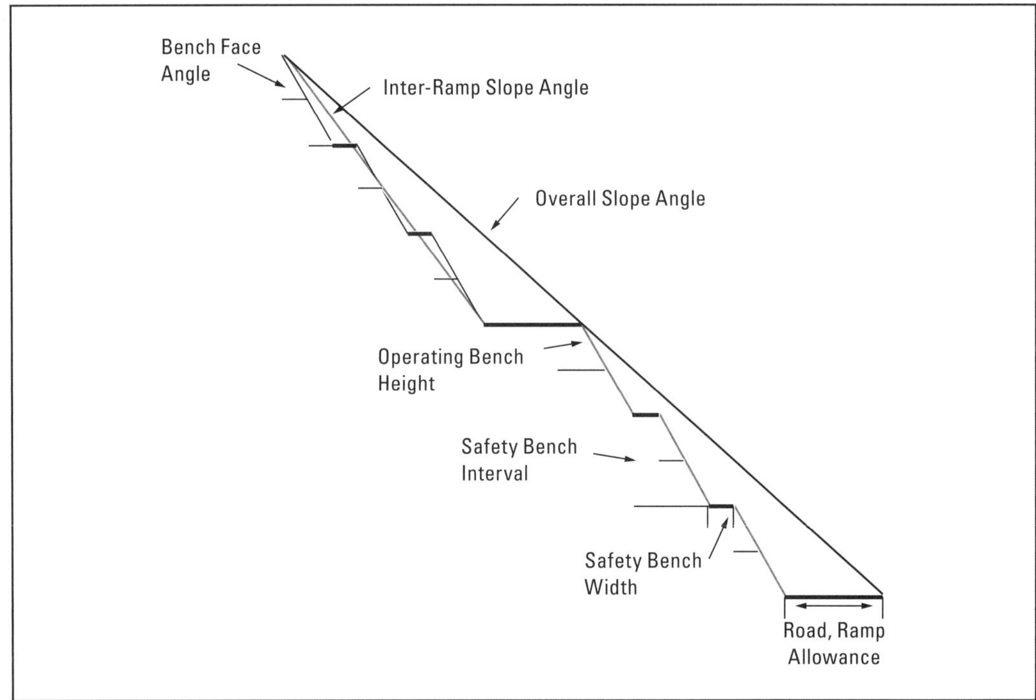

FIGURE 5.1 Schematic cross section of overall pit slope

is a primary consideration in determining the interval and width of the safety benches. If the composite angle of the bench height and face angle and the safety bench interval and width is flatter than the engineered value for the inter-ramp slope angle, then the composite value should be used in place of the engineered value. Sometimes there is reluctance to a use a slope angle flatter than the engineered value for the inter-ramp slope angle even if the composite of the individual components is flatter.

There are conditions when the use of presplit blasting is advisable to achieve a more stable bench slope and perhaps a steeper angle. This type of blasting is applied when a bench is in its final location or will not be mined for a long period of time.

Another factor in the determination of the inter-ramp slope angles is the profile of the overall composite slope. If the profile flattens from upper benches to lower benches as a "coffee cup shape," there is no need to adjust the inter-ramp slope angles from what has been discussed previously. If, however, the profile is steeper for lower benches than for upper benches as a "belly-in shape," adjustments to flatten the inter-ramp slope angles may be appropriate because of potentially adverse concentrations of stress in the steeper portions of the slope.

An important component in determining overall pit slopes is the allowances for road and ramp widths. In some instances, the slopes used to design the initial pit ultimate limit surface are based on the inter-ramp angles and exclude allowances for the road and ramp widths. When this is done, the resulting overall pit slope is too steep. The resulting pit ultimate limit is too large, the ore tonnage is overstated, and the stripping requirements are understated. Adding roads and ramps to the design results in flattening of the overall pit slope, or may inappropriately steepen the inter-ramp angle.

Allowances for the widths and number of roads and ramps in a sector of an open pit must be included in the slope angles used for designing an ultimate pit limit for the following reasons:

- The limit will be in the correct location.
- There will be sufficient room in the slope profile to add road and ramp designs without violating the inter-ramp angle.
- Ore tonnages will not be overstated.
- Waste tonnages and stripping ratio will not be understated.

The typical road and ramp width for using modern, large haulage trucks is 30–40 m (100–130 ft) including the running surface, safety berm along the open edge, and water drainage ditch along the bank toe. Changes in road and ramp allowances because of different size haulage trucks will require redesign of the pit ultimate limit and pushbacks.

Operating pit slopes for pushbacks will be flatter than ultimate pit limit slopes if interim safety bench widths are greater and there are more interim roads and ramps. Ideally, the inter-ramp slope angles for the pushbacks and the ultimate pit will be the same.

There are no significant differences in designing slope angles for open pit and strip mines. The data and analysis involved in designing slopes for shallow surface mines are generally less complex than for deep ones, but the principles involved are the same.

In-Situ and Broken Densities

Densities of ore and waste material are critical in mine planning because they are the link between volumes and weights. It is necessary to distinguish between wet and dry basis. For equipment analysis, it is also necessary to know the densities of broken material associated with various blasting or other material breakage methods. The broken density may affect the sizes of loading buckets and haulage truck payload boxes selected.

In-situ densities are usually determined by laboratory measurements on drill-core samples taken from the various ore and waste types within and adjacent to the ore body. In some instances, they may be based on the mineral composition of the material or from operating experience with a similar material in a similar geologic environment. The latter is the common approach for estimating the swell factors for blasting and material handling to determine the broken density for a given in-situ density. To convert between wet and dry basis, the moisture content of the material must be determined and it may differ between in-situ and broken material conditions.

Results of laboratory determinations for in-situ densities may be a source of errors in the true density because of overstatement. The reason for this is that, depending on the testing method used and the condition and disturbance of core samples removed from the in-situ environment, there may be in-situ voids that are not measured. Once a mine is in production or data from comparable mines are obtained and analyzed, this can usually be resolved through production/reserve reconciliation analyses.

Generally, in-situ densities are assigned to the ore-body block model blocks on the basis of the material type. Often the densities of ore are higher than for waste. In some instances, there may be sufficient data to relate the in-situ density for a given material type to a characteristic such as mineral distribution or ore grade. This is useful for deposits where the in-situ density varies significantly in relation to another measured characteristic, for example, iron ore or some polymetallic ores. In these instances, density data may be assigned to blocks by interpolation or extrapolation, but, more often, regression equations are used.

It is necessary that all density data used in mining and processing for ore-body analysis and evaluation, mine production and processing planning, and operations reporting be specified as wet or dry basis. The use of wet or dry basis for mine data depends on the commodity involved. Usually the basis in processing plants and salable product is dry. If the mine basis is dry, then it is necessary to be sure those equipment productivities are estimated and evaluated on both wet and dry bases. For iron ore, the mine data are often stated on a wet basis and the processing plants use a dry basis. In coal and Florida phosphate strip mines, it is normal to measure waste by volume and ore in tonnages.

Metallurgy as Applicable to Surface Mine Planning

Metallurgical response characteristics provide the linkage between ore characteristics, process characteristics, and salable product as concentrate or refined product, including co- and by-products and minor element qualities. They also provide the bases for selection among multiple process options. The necessary information is obtained from laboratory and pilot-plant testing of drilling, bulk, and other samples. Generally, it is collected by ore type because of the different process tests and response characteristics involved. It is necessary that metallurgical balances of weights and grades for test samples and processing products satisfy closure criteria based on Richard's Law (for ore processing material balances) for the data to be valid.

Normally the quantity and distribution of metallurgical testing samples taken for an ore body are much less than for ore grades. Samples of various types and sizes may be gathered for bench and pilot-plant testing. They are generally insufficient to permit assigning metallurgical values to ore-body model blocks by interpolation and extrapolation methods. A practical method is to develop regression equations between test sample ore characteristics and the resulting response characteristics and use the derived equations with the block ore characteristics to provide response characteristics for the blocks. It is necessary to have test data covering the full range of characteristics for a given type of ore. In many situations, this should include very low-grade and waste materials.

The metallurgical information that is needed for mine planning is ore grindability, recoveries, concentrate qualities, and criteria for selection between multiple process options when applicable. It is important that the concentrate qualities include minor elements that could affect further processing and product salability. Ore grindability data are used for processing plant design and for varying processing rates in production schedules and related costs, depending on the mix of ore types being processed.

Many ore bodies involve multiple processes and products. These can include gravity, flotation, leaching and magnetic separation processes, and concentrates and refined products, including co- and by-products. It is necessary to determine the ore characteristic criteria by which the appropriate process is selected. When multiple concentrates are involved, a payable element in one concentrate may be a contaminant in another. It is important that the recovery data separate payable components from contaminants.

The recoveries from most processes, such as flotation and magnetic separation, are instantaneous. In dump or pad leaching, however, recovery is cumulative over time. To apply a leaching recovery value to an ore-body block, it is necessary to derive an equivalent instantaneous recovery value from the recovery–time relationship. This is done using a discount factor that is compatible with the production schedule and financial analysis. Incremental recovery factors by time period are used in production schedules without using a discount.

In many situations, it is appropriate to incorporate in a mining plan the processing and recovery of payable values from material normally considered as marginal ore or waste. To do this, it is necessary to determine the metallurgical characteristics for this material. For many operations, the process involved is dump leaching. The importance of evaluating the metallurgical characteristics is because recovering values from some low-grade or waste material may be more economic than treating it as throw-away barren waste material. Where the favorable nature of recovering payable values is evident, it should be included in the mine design and planning because the benefits may be significant. It is necessary, however, to give proper regard to the risk of predictability. Including the recovery of payable values from marginal ore and waste material has the potential for a larger mine, longer mine operating life, different production rates and cut-offs, and higher economic value of the ore body.

Another metallurgical factor that must be considered is the impact of stockpiling low-grade or other ore material for long periods between mining and processing. The concern here is the affect that such things as oxidation may have on causing lower recoveries and concentrate qualities. These potential process response deteriorations need to be incorporated in evaluating and setting cut-off criteria and stockpiling strategies.

A related metallurgical matter is downstream processing (such as smelting and refining) of intermediate products. Very large ore bodies may support internal vertical integration for conversion or refining these products into final salable products. Here it is necessary to determine processing options and characteristics, and operating and capital costs.

Normally, however, the downstream processing is provided by an outside supplier such as for nonferrous concentrates. In this case, it necessary to thoroughly understand and negotiate treatment and selling terms with available suppliers. These include the following:

- Transportation and processing charges
- Payable value deductible and price reference for primary salable product
- Charges and deductible terms for non-payable quality characteristics of intermediate and salable products
- Charges, deductibles, and selling price terms for payable co- and by-products

Multiple processes, such as smelting and refining, may be involved. There may be circumstances where there are different processing options available.

Adjustments to Ore-Body Block Model Input Data

Ore grades, densities, metallurgical characteristics, and other pertinent values assigned or associated with blocks in the ore-body block model may be adjusted to reduce the impacts of selected high and low values interpreted as being operationally unrealistic or to add an element of conservatism. This is usually done by using floor and ceiling values and reducing a range of high values by a percentage for selected variables. Typical situations where such adjustments are made is when it is not practical to define an appropriate domain of influence where very high ore-grade values should apply or estimate associated localized dilution, and when it is expected that high metallurgical test values would not be realized in plant operations.

Mining Losses and Dilution

Mining losses and dilution are most often applied to ore bodies that are bedded or seam deposits or that have ore–waste boundaries that make it difficult to cleanly separate ore and waste during mining operations. They are more likely to be used for high-value ore bodies or where the distinction between ore and waste is defined by geologic rather than ore-grade boundaries. They are seldom applied to large, disseminated deposits. The calculations of mining losses and dilution are linked and must be balanced with each other and the total ore and waste in the ultimate pit design for both tonnages and grades.

Mining Losses

Mining losses are applied to ore and waste within the ultimate pit design that may not be mined because of such factors as slope failures or excess water. The ore that is not mined is excluded from the reserve for the pit. The unmined waste is excluded from stripping requirements.

Ore mining losses also occur when ore is mined with waste. In this case, the ore is excluded from the reserve for the pit, but it is added to the stripping requirements.

Dilution

Dilution is the mining of waste with ore. It occurs because of the inability to cleanly separate ore and waste along geologic or cut-off boundaries. Blasting and loading operations cause mixing of ore and waste along these boundaries. The choice of bench interval may be partial mitigation for some ore bodies. The dilution waste should be added to the ore reserves with adjustments to tonnages, ore grades, and metallurgical characteristics. Ore can also be mined with waste and thereby be lost. When planned, this lost ore is removed from reserves.

Defining reliable parameters of mining losses and dilution for an ore body is dependent on the availability of suitable operating data for the ore body or mine being evaluated or for a comparable property. Reconciliation of production data to reserves and production estimates is the essential part of this process. There could be different mining loss and dilution factors for different ore domains, but these are usually difficult to define.

Once the mining loss and dilution adjustment factors have been defined, the handling of the ore-body block economic values must be adjusted. The values of ore and waste blocks with mining loss should be reduced. The value of ore blocks mined as waste should be changed to a waste value. The value of waste blocks mined as ore should be changed to an ore value, even though negative. These adjustments may apply to individual ore-body model blocks or to the mine planning functions that use their values such as floating cone (Carlson et al. 1966) or Lerchs–Grossmann (1965) methodologies for the ultimate pit or mining area limits and the nest of pits and the commercial computer software routines that use ore-body block economic values for production scheduling and optimization.

Block Economic Value Modeling

In mine planning, economic models are used to determine economic values for blocks in the ore-body block model and for life-of-mine cash-flow financial analysis of the mine and processing production schedules. The block economic value model is less comprehensive than the life-of-mine cash-flow financial analysis model because of constraints in the block model structure and the floating cone (Carlson et al. 1966) and Lerchs–Grossmann (1965) methodologies used to generate ultimate pit designs with computers. It includes revenues and costs that are directly related mining and for ore blocks, processing for each individual block resulting in a positive or negative net economic dollar value (revenues minus costs) for each block. It is probable that the information used in the block economic model will have been derived from the life-of-mine cash-flow financial analysis model so they are consistent and compatible with each other.

The revenues and costs used in the block economics model are normally that which can be directly associated with a block's location in the block model and its geologic, grade, and metallurgical characteristics. There will be more than one revenue for a block when it will yield more than one salable product, such as a copper ore with by- or co-products. It is necessary that the salable product have a recognized market reference to estimate the selling price or net return.

Following is a list of the major revenue and cost categories that should be included in the block economic model:

- Selling prices or net returns for primary products, co-products, and by-products
- Mine operating, maintenance, and support costs covering
  - Production systems, equipment, and labor—drilling and blasting, loading, draglines for strip mining, haulage (trucks, crushers and conveyors, slurry pumping for Florida phosphate mining), roads, and dumps;
  - Mine management, technical and administrative staff, and general and administrative (G&A) costs allocated to total material mined, and ore and waste production quantities; and
  - Property and ore reserve leases and royalties, if they can be tied to specific locations and tonnages of ore and salable product
- Transportation of mined ore to processing
- Process operating, maintenance, technical, and administrative
- Overall property management, technical and administrative staff, and G&A costs allocated to mining and processing production quantities
- Transportation of intermediate product to downstream processing
- Treatment and selling terms for downstream processing of intermediate products to salable primary products, co-products, and by-products

The following costs cannot be readily incorporated in the block economics but must be included in the life-of-mine cash-flow financial analysis model:

- Initial, expansion, replacement, and sustaining capital for infrastructure, mine, processing, support, and G&A
- Pre-mining clearing, drainage, and development
- Mine and processing plant reclamation and abandonment

Because the revenue and cost structure can be expected to change through the years during the operating life of a mine, the revenues and costs used for calculating the block economic values for the ultimate pit limit generation should reflect the later years of operation. This is because most of the production at the margin of the pit or mining area limit occurs late in the life of the operation.

The costs used, whether based on engineered estimates or data from existing operations, should reflect expected operating methods, equipment, head count, and production rates. The composition of costs is unique to each combination of method, equipment, and operating schedule and rate. They need to be modified as each combination is evaluated.

The mining costs must be determined for both ore and waste. They need to reflect the differences in costs for different sectors and levels of the mine, such as the increased haulage costs for deeper levels.

PIT DESIGN FROM THE DATA OUTPUT

Ultimate Pit Limit Design

The ultimate pit limit or mining area defines the maximum extent of area and depth that an ore body can be economically mined. The ore-body block economic values, ultimate overall slope angles, surface topography and boundary controls, and the minimum pit bottom dimension criteria are used with floating cone (Carlson et al. 1966) or Lerchs–Grossmann (1965) methods to generate the ultimate pit or mining area limit. The minimum-sized pit bottom is based on the size of equipment and operating conditions expected. The ore-body block economic values are based on the near pit limit economic structure described earlier. The most practical cut-off to use at this stage for determining ore and waste is the dollar value of the blocks. The blocks in the demonstrated resource are in three groups:

1. Those that have positive values
2. Those that have a higher value if treated as ore than if treated as waste, even though the value is negative
3. Those that have a higher value, less negative, if treated as waste than if treated as ore

The blocks that are not part of the demonstrated resource are in a fourth group and must be treated as waste with negative values.

The reason some blocks in the demonstrated resource could have negative economic values as ore is because the demonstrated resource as described in this chapter does not have an economic component.

In recognizing payable values (which may be referred to as leach credits) from material normally handled as waste, the ore-body block economic value is adjusted for changes in mining costs and revenues and the costs of recovering the values. This is unlike treating waste blocks as ore. Selecting waste blocks for leach credits would require knowledge of their leaching characteristics on leach pads or in large waste dumps and the economics involved. An example of the use of leach credits is in evaluation of copper deposits using open pit mining.

The floating cone method uses the pit bottom placed in various locations and elevations of the ore-body block model by a search method. A pit removal increment is generated at a particular location and elevation with the ultimate pit slopes. The cumulative value of the unmined block economic values is calculated. If the cumulative value is positive, all of the blocks in the increment are designated as mined. The process involves generating and evaluating pit bottoms at different locations and elevations until all possible pit removal increments have been tested. The resulting composite pit generated from all of the positive-value increments is an ultimate pit design, and the ore contained within it is an ore reserve. For clarification, although this is an economic pit design, it is not the optimum because of the lack of detailed road and ramp design, economic parameters that could not be included in the block economic value, and the effect of timing that normally result in a smaller optimum pit size.

A nest of pits is generated for product selling prices ranging from a low level up to the base case reference price being used for the financial analysis of the deposit. The nest may be expanded using prices above the base-case reference for sensitivity purposes. The pit limit that becomes the basis for designing pushbacks can be selected in one of two ways. The pit limit generated for the base-case reference price is generally considered the largest allowable pit. Given that the economics of each pit in the nest is controlled by pit slope geometry and because timing of ore and waste mining is excluded, it is probable that the optimization process that uses cash-flow net present value, or NPV (which is discussed later) will result in a smaller final pit limit.

Another method for selecting an ultimate pit limit is to plot the cumulative increasing value at the base-case reference price for each successive pit limit in the nest. Typically, as the pit limit expands and deepens, the incremental value of each successive pit limit increment is larger or stable until an increment is reached where the incremental value becomes less because of progressively increasing stripping requirements and mining costs and possibly decreasing ore grade and metallurgical response. It may be preferable to select a pit limit that is smaller than the one using the base-case reference price based on the pattern of successive incremental values. This could be considered to be a subjective assessment of risk by avoiding exposure to increments having increases in stripping requirements and mining costs or decreases in ore grade and metallurgical response with only small increases in economic value. This would be verified during the financial analysis cash-flow portion of the evaluation.

The pit selected from the nest of pits serves as the trial ultimate pit limit for the subsequent mine planning steps from pushback design through optimization. It should be kept in mind that the final pit limit could be smaller than the first trial limit. Consequently, selection of the largest reasonable pit in the nest may be preferred as the trial pit limit so that no reasonable increments of value are omitted unnecessarily from evaluation and inclusion in the final mine plan.

In addition to analysis and selection of the initial ultimate pit limit, the nest of pits provides a sound economic and geometric guide for the design of pushbacks, which is discussed in the next section. In this regard, it is helpful to select intervals between prices so that incremental widths of the pits in the nest are narrower than the minimum acceptable width of a pushback.

As experience is gained regarding the economics of a deposit during successive planning cycles, it may be desirable to alter mining and processing costs and product prices as successive pits in a nest are generated to reflect changes in economics over time and for different zones of the mine.

There are two variations to typical ultimate pit limit determination procedures. One is inclusion of recovering payable values from material normally handled as waste. This was discussed earlier in defining the dollar-value cut-off criteria and may be referred to as leaching credits. The other is the open pit–underground trade-off.

The open pit–underground trade-off is treated by comparing the economics of mining a segment by open pit to mining it by underground methods. Ore zones that are minable by open pit but are more economic if mined by underground methods would be excluded from the open pit design. These circumstances are more likely to occur for a high-value, high-stripping-ratio ore body. This approach is appropriate when there would be sufficient ore mineralization at the conclusion of open pit operations to develop economic underground operations. Comparison of open pit and underground mining economics must include differences in mining costs, cut-off criteria, mining losses and dilution, and effects on processing performance and costs to obtain correct results. In addition to comparative economics, the open pit–underground trade-off may be affected by pit slope conditions limiting the pit depth, which was discussed earlier in this chapter.

The floating cone method (Carlson et al. 1966) has been adapted to single-bench ore-body block model used in phosphate strip mining by using a mining unit floating template to provide the equivalent geometric and economic analyses and output results (Crawford 1997).

The primary difference between designs of ultimate open pit limits and ultimate mining area limits for strip mining is their shape. Most open pits have a rounded character, whereas strip mines tend to be rectangular because of the configuration needed for effective dragline operations. The ultimate mining areas in Florida phosphate may be discontinuous because of irregularities in mining characteristics and property boundaries.

Pushback Design

Designing pushbacks is the process of breaking down the ore and waste volumes of an ultimate pit limit into operationally realistic production mining segments or units. This process includes

- Designing pushback bench widths;
- Using the nest of pits to define pushback configurations;
- Designing road and ramp systems; and
- Designing waste rock disposal areas, ore stockpiles, and surface facilities.

Pushback Bench Widths

Pushback bench widths are based on the sizes of the loading and haulage equipment being used and the necessary operating flexibility. The bench height is defined when the ore-body block model is set up. Pushbacks are generally in the range of 46 to 91 m (150 to 300 ft) wide for the sizes of loading and haulage equipment currently being used. There are circumstances when the width may be narrower or wider. The width for the narrowest part of pushback should be based on a combination of road and ramp design criteria, blasting drill access requirements, and single spot loading requirements. The normal width should be sufficient for accommodating double-spotting to maximize loading and haulage productivity. Making pushbacks as narrow as practical minimizes the amount of advanced stripping required to reach ore in a pushback and maximizes the benefits from economic resolution for optimization analyses of cut-off criteria and stockpiling strategies.

After the narrow and normal bench pushback-width criteria have been determined, these design criteria are used with the nest of ultimate pits to determine the configuration, location, and size of the pushbacks. If the pushback slopes differ from those of the ultimate pit limit, it will be necessary to generate a nest of pits using pushback slopes within the ultimate pit limit selected. It may also be desirable to verify whether the configuration of pushbacks is affected by the changes in the economic structure that vary with time and would be applied to the pushbacks for optimization of production scheduling and cut-off criteria.

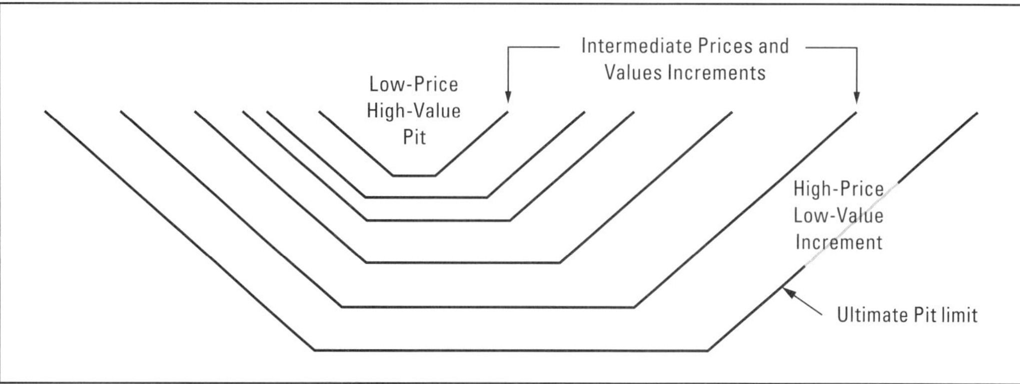

FIGURE 5.2 Schematic cross section of nest of pits

Nest of Pits

The nest of pits within the ultimate pit limit provides an economic guide for positioning pushbacks with respect to area of the pit and depth. Each increment of nest of pits from inside to outside and around the pit circumference is in an economic preference order. Each pit increment in the nest will become part of a pushback. Within the constraints of the bench pushback criteria, each pit increment in the nest should be included in a pushback in economic order before inclusion of the next pit increment is begun. Areas of a pit nest increment that are too narrow to fit into a pushback are included in the next pushback, or the width of the final pushback at the ultimate pit limit must be widened. The result of fitting the bench width criteria to the nest of pits is an economically ordered set of pushbacks that are ready to have roads and ramps added. A schematic cross section of a nest of pits is shown in Figure 5.2.

It may be desirable to modify the design of pushbacks in conjunction with pushback sequencing and production scheduling when feasible to smooth the avoidable, abrupt changes in ore characteristics and stripping ratio.

Pushbacks or mining cuts in coal strip mines are frequently rectangular with the long axis parallel to the strike. Generally, they are laid out in uniform widths from outcrop or subcrop to the high wall of the ultimate mining limit. The width is determined by the dragline digging pattern or the working room required for efficient loading and haulage operations.

In Florida phosphate strip mining, mining units for draglines are laid out in rectangular shapes of varying dimensions. The size and orientation of the mining units are related to economics, trends, and patterns of mining characteristics; matrix pumping system routings; and regulatory and environmental requirements.

Roads and Ramps

Roads and ramps are added to pushbacks using the following considerations:

- Width of roads and ramps
- Grade of ramps
- Selection of a spiral or switchback ramp system within the pit
- Providing access to loading faces on pushback benches

The width of roads and ramps is normally based on haulage unit operations. The width of the running surface (portion actually used by haulage trucks or other mobile equipment) of a road or ramp is generally three to four times the width of the haulage trucks being used. The total width is the running surface plus allowances for a safety berm on one side (the edge side within the pit) and a drainage ditch on the other side (the toe side within the pit) of the road or ramp. The running surface of main roads and ramps should be constructed and maintained with a crown to facilitate drainage and to sustain stability of the surface. The selection of subgrade and surface capping materials is critical for road durability. Some mines have crushing and screening plants for preparing these materials.

The maximum grade of ramps is normally between 8% and 12%. The flatter grades are generally preferred for electric-drive trucks. Mechanical-drive trucks tend to use the steeper grades. If there are sharp curves on ramps, it is desirable to flatten the grade of the ramp through the curve to less than 5% to compensate for the differential effect on the truck drive system and increased rolling resistance.

Location of the road system within the pit depends on the configuration of the pushbacks and provision of necessary access to loading faces as the sequence of pushbacks progress. It is best if the road system can be in a relatively permanent location except for when being pushed back as the parts of pushbacks that include the roads and ramps are mined. It may be preferable to have more than one road system within the pit if there are significant periods of concurrent mining in pushbacks in dispersed locations. Providing more than one road system out of the mine also increases the safety of operations and provides a contingency plan for sustaining operations if a road caves because of slope failure or flooding. Although additional roads may provide for additional safety and operating flexibility, the cost associated with them may be flatter overall pit slopes and greater stripping requirements.

A common concern about designing roads and ramps within a pit is whether long ramp sections should be switchbacks or spirals. The factors relating to switchback ramps are as follows:

- Ramps can be located in an area of the pit that is inactive for extended periods of time.
- Access roads to operating loading faces may be longer.
- Overall pit slopes in the pit zone of the switchbacks are flatter because of added allowance for 180-degree switchback curves.
- Ramps are longer because of the flattening ramp grade through the switchback curves.
- Lower haulage truck performance and high operating costs are likely because of slower speeds in switchback curves and greater wear and stress on haulage trucks from sharp curves.

The following factors relate to spiral ramps:

- Portions of ramps are located within the active mining areas of the pit.
- Some access roads to operating loading faces may be shorter.
- Normally there are no sharp curves that would affect overall pit slopes.
- Overall ramp lengths are shorter than for switchback ramps.
- Haulage truck performance and operating costs are better than for switchback road systems.

It is generally preferable to use spiral ramps within pits. The ramp grades should be flattened through sharp curves and switchback reversals to compensate for effects of the radius of curvature.

The overall road system should include roads outside the mine area to waste dumps, ore crushers, and ore stockpiles. There will also be access roads to the property and service roads within the property. Their locations should not conflict with the mining plans and potential future ore-body reserve and related mine design expansion.

In coal strip mines, main roads and ramps may be constructed on spoiled overburden instead of on advancing mining faces.

Waste Rock Disposal Areas

The following are important factors in designing waste rock disposal areas:

- The haulage distance should be minimized.
- The angle of repose of rock types and size distributions should typically be about 35 degrees.
- Ensure that they are not located on top of ore or encumber future pit expansion because of insufficient room for pit wall advance or for additional waste disposal.
- The location and design of the dumps need to properly control surface water and runoff drainage.
- They should not conflict with the location and layout of other surface mine, processing, and support facilities.

Ore Stockpiles

The design of ore stockpiles needs to include the following factors:

- Location with respect to mine, processing, and support facilities
- Location and design to properly control surface water and runoff drainage
- Maximum ore tonnages to be stored
- Need for separate stockpiles by ore type, grade, and metallurgical characteristics
- Haulage routes from mine to stockpiles and stockpiles to processing primary crusher(s).
- Stockpile building and rehandling method (trucks, conveyors, stackers, etc.).

Surface Facilities

Mine planning needs to be involved in the assessment of locations and layouts of mine surface, processing, and support facilities to ensure that they do not conflict with the mine design and mining plan. For the mine, these facilities include maintenance shop, tire changing and repair, equipment servicing and washing, parts and materials warehousing, diesel fuel and gasoline handling and storage, explosives handling and storage (ammonium nitrate), and offices. There are many Mine Safety and Health Administration (MSHA) and other government regulations and standards that need to be followed in planning, designing, and constructing these facilities. There need to be sufficient geologic evaluation and sampling to provide assurance that the surface facility sites do not overlay zones of known ore or of potential future exploration. In addition, consideration must be given to access roads and utilities from outside the boundaries of the property.

Routing and siting of potable, mine drainage, and waste waters; electric power and natural gas lines; and handling and distribution facilities must be considered. Their locations should not conflict with the mining plans and potential future ore-body reserve and related mine design expansion.

Other surface facilities that may be involved are as follows:

- Gasoline and diesel fuel storage
- Fueling, lubrication, and fluids checking stations
- Crushing and screening plant for preparing road subgrade and capping materials
- Explosive and related supplies handling, mixing, and storage facilities

Mining Sequence of Pushbacks

After the pushbacks, operating mining segments, or units have been designed, it is necessary to determine the sequence for mining them. In most open pits, sequencing is directed and controlled by the geometry of the pit and the pushbacks. This means that all of the pushbacks overlap each other in an onion-skin-layer order and that they must be mined in geometric order from inner to outer pushbacks. In situations where two or more pushbacks are not overlapped, selection should be made on the basis of the economic preference of the un-overlapped pushbacks and their geometric and economic relationship to subsequent overlapped pushbacks.

Generally, pushbacks are designed and sequenced to be mined from the top down. It would be unusual for a pushback to be mined partially and then combined with an adjacent or following pushback. A pushback may be mined with more than one loading face to meet production rate requirements

- By mining from both ends of each bench;
- If the pushback is wide enough to be mined with two or more loading faces from the end of each bench; or
- If a large pushback is divided into sub-pushbacks, each of which can be mined independently.

If a large pushback is divided into sub-pushbacks, it should be split to the extent practical using economic preference analysis and be operationally realistic. The nest of pits can be useful in helping determine the location of a split. The sequencing of the sub-pushbacks should be based on economic preference and their geometric and economic relationship to subsequent pushbacks. Subdivision of a large pushback should be considered as part of the design of pushbacks.

The final set of pushbacks and sequence should be such that each pushback is mined in order from inside to outside in economic, geometric preference, and top to bottom by complete benches in scheduling production. It is possible that as a result of financial analysis and optimization, one or more pushbacks will be redesigned or one or more of the outermost pushbacks will be deleted. If the latter happens, the ultimate pit limit and the resulting minable reserve will be reduced in size. The redesign of the ultimate pit limit and pushbacks gives rise to the iterative nature of mine planning because of the interactions between geometry and economics.

In coal strip mining, the sequence of mining cuts normally progresses from outcrop or subcrop toward the headwall. In Florida phosphate strip mining, the logistics of dragline moves and matrix pumping system routings frequently takes precedence over economic preference in sequencing mining units.

Production Scheduling

The pushback data and sequencing are used to generate mine ore, waste, and processing production schedules by time period over the life of the deposit. The time periods of interest may be days, weeks, months, or years. The shorter time periods are of more importance in the early years of the schedule because they may become operating budgets. The production schedule should include the effects of ore going into and out of stockpiles and flows through processing inventories if appropriate. Ore and processing product quantity and quality characteristics and recoveries should be included by ore type and process stream.

Several factors must be considered in transforming pushbacks into production schedules:

- Production rates and bottlenecks for mine ore, waste, and process facilities
- Operating schedule (number of operating hours and days in operating months and years)
- Cut-off and selection criteria for ore processing facilities
- Operating flexibility
- Preproduction development
- Blending and stockpiling strategies
- Mining losses and dilution

Production Rates

Production rate targets or goals used in generating a production schedule are normally referenced to daily or annual ore processing rates or annual salable product output rates. It is generally easier to work with ore processing rates as the primary reference. Determining desired production rates will depend on the following considerations:

- **Product marketability and sales price.** Market analysis may identify the maximum product output that can be sold at a reasonable price. This could change over time because of changes in market conditions and projections.
- **Capital requirements.** Limitations on available capital and financing may constrain production rates. The relationship between capital requirements and production rates may be partially a step function that would affect the optimum production level in some time intervals differently than gradational changes in capital–production rate relationships.
- **Optimization analyses.** Optimization analyses will yield the production rate that is associated with maximizing the economic value of the ore body. The process will balance all of the factors, including production rate, that affect the economic value.

- **Mine life.** Company objectives or regulatory considerations may establish the minimum mine life and therefore the maximum allowable production rate for the ore reserve.
- **Operating constraints.** Bottlenecks of existing mining and processing methods and equipment will affect the production rate depending on ore-body and product characteristics. They also may be adversely affected as equipment ages and performance declines.
- **Production efficiencies and operating costs.** The relationship between production efficiencies, operating costs, and production rates may be partially step functions that would affect the optimum production level in some time intervals differently than gradational changes in efficiencies and costs. They may change as equipment ages.
- **Regulatory requirements and limitations—governmental and environmental.** These requirements and limitations may constrain mine life or allowable production rate(s).

The most common production schedule target is the ore processing rate. The remaining production elements from mine through salable product are related to this rate, through the production flow material balance. Over the production life of a mine, the ore processing rate may be increased in expansion stages depending on performance of the production flow material balance. Most common is increasing the ore processing rate to offset declining ore grade, thereby sustaining the level of salable product output.

Generally, it is not practical to use mining rate as a scheduling target for open pit mines because of constantly changing operating conditions, such as increasing stripping requirements and lengthening haulage distances.

Operating Schedule

The operating schedule is the number of operating hours and days in operating months and years over the production life of the property. Generally, setting an operating schedule involves considering these parameters:

- The number of operating shifts and hours per day
- The number of operating days per month and per year
- Allowances for reduced operating schedules during periods of poor product demand or price, holidays, vacation, and maintenance shutdowns

Cut-Off Criteria

Cut-off criteria are used in the control of mining and processing operations to

- Distinguish between ore and waste during mining;
- Segregate ore between immediate processing and stockpiling for processing in the future;
- Provide a basis for selecting process options for different ore types and grades; and
- Provide a basis for determining when to process stockpiled ore.

Cut-off can be expressed in three ways:

1. Grade of the primary payable constituent of the ore (copper percentage)
2. Equivalent grade expressed in terms of the primary payable constituent of the ore for two or more payable constituents of the ore (equivalent copper percentage for copper, molybdenum, gold, and silver)

3. Dollar value of the block, discussed earlier in this chapter

Cut-off grades can be defined by three methods:

1. Minimum thresholds of ore grade and process recovery and product characteristics to achieve reasonable and economic processing performance levels and meet product characteristics in the marketplace
2. Classical breakeven economic calculations for the mine and processing options
3. Result of economic optimization analyses

An economic model is used to calculate the cut-off based on the constituents of the ore, location in the mine, and processing method used. It is proper to say there are mining and processing cut-offs. Generally the cut-offs can be expected to vary during the production life of a mine. For large mines, they can differ between mine locations. Management of cut-offs as an active part of operations production control requires a suitable model on current economics to evaluate mining location, ore grades, and processing characteristics of production blasthole and other mine operations mining area samples.

Cut-off parameters can be expressed in terms of physical and chemical ore, processing and product characteristics, or in economic (dollar value) terms. Modern computers and economics modeling software programs have greatly facilitated the use of the dollar-value cut-off concept. Parameters for ore and product quality often include minor elements and contaminants along with the payable elements. It is important that sufficient information be obtained about them during sampling and testing. Mining and process cut-off criteria may differ between areas and benches of the mine. In addition, they may change over time because of differences in ore and process characteristic distributions, and revenue and cost structures between areas and benches of the mine, and changes in them through the life of the ore body. The selection and use of variable cut-off criteria have become more sophisticated and more widely used as a result of applying advancing computerized techniques to mine planning, production scheduling control, and related economic and financial analyses to maximize the value of the ore body.

A typical practice exists to have high cut-off criteria for immediate processing early in the operating life of an ore body and then decline through time toward breakeven levels. This is done to achieve higher product output with less ore feed in the early years and to accelerate the payback of initial capital, sometimes because of the terms of financing or to obtain better terms. Reduction in cut-off and correspondingly lower average ore grade may result in staged equipment and facility capacity expansions to relieve production bottlenecks.

Operating Flexibility

Incorporating operating flexibility in a mine design, plan, and production schedule is meant to provide capabilities necessary to accommodate and adapt the resulting operations to changing conditions related to

- Errors in estimates related to ore-body and product(s) characteristics
- Errors in estimates related to operating performance and costs
- Changes in market conditions and product prices
- Compensation for lead time constraints on changes to production rates for mined ore, waste, and processing

Providing sufficient operating flexibility is a key element in designing pushbacks, sequencing them, and generating production schedules. Incorporating flexibility in the design and sequencing steps has been discussed previously. The key expression of operating flexibility in production scheduling is uncovering ore in later pushbacks with planned lead time before running out of ore in earlier ones. This means dealing with scheduling overlaps between pushbacks and the resulting impacts on waste removal and mining equipment requirements and being able to quantitatively express the amount of ore available at a given time requiring very little or no stripping. The terms *stripped ore reserve* or *ore cushion* are used to describe this concept.

Stripped ore reserve. Stripped ore reserve is the geometric measure of the current amount of ore that is available and assessable for mining in one or more operating pushbacks requiring very little or no additional stripping. The ore is material meeting the current or near-future operating mining cut-off criteria. Typically, the target amount of stripped ore reserves is 6 to 12 months of mined ore production. Setting up criteria for and measuring stripped ore reserves should take into account multiple ore types and blending distributions needed to sustain ore processing performance levels. Depending on processing production rate and performance sensitivities, ore in stockpiles may be considered as part of the stripped ore reserves.

Calculation of stripped ore reserves is based on the current pit topography and determining from the pit geometry the amount of ore that can be accessed within pushbacks without removing more than minor amounts of interstitial waste in the ore zones and without violating pit slope inter-ramp angle design criteria and parameters. This process can be tedious when being done in conjunction with scheduling production for a new operation. It is more likely that stripped ore reserves will be calculated periodically at an operating mine using a current mine topography map.

Ore cushion. The concept of an ore cushion is a convenient and effective tool to use for providing operating flexibility in generating production schedules. The approach works better when most of the ore lies below the waste in a pushback. Its effectiveness depends on the stripping ratios for the lower benches in a pushback being substantially lower than for the upper benches.

Applying the ore cushion concept involves determining where sustained ore is reached in a pushback and the amount of ore remaining to be mined in a pushback below a given bench. Sustained ore is reached when the bench and pushback remaining stripping ratios are significantly below those of the benches above. Preferably, the maximum stripping ratio for reaching sustained ore would be 1.0–2.0 to 1 or lower. A higher bench stripping ratio for sustained ore may yield meaningful results for high stripping ratio deposits.

There are circumstances where the pattern of bench stripping ratios does not allow a meaningful point of sustained ore to be reached. This occurs when there is a large amount of interstitial waste on the lower benches of a pushback. If the point of reaching sustained ore in a pushback cannot be determined because of the bench stripping ratio pattern, the stripped ore reserves approach discussed previously should be used to determine a measure of operating flexibility. Once sustained ore is reached in a pushback, most of the stripping emphasis and waste production capacity in a production schedule can be moved to the next pushback.

There are situations when more than one pushback may be being stripped down to sustained ore at the same time while most of the ore is being mined from the sustained ore zone of the current pushback. Generally, it is desirable to achieve desired operating and blending flexibility while operating in the fewest number of pushbacks practical. The best efficiencies are achieved in compact operating areas while at the same time having sufficient working sites to avoid operating and maintenance (O&M) congestion and meet ore blending needs.

It is also necessary to establish the target amount of ore remaining in the lower part of a pushback when the next pushback reaches sustained ore. Similar to stripped ore reserves, this could be 6–12 months of mined ore production. This target amount may include stockpile ore. Because it is usually a measure for only one pushback, it is normally lower than the amount for stripped ore reserves.

Scheduling an amount of ore remaining in the lower part of a pushback when the next pushback reaches sustained ore provides an overlap of ore availability between pushbacks in the event there are ore-to-waste reversals, blending difficulties, or a shortfall in the mining rate of a pushback being stripped to sustained ore.

One of the difficulties in achieving operationally realistic production schedules with commercial mine planning software is there may be insufficient capabilities and options to set up operating flexibility constraints. Being able to set up effective operating flexibility constraints provides the mine planner with a tool to ensure that the stripping schedule in a production plan is not shortsighted. This is an important capability to consider in selecting commercial mine planning software.

The concepts of stripped ore reserves and ore cushion can be applied in coal strip mining where there are separate operations for stripping waste and mining ore with either draglines or truck and shovel methods. They cannot be applied in Florida phosphate strip mining with draglines because waste and ore (matrix) are mined from the same setup position for the dragline. Here it is not practical to strip waste more than a few hours ahead of mining the related ore (matrix).

Preproduction Development

An important part of developing a new open pit mine is determining the preproduction stripping and ore stockpiling to be done before commencing processing operations. It is desirable to do enough preproduction mining of ore and waste to provide in stockpiles sufficient ore and stripped ore reserves or ore cushion for the initial several months of processing operations and to expose ore with a reasonable bench stripping ratio. Depending on the timing required for scheduling production in pushbacks after the first one, it may be necessary to schedule production from an additional one or more of the pushbacks during the preproduction period. These additional pushbacks may provide some ore for initial processing operations. It is often desirable to start up processing facilities on lower-quality ore and use it until operations have reached a level of productivity and efficiency to achieve the expected results from normal planned ore production.

Another element in production scheduling for strip mining, and perhaps for mining cut design and sequencing, is waste or overburden that is cast blasted or pushed by bulldozers into an open mining cut. It must be correctly accounted for in production schedules, including any portion that is rehandled by draglines or loading equipment. Some overburden originally spoiled by dragline may have to be rehandled, depending on the dragline digging pattern.

Blending and Stockpiling Strategies

Blending and stockpiling of ore is a common operating tool used for the following reasons:

- It provides predictable and stable ore feed characteristics for processing operations.

- It sets aside lower-quality ore that must be mined "now" for processing later in the life of the operation.

The optimization analyses that are discussed later in this chapter may result in the ore processing cut-off in the early and middle-life years of an operation being higher than the mining cut-off. Under these circumstances, the ore with a grade between these two cut-offs is stockpiled. Ore in stockpiles is processed when its grade is higher than newly mined ore. It may be that most stockpiled ore is processed after mining operations are complete.

It may be necessary to have a number of stockpiles for a variety of ore grades and characteristics or to build a stockpile(s) comprised of blended ores to meet processing plant requirements.

The nature of some mines and processes is such that run-of-mine ore from a given location is too variable in quality characteristics to be used alone for significant periods as feed to processing. In such circumstances, it is necessary to provide ore to processing from several loading locations in the mine concurrently to achieve necessary quantity of ore and control the variability of the quality characteristics. This may affect the sequencing and scheduling of pushbacks to provide a sufficient number of ore-loading locations. This usually affects the amount of waste stripping. There are circumstances where the necessary control can be obtained only by blending run-of-mine ore from different mine locations in large stockpiles before it is processed. It may be necessary to have several stockpiles to provide either the processing quantities or ranges of quality required. They also provide a measure of operating flexibility and a buffer between mining and processing operations. Blending and stockpiling will increase the costs of ore delivered to processing because of potentially lower productivity of mine equipment and the cost of operating stockpiles and rehandling material from them.

Stockpiling of lower-grade ore for later processing may provide a means of increasing value for mines with high-value ore such as gold. For such ore bodies, it is generally accepted that it is preferable to process higher grades of ore in the early years of mine life. Low-grade ore is stockpiled as it is mined and then processed later when higher-grade ore is not readily available or processing rates are increased. There may be several stockpiling options that need to be evaluated as part of optimizing the mine design and production plan. The potential contribution to added economic value from stockpiling low-grade ore may be influenced by several factors:

- Mining costs are incurred before ore values are recovered.
- Waste removal requirements may be increased.
- Stockpiling and rehandling costs are incurred.
- Metallurgical response characteristics may deteriorate because of stockpile exposure.

The use of stockpiling for ore blending has become more important as mining equipment has become larger. There are generally fewer ore-loading positions in the mine as a result of the larger equipment, and emphasis on achieving unit productivity has decreased the feasibility of in-pit blending. In addition, attempts are being made to make processing steps more adaptive and more forgiving in handling variable ore characteristics while still meeting product quantity and quality specifications and achieving reasonable recovery yields. An example for handling low-grade material through stockpiles for future processing is to stockpile ore having a lower grade than the cut-off (s) for being mined directly to processing. The cut-off grade for the stockpile would be the lowest grade that can bear the costs of rehandling from the stockpile and processing and make a profit.

Because of tight specifications on product quality, the use of stockpile blending is a common practice for coal strip mines.

Mining Losses and Dilution

Production schedules account for mining losses and dilution if their occurrences are expected to have a meaningful impact on the realized quantities of ore and waste produced and the quality of ore fed to the processing plant. In the case of mining losses, some ore may not be mined or it may be lost by mixing with waste material. Dilution occurs when a quantity of waste material becomes mixed with ore going to the processing plant, adding to the quantity being processed and thereby reducing the feed grade and perhaps salable product quality characteristics.

Financial Analysis

The financial analysis of a mine design and production schedule uses a life-of-mine profitability and cash-flow model. This model encompasses the periods of preproduction design, construction and development investment, life of production, and postproduction reclamation and abandonment. It should include all revenues, costs, taxes, and possibly inflation. Because of limitations in the capabilities of the ore-body block economic value model, discussed earlier, this model is used to validate and guide refinement and finalization for aspects of the mine design and planning that make use of that economic model. The cash-flow financial analysis model is a key tool in optimization and sensitivity analyses because it includes all revenues and costs and is the model that handles timing of revenues and costs. Chapter 20 contains a complete description and discussion of applicable and preferred financial analysis techniques.

The cash-flow model is used to generate and analyze yearly profitability and cash flow over the life of the ore body and mine being evaluated and to provide insight about year-to-year trends, patterns, and aberrations. The cash flow must be on an after-tax basis to properly combine operating costs and capital costs. It should be made clear whether inflation is included in revenues and costs through the years. In addition to generating annual profitability and cash flow, it is capable of providing NPV, discounted cash-flow rate of return, and payback statistics for optimization and sensitivity analyses and interpretation. It is useful to generate working capital patterns and requirements based on the behavior of inventories and the time lags between incurring costs and receiving revenues.

All revenues and costs that could affect the determination of the optimum pit limit, pushbacks, mining sequence and production schedule, and optimum cut-off and stockpiling strategy over the life of the ore body must be included. Revenue and cost elements are attached to segments or nodes of the production flow material balance from the mine through stockpiles and processing to salable product. It is necessary that the salable product have a recognized market reference to estimate the selling price or net return.

Replacement and sustaining capital covers equipment replacement cycles and additions to maintain productive capacity and recoveries. During the mid-years of a property, it may be sufficiently repetitive to take on the characteristics of an operating cost even though the accounting principles for capital costs and operating costs are different. Many evaluations do not give this cost category enough attention.

The structure of revenues and costs will change over time from the early years of operation through mid-years to late life. Some of the characteristics of these different time periods are described as follows:

- Early operating years
  - Low O&M costs
  - Low replacement and sustaining capital costs
- Middle operating years
  - Rising O&M costs
  - Normal level of replacement and sustaining capital costs
- Late operating years
  - Continued rising O&M costs
  - Lower management, technical and administrative staff, and G&A costs
  - Reduced replacement and sustaining capital costs
- Near pit limit (special case of late operating years, used for calculating economic values for the ore-body block model)
  - High O&M costs
  - Low technical and administrative staff and G&A costs
  - No sustaining capital costs

The costs used, whether based on engineered estimates or data from existing operations, will reflect expected operating methods, equipment, head count, and production rates. The composition of costs is unique to each combination of method, equipment, and operating schedule and rate. They need to be modified as each combination is evaluated. The composition of the costs for a given combination can be expected to change over the life of the operation between operating, maintenance, capital, and overhead. The ore-body block economic values will use the economic structure expected near the pit limit, and the cash-flow model will include the changes in composition over time.

The mining costs need to be separated between waste and ore, and among mining sectors, zones, or levels. The waste should carry its share of capital and overhead costs. It is important that all costs are properly distributed between waste and ore mining and processing to obtain the correct response and balance for changes in volumes between them in determining the optimum ultimate pit limit, production rate, cut-off, and stockpiling strategy. It may be preferable to separate labor-related (salaried, hourly, and vacation/sickness/accident) costs, wages, salaries, and benefits from equipment performance and costs to facilitate analyses of them over time and under different operating schedules and production rates.

Most of the costs can be included in the model for calculating block economic values, while others can only be handled in a life-of-mine cash-flow financial analysis model. This is because some costs that are more related to time and other non-volume-related factors can be modeled for the ore-body block model economic value only in an approximate fashion. The model for the block economic value does not include pre- and post-mining costs because of problems with timing and location. This is the reason that economic models for both block economic values and life-of-mine cash-flow financial analysis are needed as an integral part of mine planning. The life-of-mine cash-flow financial analysis model is necessary for optimization and validation of the mine design and production schedule.

Equipment Performance and Costs

During estimation and evaluation of equipment performance and costs, attention needs to be given to the following factors:

- Equipment operating schedules
- Equipment availability and utilization
- Fill factors for loading and dragline buckets and haulage trucks
- Wet and dry payloads for loading and dragline buckets and haulage trucks
- Haulage truck spotting and loading times, single or double spotting, frontal or drive-by loading
- Haulage truck turn-and-dump time at waste dumps, ore stockpiles, and crushers
- Rolling resistance and speed limits
- Productivity efficiencies
- Changes in ore and waste haulage profiles over time
- Haulage unit travel times (loaded and empty, over various profiles) for ore, waste, and ore stockpile rehandling

Haulage truck profile simulators estimate performance on the basis of gross vehicle weight and wet payloads. It is necessary to state the hourly productivities of loading and haulage equipment in the measure of the production schedule, volume or tonnage, and wet or dry basis to obtain the correct fleet requirements and operating costs. To the extent practical, data on payloads, fill factors, rolling resistance, speed limits, loading cycle times, and productivity efficiency factors should be obtained from existing or comparable mining operations using similar equipment.

In estimating payloads, it is necessary to know the density of the broken material being handled. There may also be a question regarding the capacity volume of buckets and haulage truck beds as to using an SAE (Society of Automotive Engineers) 2-to-1 or 3-to-1 heap rating. This can also be handled through the choice of fill factor. Normally, fill factors are 85%–95% depending on the material being handled and the design of the buckets and beds.

Depending on road surface conditions, tire pressure maintenance, and moisture, the rolling resistance for haulage trucks will be 3%–5%. Caution should be observed in allowing speed limits above 48 km/h (kilometers per hour) (30 mph [miles per hour]) for off-road haulage trucks on flat segments of haulage profiles. Higher speed limits depend on safety related to road conditions and visibility. Speed limits should be lower for ramps, particularly for downhill loaded travel, where they should not exceed 24–32 km/h (15–20 mph) on 8%–10% grades. Reduction in speed limits may be necessary for sharp curves. Computer programs, such as Caterpillar Fleet, Production and Cost Analysis (Caterpillar 2012), are an excellent source for estimating productivity of mining equipment. Most of the large equipment manufacturers have similar programs. Spreadsheet software can also be very useful.

Double-spotting for loading haulage trucks should be planned because the productivity for both loading and haulage can be higher than for single-spotting. A schematic diagram of

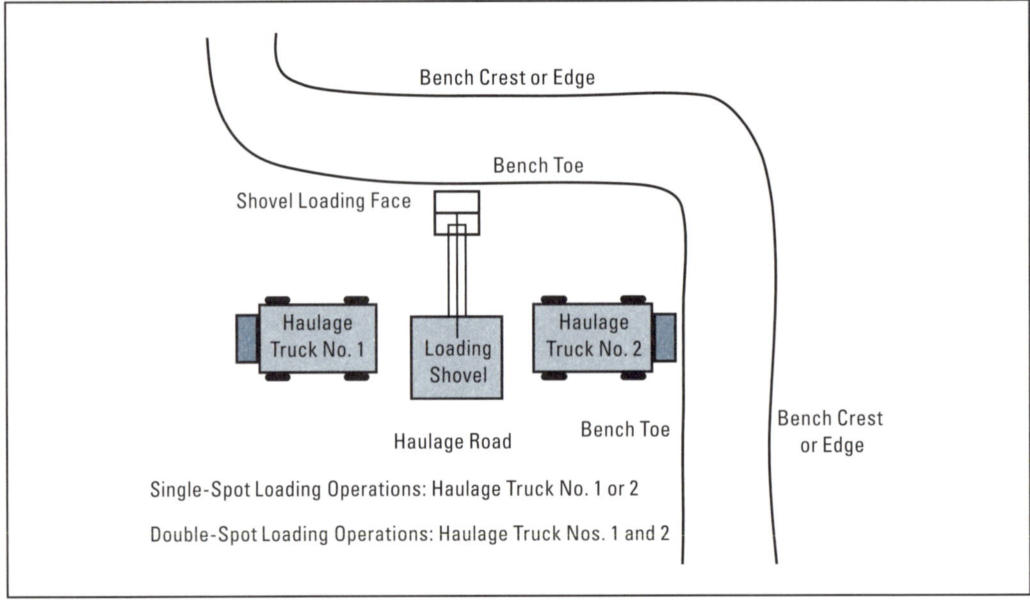

FIGURE 5.3 Schematic diagram of single- and double-spot loading operation

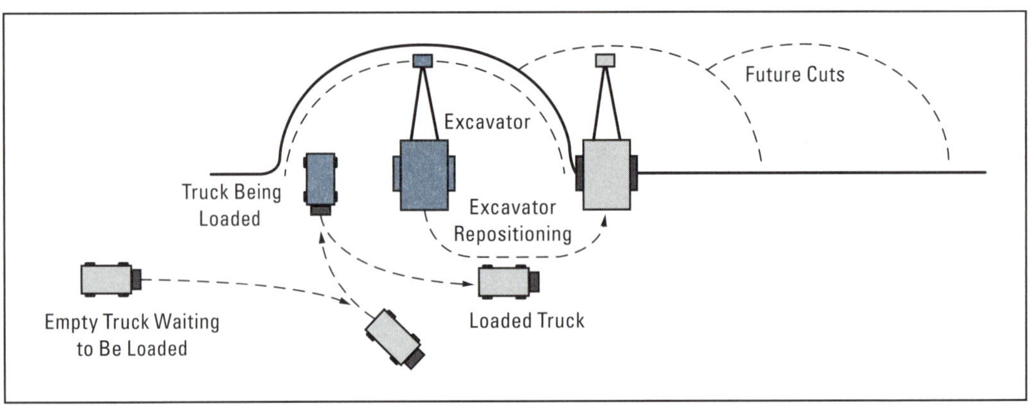

Adapted from Hustrulid and Kuchta 2006

FIGURE 5.4 Truck and shovel positioning for frontal cut

single- and double-spot loading operations is shown in Figure 5.3. The amount of productivity is influenced by spotting conditions and is related to the amount of time haulage trucks have to wait at the loading equipment to get into a loading spot. A factor involved here is the over-covering or under-covering of the loading equipment by the haulage trucks. Generally, the influence on haulage truck productivity is rather small. But the loading productivity improvement of double-spotting over single-spotting can be about 15%–20%. Other systems of truck and shovel spotting, which may be used for frontal advance of the pit face, are well defined by Hustrulid and Kuchta (2006) and Wetherelt and van der Wielen (2011) and shown in Figures 5.4 through 5.6.

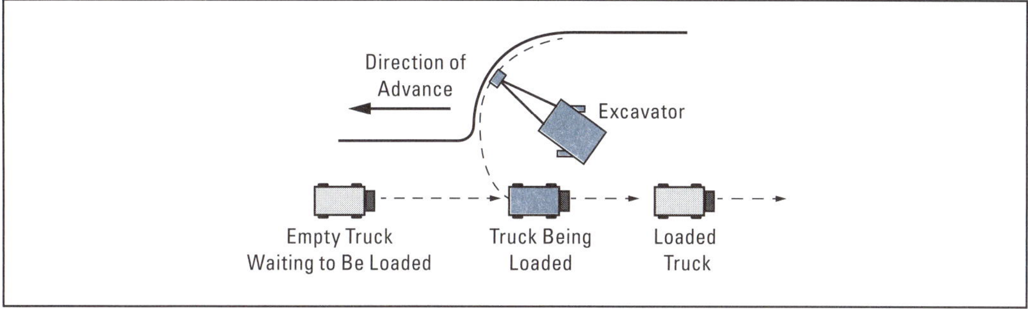

Adapted from Hustrulid and Kuchta 2006

FIGURE 5.5 Truck and shovel positioning for drive-by loading

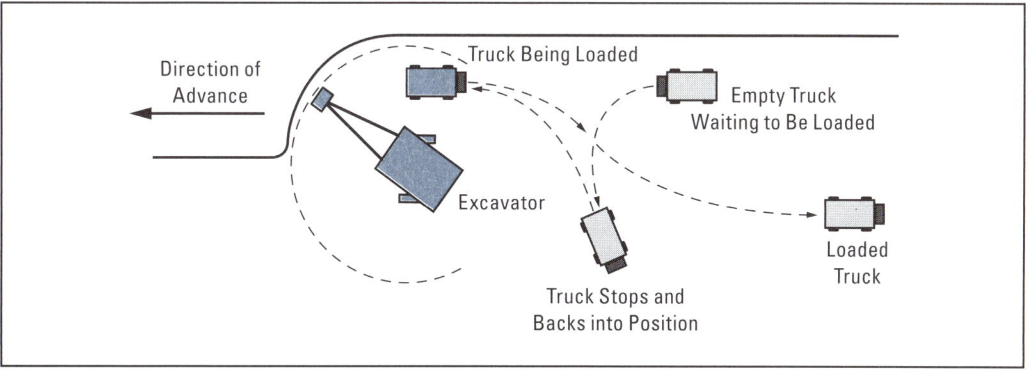

Adapted from Hustrulid and Kuchta 2006

FIGURE 5.6 Truck and shovel positioning for stop-and-reverse parallel operation

According to Hustrulid and Kuchta (as cited in Wetherelt and van der Wielen 2011),

> *...available working space, the necessary swing angle of the excavator and truck positioning time are the major considerations in the selection of the type of operation. Drive-by operations are most suitable for parallel cuts, optimizing efficiency by reducing positioning time of truck. Frontal cuts can require excessive swing angles by the loader, making it inefficient. The disadvantage of drive-by operations is that they require larger working areas and ideally a separate ingress and egress route from the loading area. Stop-and-reverse operations can be employed in combinations with both parallel and frontal cuts. They require less operation space and are more efficient from the excavation point of view. When there is sufficient space on the bench, a truck can turn without the need for reversing.*

Loading equipment should be slightly under-covered by haulage trucks to obtain the lowest combined loading and haulage costs. This is because the cost of lost production from waiting time is typically greater for haulage trucks than for loading equipment. This balance could be altered if haulage trucks are transferring material to a crushing and conveying transportation system, depending on relative costs of lost production from waiting and whether or not it is being fed from more than one source. This could also be done when it is necessary to maximize the loading rate of an ore type critical to meet blending requirements for processing.

O&M performance and cost data supplied by manufacturers are often overly optimistic. Equipment planning early in the mine planning of an ore body may tend to understate fleet requirements and cost projections.

Blasthole drilling and explosives. Drilling and blasting are required in most open pit and strip mines to break ore and waste rock to sizes that enable them to be handled for removal from the mine, or in the case of strip mining, for draglines to place waste stripping into windrows. The ore sizing must be small enough for pieces to fit into the primary-stage coarse crushers.

Blasthole drill size and bench height are normally selected to facilitate single-pass drilling without having to add drill pipe. The drills may be crawler or truck mounted depending on their size. Crawler-mounted drills may be towed by crawler bulldozers for long moves between work areas. Blasthole pattern dimension and hole diameter are selected for control of ore and waste rock breakage size distribution to reduce the amount of oversized material for loading and haulage handling and to reduce the amount of energy used in crushing and grinding the ore. The hole depth is for the height of the bench involved and may include some additional length for subgrade depth. Depending on the blasting characteristics of the rock being blasted, subgrade drilling of a short length may be included to minimize "hard bottoms" for less abuse to loading equipment while digging the bottom of a cut and for maintaining a more accurate bench elevation.

Blasthole diameters for copper and similar open pit mines are generally in the range of 22.5 to 35 cm (8⅞ to 13¾ in.) depending on bench height and rock blasting characteristics. For taconite, typical hole diameters are 38–43 cm (15–17 in.). In some instances, smaller-diameter holes and shorter-distance spacing are used for ore than for waste rock to achieve smaller-sized broken pieces to reduce process crushing and grinding requirements. Initial and periodic replacement capital or leases and operating costs for drills and related labor must be included in mine economics estimates.

The most common explosives used for surface mine blasting are ammonium nitrate and fuel oil (ANFO) and high energy and aluminized emulsions. Prima-cord, electric blasting caps, and down-the-hole delays are used to ignite and control the sequence and progression of a blast pattern. The storage, preparation, loading, and ignition of explosives may be performed by in-house or contract crews. The explosives may be stored and prepared at on- or off-site facilities. Such facilities are tightly controlled by MSHA and other government regulations. The size of blasts may also be controlled by MSHA and other government regulations regarding off-site noise levels and ground-movement limits.

A special situation is dealing with wet blastholes. Under such conditions, suitable emulsions may be used. A common method that enables ANFO to be used with wet conditions is to dewater a blasthole with a pump, insert a plastic sleeve into the hole, and then fill the sleeve with explosives. The time between loading a wet hole and igniting it should be minimized to reduce the likelihood of water infiltration causing a misfire.

Secondary blasting or hydraulic breakers are used to handle oversize boulders and "hard bottoms" in mine loading areas.

Service and other support equipment. In addition to primary production equipment, a productive and efficient mine requires numerous pieces of service and other support equipment. Many of the major items in this category are summarized as follows:

- Track and rubber-tired bulldozers for shovel pit, road, and dump maintenance and construction
- Motor-graders for road and dump maintenance and construction
- Scrapers for road and dump maintenance and construction
- Front-end loaders for production and miscellaneous loading requirements
- Hydraulic excavators for construction
- Water trucks for road, dump, and shovel pit dust control
- Diesel and gasoline trucks for fueling equipment in the field
- Explosives trucks for loading blastholes; in some instances, these are provided by contractors
- Tire-handling equipment
- Field maintenance trucks for shovel and drill repairs
- Wrecker truck for small rubber-tired equipment and service truck retrieval to maintenance shop or outdoor repair area
- Cranes for maintenance and construction
- Various vehicles for transporting personnel

Initial and periodic replacement capital or leases and operating costs for this equipment and related labor must be included in mine economics estimates.

Actual Operating Performance and Cost Data

Use of actual operating performance and cost data, when available, is essential to achieving a realistic analysis and evaluation of an ore body. These data serve as a basis for various estimates used in developing a mining plan and for validating the resulting plan. Suitable data can come from an operating property being evaluated or from a mining and processing operation comparable to the ore body being analyzed, planned, and evaluated.

Actual operating performance and cost data can be used in the full spectrum of mine planning data, criteria, and parameter requirements discussed previously. This is particularly necessary in performing studies on an existing operation. Key examples of this data are ore reserve, production reconciliation throughout the material balance, detailed comparisons of actual operating performance and costs with past plans and competitor data, and development and validation of future plans. An effective data resource is the previous three years of data interpreted and adjusted for major operating changes and inflation that have taken place over the three-year period. Using three years of data normally enables erratic patterns, aberrations in data, and favorable and unfavorable trends to be more readily identified and interpreted with regard to their impacts on current operations and planning for the future. In some situations, using five years of data may be appropriate, but excessive variability in parameters must be carefully analyzed and interpreted for their use in planning for the future.

Cost allocation methods used in accounting are often inappropriate for use in mine planning economics analyses. This is because the allocations used in accounting are often oversimplified for profit-and-loss purposes and do not adequately associate costs with the proper production material balance segment quantity. Most of the problems are in allocating G&A and some maintenance and labor costs. Often these costs are allocated to ore or final facility product rather than allocating to waste and other intermediate production volumes. It is necessary for mine planning personnel to work with accounting and operations to ensure that all costs and production material balance segments are properly matched and modeled to correctly handle changes in relevant production quantities and their effects on costs. Additional open pit cost information can be found in Chapter 15.

Optimization Analyses

The objective of optimization analyses is to determine the combination of ultimate pit limit and pushback designs, mining sequence, process selection, production rates, cut-off and stockpiling strategy, and production schedule, which will result in maximizing the value of the ore body or mine. The preferred value measure for optimization is normally the NPV of the annual after-tax cash flows. It is important, however, to look at year-to-year profitability and cash-flow patterns in addition to the total value in interpreting the results. If there are extended periods of erratic or unsatisfactory profitability, cash flow, or changes in cut-off and production rates, the resulting optimized plan is probably unsatisfactory for realistic execution in the field and requires further study.

Commercial mine planning software performs optimization analyses using the block economic values and internally generated production schedules with stockpiling being optional. These packages provide useful insights and reasonable first-approximation analyses. They may have several shortcomings, however, in their capabilities to produce some of the detail needed for refined analyses:

- Insufficient control of stripping lead time and pushback overlap
- Inability to change cost structure over the life of the production schedule
- Improper handling or exclusion of certain costs, such as capital and reclamation
- Inability to properly handle taxes and inflation, if used

These shortcomings can only be overcome through the use of a correct mine and processing production schedule and conventional cash-flow financial analysis model that can properly handle all necessary production, revenue, and cost data, and related calculations (as discussed previously in the "Financial Analysis" section).

Sensitivity Analyses

Sensitivity analyses provide insight to the impacts on the value of the ore body or mine from changes in engineering, production, and financial parameters. This is typified by determining the change in total economic value for a given change in a sensitivity variable, such as a capital overrun, change in product selling price, or a reserve or production shortfall.

A useful part of optimization and sensitivity analyses is to determine the optimum mine design and production rate that achieves the minimum required financial requirements at the lowest likely market price and output, and then determine the value of that mine design at the reference selling price and optimum production rate. The difference in values provides a sense of risk inherent in the optimum design—the greater the difference, the greater the downside risk.

Given that mine planning is a tightly integrated process of many steps and iterations to achieve operationally realistic and optimum results, it must be remembered that changing a parameter or value in any step in the process may require revision and reanalysis of other parts of the plan to achieve correct results.

In some instances, sophisticated operations research techniques may be used for optimization and sensitivity analyses. These methods include linear and dynamic programming, risk probability analysis, and Monte Carlo simulation.

Operations Monitoring and Reconciliation

Analyses and evaluations of operating mines and processing facilities involve detailed reviews and interpretations of performance and cost data for current operations and future plans. The primary purpose is to formulate or validate future plans. This receives a great deal of attention during due diligence. Reconciliation of ore reserves, production estimates, and mine production is a special part of this. It is often desirable to perform such studies using three years or more of actual data to identify, interpret, and possibly reduce the impact of short-term aberrations in the data.

It is necessary to reconcile mine ore reserves, production estimates, and actual production for mine and processing volumes, densities, ore grades, and metallurgical performance. The objective is to assess the reliability of production projections and the operations capability to achieve them. In this regard, the reconciliation and operating assessment is in two parts.

One part is to compare actual production data with the ore-body block model for volumes, tonnages, and grades. This is used to validate the reliability of the block model and determine factors that may be needed to adjust model data for production estimating. These comparisons are valuable in assessing model bias; ore to waste reversals; errors of grade and density estimates, and mining losses and dilution; and determining adjusting factors to use in mine planning.

The second part is to assess the actual performance and costs of operating equipment against budgets and plans. The primary concerns here are, are the production rates for ore and waste being achieved at the grades and from the locations in the mine as planned, and are planned costs being achieved?

The combined results of these two assessments of current operations serve as the basis to assess and validate or modify future plans and projections to achieve goals and objectives. They serve as the basis for determining adjusting factors for use in mine planning.

SUMMARY

Mine planning is a continuous, iterative, multidisciplinary process that must be performed in a systematic manner. It plays a powerful role in transforming an ore body from a mineral deposit into an economically viable business entity. It is important that people with field and operations experience be on the mine planning team and that overreliance on people with primarily computer expertise be avoided. Operating input is critical for the mine planning elements that have the important impacts on operating flexibility and field execution of the plan, and operations responsibility and accountability. A mining plan must be operationally realistic and executable in the field.

Table 5.2 provides a list of the steps and elements for preparing a sound, effective mining design and plan for a typical surface mine. It is assumed that most of the mine planning process is computer based.

TABLE 5.2 Steps for preparing an effective mining design and plan for a typical surface mine

1. Choose whether an ore deposit is to be planned and evaluated as a surface or underground mine.
2. Collect and compile data for surface topography, property boundaries, deposit and surrounding geology, and drilling and sampling for deposit mineralization grades, geotechnical characteristics, metallurgical characteristics, mining, internal processing, and outside treatment economics.
3. Generate an ore-body gridded block model encompassing the mineral deposit and surrounding zone likely to be included in the mining plan as waste, considering block size, geological and metallurgical domains, and value assignment methodology that could be by direct assignment, polygons, contouring, inverse-distance weighting, and geostatistical methods.
4. Define the demonstrated resource base within property boundary limits. Only the measured and indicated material meeting cut-off within a mining plan can be considered as ore. According to U.S. regulations, all inferred material must be treated as barren waste. Each block should be assigned a resource classification code.
5. Obtain accurate surface topographic and boundary data from reliable sources in digitized format. These data need to be ground-truthed to the extent practical to avoid major errors.
6. Determine ultimate and operating pit slope angles to be used. The geotechnical data are used to estimate the inter-ramp slope angles. The resulting slopes used for mine design and planning must be flatter than the inter-ramp slopes to incorporate the added setback required for roads within the mine and safety benches. Slope angles may differ between different parts of a mine.
7. Determine in-situ and broken densities of ore and waste to be handled from samples for various ore and waste rock types and swell amount by breakage from blasting designs and material handling methods. Assign appropriate values to each block. Distinguish between wet and dry densities from expected moisture content of ore and waste rock types.
8. Determine metallurgical behavior values for each block from metallurgical testing. It is necessary that metallurgical balances of weights and grades for test samples and processing products satisfy closure criteria based on Richard's Law for the data to be valid. Assign values to blocks by direct assignment or use equations based on block mineral characteristics and grades. For some deposits, there may be more than one set of values for multiple processing options. Potential process response deteriorations need to be incorporated in evaluating and setting cut-off criteria for stockpiling strategies.
9. Adjust block model data. This is done when necessary to reflect conservatism and concerns regarding some ranges of values being unrealistic from subjective assessment by qualified personnel.
10. Assess mining losses and dilution. Mining losses and dilution are always a concern when developing a mining plan. Estimating the amount and impact of these factors is generally difficult to assess. An empirical assessment of them from operations monitoring and reconciliation of production and reserves may be the only practical approach in most situations.
11. Develop a block economic value model for ore and waste material. Values are directly assigned to the blocks. This model is normally considered incomplete for use in financial analysis and refined optimization and sensitivity analyses.
12. Generate the ultimate pit design by the floating cone or Lerchs–Grossmann methods for the demonstrated resource material using the block economic value to define the ore cut-off.

(Table continues)

TABLE 5.2 (Continued)

13. Determine pushback design and generate a nest of pits.
    - Within the maximum ultimate pit design, generate a nest of pits related to the pushback design factors and with variable ore cut-off value control. The most important factor in pushback design is the width, which is heavily influenced by the sizes of the loading and haulage equipment being used and the necessary operating efficiency and flexibility.
    - The design of pushbacks must include the design of roads and ramps within the mine. This is also a good time to prepare estimates and designs for waste rock disposal areas, ore stockpiles if needed, related road system outside the mine, and for the wide range of maintenance, utilities, and support facilities needed for a surface mine.

14. Use pushback data and sequencing to generate mine ore, waste, and processing production schedules.
    - The sequencing of pushbacks for production scheduling is controlled by their geometry order from inner to outer pushbacks, and where they are not geometrically constrained, their economic preference order.
    - The pushback data and sequencing are used to generate mine ore, waste, and processing production schedules by time period over the life of the deposit. Some of the factors that influence production schedules are production rates and bottlenecks of mining equipment and processing facilities, operating schedules, cut-offs, operating flexibility, blending and stockpiling strategies, and mining losses and dilution that can affect realistic ore and waste production requirements. Numerous factors affect overall production targets or goals, the most common being the preferred or capacity ore processing rate.
    - One of the key elements in determining on operating cut-off strategy is practical limitations imposed by the ore processing methodology in achieving desired processing output quantities. Within such limitations, the specifics of a cut-off strategy are coupled with blending and stockpiling strategies as the result of economic optimization analyses.
    - One of the critical considerations in production scheduling is operating flexibility, which can be assessed by evaluating the patterns of the stripped ore reserve and ore cushion during the life of the production schedule and particularly as production locations make transitions between pushbacks or phases.
    - It is important for a new mine to have a plan for adequate preproduction waste stripping and ore stockpiling before commencing processing operations.

15. Perform financial analysis to validate and guide refinement and finalization for the mine design and planning.
    - The financial analysis of a mine design and production schedule uses a life-of-mine profitability and cash-flow model. This model is comprehensive including numerous items that cannot be properly represented in the model used for the ore-body block model economic values.
    - A key part of mine operating costs is the estimation of performance and costs for production and support equipment, including loading and hauling equipment, blasthole drilling and explosives, and major production support units such as dozers and graders.
    - It is important for an operating mine to have a comprehensive history of actual operating performance and cost data for use in preparing mining plans and for analysis and evaluation of progress against current budgets and plans.

16. Perform optimization and sensitivity analyses to maximize the value of the ore body or mine.
    - Optimization is the processing of using all of the mine planning and monitoring tools to typically maximize the net present value of the mine over its life while meeting other financial performance parameters needed to sustain the viability of the operations year to year. Generally, this may result in some variability in cut-offs and use of ore stockpiles for different locations within the mine and over time. Some situations result in major expansion of ore processing capacity to sustain or increase salable product output.
    - Sensitivity analyses is used to evaluate the impacts of variations in key financial revenue and cost parameters, particularly as they relate to market slumps and operating and capital cost excesses.

17. Monitor operations performance and perform production/reserve reconciliation.
    - Analyses and evaluations of operating mines and processing facilities involve detailed reviews and interpretations of performance and cost data for current operations and future plans.
    - Reconciliation of ore reserves, production estimates, and mine production is a special part of this. Reconciliation serves as the basis for determining adjusting factors for use in mine planning. There are instances when differences determined during a reconciliation process are used to make overall adjustments to ore reserves.

REFERENCES

Carlson, T.R., Erickson, J.D., O'Brian, D.T., and Pana, M.T. 1966. Computer techniques in mine planning. *Mining Engineering*, 18(5):53–56.

Caterpillar. 2012. Caterpillar Fleet, Production and Cost Analysis software. Caterpillar Inc., Peoria, IL.

Crawford, J.T. 1997. Strategic planning at White Springs, Florida phosphate mining and chemical operations. SME Preprint No. 97-058. Littleton, CO: SME

Hustrulid, W., and Kuchta, M. 2006. *Open Pit Mine Planning and Design.* London: Taylor and Francis.

Lerchs, H., and Grossmann, I. 1965. Optimum design of open pit mines. *CIM Bull.* 58(1).

Wetherelt, A., and van der Wielen, K.P. 2011. Introduction to Open-Pit Mining. In *SME Mining Engineering Handbook*, 3rd ed., Vol. 1. Edited by P. Darling. Littleton, CO: SME. pp. 867–868.

Whittle, D. 2011. Open-pit planning and design. In *SME Mining Engineering Handbook*, 3rd ed., Vol. 1. Edited by P. Darling. Littleton, CO: SME. p. 881.

CHAPTER 6

Planning the Underground Mine*

Richard L. Bullock

A discussion on planning for an underground mine is somewhat more difficult to present than describing that which goes into planning an open pit mine because there are so many different types of underground mining methods. If the person performing the mineral property feasibility study is already somewhat familiar with the various types of underground mining methods, then this will serve as a reminder of what underground configurations will be considered for the schedule of time, staffing, and material that must go into the planning for the development of each mining method. If the person performing the preliminary feasibility or intermediate feasibility (or prefeasibility) study is not familiar with the various underground mining methods, then what will be presented will only be the essence of what is done for planning each of the methods. For a more detailed description of the mining methods and many case studies on all of the methods, the reader is referred to *Underground Mining Methods: Engineering Fundamentals and International Case Studies* (Hustrulid and Bullock 2001).

ROOM-AND-PILLAR STOPING†

When one considers the number of underground mines of limestone, dolomite, coal, salt, trona, potash, gypsum, as well as all of the Mississippi Valley–type lead and zinc mines that there are, it should not be surprising to realize that approximately 60%–70% of all of the underground mining in the United States is done by some form of room-and-pillar (R&P) mining. This amounts to nearly 340 Mt/a, or million metric tons per annum (374 million stpy, or short tons per year; Zipf 2001). For the aggregate industry alone, according to a 1998 survey done by the National Stone, Sand, and Gravel Association, there were approximately 92 underground mines, all of which are R&P mines. In today's permitting environment, at any given time, there are probably between 20 and 40 more R&P underground aggregate mines being planned.

* Most of this chapter has been taken from Chapter 3 in *Underground Mining Methods: Engineering Fundamentals and International Case Studies* by William A. Hustrulid and Richard L. Bullock (Bullock and Hustrulid 2001). In some sections, the information has been summarized, and in others, it has been expanded by adding more recent data and information on coal mining.

† Both hard-rock methods (HRM) and coal mining methods (CMM) are discussed. The subject designations—HRM and CMM—are noted in the headings of each subsection of "Room-and-Pillar Stoping."

Access to the Room-and-Pillar Mine [HRM and CMM]

Although access to the mine is not always influenced by the mining method, some discussion is warranted on the various approaches to the initial mine and production opening where R&P mining will be applied:

- If it is possible to develop the resource from a hillside adit, this will obviously be the least expensive method of entry.
- If a shaft is sunk, then:
  - The production shaft should be sunk somewhere close to the center of gravity of the ore body.
  - The shaft depth should be sunk to where most of the ore is hauled downgrade to reach the shaft dump pockets.
  - The shaft depth should be sunk deep enough to accommodate adequate dump pockets, skip loading, and crusher station.
  - For aesthetic reasons, it is best to put the shaft position such that the headframe is out of site of the general public.
- If a decline is driven that will be used for
  - Trackless haulage, then 8% is the maximum grade recommended.
  - Conveyor belt haulage, but that rubber-tired trackless equipment must negotiate on a regular basis, then 15% is the maximum grade recommended.
  - Conveyor belt haulage only, then theoretically, the maximum grade could be approximately 0.26–0.44 radians (15–25 degrees) depending on the type of material. However, remember that equipment must be positioned alongside the belt to occasionally clean up the spill rock. So it is this activity that may limit the decline grade, unless hand shoveling is planned for cleanup.

Orientation of Rooms and Pillars [HRM and CMM]

Pillar Orientation Due to In-Situ Stress [HRM and CMM]

As in all mining methods, the planner of the R&P operation must be aware of the probable in-situ stress within the rock prior to mining. If indeed there is a significant maximum horizontal stress in a particular direction, then the mine planner should take this into account by orienting the room advance and the direction of rectangular pillars to give the most support in that direction. In the very early phase of development, the research should be done to determine the magnitude and direction of the inherent stress levels. When one does not know the direction of horizontal stress, at least in the mid-continent and eastern areas of the United States, pillars should be aligned at right angles with rows at N 79° E to best cope with the natural horizontal stress in the earth's crust. While this is necessary and considered good operating practice, many very shallow R&P operations may have very little horizontal stress and the direction of the orientation of rectangular or barrier pillars does not have to be of concern. This is particularly true if the mine is opened from a hillside with adits, where nature could have relieved the horizontal stresses eons ago.

However, in sharp contrast to this condition, some deep R&P mines have tremendous problems, not only with high horizontal stress levels, but with rock that will absorb a large amount of energy before violently failing. In such operations, not only is pillar orientation important, but so is the sequence of the extraction and how it takes place. Korzeniowski and Stankiewicz (2001) document such an operation in their case study, "Modifications of the Room-and-Pillar Mining Method for Polish Copper Ore Deposits."

Room-and-Pillar Orientation Due to Dip [HRM]

While most R&P mining is done on fairly flat strata, it does not necessarily have to be limited to flat horizons. With dips up to about 5%–8%, there is little difference in the layout of the R&P stope, except that the rooms should be laid out such that the haulage would be with the load going downgrade. For orientation of rectangular pillars, it would probably make sense to orient the long side of the pillar in the direction of the dip.

The other alternative is to mine a series of parallel slices in steps, following a level contour of the dipping ore body. Thus, as each round is blasted, much of the rock will cascade down to the next level where it can then be loaded; therefore, it is termed *cascade mining*. It was first reported for the Mufulira copper mine in Zambia (Anon. 1966). However, in this case, the pillars were also removed almost immediately in a second cycle of mining. This allows the hanging wall to cave as the mining retreats along the strike.

When the dip becomes very steep, say 35%–45%, and it is too steep to operate trackless equipment, some manufacturers have proposed back-mounted, cogwheel-driven jumbos that could drill stope rounds under these conditions, but then the ore would have to be removed by using a scraper. At this angle, it would not flow by gravity and would be too steep for any other type of loading equipment.

Room-and-Pillar Mine Haulage Development [HRM and CMM]

Normally, the production shaft is developed somewhere near the centroid of the ore body (unless the production opening is by adit or decline). Rail haulage should be kept as straight as possible, with long haul grades under 2%, although within the stopes themselves, grades of 5% are not unusual. The objective of the production development is to minimize the haul cost for the ore to the shaft. If this is a trackless haulage operation, other things will help minimize the haul cost:

- Keep the grades as low as possible and long hauls under 8%.
- Keep the road as straight as possible. This means keeping the pillar location from causing the road to deviate around newly formed pillars; that is, all main haul roads should be laid out prior to mining and the pillars laid out from this plan.
- Maintain the haul roads in excellent conditions with adequate crushed stone, keeping them well graded and dry. Water not only causes potholes, but water lubricates the rock, which cuts tires.

For laying out the mine development plan, keep Spearing's rules of mine openings in mind, which were listed in Chapter 4 under the "Spacing of Excavations" subsection (Spearing 1995). It is good practice in mine development to keep the intersections off of the main drift six widths apart, avoiding acute angle turnouts that create sharp "bullnose" pillars. It is also good practice to keep the nearly parallel ramps and declines apart from the main drift by at least three times the diagonal widths of the main drift.

For planning the trackless/rubber-tired haulage system, one must consider the various methods of moving rock from the working face to the crushing/hoisting facility. There are many ways to do this. With today's equipment, with the modern hydraulic excavators and powerful rubber-tired equipment, it is the very fact that equipment is so versatile and flexible that sometimes creates the dilemma as to how to be sure that the optimum method is always being used.

When an R&P mine first begins, either from the bottom of a decline or a shaft, the haul distances are short, and unless they are hauling up a decline to the surface, load-haul-dump (LHD) units with a rubber-tired loader will probably be the best way to go. As the mine gradually works out away from the dump point, there will be some juncture at which the loader should start loading a truck. But before the haul distance requires more than one truck for any given tonnage produced, the loader can load the truck, and then load itself, and then follow the truck to the dump point, that is, load and follow (LAF). Eventually, as the distance increases even more, the front-end loader (FEL) should stay in the heading and load enough multiple trucks that would keep the loader busy.

When more than one level is involved within a mining zone, and the main haulage is on the bottom level, then various combinations of LHD to the orepass or LAF or FEL with trucks hauling to the orepass can take place. At the bottom of the orepass, automatic truck-loading feeders can then transfer the ore to trucks or rail-mounted trains. If the upper ore body is rather small, and an automatic ore chute cannot be justified, then the ore can fall to the ground, and an LHD unit or FEL can load into another truck. Figure 6.1 illustrates all of the possibilities of moving the material from an R&P face to the final ore pocket at the shaft.

The mine planner needs to be aware that in an ever-expanding R&P mine, the optimum method or combination of moving the ore is constantly changing, and for every condition and distance, there is only one optimum, least-cost method. All of this can and has been accurately demonstrated using a computer simulation of the underground mine environment (Bullock 1975, 1982; Gignac 1978).

One other concept used for some R&P mines is well worth noting. In some of the older Tennessee mines, the haulage ways were driven somewhat small. Thus as the haul distance became longer, it would have been desirable to go to larger haulage trucks. However, it was not possible to go to larger trucks because of the small haulage ways. In this case, Savage Zinc, in their Gordonville mine, began hooking side-dump truck-trailers together to make a mini-truck train and increase the payload per trip for the long hauls because they could not go to a larger truck, thus optimizing their production for that load-haul condition.

If a conveyor system of haulage is used for the main distance haulage, then the preceding advice for the mine layout still applies to the general layout of the mine, except it will start from the decline of the mine entrance where the main conveyor will carry the mine production to the surface. In these cases, the mine usually has a semi-moveable crusher or breaker underground prior to the feeder to the conveyor system. The system of haulage, either LHD or FEL/truck haulage will haul broken rock from the faces to the central receiving point at the crusher. The crusher then feeds the conveyor. The crusher will have to be moved periodically, say, every 6–12 months, depending how fast the mine production moves the faces away from the central hauling point.

There was a time when rail haulage was the principal method of gathering the ore from the faces to the production shafts. In the Old Lead Belt of Missouri, there was more than 556 km

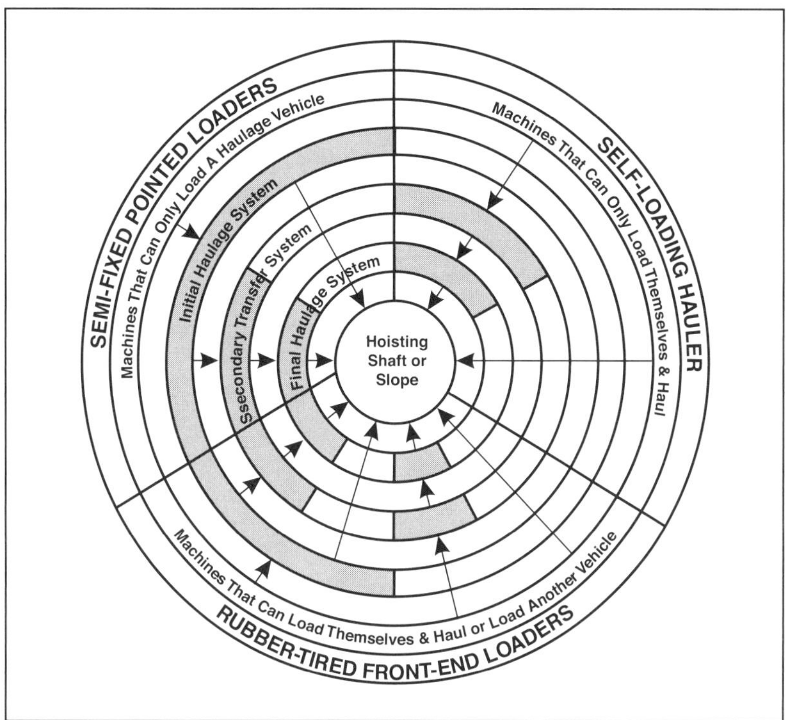

Source: Bullock and Hustrulid 2001

FIGURE 6.1 Multiple methods by which ore can be moved from the face of an R&P mine to the hoisting point

(300 mi) of interconnected railroad that was being used to bring all the ore into two main shafts, from what was originally about 15 mines. Today in the United States, there are very few R&P metal mines using rail haulage. But there are still a lot of coal mines in the United States that use rail haulage, and in other countries, there are still many metal mines using rail haulage.

Room-and-Pillar Extraction Methods [HRM]

One of the advantages of modern R&P mining systems is that every task to be done can be mechanized to some degree, provided that it is economically sound to do so. That minimizes the operating labor force and makes staffing the operation easier. The high-capacity equipment for modern R&P operations is reasonably simple to learn and operate. Anyone having operated any heavy machinery in construction work, the military, or even on a farm has little trouble adapting to loading and hauling equipment in an R&P operation.

Although most R&P mining is done by drilling and blasting, particularly in the aggregate and metal mining businesses, a very large portion of R&P mining is also done by mechanical excavation. Many trona, potash, and some salt mines perform all of their excavation mechanically. The reader is directed to the case studies in "Section 3: Room-and-Pillar Mining of Soft Rock," of *Underground Mining Methods* (Hustrulid and Bullock 2001) to illustrate those types of R&P mines that use drilling and blasting and those that use mechanical excavation. With the power of today's mechanical excavating machines and with the improvements that are being made in the tools, such as disk and pick cutters, the possibility of using mechanical excavation should be considered during the feasibility study for any rock under 100 MPa

(15,000 psi) or even up to 136 MPa (20,000 psi) if it has a lot of fractures and is low in silica content. The advantages of various types of mechanical excavation, where they are applicable, are well documented (Bullock 1994). Furthermore, for long developments, full-face tunnel-boring machines are proving their worth for certain conditions (Snyder 1994; Alexander 1999).

Where it is viable, there are advantages to mechanical excavation (Ozdemir 1990):

- Improved personal safety
- Minimal ground disturbance
- Reduced ground support
- Continuous, noncyclic operations
- Low ground vibrations and no air blast
- Uniform muck size
- Less crushing and grinding in the mill
- Reduced ventilation requirements
- Conducive to automation

Where mechanical excavation is truly viable, it adds up to higher production rates and reduces mining operating cost. Many R&P mines, however, must rely on drilling and blasting the face. The initial pass of mining by drilling and blasting is usually done by drilling "vee cut" or "burn cut" types of drill patterns. These are well documented in the literature (Bullock 1961, 1982; Casteel 1973; Langefors and Kihlstrom 1963; Hopler 1998). However, one aspect that is often overlooked is that only about 40% of the rock should be broken with drilled swing patterns (rounds), breaking to only one free face, but about 60% of the rock should be broken by slabbing, that is, drilling holes parallel to the second free faces as they are exposed (Figure 6.2). This minimizes the cost per ton of rock broken and maximizes the productivity.

Single-Pass or Multiple-Pass Extraction [HRM]

There are two approaches to mining the existing thickness of the ore body or valuable rock: taking the entire thickness in one pass (mining slice) or removing the ore body by multiple passes or slices. The overall thickness to be extracted determines which approach should be chosen.

Normally in mining bedded deposits for aggregate, the total thickness of the desired horizon is known and the decision as to how thick a slice to take can be made in advance. However, in metal ore deposits, particularly of the Mississippi Valley or collapsed breccia-type deposits, most of the time the mine planner does not know how thick the total mining horizon is going to be except where each diamond drill hole passed through the formation and identified it; a few feet away from that hole, it may very well be different. In both cases, for the aggregate producer and the metal ore producer, the best approach is to first mine what one thinks is the top slice through the ore body. As to the thickness that this should be, this depends on what equipment the producer has or is going to buy. But it also depends on what height of ground can be mined and maintained safely and efficiently. Mining the top slice first is so that whatever back and rib/pillar scaling and reinforcement is required, it can be reached easily and safely. In this author's opinion, this should not exceed 8.5–9.8 m (28–32 ft). However, there are aggregate producers that will mine 12 m (40 ft) and more in one pass by using a high-mass jumbo and extendable-boom roof-scaling equipment.

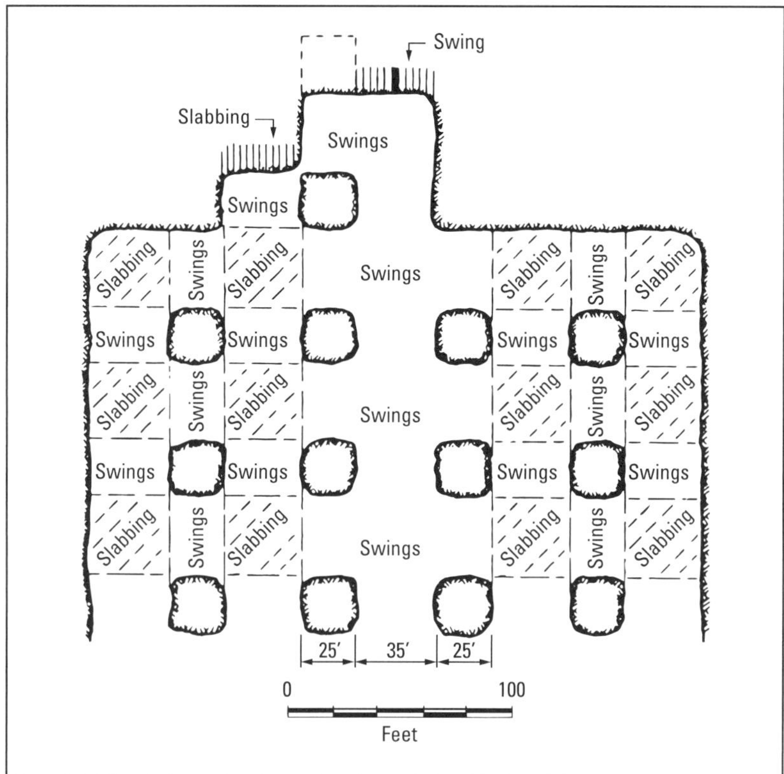

Source: Bullock and Hustrulid 2001

FIGURE 6.2 How to optimize the drilling and blasting of R&P stoping

After the first pass is completed in a metal mine, for a given stoping area, then the back and floor should be "jackhammer" (air hammer machine) prospected to identify what ore remains in the back and floor that will need to be removed by other mining slices. If ore is found in the back and floor, then the ore in the back should be mined first. After it is removed and additional back prospecting reveals no more ore, and the back is again made secure and safe, then the ore in the floor can be taken.

Methods of mining the ore in the back vary somewhat on the thickness of the ore yet to be mined and the original stope height. If the original stope height is less than 7.6 m (25 ft) and the thickness of the back slice is less than about 2 m (7 ft), then most extendable-boom face jumbos will reach this high to drill the brow with breast (horizontal) drilling. By drilling horizontal holes, the miner has a better chance of leaving a smoother back, requiring less maintenance than if the miner had drilled upper holes to break to the free face of the brow. This is especially true if smooth-wall blasting is practiced. This author does not recommend tilting the jumbo feed up and drilling nearly vertical holes and drilling uppers to break to a brow, particularly if working in a bedded deposit as are so many R&P mines.

If the ore thickness in the back is greater than the conditions just mentioned, or the room height is already at the maximum height that can be safely maintained, then the approach to mine the back slice is that of first cutting a slot in the back at the edge of the entrance to the stope. This slot should be between the pillars and reaching the height of the next slice or the

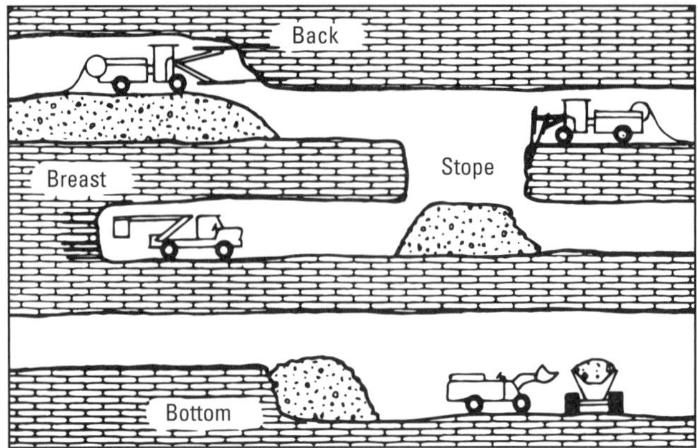

Source: Bullock and Hustrulid 2001

FIGURE 6.3 Various types of stoping action involved in many R&P metal mines

top of the ore body, being very careful not to damage what will be the rock that forms the top of the new pillars. Smooth-wall blasting could be used to advantage in this area. A mine dozer (usually a small dozer, such as a D-4 or D-6 size) should begin pushing the rock up and making a roadway for the jumbo to travel upon the rock pile. An FEL can also do a reasonable job of building the roadway if no dozer is available. From here, the jumbo can drill breast (horizontal) holes in the brow and follow the ore zone throughout the stoped area. Even if the ore zone goes into the solid beyond the original stoped area, this will not be problem, except that a loader will now have to load out the ore as it is broken. About 75% of the ore will have to remain in the stope rock pile until the back-mining job is completed. This can be a disadvantage if the mine needs the production immediately, or it can be a big advantage if the ore can be left to be moved when many of the miners take off on summer vacations. When the top of the ore is finally reached, the back can be made safe because the miners are still up working very close to the back. This type of mining has been practiced in both the Tennessee zinc and the Viburnum lead/zinc districts. It is not uncommon for several passes of the ore to be mined from broken rock piles. In these cases, ore will need to be loaded from the edges of the rock piles at the bottom of the rock piles to make room for new broken rock, or a loader will have to go upon the rock pile and load out the excess rock (Figure 6.3).

If after the first slice is taken through the stope it is then discovered that a very thick, continuous ore zone lies above this first slice, say 15–20 m (50–65 ft) or more, then an entirely different approach may be taken. In this case, it may be better to drive a development ramp to what is now known as the top of the ore and mine out the top slice from this new ramp. Then put through a slot raise from the bottom level to the top level, which can then be slabbed out to make room for long-hole drilling and blasting the ore body down to the level below. Again, the pillars should be presplit or smooth-wall blasting used to protect them.

Once all of the back ore is removed from the stope and the final back is made completely secure, the bottom ore can then be removed from the floor. This is best done by first cutting a ditch or short decline in the floor at the entrance to the stop to the necessary depth to carry the bluff. Bluffs can be carried very thick, only limited by the height that it is safe for the equipment doing the loading to work. It is common practice to carry up to 9-m (30-ft) bluffs in the lead mines of the Viburnum area, using 7- to 11-metric ton (8- to 12-ton) size loaders. However, beyond this height, safety to the loader operator may become an issue.

The drilling and blasting of the bluffs is usually done with downholes by small-surface quarry-type drills if the bluffs are at least 4-m (13-ft) thick. If they are less thick, they may be drilled out with the face jumbos, drilling breasting or horizontal holes (sometimes termed *lifters* and *splitters*).

This procedure can be repeated over and over until the bottom to the ore is reached. However, in areas where this type of mining is suspected to take place, for the initial pass, the pillar width *must* be large enough to accommodate considerable height if it should be needed. Then if additional ore is not found, the pillars can be slabbed down to a smaller size. However, one precaution is that in both removing the back ore and taking up the bottom, the pillars must be protected with smooth-wall blasting or presplitting around the pillars. The pillar design is briefly discussed in "Pillar Width," but the point here is that for any given pillar width that the mine planner may have assumed, unless taking down the back and taking up the bottom is planned or there is a very large safety factor in the pillar design, the designed safe width-to-height ratio will be exceeded by the multiple-pass mining that may need to be made. It cannot be overemphasized that everything should be done to protect the integrity of the pillars during the first pass of what may become multiple passes. Remember that unless caution blasting is observed, blasting fractures may extend into the pillar 1–2 m (3–6 ft) or more. As more slices continue to be removed, these fractures will begin to open, revealing a much reduced size of pillar that could fail.

Room-and-Pillar Extraction Methods [CMM]

A typical layout of an R&P mine used in the mining of bituminous coal is shown in Figure 6.4. In this case, a set of five main entries allows access to the production panel through panel entries. The entries in coal mining are limited to 6 m (20 ft) in width and are generally driven about 18–31 m (60–100 ft) apart, center to center. The panel itself may be approximately 120–180 m (400–600 ft) in width, limited primarily by the cable reach of the electric shuttle cars that are usually used to move the coal. The length of the panel commonly varies from about 600–1,200 m (2,000–4,000 ft) but could be greater. In the case of Figure 6.4, note that the panel pillars are being mined. This is called *pillaring followed by caving*. The normal practice in pillaring is to drive the rooms and crosscuts on advance (first mining into the virgin coal seam in that panel) and to pillar on the retreat (second mining when moving back out of the panel). The caved area then becomes known as the *gob*. If the surface must be protected from caving, then the pillars are not removed.

There are two methods of R&P coal mining: conventional and continuous. The conventional method, using drilling and blasting of the undercut coal seam, is an antiquated method and is only practiced in less than 5% of the coal mining in the United States, thus it will not be discussed here. The continuous method gets its name from the mechanical excavating drum miner that continuously extracts the coal. In this method, the coal is continuously mined and hauled from the face, usually with electric shuttle cars, although it may be done by conveyor belt or diesel scoop machines. There is usually some waiting between the loading of each shuttle car or scoop. Both in conventional and continuous mining, the auxiliary operations of roof control, ventilation, and cleanup must also be performed.

The plan view is shown in Figure 6.4 (which represents the section entries, room entries, and rooms associated with a production panel within the mine). The room entries, rooms, and the associated crosscuts are mined on advance, and the pillars are mined on the retreat. This or a similar system is used with continuous miners. Note also the numbering on the entries in the section entry set. It is common practice in coal mining that the entries in any set are numbered from left to right as one looks *inby* (toward the faces and away from the outside of the mine). Also note that if someone is looking at the face, then that person is looking inby; but if turned in the opposite direction and facing toward the outside of the mine, the person is looking *outby*. Another example is shown in Figure 6.5, which illustrates the sequence of cuts in developing the mine on the advance.

Room Width [HRM]

For productivity reasons, the room widths should be as wide as is practical and safe. The wider the rooms, the more efficient the drilling and blasting and the larger and efficient can be the

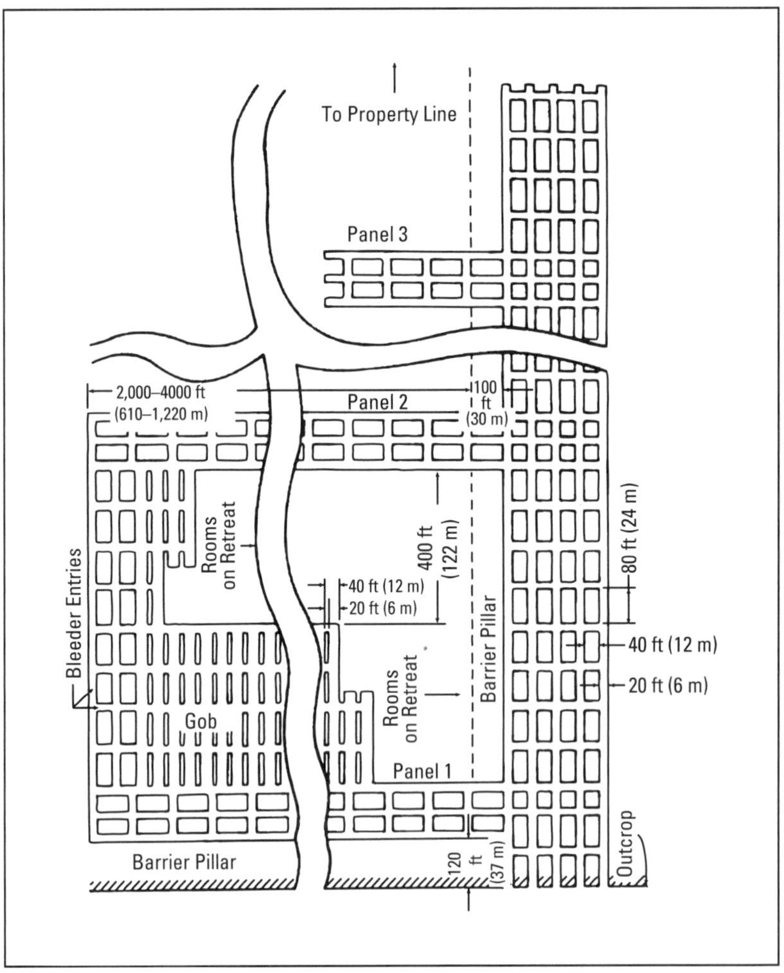

Source: Stefanko and Bise 1983

FIGURE 6.4 Typical soft-rock or coal R&P five-entry mining system

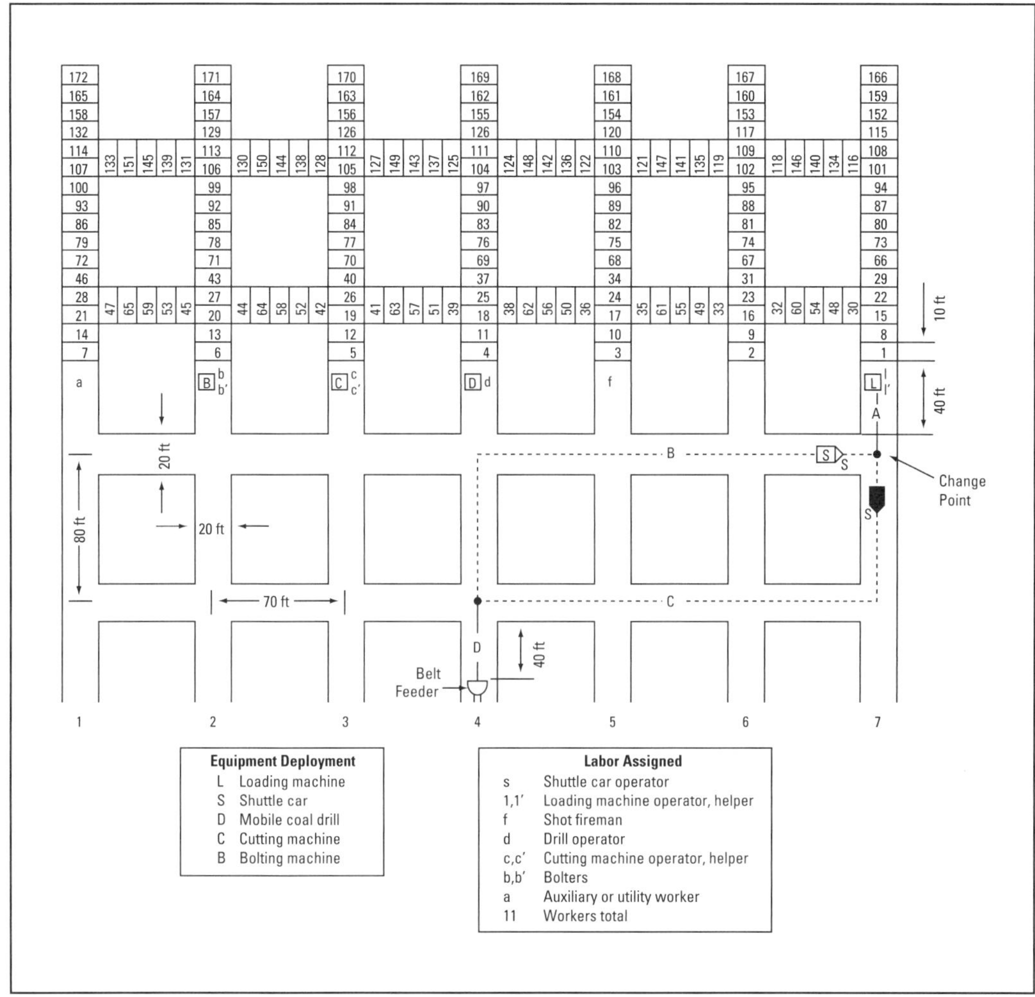

Source: Stefanko and Bise 1983

FIGURE 6.5 Advancing R&P, seven-entry coal mining, showing the sequence of cuts

loading and hauling equipment. However, room width for any given mine environment will be limited by the rock mass strength of the ore body as well as the rock mass strength of the back and floor, compared to the stress levels induced into the rock. It is inappropriate to simply try to design the R&P widths from elastic theory without taking into account the rock mass strength. However, because the rock and pillars can be reinforced and thus increase the affective rock mass strength, it may become a matter of economics as to how wide to extend the room width.

There have been many discussions written on how to design the roof span. For a complete discussion on all the rock mechanics aspects of this problem, the reader is referred to "Section 10: Foundations for Design," in *Underground Mining Methods* (Hustrulid and Bullock 2001). At this point, it is important to consider what information will be needed for the design and how much of the needed information is already at hand. In Chapter 4 of this handbook, there is a general summary of the geological and structural information that should have been determined during the exploration of the ore body. Unfortunately, most exploration groups spend little time and money trying to determine the information that is needed to construct a rock mass classification of the mineralized areas and rock surrounding the mineralization. It may be that a "best guess" rock mass analysis has to be done with nothing but the exploration information. In any case, it is hoped that there would be enough mapping of the underground structures from core logs, surface mapping, mapping of surface outcrops of the same underground formation, and geophysical information that a crude rock mass classification could be constructed.

If it is a new mining district, then what is really needed is an underground test mine. In Chapter 11, it is noted that there are more than 20 good reasons for developing a test mine during the intermediate feasibility (or prefeasibility) study. One of the most critical reasons is to obtain better geotechnical information on which to base the mine planning, which in this case are the room widths and pillar widths and height, which will greatly affect the mine operating cost. Structural information can then be seen and accurately mapped, and all of the data needed for the joint and fracture information can now be accurately measured. Likewise, in-situ stress measurements can then be taken, as well as larger core samples for laboratory testing.

Pillar Width [HRM]

One cannot discuss pillar width without relating it to pillar height. The overall strength of the pillar is related to the height of the pillar; that is, it is a matter of a ratio of pillar width w to pillar height h. The amount of load that the pillar can safely support is proportional to the "w/h" ratio. Thus it is that a pillar of a 4:1 ratio has a much larger safety factor than pillars of 1:1 or 1:5 ratios.

The actual load that the pillars can really carry can only be measured. The theoretical load as calculated by the overburden load distributed to the pillars may or may not be the load that is actually being carried. There is a good chance that the load may be arching over some of the interior pillars of the stope and transferring the load to barrier pillars or waste areas. In such cases, it may be that the interior pillars can be made smaller as yielding pillars. In such cases, if the stopes mine very wide, then on a regular interval, a row of large rectangular barrier pillars should be left. In areas of very large lateral extent, this will prevent cascading pillar failure of the entire area in a domino effect (Zipf and Mark 1997).

The reader is referred to the case studies on R&P stoping to see examples of how each mine approaches this problem, to the design theory expressed in "Section 10: Foundations for Design" in *Underground Mining Methods* (Hustrulid and Bullock 2001), and particularly to the chapter by Zipf (2001) on catastrophic failure of large areas of R&P mining where proper precautions have not been taken.

Pillar Robbing [HRM]

Pillar removal should be a planned part of the overall mining of areas where the economic value of that which remains justifies and warrants the extraction of some or all of the pillars. As an example, it is not uncommon for some very high-grade pillars in the lead/zinc/copper mines of the Viburnum Trend to have a value of more than $1 million per pillar. For more on pillar removal in this area, the reader is referred to a case study by Lane et al. (2001) in *Underground Mining Methods*.

If future pillar extraction is planned, whether it be partial slabbing of pillars, removal of only a few high-grade pillars, or complete removal incorporating some system of backfilling, then what is designed and left from the initial mining highly influences what can be done in the future. From the experience of this author in planning and supervising the slabbing and removing of hundreds of pillars, the first and most important thing to do is to install a complete network of convergence stations throughout the area that will be affected. Regular monitoring of such a network over time should allow the operator to learn what an acceptable amount of convergence is and what amount of convergence is leading to massive failures. Some mines have back–floor convergence that can move as much as an inch per month without the back breaking up and failing (Parker 1973). However, if the back-floor convergence in some of the mines, such as those of the mines of the Viburnum Trend, have convergence in the order of a few thousandths a month, this becomes significant. Convergence of 0.0254 mm (0.001 in.) per month is not significant. Convergence of 0.0762 mm (0.003 in.) per month indicates a serious problem but is controllable with immediate action. Convergence of 0.1778 mm (0.007 in.) per month indicates the acceleration is getting out of control and the area may be lost.

There are several methods to remove pillars from R&P stoping. In broad terms, they are as follows:

- Slab some ore off of each pillar containing the high-grade portion of the ore during retreat from the area.
- Completely remove a few of the most valuable pillars, but leave enough pillars untouched to support the back.
- In narrow stopes, completely remove all of the pillars in a controlled retreat.
- Use massive backfill methods to remove all or some of the pillars.

Placing cemented backfill around pillars, all the way to the back, to give the proper support of the stope between the solid and the backfill creates a "pressure-arch" over the subsequent pillars in between, and then these pillars can be removed. Economics permitting, this area could then be backfilled if necessary, and the pillars encapsulated or trapped in the original fill could then be mined from a sublevel beneath the pillar, with long-hole blasting of the pillar into the sublevel area (Lane et al. 1999). The total backfill would prevent any future subsidence.

All of the preceding methods took place in the final mining of the Old Lead Belt of Missouri over a period of approximately 25 years, but much more intensely over the last 10 years. This was an R&P mining district that was mined for 110 years before finally shutting down.

All of the preceding methods described have been applied and have been proven to be profitable. Each method must be analyzed from an economic feasibility point of view, as well as from a technical ground control/rock mechanics point of view before planning and executing the pillar-removing practices.

Sometimes when partial pillar removal takes place without the proper planning, or when pillars that are too small are left in the first pass of mining, the pillars begin breaking up and serious convergence begins to accelerate that cannot be controlled. One of two things will have to take place to save the area: Either a massive pillar reinforcement will have to take place (if there is time prior to collapse) or massive amounts of backfill will have to be placed in the entire area. The author has been involved with both methods of stopping convergence and eventual catastrophic collapse of large areas.

For the first solution, fully grouted rebars were placed in more than 300 pillars in the R&P mines of the Viburnum Trend. This major project is well documented by Weakly (1982) and covers the method employed, reinforcing pattern, cement grout mixture, convergence instrumentation, and results. In the Old Lead Belt areas where the pillars needed reinforcing, they were wrapped with used hoist cable with a load of 6 tons placed on each wrap (Wycoff 1950). But this method of reinforcing was not as fast, economical, nor as effective as the fully grouted rebars in the pillars.

The second solution—the massive backfill system—was used to mitigate a potential massive catastrophic failure that nearly took place at the Leadwood mines of the Old Lead Belt that was only about 132 m (425 ft) deep. However, the back was thin-bedded dolomite, interbedded with shale and glauconite, and was also badly fractured and leached. The "roof bolt" was originally developed in these St. Joe mines in the early 1930s as a means of tying these layers together to act as a beam (Weigel 1943) and using the roof bolts with channel irons, which formed a crude truss. Even though the rock in the pillars was equally as bad for support because they also contained bands of high-grade galena, as a means of economic survival, pillar removal and slabbing took place over a period of 25 years. Occasionally, local cave-ins would occur after an area had been "pillared." But since these cave-ins were beneath uninhabited St. Joe–owned land, they were of no real concern. However, when slabbed pillars between two of these smaller cave-ins began to fail and a third and fourth cave-in occurred in more critical areas, there was a considerable amount of concern. The initial extraction of some of these areas involved multiple-pass mining and room heights that were mostly 6.1–12.2 m (20–40 ft). Final mining of the area resulted in approximately 95% ore extraction. To stop the caving in the third and fourth areas in about 1962, more than a million tons of uncemented, cycloned sand tailings were put into the mines, filling the rooms nearly to the back. The results were very successful in controlling the converging, failing ground.

In retrospect, the end result might be compared to the overhand cut-and-fill practice of deliberately mining the pillars very small and immediately filling in around them in what is known as *postpillar* mining. This system was used by Falconbridge Nickel Mines (Cleland and Singh 1973) and the Elliot Lake uranium mines (Hedley and Grant 1972). The author also observed its use by San Martin and Niaca in Mexico and San Vincente in Peru. The two mining systems, with small pillars encapsulated in sand tailings, look similar.

Pillaring Coal on Retreat [CMM]

Four basic pillaring methods are illustrated in Figure 6.6 and are described in detail by Kauffman et al. (1981). They are (A) split and fender, (B) pocket and wing, (C) open ending, and (D) outside lift. The numbers on each diagram represent the successive cuts by the continuous miner. The temporary prop supports are shown as rows of black dots. For more details on these methods, the reader is directed to Hartman and Mutmansky (2002).

Ventilation

It is not the intent to go into the design of a noncoal R&P mine ventilation system in this chapter. There are current books written on the subject of mine ventilation, which the mine planner should refer to if need be to learn mine ventilation planning (Hartman et al. 1997; Ramani 1997; Tien 1999). However, most of the literature on ventilation design for R&P

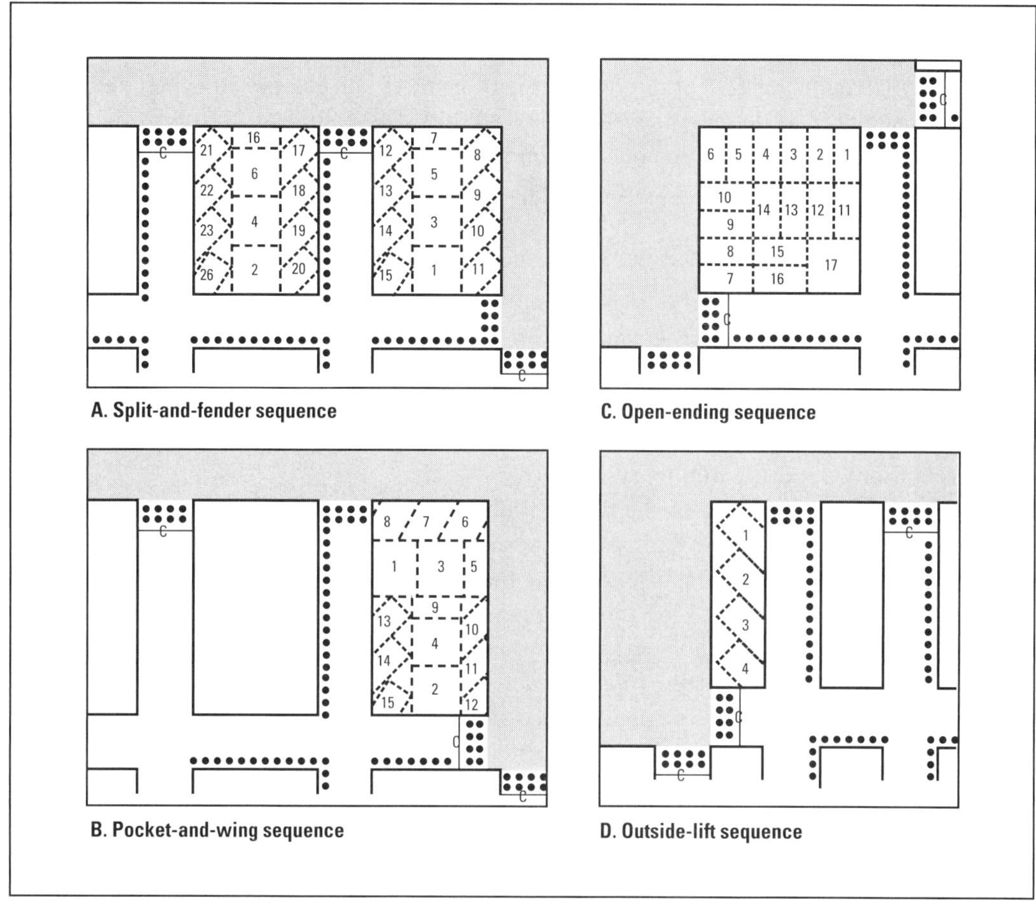

Source: Kaufman et al. 1981

FIGURE 6.6 Examples of cut sequence for pillar extraction during retreat mining

mining has been written for coal mines. Thus it may be advisable to mention a few reminders of details that are unique to many of the metal and aggregate R&P mines.

Metal or aggregate R&P mine ventilation is comparable to R&P coal mine ventilation:

- Everything is larger, and it is not uncommon to have 9 × 9 m (30 × 30 ft) entry drifts and rooms stoped out to 12 × 21 m (40 × 50 ft). It will take a lot more air or a stream of high-pressure, directed air to meet the minimum velocity across the working face in these conditions.
- Stoppings are difficult to build and the total force against such a stopping can be enormous. Thus many R&P mines rely on auxiliary fans to pick up air from the main ventilation drifts and carry enough air through vent tubing to serve the needs of the active faces.
- Ventilation doors are more like airplane hangar doors than small ventilation doors. Again, the total force against these doors that must be controlled automatically is enormous.
- Air stratification in large stopes can be a problem.

- Diesel equipment is extensively used in these mines, the exhaust of which must be diluted. Current Mine Safety and Health Administration (MSHA) regulations for diesel particulate material (DPM) are very difficult to meet and require an expert ventilation mine engineer to design the system. The MSHA DPM level is set at 160 µg of total carbon standard as measured on a personnel dosimeter.
- Main ventilation fans can be placed underground, if it is beneficial to do so.

Changing Market Conditions and Room-and-Pillar Mine Planning

Fluctuating market conditions should be considered during the R&P stoping feasibility study. This is because the commodity is gradational in value related to changing market conditions. The flexible mining conditions of R&P stoping can usually be adjusted to the elastic nature of markets if the mine planners are always aware of the current market trends. The extremely modifiable conditions of the R&P mining method allow the mine operation to react to market needs faster than with other mining systems.

Another of the advantages to the method of R&P stoping is that new faces to work are continuously opening if the ore body is continuous. Even if the stope is only four pillars wide (see Figure 6.2 as an example), at any one time that stope may have as many as 12 to 15 faces open for drilling and blasting. One can imagine if the stope were 10 pillars wide how many faces would be exposed to drilling and blasting. For metal mines, this offers a lot of flexibility to mine the grade of ore that is the most desirable for any current price of the metals that are being mined. For short periods of time in each stope unit, it is usually possible to work only the higher- or lower-grade faces of a particular mineral depending on the market. This usually has a drastic effect on the grade within a few days. After the high-grade Fletcher mine in the Missouri Viburnum Trend had been operating for about three to four years, there were approximately 50 to 70 faces that were open for mining on any given day, when only 10 to 12 would actually be worked. It becomes a matter of face selectivity to maintain a grade of lead, zinc, or copper ore that can best be handled by the concentrator and still optimize the financial objective of the mine.

Similarly, spare equipment can be put into reserve stopes to increase production if the remaining materials' flow can take the added capacity. However, if these practices are carried on too long or too often, mine development must also be accelerated. If maintained, old stopes can be reactivated quickly to mine lower-grade minerals, which become minable because of economic cycles.

Another aspect of the R&P mine is that often lower-grade resources are left in the floor or the back of the stopes. When the price increases, new reserves are readily available for quick mining. This technique is often overlooked by individuals not accustomed to planning R&P metal mining where the mineral values are gradational. There is the option of mining through the better areas of the mineral reserve and maintaining a grade of ore that satisfies the economic objectives at that time. At a later time, when the mining economics may have changed, the lower-grade areas left as remnant ore reserves can be mined.

However, despite the preceding discussion, even in R&P mining, drastic changes in the rate of mining (momentum) cannot be assumed to be free. It often takes several months with an increased labor force to regain a production level that seemed easy to maintain before a mine production cutback. If spare equipment is used to increase production, maintenance probably will convert to a breakdown overtime schedule compared to the previous preventive maintenance schedule on shift, at least until permanent additional equipment can be obtained. Nevertheless, the necessary changes can be made.

As discussed in the "Pillar Robbing" section, the other technique is that of slabbing or removing high-grade pillars. Thus, even in the latter years of the mining operation, some of the "sweetener" is left to blend with the lower grade. Although not unique to R&P mining, this technique certainly is easier to accomplish in an R&P operation than in other more complex mining systems.

SUBLEVEL OPEN STOPING MINING

As discussed in the previous section, the R&P mining system is applied to subhorizontal ore bodies often of relatively uniform thickness. A portion of the ore body is removed in the form of rooms, and pillars are left to support the overlying strata. The mining may be done in a checkerboard pattern of rooms and pillars, or long rooms may be created with rib pillars left between. The strata making up the roof and floor are competent as is the ore. The deposits mined with this technique range from thin to moderately thick. As the dip of the strata increases, and/or the ore-body thickness increases, other methods must often be employed. Consider a moderately thick deposit that would be mined by the R&P mining system if flat-dipping, but now the dip increases to 90 degrees. In this case, the loading on the pillars would come from the horizontal direction and the blasted ore would fall downward to be collected at bottom. Although the general geometry is the same as the R&P method, the generic name given the system is *sublevel open stoping.* Blast-hole stoping, vertical crater retreat (VCR) mining, and vein mining also fall under this general heading. The shrinkage stoping method is a special form of sublevel stoping. In general, the method is applied to ore bodies having dips greater than the angle of repose of broken material (greater than about 50 degrees), so that material transport to the collection points occurs by gravity. For massive deposits, stopes with vertical walls are created, and the overall dip of the deposit is immaterial. The criterion for applying the method is that the openings created remain open during extraction. These openings may be later filled or left open. The pillars left between stopes may be extracted at a later time or left in place. Some typical layouts used for extracting the ore are briefly presented in the following subsections. It is assumed that mobile equipment is used with ramp access.

Throughout the planning during the feasibility study, one must be aware of the potential ore dilution or ore loss when one is drilling long holes and the vein being mined is an undulating type vein, as there is always the risk of getting more ore dilution and the possibility of more ore loss than might first might be expected. In feasibility studies, the value of dilution for sublevel long-hole stoping should be at least 15% and ore loss at least 5%. However, for short blast-hole methods, such as shrinkage methods, dilution and ore loss might well be less than 5%.

Extraction Principles

Consider the ore block of width w = 1–10 m (1–33 ft), length l = 10–40 m (10–131 ft), and height h = 20–30 m (65–98 ft). For simplicity let us consider that ore block is vertical, although for this method, the dip of the block is immaterial. If the block is thick enough, a stope can be developed that will flow by gravity and can be mined by using a number of sublevel stoping techniques. The blasted ore will fall to the bottom of the block and be removed using LHDs. There are various designs for the extraction level. Here it is assumed that a trough is created

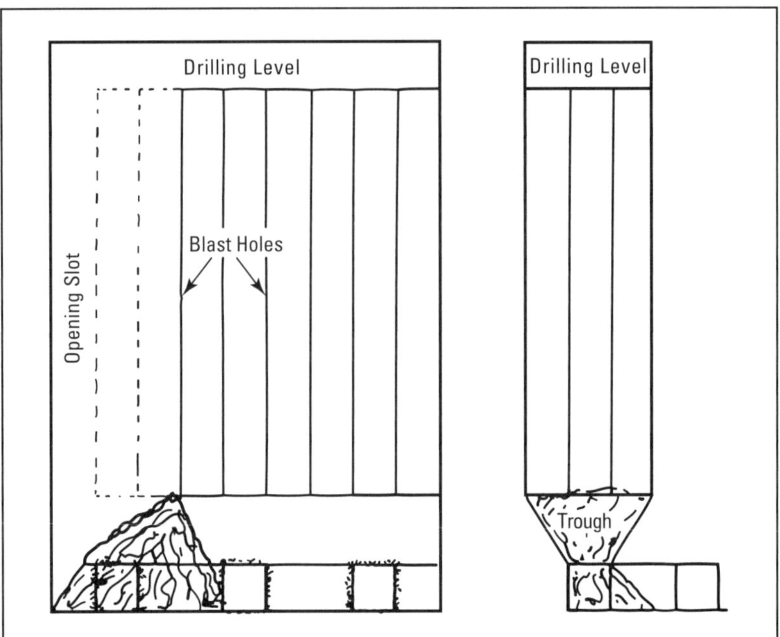

Source: Bullock and Hustrulid 2001

FIGURE 6.7 Blast-hole stoping, starting with a slot opening

using fans blasted toward an opening slot. The LHDs travel in a footwall haulage drift running parallel to the trough. Access to the trough is from the side. The location and number of access points (drawpoints) are such as to provide full extraction coverage.

Blast-Hole Stoping

Blast-hole stoping will be the first method considered for mining the block. From the drilling level located at the top of the block (Figure 6.7), rows of parallel blast holes are drilled down to the top of the extraction trough. A raise is driven at one end of the block, and it is slashed to full stoping width to form a slot. The rows of blast holes are now blasted one row or several rows at a time toward the open slot. The blasting design and layout is very similar to that used in bench blasting. The hole diameters used vary widely but typically would lie in the range from 76 to 165 mm (4.5 to 6.5 in.). For wide blocks, 165-mm (6.5-in.) diameter holes are often used. Hole straightness is an important design consideration that affects fragmentation, ore loss, and dilution. In general, one would select the largest hole diameter possible for the stope geometry since straight hole length is strongly dependent on hole diameter. The specific development (amount of development required to exploit a certain volume of ore) is inversely proportional to block height. Because the cost of development is significantly higher than that of stoping, one wants to have the highest possible extraction blocks associated with a given extraction and a given drilling level.

Sublevel Stoping

If geomechanical studies indicate that very high blocks (heights exceeding the straight drilling length from one drill location) can be extracted using the same extraction level, then several drilling levels at various heights within the block must be created. Because of the multiple drilling levels (sublevels), this method is called *sublevel stoping*. The layout is very similar to blast-hole stoping with an extraction level and an opening slot, but now there are multiple drilling levels. The mining takes place overhand in which the lower drilling blocks are extracted before the upper or underhand in which the extraction of the upper drilling blocks precedes those underlying. Here it is assumed that overhand stoping is employed. The simplest approach is to repeat the drilling layout for single-level blast-hole stoping. This is shown in Figure 6.8. The ore-body thickness is assumed to be such that the full width is undercut and becomes available for drilling access. Parallel holes can be drilled in this case. An alternative is to drill fans of holes (Figure 6.9) rather than parallel holes from the sublevels. Furthermore, there may be one or multiple drill drifts on each sublevel, and the rings may be drilled downward, upward, or in full rings. The selection is based on a number of factors, a full discussion of which is beyond the scope of this section. As indicated, the application of the sublevel stoping technique assumes good stability of the openings created. Stability surprises can mean the partial collapse or even full collapse of partially extracted stopes. Production may be stopped fully because of the presence of large blocks in the drawpoints. Even in the best case, there is ore loss and dilution. Reinforcement of the footwall, hanging wall, and the roof can be done prior to or during

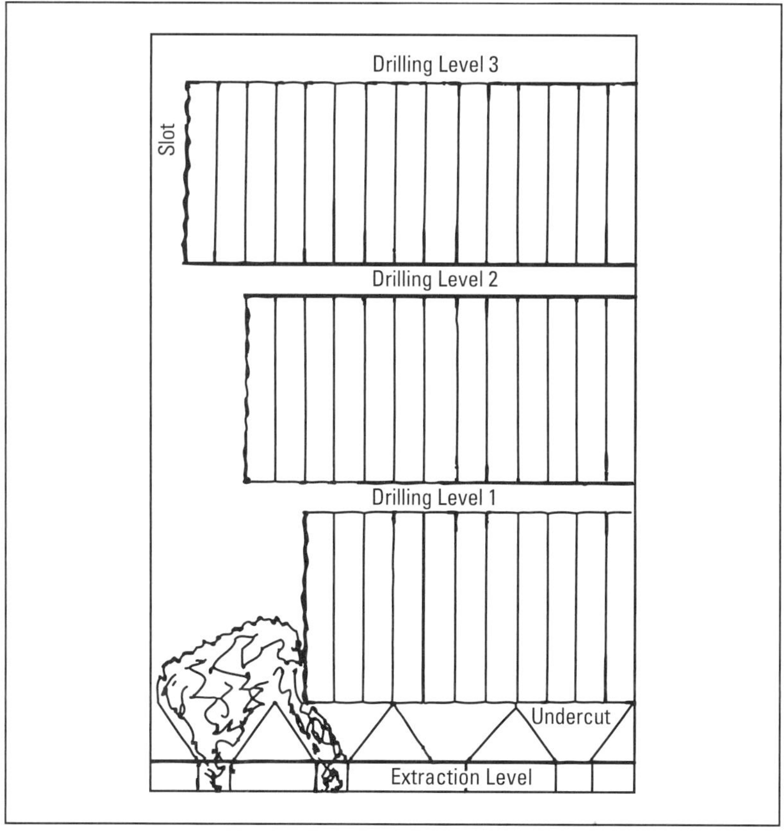

Source: Bullock and Hustrulid 2001

FIGURE 6.8 Multilevel sublevel stoping

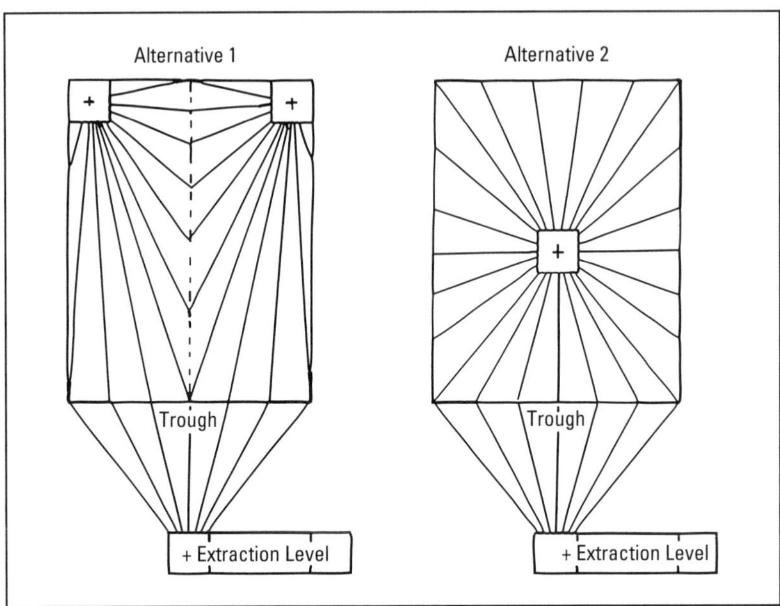

Source: Bullock and Hustrulid 2001

FIGURE 6.9 Two typical drill patterns for sublevel stoping

mining. These extraction blocks can be oriented along (longitudinally) or transversely to the strike of the ore body.

Vertical Crater Retreat

In the cases just discussed, rings of holes were blasted toward a vertical slot. In the VCR or vertical retreat mining systems, the need for the slot connecting the drilling and extraction level has been eliminated, thus simplifying the development. It is replaced by a horizontal slot (undercut) created at the bottom of the block on the extraction level. Although a real trough may be created, it is not necessary. From the drilling level, large-diameter (approximately 165 mm [6.5 in.]) parallel holes are drilled downward to the undercut level (Figure 6.10). Short explosive charges (length = six hole diameters) are lowered to positions slightly above the top of the undercut. These spherical charges are detonated, dislodging a crater or cone-shaped volume of rock into the underlying void. As each layer of charges is placed and detonated, the mining of the stope retreats vertically upward; hence the name *vertical crater retreat* mining. The design of the blasting pattern is based on full coverage of the block cross section by the adjacent craters. Normally, the blasting pattern is tighter (holes spaced closer) than would be the case in large-hole blast-hole stoping, and hence the powder factor is larger. When blasting under these confined conditions, the fragmentation is generally finer than with blast-hole stoping. Prior to charge placement, care must be taken in determining the location of the free surface. Special tests are performed to determine crater dimensions. In this system, the level of broken rock remaining in the stope can be controlled to provide varying levels of support to the stope walls. If the stope is kept full except for a small slot to provide a free surface and swell volume for the blasted rock in the slice, it is classified as a shrinkage method. In this case, the remaining ore would be drawn out at the completion of mining.

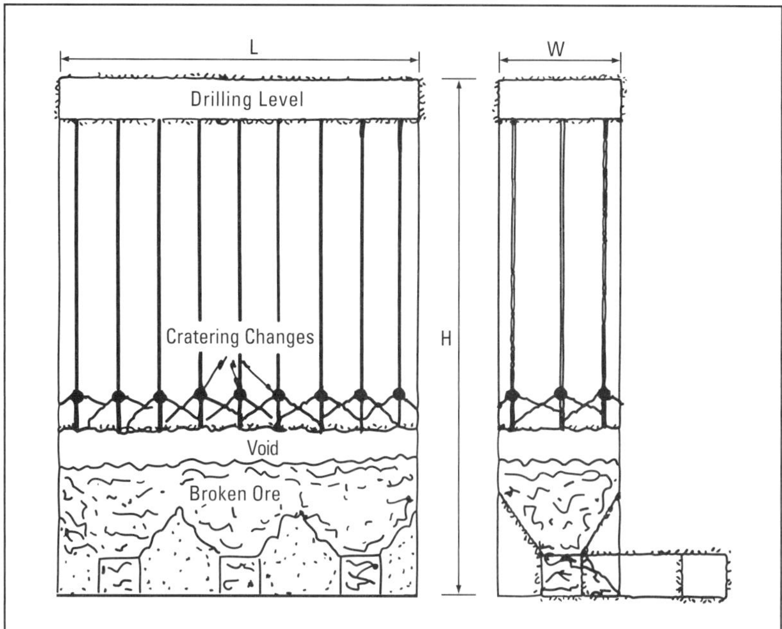

Source: Bullock and Hustrulid 2001

FIGURE 6.10 Vertical crater retreat stoping

Vein Mining

Another approach to extracting the ore block is called vein mining. At the highest level of the block to be extracted, a connection is made to the ore body. It is assumed that the access is located on the footwall side and the connection is made in the middle of the extraction block. On the extraction level, an undercut or an extraction trough is prepared. A raise is driven between the extraction level and the upper access point, from which long blast holes can be drilled. In the United States, this was first tried at Pea Ridge Iron Ore Company (Ovanic 2001) using a Boliden-type cage. More recently, using the Alimak technique (Figure 6.11), the raise is located in the footwall, a small distance from the ore–footwall contact. The next step in the process is the drilling of subhorizontal fans of blast holes from the Alimak platform (or Boliden-type cage) in such a way that the plan area of the extraction block is fully covered. The hole diameter is determined by the capacity of the drilling machine but should be as large as possible since the toe spacing and the burden (distance between fans) is determined by the hole diameter and the explosive used. Once the drilling of the entire extraction block has been completed, the fans are charged and blasted one or more at a time working off of the raise platform. The access to the block is now only from the upper level as the stope is retreated upward. The ore in the stope can be removed after each blast or it can be left in place, removing only enough to provide swell volume for the next slice(s). Rock reinforcement can be installed in the hanging wall if required from the raise platform during the drilling of the production holes. The method allows the extraction of very high ore blocks with a minimum of development (upper access point, the extraction level, and the connecting raise). The overall length of the extraction block is determined by the straight-hole drilling length of the available drilling equipment. However, if larger blocks are more economically mined, then multiple raises can be used, and

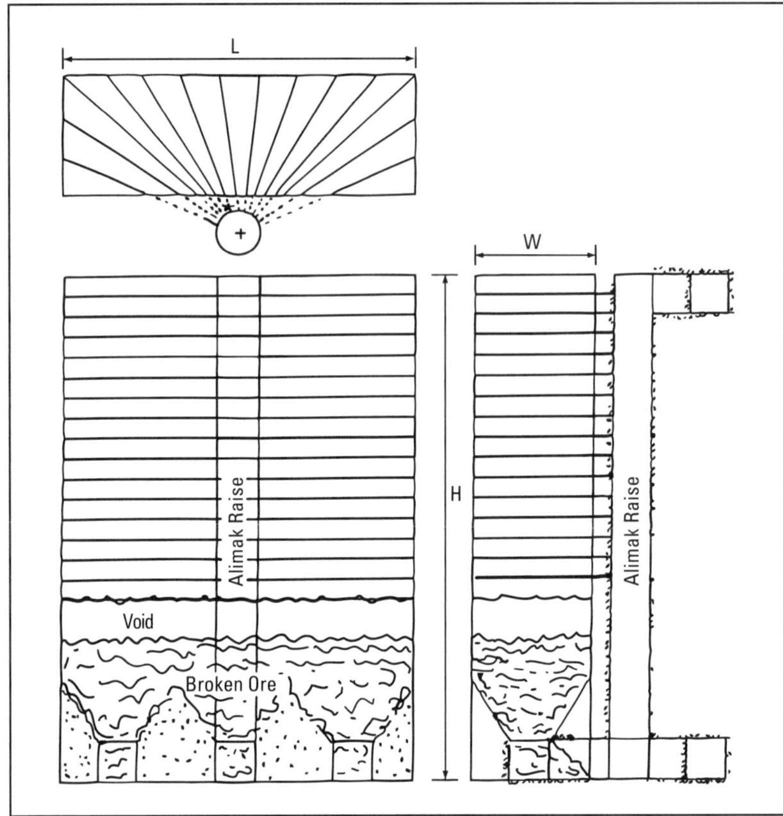

Source: Bullock and Hustrulid 2001

FIGURE 6.11 Sublevel stoping a vein with long holes

this should be a factor considered in the feasibility study. The disadvantage of the method is that the drilling and charging must be done from a raise environment, which traditionally has not been pleasant. However, major advances have been made in the mechanization/automation of the rigs used for the drilling.

Shrinkage Stoping

The final method to be considered under this category is shrinkage stoping. Although normally considered a separate method, it is logical to include it here, since it is an open stope method. The method is generally applied to very narrow extraction blocks, which have traditionally not lent themselves to a high degree of mechanization, but at the same time has been applied successfully in high-grade precious metal mining because of its low dilution and low ore loss. Here a very simplified layout (Figure 6.12) is illustrated. The extraction block is laid out longitudinally because of the very narrow nature of the ore body recovered. An extraction drift is located in the footwall with loading crosscuts positioned at regular intervals. Raises are driven at each end of the extraction block connecting to the above-lying level. An initial horizontal extraction slice is driven across the block from raise to raise. Extraction troughs are created by drilling and blasting the rock between this level and the underlying extraction points. When the extraction system has been created, short vertical holes are drilled into the roof of the first

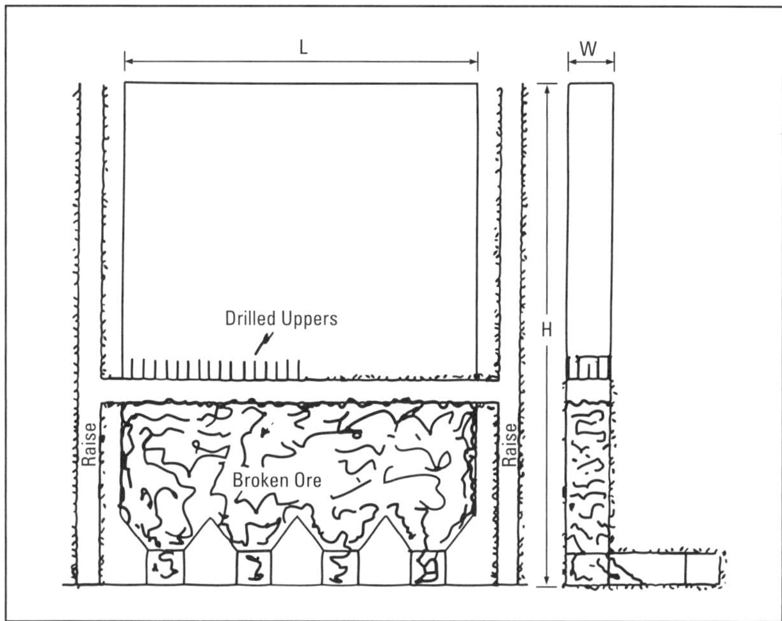

Source: Bullock and Hustrulid 2001

FIGURE 6.12 Shrinkage stope

extraction slice using the raise access. The miners stand on the broken ore, which forms the working floor. Jackleg or stoper drills are used for the drilling of the small-diameter holes. The holes are charged, and then ore is extracted from the stope to provide room for the blasted material. The blast is initiated, and the miners reenter the newly created void to drill out the next slice. The process continues working upward one slice at a time. On reaching the upper end of the extraction block, the ore is drawn out. Until that time, the stope is filled with broken ore.

Summary

Depending on the geometry of the ore body, several varieties of sublevel stoping can be employed. The ore bodies must have strong wall rocks and competent ore, either naturally or helped by the emplacement of reinforcement as large openings are created in the process of ore removal.

The extraction block used to illustrate the layouts for the different mining systems can now be duplicated and translated laterally and vertically in the ore body, leaving pillars to separate adjacent blocks. The size and shape of the extraction block can be adjusted to fit the orebody geometry and the mine infrastructure. The openings created during this primary mining may be filled with various materials or left unfilled. The filling materials may be cemented or uncemented depending on the next stage of recovery envisioned. Various methods are used to recover the remaining reserves tied up in the pillars. During the feasibility studies, these secondary recovery methods should be examined at the same time as the primary system is designed. Although for simplicity the basic extraction block was considered vertical, the process could obviously be repeated for ore bodies having various dip conditions.

CUT-AND-FILL MINING

In the previous section, it was assumed that the rock mass properties were such that large openings could be created. Because of the way that ores are emplaced, there are many instances where the ore and/or the wall rocks are weak, and hence both the opening size and the allowable time between ore removal and the filling of the excavation are strictly limited. There are a number of different extraction designs that can be applied, all of which fit under the general category of cut-and-fill mining. It is a very versatile method and can be adapted to the extraction of any ore-body shape. With some exceptions, all of the ore is removed via drifting and the drifts created are then filled. As a result, the mining costs are high compared to the other methods. But with these systems, when the methods are applied correctly, the recovery is high and the dilution is generally low. Thus it is an appropriate approach to the extraction of high-grade ore bodies.

Extraction Principles

For simplicity, an extraction block of the same type used in the previous section is assumed. The access is via a ramp driven in the footwall, and mobile equipment is used. Typically, the drifts used in mechanized cut-and-fill mining are on the order of 5 m (16 ft) high. To begin the discussion, it is assumed that the ore block to be extracted is vertical and has a width that can be removed during normal drifting. When the ore-body strength is fairly good, overhand cut-and-fill mining is normally applied (Figure 6.13). This means starting at the bottom of the block and working upward.

Ideally, the access to each level is via crosscuts originating at the midlength position of the block. In this way, two headings can be operated at any one time. Typical drift rounds consisting of drilling, blasting, loading, scaling, and the installation of rock reinforcement are used. This progression of operations can lead to delays unless carefully planned. The drilling of the second heading is carried out while the other operations are being done at heading number 1. When the slice has been completed, filling is done. The fill is placed leaving a small air gap to the overlying ore. On the next slice, this gap forms the free surface for the blasting. The process continues upward slice by slice to the top of the block. Several such extraction blocks can be in operation at any one time to meet the production requirements. The horizontal pillar created between two such stacked extraction blocks is called the *crown pillar* with respect to the underlying stope and the *sill pillar* for the stope above. Normally the first cut of the extraction block above the sill is filled with a cemented fill to facilitate later extraction of the pillar.

The access to this one-drift-width cut-and-fill stoping is from an access ramp located in the footwall. Often four slices are accessed from a given point on the ramp. This is shown in Figure 6.13. In an overhand cut-and-fill method, crosscut 1 is made first. When the slice is completed, the roof of the crosscut is slashed down to form crosscut 2. This continues for the four slices at which time a higher point on the ramp is selected as the origin of the crosscuts. Generally, the maximum crosscut grade is limited to about 20 degrees. This sets the position of the ramp with respect to the ore body.

In some cases, the wall rock is strong enough to allow a double slice to be open at any given time. Here the first slice is mined by drifting (Figure 6.14A) and then rather than filling directly, uppers (upward-oriented drill holes) are drilled the length of the slice (Figure 6.14B). Once the drilling has been completed, the several rows of holes are charged and blasted beginning at the ends of the extraction block and retreating toward the access. The ore is extracted by

LHD after each blast and transported to the orepass. In this way, efficiency can be improved by changing the typical cycle to one in which all of the drilling is done first, followed by the charging and loading. Then one lift can be filled, followed by the drilling of uppers, and so forth, or both lifts can be filled followed by drifting, followed by the drilling of uppers.

If the strength of the wall rock and the ore is quite good, then spans of more than two lifts can be created. Figure 6.15 shows an example where slices 1 and 4 are extracted by drifting (stage 1). Rows of vertical blast holes are then drilled from the floor of slice 4 to the roof

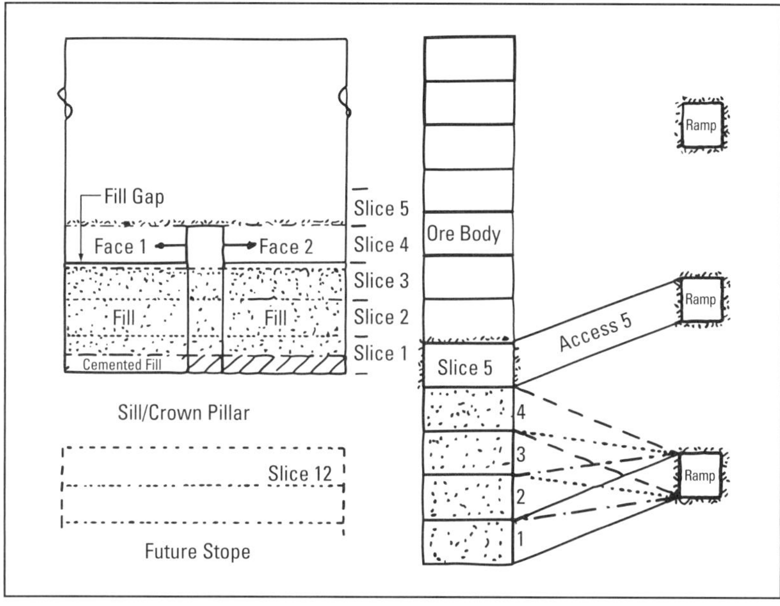

Source: Bullock and Hustrulid 2001

FIGURE 6.13 Overhand cut-and-fill stope

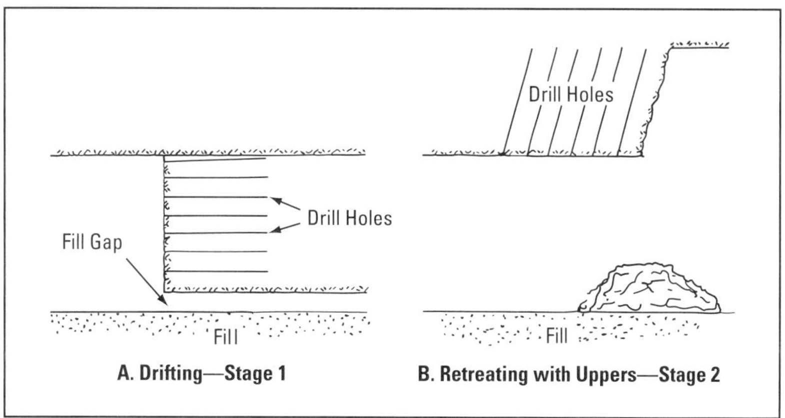

Source: Bullock and Hustrulid 2001

FIGURE 6.14 Two-pass cut-and-fill stoping

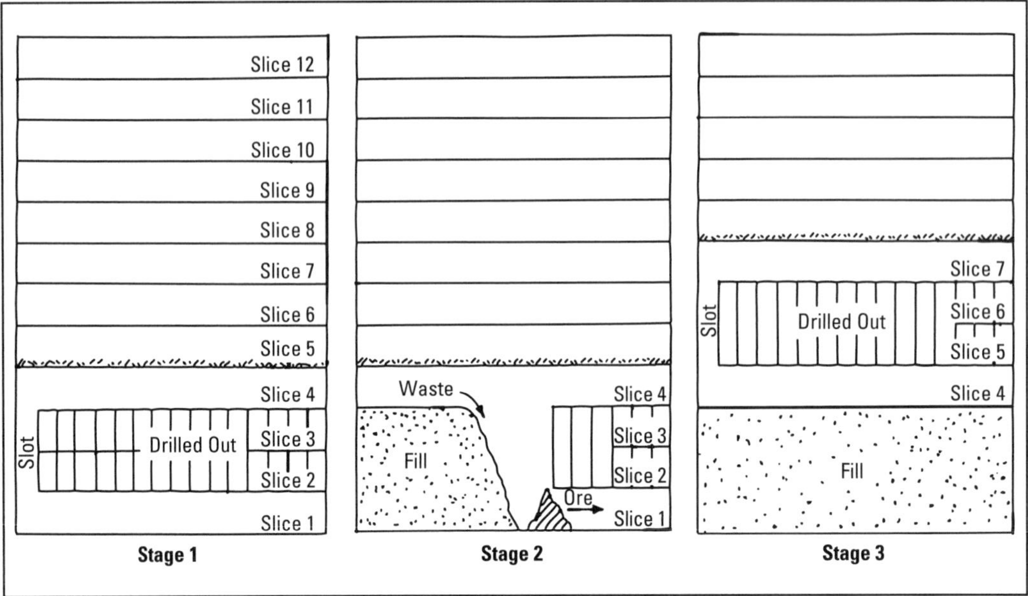

Source: Bullock and Hustrulid 2001

FIGURE 6.15 Mining sequence of Avoca cut-and-fill system

of slice 1. A vertical slot is created, and the rows of holes are blasted one or more at a time toward the slot. The ore is extracted by an LHD operating in slice 1. At the same time that retreat extraction is underway, filling is being conducted from the opposite end of the stope (stage 2). A gap is maintained between the extraction and filling fronts to minimize dilution. When completed, slice 7 is removed by drifting (stage 3). Slices 5 and 6 are now removed using slice 7 as the drilling level and slice 4 as the extraction level. The method is called *rill mining* or the *Avoca* method.

If the extraction block is quite wide, then the cut-and-fill method can still be used, but now several drifts are driven side by side (Figure 6.16). This is similar to R&P (rib pillar) mining with the rooms being filled and the pillars then being extracted. As shown, there are various techniques used to shape the drifts, but the most common is shown in Figure 6.16C. Here straight walls are used, and every other drift is removed in a primary mining phase. Cemented fill is used to avoid dilution during the removal of the interlying drifts. One option is to make the primary drifts narrow and the secondary drifts wide to minimize the use of cemented fill.

If the strength of the ore or hanging wall is very poor, then the undercut-and-fill method may be used (Figure 6.17). The first slice is taken and then various techniques are used to prepare a layer that will become the roof when extracting the slice below. In the past, a timber floor pinned into the walls was the main technique. Often, it is more common to pour a layer of cemented fill with or without reinforcement. The remainder (upper portion) of the drift may be left open or filled with uncemented fill. The next slice is then extracted under the constructed roof. A more common practice in the United States and Canada is to use engineered cement or past fills. When they are used for mining wide ore bodies, the material must be jammed tight to the back or previous floor.

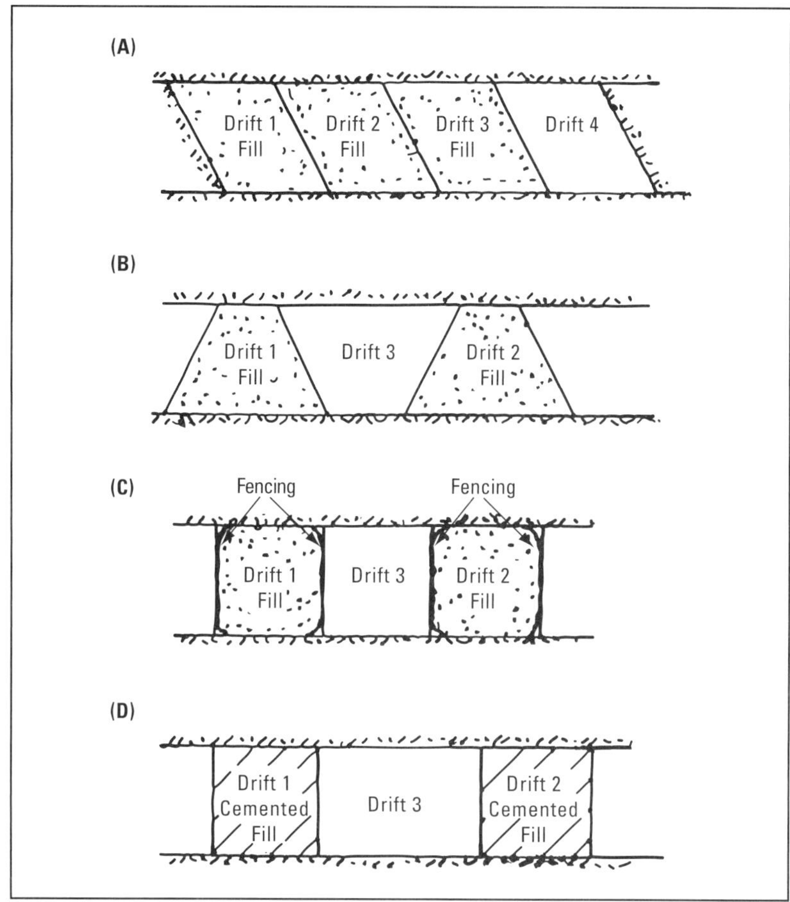

Source: Bullock and Hustrulid 2001

FIGURE 6.16 Drift shapes and sequence for drift-and-fill method

From the same development access level, some mines use overhand cut-and-fill, working upward from this level while employing an undercut-and-fill method, working downward. This doubles the number of working faces in operation at any one time from a given level. Whether or not this is advisable will depend on the ore/rock strength. Wide ore bodies can also be mined using the underhand cut-and-fill method (Figure 6.18A). The process is the same as described earlier, but now cold joints occur between the individual drift floor pours. Generally, whether it be engineered fills, cemented fill, or a paste fill, it is better to avoid positioning drifts of the underlying layer directly under those above. The drifts may be shifted sideways or driven at an angle to those above. In the latter case, one gets a basket-weave pattern (Figure 6.18C). Wide precious metal ore bodies can also be mined by undercut drift-and-fill methods, using a basket-weave approach, in this case using an engineered cemented fill. Undercut drifts can be cut on a 45- to 60-degree angle in a variation of the basket-weave pattern, such as is shown in Figure 6.19 of the Murray mine at Jerritt Canyon, Nevada.

In thick, inclined ore bodies or in very wide ore bodies, which are appropriate for overhand cut-and-fill mining, vertical pillars are sometimes left to provide additional support between the hanging wall and footwall (Figure 6.20). If possible, these pillars are located in the internal

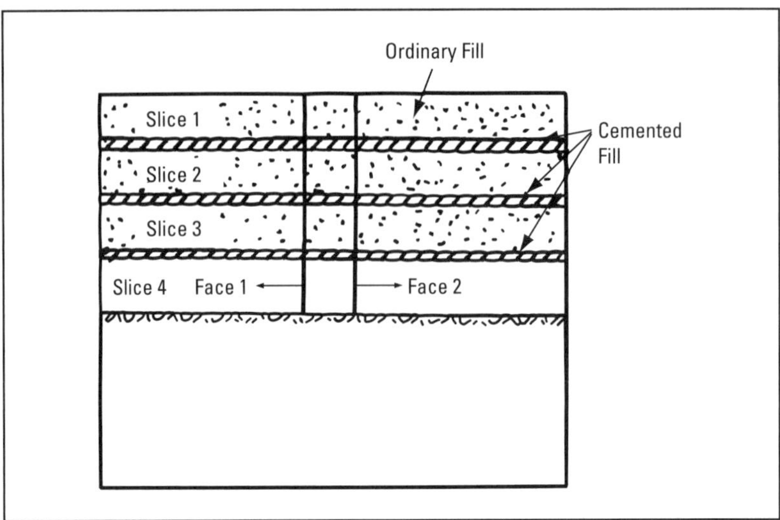

Source: Bullock and Hustrulid 2001

FIGURE 6.17 Undercut-and-fill mining sequence

waste or low-grade areas. On the lowest slice, an R&P mine is created. The rooms are then filled. A second slice is taken, continuing the vertical upward extension of the pillars. This level is then filled and the process continued. The mining system is called postpillar mining because the pillars appear as vertical posts surrounded by fill. Some authors include the method under R&P mining and others under cut-and-fill mining. Because of the presence of the surrounding fill, even very tall and slender pillars can be quite strong.

SUBLEVEL CAVING

Sublevel caving was initially applied in extracting the soft iron ores found on the iron ranges of Minnesota and Michigan. The sublevel caving as practiced today is significantly different from this early version and should probably be given another name such as sublevel retreat stoping, continuous underhand sublevel stoping, or something similar that better reflects the process. Sublevels are created at intervals between 20 and 30 m (65 and 98 ft) beginning at the top of the ore body and working downward. On each sublevel, a series of parallel drifts are driven at a center-to-center spacing, which is of the same order as the level spacing. From each sublevel drift, vertical or near-vertical fans of blast holes are drilled upward to the immediately overlying sublevels. The distance between fans (the burden) is on the order of 2–3 m (6–10 ft). Beginning typically at the hanging wall, the fans are blasted one by one against the front-lying material consisting of a mixture of both ore from overlying slices as well as the waste making up the hanging walls and/or footwalls. The extraction of the ore from the blasted slice is continued until the total dilution reaches a prescribed mineral cut-off level. The next slice is then blasted and the process continued. Depending on ore-body geometry, the technique may be applied using transverse or longitudinal retreat.

Today the sublevel caving technique is applied in hard, strong ore materials for which the hanging-wall rocks readily cave. The key layout and design consideration is to achieve high recovery with an acceptable amount of dilution and ore loss. The uncertainties of fragmentation

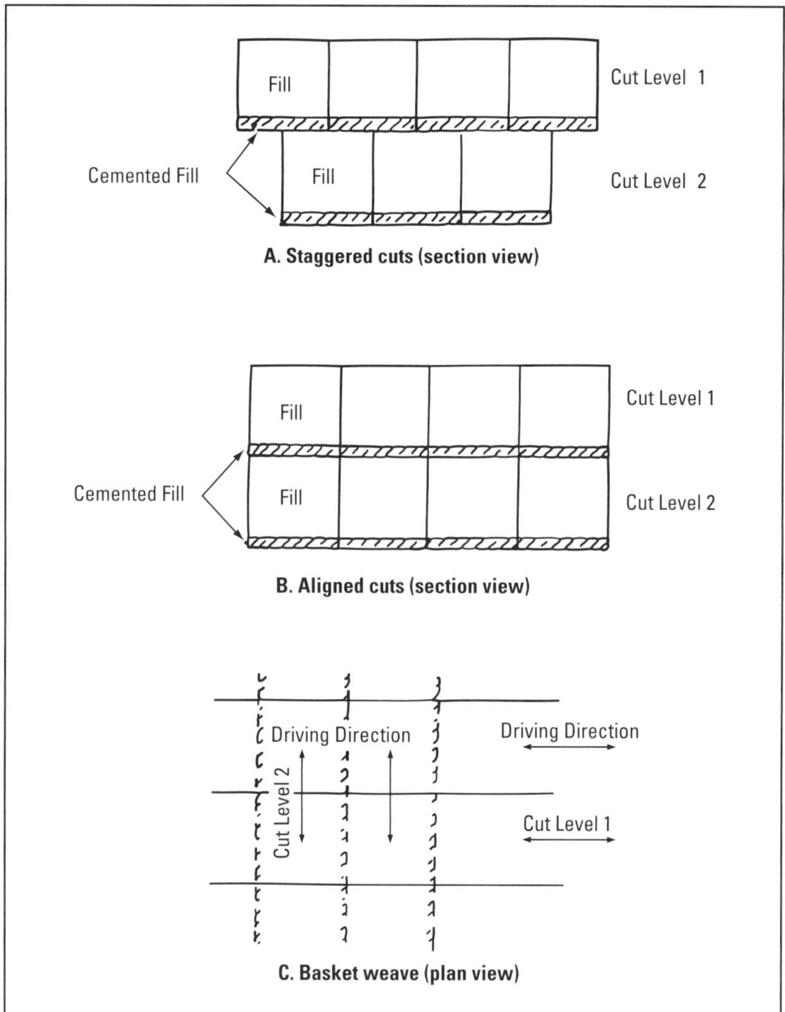

Source: Bullock and Hustrulid 2001

FIGURE 6.18 Drift development pattern for undercut and fill levels

and ore cavability present in panel caving (discussed in the following subsection) are removed, because each ton of ore is drilled and blasted from the sublevels. The method has been used most for the mining of magnetic iron ores that can be easily and inexpensively separated from the waste. However, it has been and can be applied to a wide variety of other ore types.

Sublevel Caving Layout

As indicated, the ore is recovered both through drifting and through stoping. Because the cost per ton for drifting is several times that for stoping, maximizing the stoping and minimizing the drifting is desired. This has meant that through the years the height of the sublevels has steadily increased until today they are up to 30 m (98 ft). Whereas approximately 25% of the total volume was removed by drifting in the early designs, today in the largest scale sublevel caving designs, that value has dropped to about 6%. The sublevel intervals have changed from

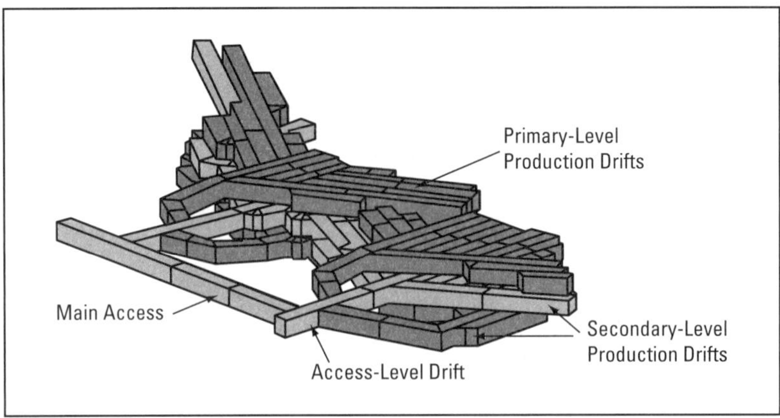

Source: Brechtel et al. 2001

FIGURE 6.19 Undercut drift-and-fill method used at Murray mine, Jerritt Canyon, Nevada

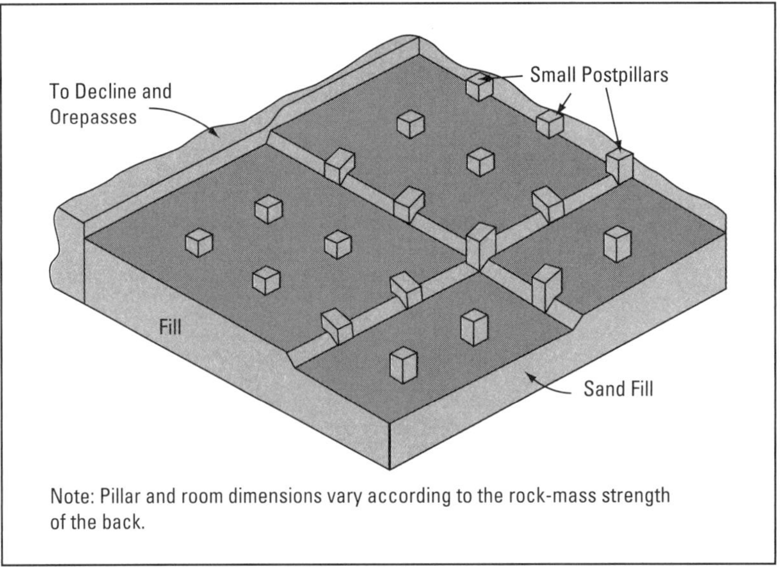

Source: Bullock 2011

FIGURE 6.20 Mechanized postpillar cut-and-fill method

9 m (29 ft) up to nearly 30 m (98 ft). The key to this development has been the ability to drill longer, straighter, and larger-diameter holes. Sublevel caving is an underhand method with all of the blast holes drilled upward. The ore moves down to the extraction/drilling drift under the action of gravity. Several factors determine the design. The sublevel drifts typically have dimensions (w/h) of 5 × 4 m (16 × 13 ft), 6 × 5 m (19 × 16 ft), or 7 × 5 m (23 × 16 ft) to accommodate the LHD loading equipment. In the example used to illustrate the layout principles, it is assumed that the drift size is 7 × 5 m (23 × 16 ft). The largest possible blast-hole diameter from the viewpoint of drilling capacity and explosive charging is normally chosen. The maximum hole size in use today is 115 mm (4.5 in.), based largely on the ability to charge and

retain explosive in the hole. These large holes may be drilled using either in-the-hole (ITH) or top-hammer machines. The large diameters and large drift sizes permit the use of tubular drill steel of relatively long lengths (thereby minimizing the number of joints and maximizing joint stiffness) so that the required long, straight holes can be produced. The largest ring designs incorporate holes with lengths up to 50 m (164 ft).

The distance between slices, the burden (*B*), depends both on the hole diameter (*D*) and the explosive used. For initial design when using ANFO (ammonium nitrate and fuel oil) as the explosive, the relationship is $B = 20D$. For more energetic explosives (bulk strength basis), the relationship is $B = 25D$. Assuming that the hole diameter is 115 mm (4.5 in.) and an emulsion explosive is used, the burden would be about 3 m (10 ft). Typically the toe spacing (*S*) is 1.3 times greater than the burden. Hence the maximum toe spacing would, in this case, be 4 m (13 ft). To achieve a relatively uniform distribution of explosive energy in the ring, the holes making up the ring would have different uncharged lengths. Both toe and collar priming initiation techniques are used.

The sublevel drift interval is decided largely on the ability to drill straight holes. In this example, it is assumed that the sublevel interval based on drilling accuracy is 25 m (82 ft) (Figure 6.21). Once the sublevel interval has been decided, it is necessary to position the sublevel drifts. In this example, the drifts are placed so that the angle drawn from the upper corner of the extraction drift to the bottom center of the drifts on the overlying sublevel is 70 degrees. This is approximately the minimum angle at which the material in the ring moves to the drawpoint. The resulting, center-to-center spacing is 22 m (72 ft). A one-boom drill is assumed to drill all of the holes in the ring. The inclination of the side holes has been chosen as 55 degrees although holes somewhat flatter can be drilled and charged. The function of the holes drilled flatter than 70 degrees is largely (a) to crack the ore, which is then removed from the sublevel

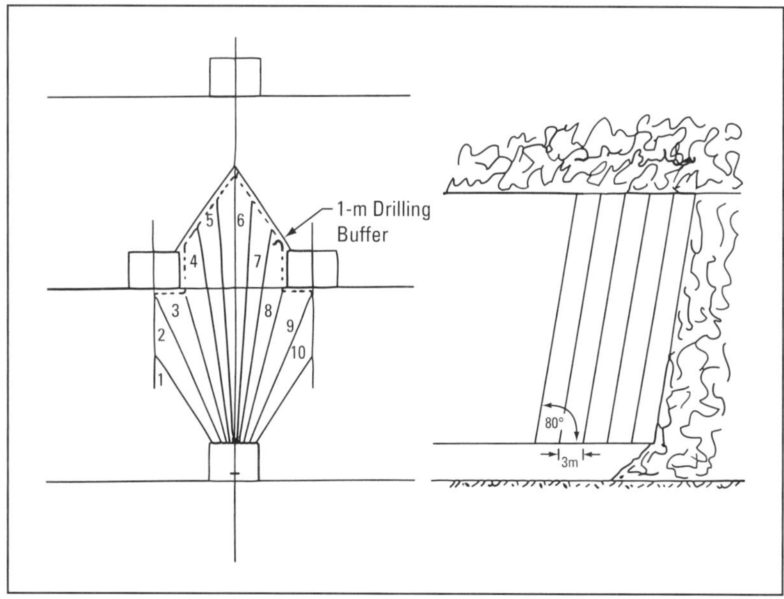

Source: Bullock and Hustrulid 2001

FIGURE 6.21 Initial stope design step for sublevel caving

below; and (b) to reduce the maximum drill-hole length. Holes flatter than 45 degrees are difficult to charge because of the angle of repose of the ore at the extraction front.

In Figure 6.21, an extraction ellipse has been developed. The fans may be drilled vertically or slightly inclined from the horizontal at an angle, typically 70–80 degrees. Inclining the fans improves brow stability and access for charging the holes. To initiate mining of a new sublevel, an opening slot must be made toward which the fans can be blasted.

Recovery and Dilution

Sublevel caving lends itself to a very high degree of mechanization and automation. Each of the different unit operations of drifting, production drilling, blasting and extraction can be done largely without disturbance from one another (Figure 6.22). Specialized equipment and techniques can and have been developed leading to a near factory-like mining environment. As indicated earlier, because every ton of ore is drilled and blasted, there are not the same uncertainties regarding cavability and fragmentation present with block caving. However, a very narrow slice of blasted ore surrounded by a mixture of waste and ore must be extracted with high recovery and a minimum of dilution. As can be easily visualized, the ore at the top part of the ring in the example is more than 40 m (131 ft) away from the extraction point whereas the waste–ore mixture lies only the distance of the burden in front of the ring (on the order

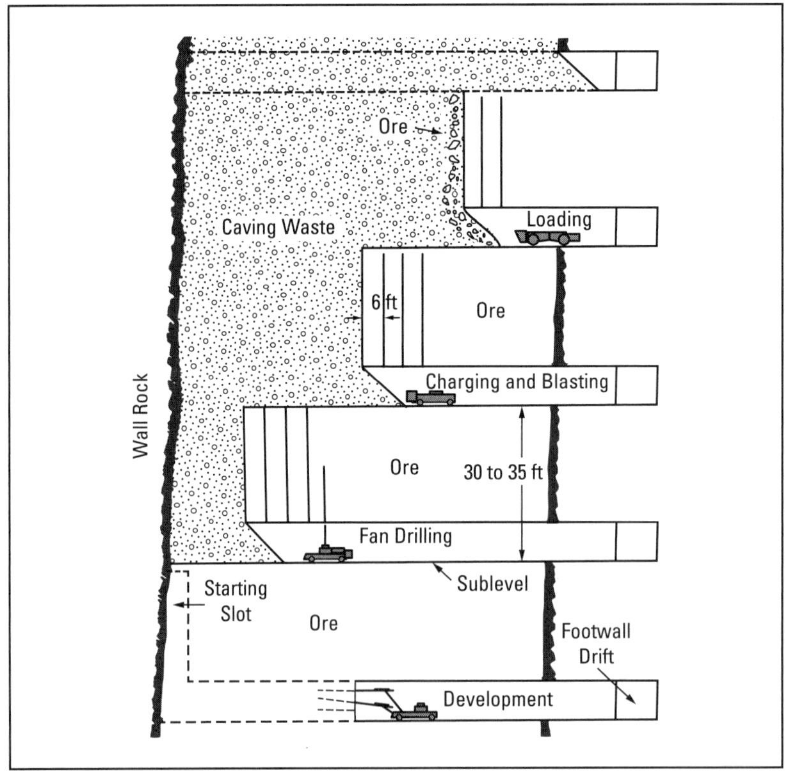

Source: Cokayne 1982

FIGURE 6.22 Typical sublevel caving showing four simultaneous operations

of 3 m [10 ft]). With care, recoveries on the order of 80% with dilution held below 25% can be achieved.

PANEL CAVING

In this section, the term *panel caving* is used to represent both *block caving*, suggesting the mining of individual blocks, and *panel caving*, which is mostly used today to indicate a laterally expanding extraction. There are a great number of variants of this system and it is impossible to do them all justice in a very short discussion such as this. The intention is to provide the reader with an introduction to some of the more important layout considerations. The emphasis is on development and extraction. For panel caving, the three most important elements of the extraction system are the undercut level, which removes the support from the overlying rock column, the funnel through which the rock is transported downward to the extraction level, and the extraction level itself. The basis for system design and performance is the degree of fragmentation present as the rock blocks enter the top of the funnel. The impact of the fragmentation is discussed in more detail as the section proceeds. In the early days of block caving, the materials were soft and caved readily. Today the trend is to use cave mining on ever harder and tougher ores. The result is that the engineer must thoroughly evaluate the ore body and tailor the design so that a successful extraction will result. This is the least expensive of the mining systems as measured on an extracted ton basis. This requires a large amount of testing, which must take place during the feasibility study, and then only an engineer experienced in panel caving design should attempt to design the configuration of the panel caving system.

Extraction Level Layout

Assuming the use of LHDs, the major development on the extraction level consists of extraction drifts, drawpoints, and the extraction troughs/bells. To simplify the discussion, it is assumed that all of the drifts have the same cross section. Design is an iterative process and it is always a question as to where one begins. In this case, one begins with knowing or assuming the size of the material that must be handled. The physical size of the loading equipment is related to the required scoop capacity that, in turn, is related to the size of the material to be handled. If the fragmentation is expected to be coarse, then a larger size bucket and a larger machine would be required than if it is fine. Knowing the size of the machine, one arrives at a drift size. In the sizing of orepasses, it is expected that the orepass diameter should be three to five times the largest block size to avoid hang-ups. If this same rule is applied to the sizing of extraction openings, then the size of the extraction opening should be on the order of 5–7 m (16–23 ft) for block sizes with a maximum dimension of 1.5 m (5 ft). Depending on the density and the shape, such a block would weigh 5–10 t (5.5–11 st). A large piece of equipment is required to be able to handle such blocks. It is typical for extraction drifts to be sized (width-to-height ratios expressed in meters) according to the ratios 4:3, 5:4, or 6:5. For the machine in the example used in this section, the drift size would be on the order of 5 × 4 m (16 × 13 ft) or larger.

To begin the design of the extraction level, one creates the grid of extraction drifts, which are to be traversed by the LHDs and the lines of associated drawpoints. In practice, a series of circles of radius R corresponding to the draw radius of influence on the undercut level are first drawn. Figure 6.23 shows one such pattern for staggered coverage with the locations of the extraction drifts superimposed. It has been found that the value of R depends on the degree of fragmentation. The radius will be larger for large fragmentation than for finer fragmentation.

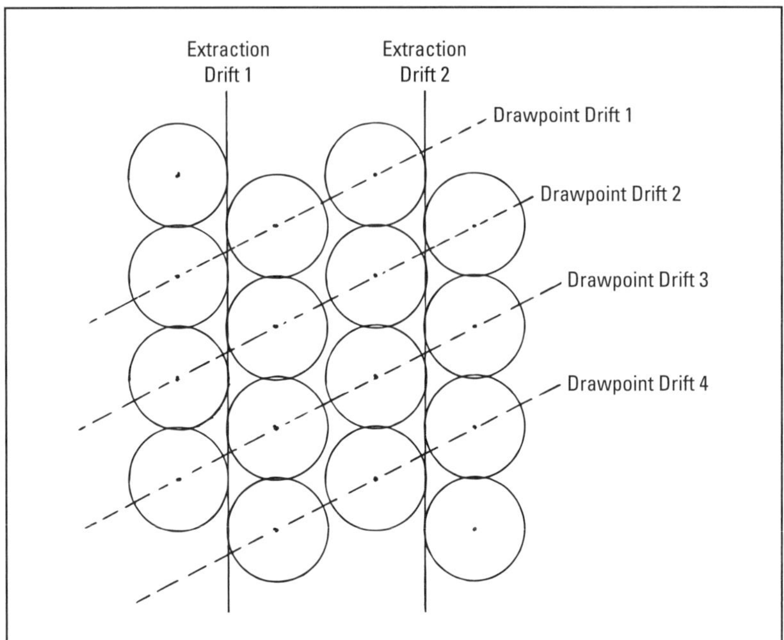

Source: Bullock and Hustrulid 2001

FIGURE 6.23 Initial panel caving design with staggered draw circles

This presents a design problem, because in the initial stages of draw, the fragmentation will generally be larger than at later stages. The degree of desired coverage is one of the design factors. The "just-touching" case shown in Figure 6.24 shows triangles between circles not covered. However, moving the circles to a condition of even more overlap would finally give complete total coverage. In the example, it is assumed that $R = 6$ m (20 ft) and a square, just-touching drawpoint pattern is used. Shown in Figure 6.24 are the locations of the extraction drifts, the drawpoint drifts, and the drawpoints on the extraction level. Figure 6.25 indicates that there are two draw circles associated with each drawpoint in this herringbone design. For drawpoint 1, the draw circles are 1 and 4, whereas for drawpoint 2 they are 2 and 3. In continuing the design example, one must decide the orientation of the drawpoint drift with respect to the extraction drift. Figures 6.25 and 6.26 show two possibilities involving the use of a 45-degree angle. A careful examination of these figures reveals that the choice affects both loading direction and the ease with which the openings can be driven.

A drawpoint entrance made at 60 degrees to the axis of the extraction drift is a very convenient angle from the loader operator's point of view. Some designs involve the use of 90-degree angles (square pattern). In this case, loading can be done from either direction. The 90-degree pillars provide good corner stability, but the loading operation is more difficult. When considering the different drawpoint design possibilities, loading machine construction must be taken into account. It is important for the two parts of the LHD to be aligned when loading to avoid high maintenance costs and low machine availability.

As indicated, the design of the extraction level is made in response to the type of fragmentation expected. For coarse fragmentation, the openings have to be larger to permit extraction of the blocks. However, the larger openings present the possibility for stability problems, and

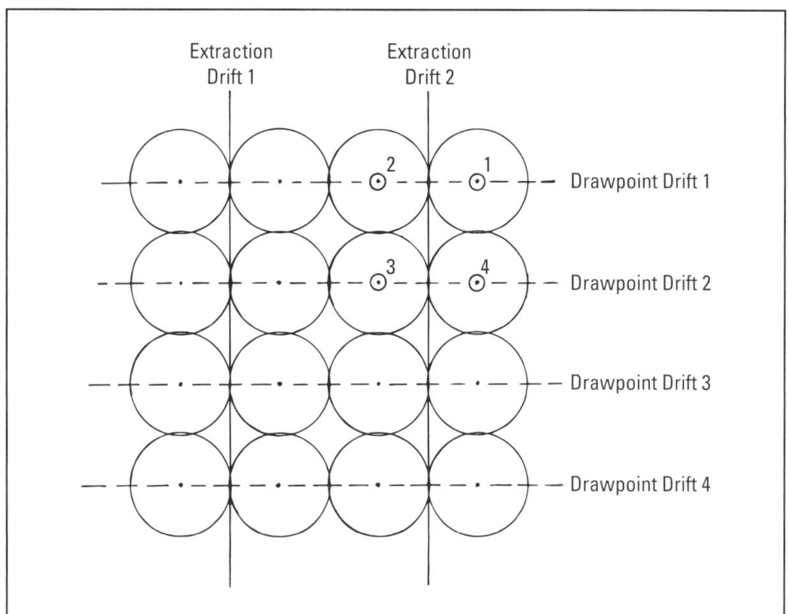

Source: Bullock and Hustrulid 2001

FIGURE 6.24 Square layout of extraction and drawpoint drifts

because these openings must last for the time required to extract the overlying column of ore, the design, creation, and reinforcement of the openings must be carefully made. Fortunately, in the rock in which one expects coarser fragmentation, the rock is also stronger, providing a better construction material. In softer rocks yielding a finer fragmentation, the openings can be smaller. The need to protect the integrity of the openings is of highest importance. This is discussed in more detail in the following subsection.

There are, as indicated, a great number of different design possibilities for the extraction level. All involve the basic components of fragmentation, radius of influence, draw coverage, machine size, and drift size examined roughly in that order.

Undercutting and the Formation of the Extraction Trough

In the undercutting process, a slice of ore forming the lower portion of the extraction column is mined. As this drilled-and-blasted material is removed, a horizontal cavity is formed beneath the overlying intact rock. Because of the presence of this free surface, the subhorizontal side stresses, and the action of gravity, the intact rock undergoes a complex process involving loosening, crushing, and caving. The ease by which the intact rock transforms into a mass of fragments is reflected in its characteristic *cavability*. One approach to addressing a material's cavability is to describe the size and shape of the area that must be undercut to promote caving. The other, and more important part, of cavability is the description of the fragment size distribution. This is much more difficult to predict but ultimately more important from a design viewpoint. In this section, both the undercutting process and the design of the trough required to deliver the resulting fragments to the extraction level are described.

The simplest design is to combine the undercutting and the trough formation process into a single step. As described in the previous section, a series of parallel extraction drifts are driven.

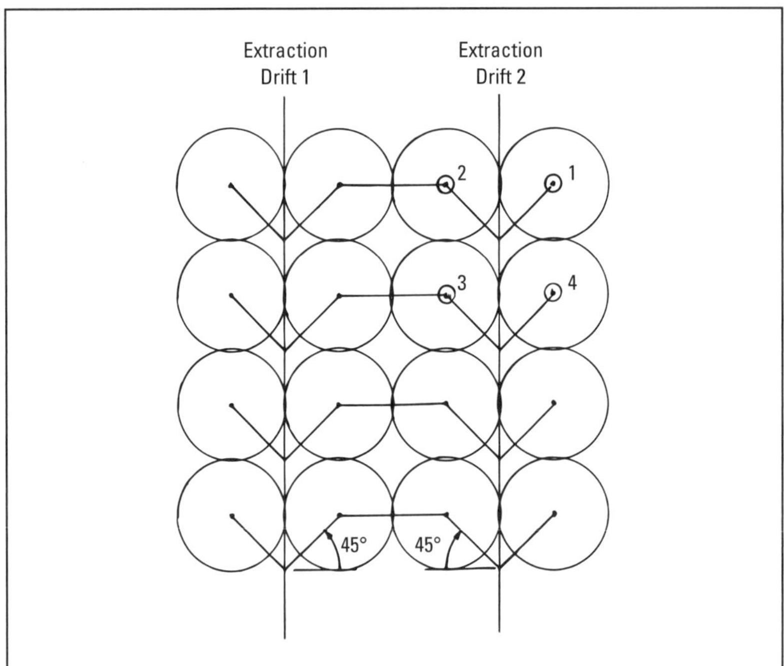

Source: Bullock and Hustrulid 2001

FIGURE 6.25 Herringbone pattern of drawpoint drifts

The center-to-center spacing of these drifts is determined by the size of the influence circles. For this example, the square layout plan shown in Figure 6.24 is used. The center-to-center spacing of the extraction drifts is 24 m (80 ft) (4*R*). A series of parallel trough drifts will now be driven between the extraction drifts. Starting at the far end of the extraction block, fans of holes are drilled and then blasted toward opening slots. In the case shown in Figure 6.27, the side angles of the fans have been chosen as 52 degrees, and the resulting vertical distance between the extraction level and the top of the undercut is 18 m (60 ft). It is noted that the trough drifts and the troughs can be created either before or after driving the extraction drifts. In the latter case, this would be termed *advance* or *pre-undercutting*. An advantage with this design is that all of the development is done from one level. An example of the use of this design has been presented by Weiss (1979).

Most mining companies using panel caving have separate undercut and extraction levels. Figure 6.28 shows the same cross section as in Figure 6.27, but now a separate undercut will be constructed. As seen in Figure 6.29, the undercut level has been designed as a rib pillar mine. The rooms are 5 × 5 m (16 × 16 ft) and the room center-to-center spacing is 12 m (40 ft). In step 2 of this design, the interlying pillars are drilled and blasted. In step 3, the extraction troughs are created to complete the undercut/trough development. It is possible and often desirable to develop the undercut level first followed by the development on the extraction level.

Figure 6.30 is an alternative design for the same basic extraction level layout. A separate undercut level has been used with the undercut drifts spaced on 24-m (80-ft) centers. From these drifts, fans of holes are drilled to form a trough. The angle of the side holes has been

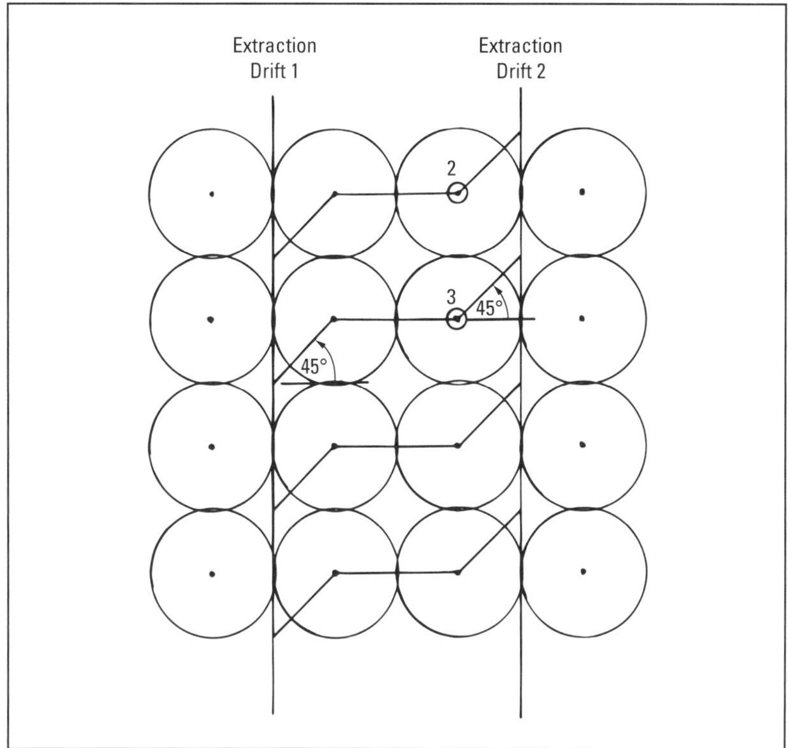

Source: Bullock and Hustrulid 2001

FIGURE 6.26 Flow-through pattern of drawpoint drifts

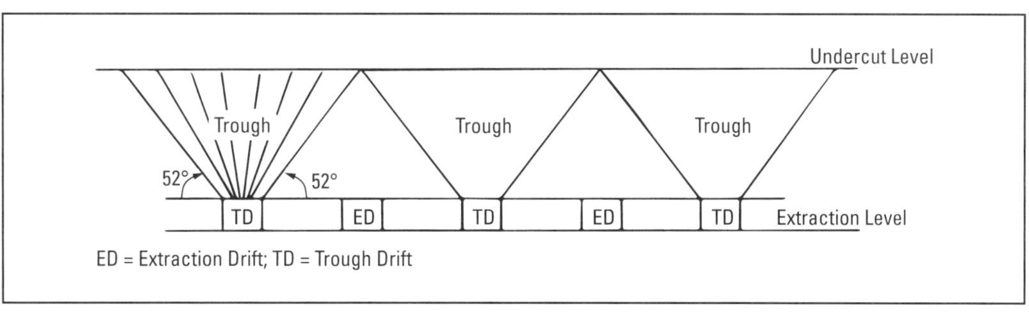

Source: Bullock and Hustrulid 2001

FIGURE 6.27 Single-level extraction and undercutting

chosen as 52 degrees. As can be seen, the undercut drifts are positioned directly above the underlying extraction drifts. Once the undercut has been created, a sublevel caving–type of fan pattern is drilled from the trough drifts on the extraction level. This completes the development. The total height of the undercutting in this case is 36 m (118 ft), which has some advantages in the caving of harder rock types.

Figures 6.31 and 6.32 are the plan and section views of a more traditional undercutting and bell layout for panel caving. In the previous examples, an extraction trough has been used, primarily to demonstrate the principles involved. A trough has the advantage of simplicity of

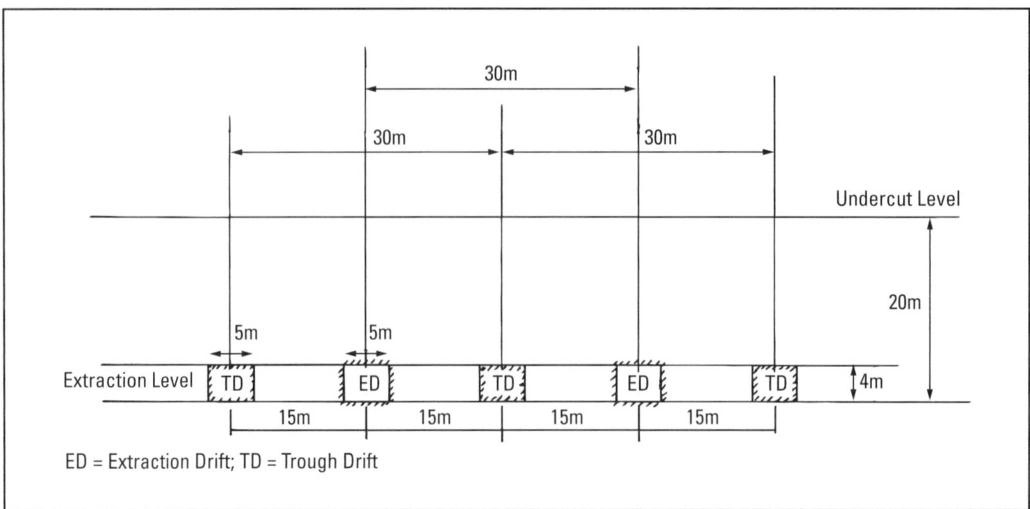

Source: Bullock and Hustrulid 2001

FIGURE 6.28 Separate-level undercutting and extraction

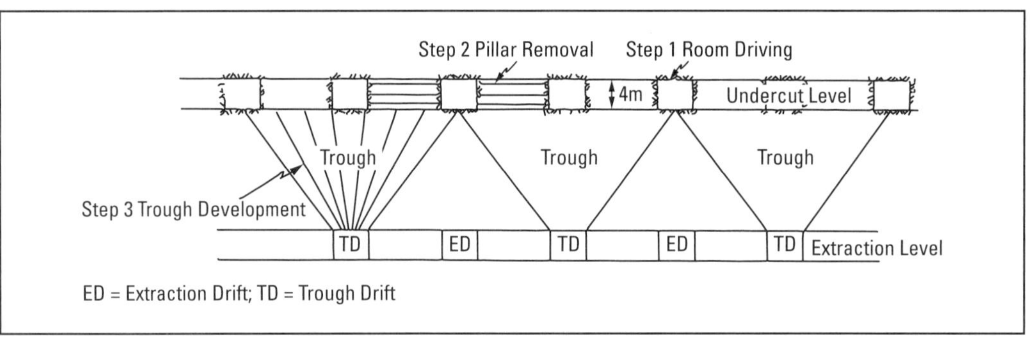

Source: Bullock and Hustrulid 2001

FIGURE 6.29 Undercutting and trough development

construction but the disadvantage that additional rock is extracted during the development process. This rock, if left in place, could provide extra stability to both the extraction drifts and the drawpoints. Drawbells are created rather than troughs. The first step in the drawbell construction is the driving of a drawpoint drift connecting adjacent extraction drifts. A raise is driven from this drift up to the undercut level. Fans of drill holes are then drilled from the drawpoint drift around the opening raise to form the bottom of the drawbell. Fans of holes are also drilled from the undercut drifts to complete the bell formation. A disadvantage with this design is that the amount of development and the level of workmanship required is higher than when using the trough design. As a result, it is more difficult to automate.

For all of the designs, it is important that a complete undercut be accomplished. If this is not done, then very high stresses can be transmitted from the extraction block to the extraction level causing major damage. Traditionally, the extraction level has been prepared first, followed by the creation of the undercut and the completion of the drawbells. This procedure does have many advantages.

Unfortunately, very high near-vertical stresses are created just ahead of the leading edge of the undercut. These stresses are transmitted through the pillars to the extraction level and can induce heavy damage to the newly completed level. The result is that repairs must be made before production can begin. The concrete used for making the repairs is generally many

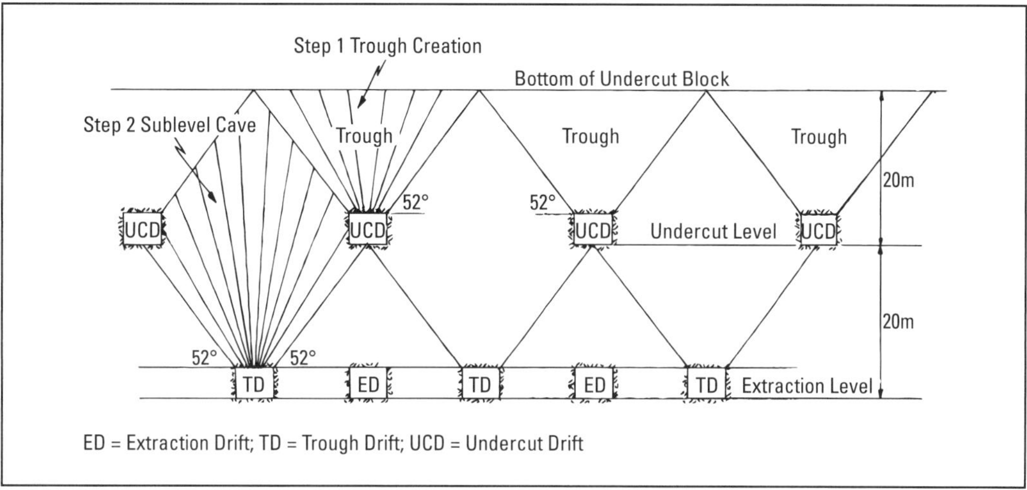

Source: Bullock and Hustrulid 2001

FIGURE 6.30 Two-level extraction and undercutting

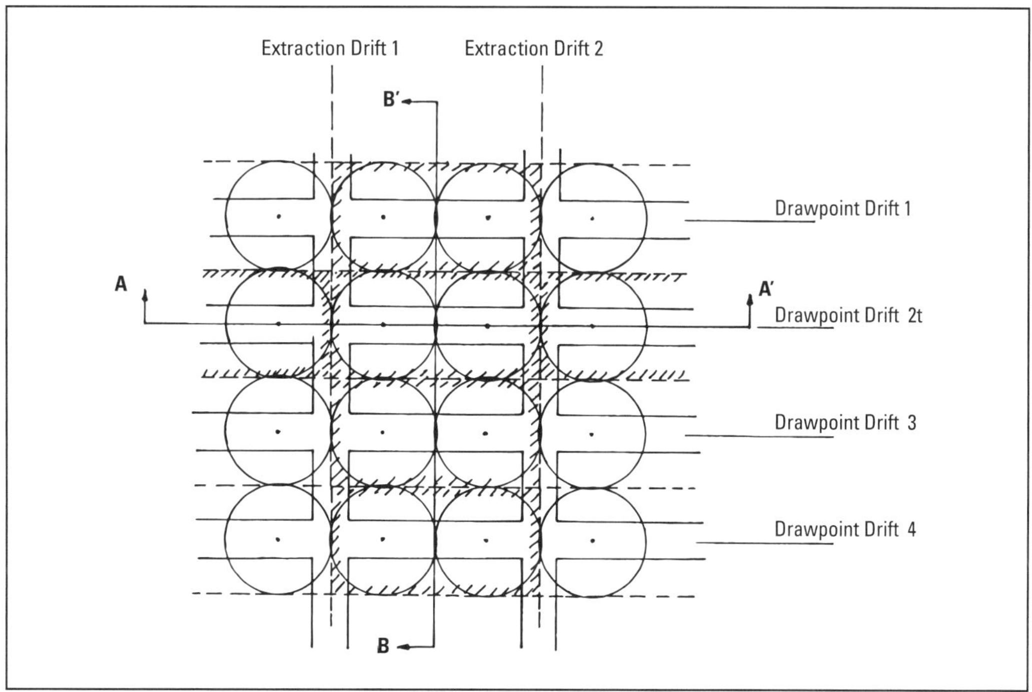

Source: Bullock and Hustrulid 2001

FIGURE 6.31 Plan view of traditional bell development

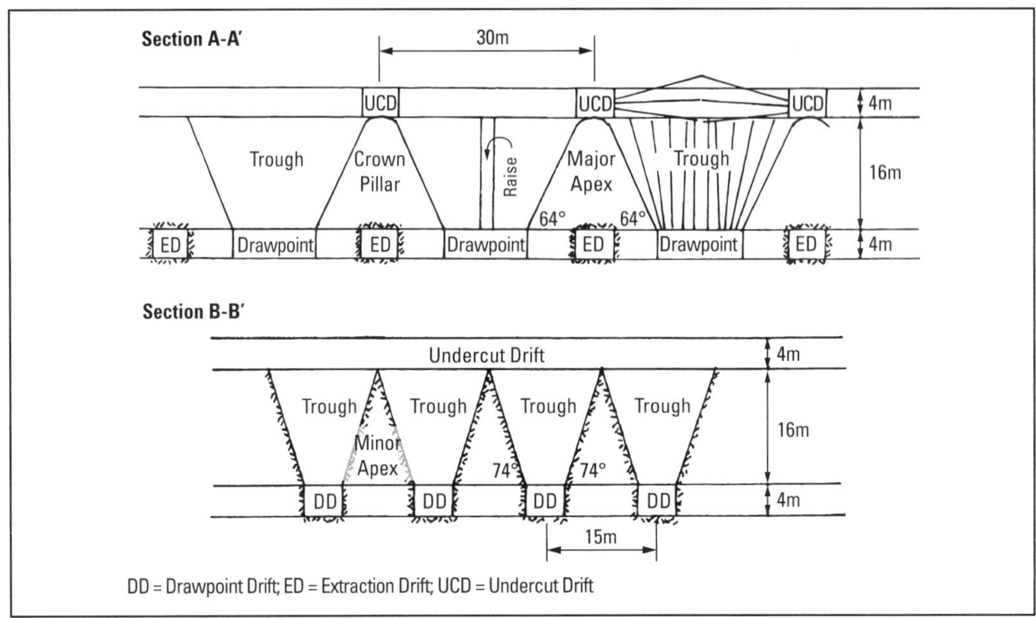

Source: Bullock and Hustrulid 2001

FIGURE 6.32 Sectional view of bell development

times weaker than the rock that has been broken, and the structural strength can never be completely restored.

An alternative to this procedure is to create the undercut first (advance undercutting), thereby cutting off the vertical stress. The extraction level is then created under this stress umbrella. Where this has been done, the conditions on the extraction level are markedly improved over those in which undercutting has been done afterward. There are pros and cons with both techniques, but advance undercutting will be the way of the future for most mines.

The Size of the Block

The size of the block refers both to the height of the extracted column and to the plan area. In the early days of block caving, the height of the blocks was on the order of 30–50 m (98–164 ft). With time, this has progressed to the point that extraction heights of several hundred meters are being used or planned. Obviously, as the specific development is inversely proportional to the height of the block, there are pressures to make the extraction units as high as possible. Naturally there are limits imposed by ore-body geometry, mineral types, and so forth. There are also limits imposed by the life of the extraction points. If the reasonable life of the extraction point is, for example, 100,000 t (110,200 st), there is no point in selecting a block height yielding 200,000 t (220,400 st) per drawpoint. Drawpoints can and are rebuilt, but it is best if they can last the life of the draw. As indicated in the introduction, most caving today is done in the form of panel caving rather than the caving of individual blocks. Once the initial cave is started, the lateral dimensions are expanded. Cavability is an issue affecting the size of the undercut that must be created to get a sustainable cave. Relationships have been developed relating the rock mass characteristics, the hydraulic radius (area/perimeter), and the ease of caving. It is possible, unfortunately, to begin initial caving and then for a stable arch to form. The undercut area must then be expanded and/or other techniques, such as boundary weakening, used to get the cave started once again. With a large enough undercut area, caving can be induced in any rock mass. Although necessary, it is not sufficient for successful block caving. The other factor is the degree of fragmentation that results. As the method is being considered for application to ever stronger rock types, both of these factors—cavability and fragmentation distribution—must be satisfactorily addressed prior to any method selection decision. Unfortunately, the database on which such decisions must be taken is very limited.

Cave Management

Cave management refers to keeping control over how much is extracted from each drawpoint each day. It involves many different factors. The rate of draw is an important parameter in planning the required area under exploitation. As the loosening of the fragments appears to be a time-dependent process, this must be recognized in planning the draw. The rate must not be so rapid that a large gap results between the top of the cave and the bottom of the block. A sudden collapse of the rock above can result in disastrous air blasts. In high-stress fields, it has been observed that too rapid a draw can result in rock-bursting conditions being created. In one section, there is a zone in which the height of the column under the draw area increases from near zero (where extraction is just beginning) to the full column height. This is followed by a zone in which the column height of the ore column decreases to near zero where extraction is complete. It is important to maintain the proper height of draw versus distance slopes in these two sections to avoid the early introduction of waste from above. Poor cave management can also mean the buildup of high loads in various areas and subsequent stability problems. Typical rates of draw as taken from the available literature are on the order of 0.3–0.6 m/day (1–2 ft/day).

The proper sequencing of undercut and extraction is a very important aspect of cave management. Unfortunately, design guidelines are difficult to obtain from the literature in this regard.

An important design consideration for the extraction level is the means by which oversize will be handled. There are several different problems to address. The first concern is the management of true hang-ups at the extraction points. Sometimes these can simply be blasted down by the careful placement of explosives. At other times, the boulders must be drilled first. This is not a simple procedure and involves dangers to workers and machines. The second concern is where and how to handle the *movable* oversize. These blocks can be (1) handled at the extraction points, (2) moved to a special gallery for blasting, (3) moved to an orepass equipped with a grizzly and handled there, or (4) directly dumped into an orepass for later handling. All variations are used, and each company has its own philosophy in this regard.

Initially, the sizes of the blocks arriving at the drawpoints are defined by the natural jointing, bedding, and other weakness planes. As the blocks separate from the parent rock mass, they displace and rotate with the loose volume occupying a larger volume than the intact rock. The swell volume is extracted from the extraction points, thereby providing expansion room for the overlying intact rock. Loosening eventually encompasses the entire column. As the column is withdrawn, the individual blocks abrade and split, resulting in a finer fragmentation than in the early part of the draw. The initial fragmentation corresponding to that caused by the initial fractures in the rock is termed *primary fragmentation*. As the column moves downward and new breakage occurs, the resulting fragmentation is termed *secondary fragmentation*.

Data concerning this transition from primary to secondary fragmentation are very difficult to obtain.

SUMMARY

This chapter has presented some of the design/layout aspects of the major mining systems used in underground mining. With this background, it is hoped that readers will better understand how to apply planning of time, materials, and workforce to develop costs and schedules of the appropriate design to the particular feasibility study on which they are working. All of these methods are discussed in much more detail in *Underground Mining Methods* (Hustrulid and Bullock 2001) and in the *Design and Operation of Caving and Sublevel Stoping Mines* (Stewart 1981), where there are many case studies of the various mining methods. The methods are again discussed in the *SME Mining Engineering Handbook* (Darling 2011), with many chapters written on the mine design and planning aspects.

REFERENCES

Alexander, C. 1999. Tunnel boring at Stillwater's East Boulder project. *Mining Engineering,* 51(9):15–24.

Anon. 1966. Cascade mining. *World Mining,* (June):22–23.

Brechtel, C.E., Struble, G.R., and Guenther, B. 2001. Underhand cut-and-fill mining at the Murray mine, Jerritt Canyon joint venture. In *Underground Mining Methods: Engineering Fundamentals and International Case Studies*. Edited by W.A. Hustrulid and R.L. Bullock. Littleton, CO: SME. p. 334.

Bullock, R.L. 1961. Fundamental research on burncut drift rounds. *Explosive Engineering,* 1 and 2 (January and March).

Bullock, R.L. 1975. Optimizing underground, trackless loading and hauling systems. Ph.D. dissertation, University of Missouri-Rolla.

Bullock, R.L. 1982. General mine planning. In *Underground Mining Methods Handbook.* Edited by W.A. Hustrulid. New York: SME-AIME. pp. 113–154.

Bullock, R.L. 1994. Underground hard rock mechanical mining. *Mining Engineering,* 46(11):1254–1258.

Bullock, R.L. 2011. Comparison of underground mining methods. In *SME Mining Engineering Handbook*, 3rd ed. Edited by P. Darling. Englewood, CO: SME. p. 397.

Bullock, R.L., and Hustrulid, W.A. 2001. Planning the underground mine on the basis of mining method. In *Underground Mining Methods: Engineering Fundamentals and International Case Studies.* Edited by W.A. Hustrulid and R.L. Bullock. Littleton, CO: SME. pp. 29–48.

Casteel, L.W. 1973. Open stopes–horizontal deposits. In *SME Mining Engineering Handbook*, Vol. 1. Edited by A.B. Cummins and I.A. Given. New York: SME-AIME. pp. 12135.

Cleland, R.S., and Singh, K.H. 1973. Development of post-pillar mining at Falconbridge Nickel Mines Ltd. *CIM Bulletin,* 66(732).

Cokayne, E.W. 1982. Sublevel caving: Introduction. In *Underground Mining Methods Handbook.* Edited by W.A. Hustrulid. New York: SME-AIME. p. 872.

Darling, P., ed. 2011. *SME Mining Engineering Handbook*, 3rd ed. Englewood, CO: SME.

Gignac, L.P. 1978. Hybrid simulation of underground trackless equipment. Ph.D. dissertation, University of Missouri-Rolla.

Hartman, H.L., and Mutmansky, J.M. 2002. *Introductory Mining Engineering.* Hoboken, NJ: John Wiley and Sons. pp. 324–327.

Hartman, H.L., Mutmansky, J.M., Ramani, R.J., Wang, Y.J. 1997. *Mine Ventilation and Air Conditioning*, 3rd ed. New York: John Wiley and Sons.

Hedley, D.G.F., and Grant, F. 1972. Stope and pillar design for the Elliot Lake uranium mines. *CIM Bulletin,* 65(723).

Hopler, R.B., ed. 1998. *Blasters Handbook*, 17th ed. Solon, OH: International Society of Explosives Engineers. pp. 455–460.

Hustrulid, W.A., and Bullock, R., eds. 2001. *Underground Mining Methods: Engineering Fundamentals and International Case Studies.* Littleton, CO: SME.

Kauffman, P.W., Hawkins, S.A., and Thompson, R.R. 1981. *Room and Pillar Retreat Mining: A Manual for the Coal Industry*. Information Circular IC-8849. Washington, DC: U.S. Bureau of Mines.

Korzeniowski, W., and Stankiewicz, A. 2001. Modifications of the room-and-pillar mining method for polish copper ore deposits. In *Underground Mining Methods: Engineering Fundamentals and International Case Studies.* Edited by W.A. Hustrulid and R. Bullock. Littleton, CO: SME. pp. 103–110.

Lane, W.L., Yanske, T.R., and Roberts, D.P. 1999. Pillar extraction and rock mechanics at the Doe Run Company in Missouri 1991 to 1999. In *Rock Mechanics for Industry, Proceedings for the 37th Rock Mechanics Symposium.* Vail, CO: American Rock Mechanics Association. pp. 285–292.

Lane, W.L., Yankske, T.R., Clark, L.M., and Roberts, D.P. 2001. Pillar extraction and rock mechanics at the Doe Run Company in Missouri 1991 to 2000. In *Underground Mining Methods: Engineering Fundamentals and International Case Studies*. Edited by W.A. Hustrulid and R.L. Bullock. Littleton, CO: SME. pp. 95–102.

Langefors, U., and Kihlstrom, B. 1963. *The Modern Technique of Rock Blasting.* Stockholm: Almqvist and Wiksell. pp. 230–257.

Ovanic, J. 2001. Mining operations at Pea Ridge Iron Ore Company: A case study. In *Underground Mining Methods: Engineering Fundamentals and International Case Studies*. Edited by W.A. Hustrulid and R. Bullock. Littleton, CO: SME. pp. 229–234.

Ozdemir, L. 1990. Principles of mechanical rock cutting and boring. In *Mechanical Excavation Short Course.* Golden, CO: Colorado School of Mines. p. 2.

Parker, J. 1973. How convergence measurements can save money: Practical rock mechanics for the miner, part 3. *Engineering and Mining Journal,* 174(8):92–97.

Ramani, R.V., ed. 1997. *Proceedings of the 6th International Mine Ventilation Congress.* Littleton, CO: SME.

Snyder, M.T. 1994. Boring for the Lower K. *Engineering and Mining Journal,* 195(41):20ww–24ww.

Spearing, A.J.S. 1995. *Handbook on Hard-Rock Strata Control.* Special Publication Series SP6. Johannesburg, South Africa: South African Institute of Mining and Metallurgy. pp. 89–93.

Stefanko, R., and Bise, C.J. 1983. *Coal Mining Technology: Theory and Practice*. New York: SME-AIME. p. 102.

Stewart, D., ed. 1981. *Design and Operation of Caving and Sublevel Stoping Mines.* New York: SME-AIME.

Tien, J.C., ed. 1999. *Proceedings of the 8th U.S. Mine Ventilation Symposium.* University of Missouri-Rolla.

Weakly, L.A. 1982. Room and pillar ground control utilizing the grouted reinforcing bar system. In *Underground Mining Methods Handbook.* Edited by W.A. Hustrulid. New York: SME-AIME. pp. 1556–1560.

Weigel, W. 1943. Channel irons for roof control. *Engineering and Mining Journal,* 144(5):70–72.

Weiss, P.F. 1979. Development system for block caving under severe conditions. In *Design and Operation of Caving and Sublevel Stoping Mines*. Edited by D.R. Stewart. New York: SME-AIME. pp. 143–146.

Wycoff, B.T. 1950. Wrapping pillars with old hoist rope. *Transactions of AIME,* 187:898–902.

Zipf, R.K. 2001. Pillar design to prevent collapse of room-and-pillar mines. In *Underground Mining Methods: Engineering Fundamentals and International Case Studies.* Edited by W.A. Hustrulid and R.L. Bullock. Littleton, CO: SME. pp. 493–496.

Zipf, R.K., and Mark, C. 1997. Design methods to control violent pillar failures in room-and-pillar mines. *Transactions of the Institution of Mining and Metallurgy Section A,* 106:A124–A132.

CHAPTER 7

Planning the Mineral Processing Plant

Mark A. Anderson

Key to the extraction of minerals and metals from their ores is the geological and mineralogical foundation of their discovery. With proper identification and characterization of the potential ore types found in a mineralized body, the methods of metallurgical extraction and process plant design can proceed with varying degrees of reliability. In the early stages of project development (i.e., preliminary feasibility or order-of-magnitude study), the geological, mineralogical, and mining data may be very vague and lacking adequate data as to representation. Nevertheless, the metallurgical engineer must plan the extraction techniques appropriate for the level of data reliability, knowing full well that the eventual ore body may bear little resemblance to its original characterization. The statements within this chapter on geology or mining apply to the processing of the mineral product, even though this may not be the subject of discussion.

The engineering and design information used in the establishment of the metallurgy and process plant design can be simple or complex, inexpensive or extremely costly, and well established or on the cutting edge of technology. Regardless of the relative stage of project development, the basic building blocks of the metallurgical and plant design effort remain constant. Only the reliability of the data and associated confidence levels change. This chapter identifies the basic building blocks of a successful metallurgical and process plant design and attempts to quantify, in a very subjective manner, the level of confidence for each.

The genesis or progression to the final feasibility-level study is usually preceded by very early-stage studies typically identified as a *preliminary feasibility*, *scoping*, *order-of-magnitude*, or "*back of the envelope*" study followed by *intermcdiate feasibility* (or *prefeasibility*) *study* and then the *final feasibility study* proper (see Table 1.1). These steps are not cast in stone or even required by all institutions but rather suggest the more typical progression of projects from their discovery to approval by company management and/or financial institutions. The reader should not be confused by other levels of progression such as *indicative*, *interim*, *budget*, and so forth. These levels are all used periodically and accomplish the same end. What is important is that the reader understands the changing levels of confidence as a project advances through its stages of development.

The metallurgical building blocks of all feasibility studies include the following titles or areas of study and influence. They may involve simple statements or represent the culmination of exhaustive testing and design. In any case, each must be as fully discussed as the data suggest in each of the reports leading up to a final decision to proceed with a project. Following are the basic building blocks of a feasibility study:

- A geological description of the discovery
- A description of the exploration program as it relates to metallurgical sampling
- A brief discussion of the proposed mining plan and methods as they relate to processing (The proposed mining plan is used to schedule metallurgical recoveries and concentrate grades over the term of the project.)
- The sampling protocol and procedures used to establish representation
- Physical characterization of the known mineralization types, including mineralogy, ore hardness, specific gravity, screen analyses, liberation studies, and unique properties such as magnetic susceptibility, radioactivity, and so forth
- The metallurgical testing program, including laboratory and pilot-plant studies to establish the comminution, screening, classification, extraction, dewatering, and proposed tailings disposal techniques for the various mineral types
- The recommended flow sheet and material balance
- The recommended mineral or metal recoveries and concentrate grades for each mineralization type identified
- The operating costs associated with principal unit operations of the process and including process administration, engineering, and other overheads
- An evaluation of the infrastructure required to support the processing operations
- Capital costs for process engineering, metallurgical development, plant construction, tailings disposal, and concentrate or product transportation facilities
- A recommended process plant organization, including administration, metallurgical engineering, assaying and quality control, process operations, and process maintenance
- A description of process and maintenance control programs, including instrumentation, computerized process control, preventive maintenance scheduling, and so on
- A realistic quantification of the technical and economic risks associated with the elements of the process operations

PRELIMINARY FEASIBILITY STUDY

A preliminary feasibility study is often initiated when exploration organizations have had some success in assembling sufficient data to outline a mineral resource. Using intuitive reasoning rather than factual data, assumptions about future project success are used to generate an initial view of the prospect's expected production and economic viability. Studies of this sort are also used to advance plant expansion or modernization projects that are not exploration driven, although they could be. The levels of accuracy associated with preliminary feasibility studies are often quoted in the range of plus 50% to minus 30%. At these levels, accuracy determinations are a waste of effort. It would be better to quantify them as unknown and assume that the project is simply a good technical or economic bet with fairly long odds. A fair representation of the position of an order-of-magnitude (or preliminary feasibility) study is stated by Mular and Bhappu (1980) as follows:

> *Shortly after discovering a mineral deposit and analyzing the results of initial sampling, an economic evaluation should be made to determine if this deposit has possibilities for a*

viable project. As practically no information is available, this evaluation must be made entirely on assumptions, which should be on the optimistic side. If the economics are unsatisfactory after using these optimistic assumptions, then it is prudent not to spend any additional money on exploration on this deposit.

Projects that are not exploration driven but rather incorporate plans for modernization or expansions can be quantified in a similar manner. This step is often called the "sniff test" and can be used to save considerable, valuable time and technical effort.

Project Organization

One of most frequently overlooked keys to project development success is the formation of a project team at the earliest possible date. It is highly likely that the project is still being managed by an explorationist if it is a new discovery or within the confines of the mine site organization. In either case, it is critical that the four key legs of the project have an assigned representative to ensure the melding of environmental, geologic, mining, and metallurgical efforts. It is likely that all of the other facets of the potential project can be addressed by this group at this time.

Description of the Project Organization

A major mining company will have at its disposal the resources to staff a project at all levels, from inception to final completion. In today's rapidly changing environment, it is likely that the project is being advanced by a small entrepreneurial exploration company with no ready access to mining, metallurgical, environmental, or cost estimating personnel.

The established mining company can be expected to assign personnel to the project on a full- or part-time basis to gain a coordinated effort among all disciplines. What is important is that all of the disciplines be represented and that their tenure with the company be long term. In this manner, management can be kept well apprised of project progress within a multitude of reporting systems from all disciplines for the length of the project.

In the case of a small exploration company, or small mining company with limited staff, it is likely that outside consultants be chosen to guide the work. The project leader, usually a geological type, must be trusted to form strong technical relationships with mineral resource, mining, metallurgical, environmental, and economics personnel to obtain a project completion design. It is important that the project leader not have a vested or "founding father" interest in the project to maintain an unbiased evaluation of the mineral property.

Qualifications of Assigned Personnel

Geologic, mining, metallurgical, environmental, and economic contributors will almost certainly be persons with a depth of knowledge of the minerals industry to allow them to make long-range decisions based on what is certain to be a limited database. These individuals will have well-established track records in bringing projects to completion and will be capable of forming long-range project development plans to attain the next level of certainty.

Since the emphasis of this section is on the metallurgical component of a project study, it serves the reader to be reminded that meaningful metallurgical forecasts and predictions are the forte of experienced mill process engineers. The process engineer will be able to discern, by virtue of experience and with a high degree of certainty, the probable success or failure of a project based on what may be viewed as meager data.

Geological Information

The process personnel faced with the challenge of predicting a range of likely outcomes for a project with a very limited amount of data will require considerable insight from their geologic counterparts. Geological information, even though limited to a small number of samples, must be relied upon for ore type, mineralogy, and physical characteristics information. From this limited database, the metallurgist will predict possible treatment flow sheets and attempt to estimate capital and operating costs. If the potential process design yields a concentrate that will require downstream processing, the metallurgical engineer will scrutinize assay and mineralogical data for clues as to possible deleterious contaminants and trace minerals that could lead to adverse effects in process metallurgy and smelter penalties.

Project Location

The project location will yield certain clues as to foundation design parameters such as possible mill site locations, the availability of power, water and transportation, and the existence of other historical operations in the area. All will have a subjective use in assembling a valid prediction of future metallurgical performance. As an example, freight for both incoming equipment and outgoing product delivery can become an outsized contributor to capital and operating costs if all process units must fit in the airframe structure of a C-130 transport aircraft or be capable of being transported by small equipment over primitive roads.

Geological Description

Although to most metallurgists the geologic description of the project may be turgid prose, it will nevertheless yield significant clues to expected metallurgical results. Both mineral recovery and product quality will be affected by the relative fracturing and faulting of a resource. Intense clay alteration will serve as a warning to those contemplating both heap leaching and flotation operations. A recent project contained a reference to the mineral creedite ($2CaF_2{\cdot}2Al(F,OH)_3{\cdot}CaSO_4{\cdot}2H_2O$) in the geologic description. Previously thought to exist only in Colorado and Kazakhstan, the mineral turned up in a proposed copper heap leach in Central Nevada. The mineral is acid soluble and yielded prodigious amounts of fluorine to the leach solutions, which affected both bacteria mortality and cathode quality. Had the metallurgist been looking for problems, the mineral may have been spotted, or, better yet, the geologist may have tipped off his counterpart on the project team. In this case, the project was compartmentalized and no meaningful fatal-flaw discussions were held among geologic, mining, and metallurgical personnel.

Exploration Program

From a process design standpoint, the infancy of an exploration program will be a time of limited information. Geologists will be more interested in tons and grade than with the expectations of the metallurgist. However, if dialogue is maintained, even in the early days of the project, sampling and drilling programs can yield results that can be interpreted and forecasted by the metallurgist.

The limited metallurgical testing, usually only bench-scale tests, will nonetheless be used by the metallurgist to project success or lack thereof for the project. If the testing indicates very favorable recoveries and product qualities, the metallurgist can gain confidence in the deposit.

If mineral recovery and product quality are tenuous from the start, it is usually a sign of more bad news to come.

Sampling and Assaying

Working with a limited number of tests, the metallurgist can gain comfort from the repeatability of assays for both mineral samples and products of metallurgical tests. If sufficient samples are available, the metallurgist can work with both the geologist and mining engineer to pick samples that represent the whole of the known resource and use these samples to the best advantage in obtaining possible flow-sheet alternatives for consideration. Most resources in the present economic climate will not tolerate the costs of cutting-edge technology or complex treatment schemes. Good sample representation, even at the earliest stages of a project, can prevent future process failures and a run-up in metallurgical testing costs.

Physical Characterization

Enlightened explorationists will include the observations of both field exploration and mineralogy-based personnel in their review of a potential resource. The metallurgist can examine this work and gain insight into the possible requirements for mineral liberation, mineral recovery, percolation rates, reagent requirements, environmental requirements, and tailings disposal. The ability to delve into these observations from a geologic report requires astute observations by all of the personnel involved; quality communications between the mining, metallurgical, environmental, and geological personnel; and a depth of experience that yields wisdom from comparable past experiences. If a mining professional with more than 30 years of experience has not seen or heard of a particular problem, it can be expected to be a big one. This is analogous to a paraphrased Murphy's Law, which states that if something unknown is going to happen, it will probably be bad for all concerned (or more commonly stated as "anything that can go wrong will go wrong").

Mining

The process engineer or metallurgist must rely heavily on the prognostications of the mining engineer from the earliest stages of the project. A well-experienced mining engineer will serve as the intermediary between the geological and metallurgical engineers and must be relied upon to furnish key data to both individuals. Exploration drilling programs will be guided by the mining engineer and directed by the geologist. The metallurgist will rely on both for foundation data upon which to base predictions of recovery, product quality, and costs.

Again, it must be argued that at this level of project development, it is the sum of several professional opinions that provides the foundation for a decision to proceed or to pack up and go home. If the mining engineer deems a project uneconomic or impractical to mine, it is highly unlikely that any geologic or metallurgical wizardry will save it.

Mining Method

Although seemingly basic, the choice of future mining method by the mining engineering contributor will provide a series of limits to the metallurgist's field of play. An underground development will, in most cases, provide limited ore tonnage to treat (i.e., less than 10,000 stpd, or short tons per day; smaller sized rock delivered to the crushing circuit; higher ore grades; and a plethora of contaminants, including oil-fouled mine water, blasting remnants, tramp steel, concrete, and timber). All of the contaminants will, in one way or another, hinder the effective operation of most mill circuits and will, at the very least, contribute to an increase in capital and operating costs. In most underground operations, the mine will lag the mill in coming up to ultimate production levels and will require a high level of planning and execution to maintain even throughput rates.

Open pit operations will yield higher throughput rates, larger-sized rocks, smaller quantities of contaminants, lower ore grades, weather-induced delays, and more highly variable moisture contents. Open pit mining operations are usually capable of outrunning the mill and will come to projected operations levels much quicker.

Production Plan and Mining Rates

In an early-stage project, the mining engineer will produce a "best guess" production schedule for the resource as it is known at the time. The annual plan will likely assume a uniform ore grade to the mill with no outside influences on production tonnage rates. The metallurgist will base the annual project production on the production plan and mining rate and adjust the flow sheet and product delivery strategies accordingly.

Underground development projects, particularly vein-type deposits, have a tendency to be designed at unrealistically high levels during the preliminary stages of a project. This is due to the inherently high costs of underground mining and the attempt to mitigate those higher costs by upping the tonnage divisor. This is not as much of a problem with the development of room-and-pillar stoping, since given that this type of mining requires a minimum of mine development to reach full production very quickly. An exception is when the mine expectedly encounters large inflows of water. Open pit operations usually start out at a fairly realistic level and then are escalated to take advantage of the largest equipment units available. In either case, the metallurgical engineer will often find his or her early work unsupportable and will have to change to meet the realities of the mining situation.

It is at this time that communication between the project contributors is critical. Having a mill capacity of 50,000 stpd and a mining rate of 10,000 stpd is unproductive.

Mining–Metallurgical Interface

The mining engineer–metallurgical engineer interface is a natural one. For the most part, the two disciplines share considerable course work in their training, speak the same technical jargon, and may be involved with the project well into its design, construction, start-up, and operation. The exploration geologist, on the other hand, should be transferred to another exploration venture and be replaced by an experienced mine geologist for expert mine geology advice. From this author's experience, this transfer is needed because the founder of the resource will have difficulty remaining unbiased with the many judgment decisions that must be made and it will avoid confrontation from other members of the project team. At this stage of a project, the metallurgist seeks out the geologist to gain insight and understanding and connects with the mining engineer because they will be together in explaining the success or lack thereof for the project.

Metallurgical Testing

It must be assumed that there exists a certain number of samples from either bulk sampling or drilling from which to gain some metallurgical insight into a project. In some cases, the metallurgist will be faced with the restart of a shutdown operation and will have a wealth of experience to draw from. This is the exception rather than the rule.

The challenge to the metallurgical engineer is to use the samples available to the best advantage. This means carefully evaluating the input of the geologist, mining engineer, and environmental engineer; making a subjective decision as to what will probably work; and then designing a small test program that will provide clues as to the future flow sheet and expose as many "problem areas" as possible.

Historical Laboratory Testing

Unless the project involves a change to an existing operation, a restart of operations, or a development within an existing district, there will be no historical record to rely on when making prognostications in the early stages of the project. The metallurgist must make value judgments based on personal experience and the recommendations of geologic, mining, and environmental personnel assigned to the project.

Sampling and assaying. If historical records are available, the metallurgist should try to determine their applicability to the potential resource being studied and to weigh predictions of sample representation upon those areas that can be safely regarded as at least similar to the current resource.

Comminution. At this stage in the project, historical and district records can be used to advantage in adjusting metallurgical parameters to fit. Large, homogeneous mineralized areas, as are typified by certain iron ore formations, limestone, trona, and so on, can usually be relied on to give the metallurgist a fairly good estimate of crushing and grinding conditions and the steps required to reach liberation size. Greenfield projects are highly vulnerable to poor estimates at this stage of the project. Strong collaboration is required between the geologist, mining engineer, and metallurgist to arrive at approximations of resource crushability and grindability. Although it is patently unfair to be too conservative at this stage of the project, the metallurgist must attempt to err on the pessimistic side. The large capital and operating cost contribution of the comminution circuits demands a bit of cynicism. It is also important to remember that the review of historical data will be tempered by contemporary testing and serves as a good reference point to judge the efficacy of current work.

Screening and classification. Certain problem ores have a history of preserving their reputations from earlier experiences. If historical operations were plagued by wet and sticky ores or generated very little undersize, the metallurgist can assume that the project will be inconvenienced by the same problems. Not to be ignored is the effect of weather and its historical impact on screening operations in particular. Certain concentrator operations that are located in semiarid environments have experienced all of the trials and tribulations of an operation located in a tropical rainforest. In Arizona, the monsoon season is aptly named.

Extractive techniques. If preexisting operations used flotation or if a district is populated with flotation mills, it can be safely assumed that the project under study will probably be the same. A notable exception to this statement is the changing face of the gold industry. Historical gold operations that relied on gravity concentration, flotation, or direct cyanidation may now be economical as low-grade heap leach projects. It is also quite likely that today's proposed extraction technique will be a combination of all of the previously mentioned methods.

Dewatering. Filtration and thickening records are useful in planning for flow-sheet adaptations to handle problem ores. More than likely, this area will have the least amount of collective information and the metallurgist will need to use experience and limited current test work in preparing for dewatering.

Concentrate grade. The historical record, if available, can prove invaluable in this instance. Contemporary testing will be limited to very few tests, and strong operational or district experience is invaluable. If local ores are known to contain extraordinary levels of arsenic or other smelter penalty elements, the metallurgist must assume "bad luck" and design accordingly. A large restart of a copper porphyry operation in Nevada ignored the historic problems with concentrate grade and ended up with a marginally economic operation that no amount of metallurgical manipulation could redeem. An industry feeling expressed by a Michigan Tech professor, Duane Mark Thayer's corollary to Murphy's Law, states that "old metallurgists are considered morons" and "bad things are guaranteed to happen to the enlightened minds of today." Or restated, "Those who fail to learn from history are bound to fail."

Recovery. The same discussion of historical evidence as was made for concentrate grade can be made for recovery. Historical recovery problems have probably not gone away and will have to be addressed today. Again, comparisons of historical records with contemporary testing may reveal change or simply cause confusion. In any case, the metallurgist must use the historical data to frame his or her current-day predictions and forecasts.

Reagents and consumables. Actual plant reagent and consumable consumption rates are invaluable in forming the basis for current predictions. The metallurgical engineer can profit from historical work by using experience to provide the initial suite of chemicals proposed for current testing.

Historical Pilot-Plant Testing

The existence of historical pilot-plant results is likely to be problematic. However, the current project may have been taken through pilot testing by a previous owner or even by another generation of employees of the current company. Given adequate assurances of sample representation, the results would be invaluable in reinforcing the validity of an early-stage project study.

If pilot-plant results are available, the metallurgist can use them to confirm suspicions about a probable flow sheet; estimate power, wear steel, and reagent consumption; and obtain a reasonable fix on recovery and mineral concentrate grade.

Contemporary Laboratory Testing

Projects at this level of development are in search of funding from either a parent company or financial backers of an entrepreneurial exploration company. The results are used as a justification to spend further money on exploration and development and not to proceed with design and construction. To that end, the metallurgist is expected to be realistically optimistic, keeping in mind that money can be made or saved by either a negative or positive report.

The opportunity for metallurgical testing will be limited to a small number of samples or drill-hole intersections, and the metallurgist will be primarily concerned about the work representing the resource as it is currently known and not the fully developed ore reserve at some future date. The key again is teamwork. Working together, the metallurgist, geologist, and mining engineer obtain as good a sample for metallurgical work as possible. If different ore types have been recognized, they should be tested independently prior to compositing. The testing will likely be confined to laboratory open-cycle and locked-cycle tests to determine reagent consumption, process conditions, concentrate grade, and recovery. If the metallurgical test work is marginal or poor, it may be an indication that no more money be spent for drill holes until the causes of such problems are identified and possibly remedied.

Sampling and assaying. The metallurgist has the opportunity to participate in the selection and assembly of samples for laboratory testing. To this end, it is best to ensure that the most representative sample possible is obtained and the limited number of tests yield results that can be used for global estimates of possible mill performance. The metallurgist must rely on the geologist to ensure quality sampling and assaying techniques so that the completed metallurgical tests will have meaningful metal and material balances.

Comminution. It is highly unlikely that enough time will be available to do detailed comminution testing. If samples and time are available, first-pass crushability and grindability tests can be run. In most cases, the metallurgist will collaborate with the geologist and mining engineer to assess rock quality and probable crushing and grinding requirements. Also available will be first-pass mineralogical examinations of drill-hole footages. If performed by a qualified mineralogist, the mineralogy review will include information on fracturing, mineralization, liberation size, occurrence of free metals, levels of clay alteration, and so on. Utilizing experience and the information given by other team members, the metallurgist can make an educated projection of crushing and grinding requirements.

Screening and classification. No special screening or classification tests will be run. The exception is fractional analysis of head, concentrate, and tailings products from the metallurgical tests. Mineralogical information, particularly concerning fracturing and clay alteration, can give the metallurgist a heads-up regarding possible screening problems. At this point in the project, the classification circuits will be assumed to present no problems.

Extractive techniques. Now the metallurgist must control testing closely. Given limited budgets, samples, and time, there exists little room for research. This is mentioned only to guard against the propensity of some industrial testing facilities to try every unit operation known to mineral processing simply to come to the only practical conclusion that was available at the onset of the project.

The differentiation of gravity separation, magnetic separation, flotation, and leaching as process techniques is usually quite clear very early in the project work. It is amusing and a little scary to observe limited project funds being spent on gravity concentration of porphyry copper ores with head grades of less than 1%, yet it happens frequently.

It is more important that the limited number of tests be designed to provide the maximum amount of information. Favorable mineral or metal recoveries coupled with initially high concentrate quality provide an adequate foundation for the metallurgical forecasts required at this level of a study.

Dewatering. Filtration and thickening testing will usually not be run at this stage of a project. Clues about concentrate and tailings thickening will usually emerge from the laboratory testing. These clues, along with the metallurgist's experience, will have to suffice for the foundation design process selection at this time.

Concentrate grade. The concentrates made from laboratory tests should be submitted for a complete study using inductively coupled plasma. The results will give an early indication of the probable willingness of smelters to take the material and will provide a basis for estimating smelter terms, payables, and freight. This is probably the most overlooked area in an early-stage report. Even though final estimates are not available, the project team and management need to be alerted as to potential problems that could be revealed later. More than one project has been abruptly ended by an overlooked arsenic, fluorine, or insolubility problem.

Recovery. It is highly probable that only a few tests will be available to use in the evaluation. Consistently low tailings assays (high recoveries) are usually a good sign of probable project success. Mixed results demand more work before advancing the project. To simply say or assume that the metallurgical problems can be fixed is nonsense. Many projects in the minerals industry have been developed that failed as an operation because of overoptimistic mineral recovery projections in the feasibility studies.

Locked-cycle tests, which confirm projections from open-cycle tests, are particularly useful in predicting future mill performance. Deciding how to include the values attached to intermediate products is the forte of an experienced mill metallurgist. It is usually reasonable to assume that locked-cycle tests will improve over open-cycle tests and that pilot-plant and mill results will exceed laboratory findings. Mills that confound their designers and operators with poor performance usually have side-issue problems, such as poor sample representation, application of poorly understood "cutting-edge technology," slurry rheology, and so on.

Reagents and consumables. The limited number of testing opportunities will also demand that the initial reagent suite be functional—not optimum, but functional.

Metallurgical and Process Criteria

The preliminary feasibility (order-of-magnitude or scoping) study will be used to provide the basis for a proposed flow sheet and material balance and to support a capital cost estimate and plant operating costs. Again, limited data in the hands of an experienced engineer can be used to form a vision of what a potential operation may look like and what its potential returns may be. The process engineer will be expected to come to reasonable conclusions for each of the following parameters, depending on the type of resource under study:

- Recovery
- Concentrate grade
- Process residence times
- Abrasion index
- Crushability
- Grindability
- Magnetic susceptibility
- Reagent consumption
- Water requirement
- Electric power requirement
- Wear steel consumption
- Concentrate moisture
- Ore specific gravity
- Evaporation rates

Flow Sheet and Material Balance

At this stage of project development, the flow sheet and material balance will be very general in nature. The flow sheet will usually consist of a block diagram with certain major pieces of equipment noted. The more thought and detail that the process engineer can build into this portion of the study, the more directed future work will be. For example, if a 124-mi (200-km) pipeline is required to supply water to the project or to convey concentrates to a port, it is important to annotate that fact here.

The flow sheet, even though conceptual, will very often resemble the final version. The inclusion of as much forward thinking as possible will serve to provide the basis for planning and completing metallurgical and laboratory testing in the advanced stages of the project. Figure 7.1 is a partial example of the form a preliminary flow sheet may take.

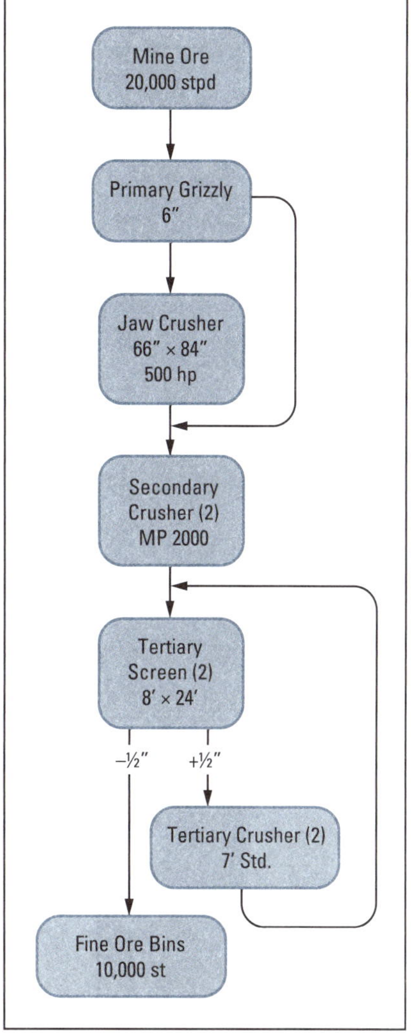

FIGURE 7.1 Simple flow-sheet example

Process Description

The process description will, as in the case of the flow sheet, be more the product of the process engineer's imagination than a detailed unit-by-unit description as is found in the final versions. At this stage, the metallurgist should describe the possible flow sheet in terms of the major units of process equipment and the materials handling systems that connect them. Given a mining rate by the geologist and mining engineer, the process engineer can illustrate a basic process, identify the major unit operations, assign a possible size to each unit, and briefly describe the results from each unit operation. Fair representations of costs can be acquired by also obtaining telephone or budget quotes on the major pieces of equipment and the materials handling systems. These costs, which are usually high, can then be factored to yield a possible range of future plant construction costs.

Process Plant Infrastructure

At this stage of the project, the process engineer should identify the concentrator infrastructure in sufficient detail to provide a basis for making cost allowances for each. Included in typical mill infrastructure are the following:

- Assay and metallurgical laboratory
- Compressed air system
- Instrument air system

- Dust collection system
- Process water supply
- Communications system
- Process control instrumentation
- Maintenance shops and equipment
- Mill transport and service vehicles

The allowances for the preceding systems and equipment may be quite definitive as in the case of vehicles, or the product of historical data for plants of the suggested size. What is important is that all areas are addressed and not necessarily the ultimate accuracy of the estimate.

Tailings Disposal

All process operations must address the subject of tailings. The tailings can range from the storage of concentration products to detoxified heaps. In the early stages of the project, the process engineer must collaborate with the geologist and mining engineer in establishing the most convenient areas for the eventual storage of waste products and attempt to visualize the eventual storage methodology. The early-stage site visit should yield a good view of the topography of the site, identify areas with sufficient suitable area for accommodating a tailings receptor, and determine the approximate size of a starter tailings facility.

Possible Design Configurations

Frequently there will exist, in close proximity to the mill operations, an area that requires minimal additional structure to be turned into a tailings repository. Beginning with a starter dike, usually of mine run materials, the project can then expand with a containment made wholly from cycloned tailings. The initial capital estimate will include the cost of the starter dikes and piping systems. Given the costs associated with closure, it may be that lining the facility will provide insurance against a long-term problem of releasing contaminated water into the groundwater system.

Tailings facilities located in arid areas may be constrained in size by filtering the tailings prior to deposition in conjunction with placement in a lined facility.

In areas of high seismicity, the range of tailings dam construction techniques will be limited. The engineered design will probably demand dam construction methods that are associated with water retainment structures, which are structures that are maintained with low phreatic levels and high tolerance for seismic events and/or large rainfall events. As an example, fully compacted tailings are required for the coarse fraction spigoted on the downstream side of tailings dams in Chile, where dam failure would result in a catastrophic event (R.L. Bullock, personal communication).

At the early stage of development, the metallurgist needs to obtain visionary consultation from well-qualified geotechnical support personnel. A simple, low-initial-cost facility may turn into a highly complex engineered structure that demands engineering and operational control over its entire life. The lack of adequate planning for tailings deposition can ruin a project operationally and economically if not adequately planned for in its early stages.

INTERMEDIATE FEASIBILITY STUDY

From the processing standpoint, the intermediate feasibility (or prefeasibility) study level of project development differs little from the final feasibility study. However, advanced pilot-plant work, perhaps some laboratory test work associated with slurry rheology (i.e., concentrate and tailings thickener design), detailed engineering, and complete ore characterization studies, are not covered in the intermediate feasibility study. In many instances, depending on the size of the deposit under study, the metallurgical characteristics of the ore body are fully understood and the final feasibility study will only represent fine tuning and optimization.

Project Organization

The project organization, at this level of development, should represent the personnel that will be charged with the design, construction, and start-up of the project. Small projects (<10,000 stpd) may only require the future project manager, chief geologist, mill manager, mine manager, environmental manager, and maintenance manager. This team will remain intact until a final feasibility study has been approved and the project brought to operational status.

Description of the Project Organization

The project organization will begin with key engineering and operations personnel, each of whom has had pertinent operational experience and can function well as a team. They will direct the remaining exploration and engineering tasks and ensure that a smooth transition to full operations is made. Smaller mineral companies usually attempt to save money by either neglecting this requirement altogether and trusting the work consultants or loading up too few individuals with an extraordinary amount of responsibility. The outcome is usually an ill-prepared management team, a poorly executed design, an extended start-up period, and below-target performance for several months of operations.

The project team must be able to visualize the future operation, interface with the engineering contractor to build in the factors necessary for a successful design, and plan for the full staffing of the project at some point in the genesis of the project. Figure 7.2 is a simplified organization chart for a client organization.

Qualifications of Assigned Personnel

The persons identified in Figure 7.2 will have senior-level experience in engineering, management, and operations. They will be able to direct the necessary in-house engineering and work to be accomplished in completing the task and will interface with the engineering and construction firm on a routine basis.

Geological Information

At this stage of the project, the geological and exploration work will be essentially complete in sufficient detail to support the initial years of the project. Exploration drilling will be directed to filling in previous work, providing geotechnical data, and creating composites necessary for metallurgical laboratory and pilot-plant testing.

Project Location

The information contained in this section will likely be unchanged from the preliminary feasibility study. However, exploration drilling may not be able to sterilize the chosen sites for

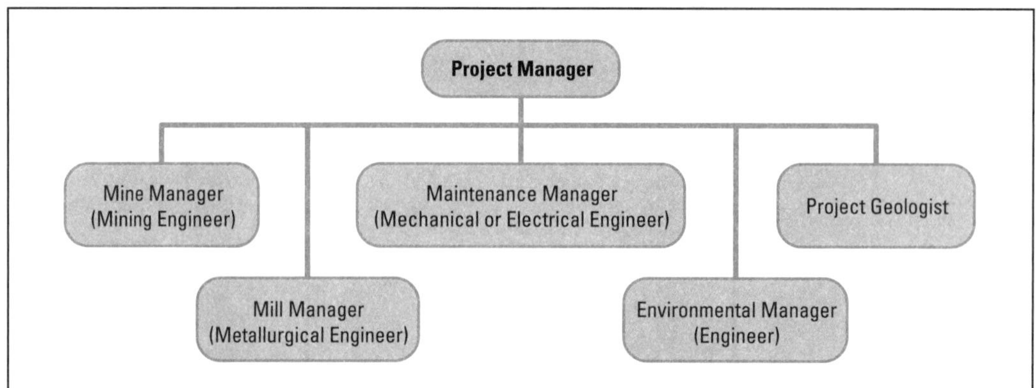

FIGURE 7.2 Project organization chart for the intermediate feasibility study

mill and maintenance facilities, necessitating major increases in costs and/or extended project schedule times.

Geological Description

One of the challenges presented by the preparation of a series of feasibility studies at varying levels of confidence is that the situation is not static. Exploration often continues through the intermediate and final feasibility studies and frequently presents different glimpses of the same ore body as geologic information is expanded. At the intermediate feasibility level, the metallurgist must be fully cognizant of the latest geologic and mineralogical data and adjust testing programs accordingly.

Exploration Program

Metallurgical data requirements are now focused on full characterization of all known ore types and on the initiation of studies to investigate the behavior of the various ore types in crushing, grinding, flotation, leaching, filtration, thickening, and so forth. In many cases, the level of technical data presented in an intermediate feasibility study will be nearly completed with only extended pilot-plant runs and basic engineering necessary for final (full) feasibility status. Again, the key is the full understanding of all known ore types and a general sense of how they will be scheduled for milling. Portions of the exploration program will often involve large core or even underground bulk sampling to give the metallurgist an increased level of confidence. Relatively evenly mineralized porphyry ore bodies require a minimum of pilot-plant work while complex ore bodies with little surface expression will require the skillful assemblage of composites from which the metallurgist can extract plant design data.

Sampling and Assaying

During the intermediate feasibility study, it is key that the full range of ore types be represented by drilling and composites of critical drill holes. The metallurgical testing program must be conducted on ores having the same mineralogical character and grade as those expected in the mining plan. Many projects have been based on high-grade metallurgical samples with inherently high recoveries. Through shoddy work and oversight, the lower-grade ores that were eventually mined and shipped to the mill gave almost identical tailings assays, along with catastrophic damage to mineral recovery. This is a very common failure of many operations to yield the rate of return on the investment as projected in the feasibility report.

Mineralogy will often lead to problems in obtaining good material and metallurgical balances for laboratory and pilot-plant tests. Assay programs, such as "metallic assays," must be developed to ensure that metallurgical recommendations are based on solid assay evidence and not widely scattered head assays for the same sample.

Physical Characterization

At the intermediate feasibility level, ore conditions, such as specific gravity, moisture, hardness, clay alteration, and so on, are almost fully developed. Additional exploration work may result in some minor changes, but the nature of the ore body and how it is going to react to handling and transport should be well understood.

Mining

It continues to be assumed that the geologic, mining, and metallurgical team is in place for the completion of the intermediate feasibility (or prefeasibility) study. Given this relationship, the project can be scrutinized by three technical specialties with different objectives and yield a sound evaluation that includes the concerns of all involved parties. The relationship between the mining engineer and the metallurgist continues to be very important and, when missing from a project, often results in poor initial results and a labored start-up.

Mining Method

In the intermediate feasibility study, the mining method is usually well developed. The mining engineer can communicate to the metallurgist the expected ore conditions to be planned for and give good approximations for variables such as ore top size and total moisture content.

Production Plan and Mining Rates

Well-developed mining and/or stoping sequences will have been developed that give the metallurgist a schedule of ore deliveries by year, ore type, and grade. The schedule of mining also serves as a foundation for the remaining metallurgical composites to be used in remaining laboratory or pilot-plant tests.

Mining–Metallurgical Interface

Invariably it will fall to either the mining engineer or the metallurgist to shoulder the responsibility for a poorly functioning project. This alone makes close cooperation between the two work centers critical. At the beginning of the project, it may be worth demanding that their qualifications be similar so that their positions become almost interchangeable for management of the project.

Metallurgical Testing

The intermediate feasibility (or prefeasibility) study will include all of the historical metallurgical work along with initial tests done to support the preliminary feasibility (or scoping) study. The historical and preliminary feasibility study data will generate a list of variables that need to be isolated and studied in the intermediate feasibility study. The intermediate feasibility study is nearing completion from the metallurgist's perspective and requires only certain esoteric data and the completion of extended pilot-plant runs to reach final feasibility status. It is important to understand that, in many instances, the intermediate feasibility study is the first report that may be circulated outside of the project company. Preliminary feasibility (or scoping) studies are used by junior mining companies to help secure private financing for progression to the next level of feasibility. These reports may also be subjected to "due diligence" scrutiny if the company is trying to secure further third-party funding for the project.

Contemporary Laboratory Testing

There are many reliable and dependable metallurgical test and research facilities in the industry, and the project metallurgist must become acquainted with the expertise of each of these facilities before deciding which ones to use. A word of caution: Stay away from fraudulent laboratories that claim "state of the art" (yet unproven) and unexplained "black-box" technologies. These types of facilities will not explain the phony results that are produced, claiming it is because of proprietary information or that the results are simply too complex for the lay metallurgist to understand.

Sampling and assaying. All sampling and assaying for laboratory test samples and the products of testing must stand up to intense scrutiny. When the project advances beyond the intermediate feasibility level, these same results will serve to provide validity to all of the metallurgical conclusions and recommendations. Quality sampling and assaying also provides the foundation for accurate and reliable material and metallurgical balances. A testing program, replete with poor metallurgical balances, will draw increased scrutiny from third parties and seriously damage the credibility of the project. Although poor laboratory procedures occasionally affect results, these instances are usually limited to highly suspect "clandestine laboratories" that make unsubstantiated claims. If highly erratic results are encountered using the assays and sampling procedures of reputable laboratories, it is often due to such variables as nugget effect, sample oxidation, and the use of small sample sizes for testing. Although expensive, the philosophy of "the larger the better" often saves time and money when contemplating the results of both single-batch and locked-cycle laboratory tests.

Comminution. It is important that the comminution tests be performed on samples that are representative of the known ore body. These reports may even be subjected to "due diligence" scrutiny if the company is trying to secure further third-party funding for the project.

Metallurgical tests at the intermediate feasibility level will have yielded quantitative results for both ore crushability and grindability. Standard Bond crushing and grinding indexes, coupled with experience, will yield sizing parameters useful for the initial selection of crushing and grinding equipment. Metallurgical personnel with a plethora of operating experience will have an intuitive feel for the probable size and power requirements for the equipment in question and will be cautious when conflicting results are obtained from laboratory testing.

Vendors of crushing and grinding equipment are frequently the best source of data for the selection of crushing and grinding units. Oftentimes, the cost of testing will be waived if the equipment is selected from the vendor who performed the sizing tests. A relationship between the comminution vendor and the project, which originates at the testing stage, can offer practical long-term cost benefits. In contrast, projects that are shopped around are often victimized by poor service and substandard ultimate results.

The subject of mill sizing begins and frequently ends with the size of the crushing and grinding circuit. Although many highly self-congratulatory technical tomes have been written about the genius involved in increasing plant throughput, it is usually a plant that was overdesigned in the first place. No bargaining should begin at this stage of a project. If an expandable plant is desired, the testing personnel should be aware of it.

There are also examples of closely designed plants that will produce at their "name plate" level and not one ton more. These project constructions are usually the product of hard-dollar, fixed-price contracts with built-in engineering and construction fees. It also requires a disciplined owner who fully realizes the cost of change orders.

Screening and classification. A fairly complete picture of mineral liberation size and the distribution of metal values over the range of sizes for heads, concentrates, middlings, and tailings should also be available at the intermediate feasibility stage of a project. Coupled with comminution testing, the data will yield an effective and efficient plant design. Undergrinding is exasperating, and overgrinding is both expensive and frequently detrimental to good mineral separations.

Good laboratory data for screening and classification will point out the need for combination circuits. Such examples include gravity concentration combined with leaching in the case of gold plants and magnetic separation combined with flotation as is the case in some iron ore and also copper circuits.

The screening and classification data can also save money in the comminution circuit by allowing for the selective comminution of only those products that require it. It makes little sense to grind thousands of tons per hour down to final mineral liberation size when a coarser primary grind will yield good recoveries and require only the regrinding of concentrates or middlings.

Extractive techniques. At the intermediate feasibility level of the project, all research should be completed, and a technique and circuit identified. It can be tempting to investigate every mineral separation technique identified in Taggart (1976), but this is usually the result of poor project management control over the laboratory doing the testing and a not a requirement of the ore. Ores with exceptionally trying metallurgical problems can bankrupt even the best of companies and several nearly have. It is best that these mineralized bodies be left to be experimented with by the well-financed upper-tier companies of the industry.

Today's metallurgist is equipped with the same separation techniques that were available 50 or more years ago. What have dramatically changed are design, construction materials, quality and process control standards, and modernized designs of equipment, such as crushers, flotation machines, and reagents.

The intermediate feasibility study will have identified one of the core separation techniques as applicable to the project ores. Among these are flotation, gravity separation, screening and classification, magnetic separation, heavy media separation, leaching, and simple hand sorting of ore. The latter is often overlooked in our zeal as metallurgists to be high tech. A common story in our industry is about an engineer faced with the occurrence of uniform baseball-sized gold nuggets in a gravel deposit. The proposed circuit included crushing, grinding, classification, leaching, filtration, and tailings disposal at an ultimate recovery of 80%. Local labor could have been employed to pick out the nuggets at a much lower cost.

The selected extraction technique will have yielded acceptable recoveries and concentrate grades and exposed no sizable ore zones that reflect refractory characteristics. The comminution requirements for achieving the expected recovery and concentrate grade are well understood and the reagent requirements quantified. In most cases, these values will be conservative with the full knowledge that actual plant practice may yield more desirable results.

Dewatering. If the laboratory testing is done on large (≥2,000 g) samples or if locked-cycle tests are run, sufficient sample will be available to make some preliminary judgments about thickening and filtering requirements. Often these tests are delayed until late into the final feasibility testing or are ignored. It is most embarrassing to be engaged in start-up and find that reclaim water is not available or that concentrate flow moistures are out of specification because of the lack of good thickening and filtration tests.

These tests are typically run in-house by individual companies or the work can be farmed out to one of several very good testing laboratories. In the case of very slimy or clayey ores, the amounts of reagents will be increased to compete as a major cost item in the mill flow sheet. This possibility, and others like it, needs to be addressed early and not when the project is about to start up.

Concentrate grade. The extensive metallurgical testing program recommended so far will have yielded ample concentrate samples from tests on representative samples of the mineralized body. Of particular importance are the presence of penalty elements such as arsenic, antimony, bismuth, nickel, alumina, fluorine, chlorine, magnesium oxide, and mercury. In addition, for example, copper concentrates may be penalized for zinc, lead, and silica.

Opening dialogues with receiving smelters should be begun no later than this intermediate feasibility stage of a project. One would hope that excessive arsenic values would show up in the preliminary-level work. In the smelting world, high arsenic content can make marginal concentrate grades unmarketable.

Recovery. The end of intermediate feasibility testing would also witness a firm estimate on the mineral recovery from each of the existing ore types. Relationships between grind, process time, and mineral recovery will be fairly well developed, at least to the point of making some intelligent estimates of equipment sizes in grinding and separation.

A common error in estimating recovery is associated with the improper selection of samples for metallurgical testing. Frequently, geologists will bias samples on the high side of the average head assay for the deposit, and, given a relatively fixed tailing, the testing will yield high recoveries. An astute project manager, of course, should not allow this to happen. It is embarrassing to find that the tail assay for a particular ore is relatively fixed at the optimum grind size and that lower head grades yield a substantially lower recovery.

Reagents and consumables. In addition to liberation size, process retention time, comminution power requirements, tailings assays, and concentrate grades, the laboratory testing, if well planned, will have also yielded expected consumption of process reagents. These include, but are not limited to, collectors, frothers, pH control agents, pulp conditioners, leaching agents, adsorption mediums, lixiviants, and other process-modifying chemicals such as wetting agents.

Contemporary Pilot-Plant Testing

Pilot-plant testing is usually not available for the intermediate feasibility stage of a project. If the deposit is easily bulk sampled or if large-diameter core samples are available, the opportunity to include some pilot-plant work may be practical at this intermediate feasibility level.

Pilot-plant testing is an extensive subject. For the purposes of this chapter, it is assumed that the reader is familiar with pilot-plant design and the difficulty of achieving representative results for some process variables when dealing with very low head-grade ores.

Pilot-plant tests can be very expensive, and oftentimes the data are no more revealing than that obtained from well-designed laboratory tests. New or novel approaches to separation or comminution may demand pilot-plant level work and may mollify some corporate boards of directors. Successful pilot-plant testing may be the catalyst for generating project funding given practical solutions to mining and environmental concerns.

The pilot plant will offer very valuable data for estimating comminution power requirements, determining near-optimum process retention time, and confirming the choice of the reagent suite. Mineral recovery will also be determined to a level very similar to actual plant operations. The term *pilot plant* is used synonymously with *large column leach tests* and the dedication of certain mill circuits as test platforms.

Sampling and assaying. The sample provided for pilot-plant work must be meticulously assembled. The pilot plant may be used to test one or more ore types, and the samples selected need to very closely characterize each ore type. Given the expense of pilot-plant test work, data achieved from biased or nonrepresentative samples are "expensively worthless."

Comminution. One of the most important pieces of information that will be garnered from the pilot-plant run is an estimate of comminution power requirements. The pilot-plant data will confirm laboratory-scale estimates of crushability and grindability and serve to provide an accurate foundation for making equipment selections. In ores, that have straightforward metallurgy, very similar ore types, and no suspect characteristics, the only data required from the pilot plant may be that associated with comminution system design.

In the latter instance, the preliminary selection of a vendor for the comminution section of the project will enable the pilot-plant work to be accomplished in the vendor's testing facility. This is usually less expensive than erecting one's own plant or contracting the testing at an industrial facility. A certain level of technical paranoia is evident in the industry, with testing on the same project being spread to several testing venues. The only thing accomplished is to confuse the report reader and offer several often-opposing answers to the same question. In most cases the "victim," at least financially, is the project.

Screening and classification. Continuous pilot-plant operations and very large-scale leaching tests will yield confirming data for liberation estimates and provide a platform for design of the grinding circuits. Ores that did not exhibit any troubling physical problems in laboratory testing may turn out to have serious pulp rheology problems when large, more reliable samples are tested. The pilot plant may also yield data useful in choosing between cyclone classification and the more traditional mechanical classifiers.

The ores may also exhibit peculiar screening problems, such as high clay content and/or sizing phenomena, that generate platey or other nonconforming particle shapes.

Extractive techniques. The pilot plant is not usually used to select between possible separation techniques. It does provide the opportunity for inserting reinforcing process steps such as gravity concentration ahead of leaching in the case of a gold ore.

The pilot plant will confirm selection of the process technique and provide information that is invaluable during start-up of actual operations.

Dewatering. Testing for the determination of the characteristics of process products for thickening and filtration will have been, up to this point, limited to very small samples from laboratory tests. This is particularly true for concentrates. The pilot plant is expected to provide data that are useful for plant scale-up based on large-sized samples, which may be collected under a variety of operating conditions.

Concentrate grade. The use of pilot-scale operations to verify concentrate-grade expectations is frequently troublesome, particularly in the case of low-grade sulfide ores. Large bench-scale tests yield small amounts of concentrates, and pilot plants are difficult to operate with the limited pulp flows associated with attempting to clean the concentrates to marketable levels.

Frother addition, in many cases, also severely limits process control. Pilot plants treating low-grade iron ores, phosphates, lead, zinc, coal, and industrial minerals usually retrieve sufficient weight into the concentrate to enable a good estimate of cleaning and regrinding requirements.

Estimates of concentrate grade from pilot-plant results using accumulated weights of rougher concentrates and then performing large-scale laboratory cleaning tests are acceptable when corroborated by initial bench-scale work.

Recovery. Tailings assays gained under steady-state operating conditions will yield mineral recoveries, which, for the sample tested, are an accurate estimate of future plant operations. The pilot plant can be expected to provide recovery data over a wide range of throughput rates, reagent conditions, and grinds.

Reagents and consumables. Reagent addition rates, which are typically lower than laboratory bench-scale tests, will be expected from the pilot-plant data. The effects of recirculated water will often enhance recoveries and provide a clearer look at what future plant performance will be.

Metallurgical and Process Criteria

The completion of an intermediate feasibility (or prefeasibility) study is similar to a final judgment. The study will be used to prompt the expenditure of additional funds for exploration, provide for the operation of a pilot plant, support mine design efforts, and form the basis for long-range financial and economic expectations. The intermediate feasibility study, particularly from the process standpoint, is essentially a mirror of the final feasibility study and only lacks the completion of all tests, perhaps a pilot-plant program, and the incorporation of ongoing exploration, geology, and mining results.

Recovery

The mineral recoveries will be based essentially on bench-scale tests performed on as large a sample as possible and with attention to representing all of the ore types present. A metallurgical forecast based on mine production from various ore zones will be developed and appropriate recoveries assigned on a monthly, quarterly, and annual basis.

Concentrate Grade

Concentrate-grade variables will have been discussed with a variety of smelters or refineries and initial terms agreed upon. The elimination of penalty elements may require additional metallurgical test work during the final feasibility phase of the project.

Comminution Variables

The parameters of comminution will include test results for the following variables:

- Abrasion index
- Crushability
- Grindability

Based on the prior data, the power requirements and preliminary equipment sizing designs can be initiated.

Magnetic Susceptibility

Ores that can be beneficiated by using their respective magnetic properties will have been thoroughly tested at several levels of magnetic intensity. Stages of separation have been evaluated and power requirements have been developed for the initial sizing of the plant.

Reagent Consumption

Reagent consumption rates have been determined for all chemicals to be used in the process and, in some cases, the infrastructure. Sources and prices have also been obtained.

Flow Sheet and Material Balance

The flow sheet and material balance will have been developed for the scale of operations recommended in the intermediate feasibility (or prefeasibility) study. The flow sheet will typify the process as it is currently understood and will include all major equipment units. Sufficient detail will be included to provide the basis for completing initial factored cost estimates and preliminary engineering.

Flow Sheet Development

The flow sheet will identify all major process equipment units, including crushing and grinding, classification, pumping, screening, conveying, and separation (flotation, leaching, magnetic separation, gravity separation, etc.). The level of detail will enable the process designer to complete several typical general arrangement and plan or section drawings. These drawings, along with a preliminary equipment list and perhaps a simple piping and instrumentation drawing, will form the basis for the initial factored capital cost estimate.

Material Balance Development

The material balance will be developed along conventional lines and will identify and quantify process flows within the process. An integral part of the balance is the development of a water balance for the entire operation. Material balances are usually done as a matter of habit. Water balances are often neglected and the results include overtaxed water supplies, limited expansion possibilities, and the elimination of certain process technologies due to inadequate water resource planning and development.

Figure 7.3 is a simple rendering of an acceptable form of flow sheet and material balance calculation format. Other flow-sheet/material-balance representations will identify process flow quantities on the flow sheet itself. There is no preferred format and either will work well.

Process and Operations Description

Having completed the metallurgical and pilot-plant testing, and the process flowchart development, the metallurgist should now proceed to describe the process plant and operations that will take the run-of-mine ore and produce a concentrate as predicted by the previous metallurgical feasibility testing and analysis.

Unit operations. A complete process and operations description should be included in the intermediate feasibility study. The description will include a step-by-step explanation of

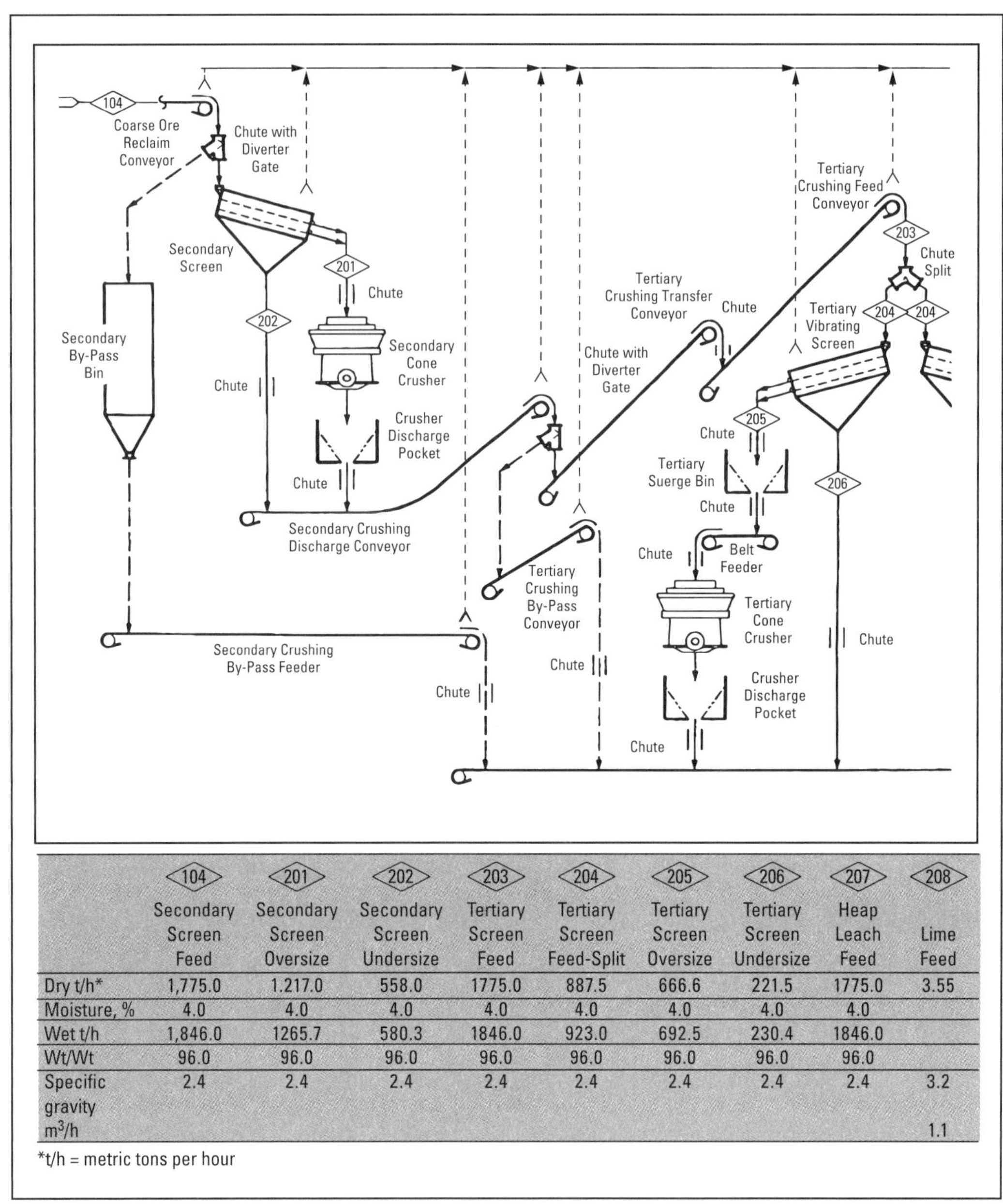

	104 Secondary Screen Feed	201 Secondary Screen Oversize	202 Secondary Screen Undersize	203 Tertiary Screen Feed	204 Tertiary Screen Feed-Split	205 Tertiary Screen Oversize	206 Tertiary Screen Undersize	207 Heap Leach Feed	208 Lime Feed
Dry t/h*	1,775.0	1.217.0	558.0	1775.0	887.5	666.6	221.5	1775.0	3.55
Moisture, %	4.0	4.0	4.0	4.0	4.0	4.0	4.0	4.0	
Wet t/h	1,846.0	1265.7	580.3	1846.0	923.0	692.5	230.4	1846.0	
Wt/Wt	96.0	96.0	96.0	96.0	96.0	96.0	96.0	96.0	
Specific gravity	2.4	2.4	2.4	2.4	2.4	2.4	2.4	2.4	3.2
m^3/h									1.1

*t/h = metric tons per hour

FIGURE 7.3 Flow sheet and material balance example

the process, giving equipment sizes, process flows, and expected product quantities at various stages of the process.

Maintenance program. Included in the process and operations description should be a preliminary statement as to the style and breadth of anticipated maintenance for the process. The section should include the philosophy of preventive maintenance, warehouse inventories, and plant operations redundancy.

Process control program. The basic instrumentation and control program should be illustrated and should include recommendations as to style and comprehensiveness. Small plants (i.e., less than 1,000 stpd) are usually operated with local controls only and a minimum of automatic control. Larger mills can involve full programmable logic control, diagnostics, and *expert* style systems to effect maximum process efficiency. It is not important at this stage of a project to be exact in detailing the instrumentation. The process designer does need to understand that one will exist and that it will require costing.

Process Plant Infrastructure

At this stage of the project, the process engineer will identify and size the concentrator infrastructure in sufficient detail to provide a basis for making factored cost allowances for each. Included in typical mill infrastructure are the following:

- Assay and metallurgical laboratory
- Compressed air system
- Instrument air system
- Dust collection system
- Process water supply
- Communications system
- Process control instrumentation
- Maintenance shops and equipment
- Mill transport and service vehicles

The cost allowances for the preceding systems and equipment may be quite definitive, as in the case of vehicles, or the product of historical data for plants of the suggested size. What is important is that all areas be addressed and not necessarily the ultimate accuracy of the estimate. The allowances will remain unchanged until definitive estimates are made in the final feasibility study.

Electrical

Process plant electrical and the electrical personnel to support administrative and maintenance operations are usually included in, or as an adjunct to, the process plant capital. The intermediate feasibility study will include an evaluation of power draw from the process operations, and the power associated with infrastructure (i.e., water wells, shops, offices, etc.) will be allowed.

Water

The process water system will be detailed with the freshwater supply characterized and the internal water conservation and recycle streams identified. The water system must be designed for the initial stages of start-up when little or no recycle water from tailings is available. In arid areas, specific allowances for evaporation must be addressed, along with the location and drawdown capability of the water resource.

Transportation

The process plant will depend on surface and/or air delivery of heavy maintenance and operating supplies. The cost structure for highway, railroad, and air freight must be evaluated and specific routings assigned to individual supplies. In cases of foreign or offshore locations, the extra burden of ocean freight, duties, and taxes must be adequately accounted for either in the process plant costs or in the general project account.

Communications

The process plant communications system must be separated from the project communications system if necessary. Frequently, the plant will incorporate the project-wide radio communications system, a system of cellular phone applications, and an internal plant intercom system.

Compressed Air

Compressed air requirements for process blowdown and instrument air must be identified and allowed. Projects anticipating wet, sticky ores should be advised to apply a generous safety factor to compressed air requirements in that "there is never enough air" when constant air usage to free chutes, screens, and bins is required.

Instrument air requirements need to be identified and given a preliminary sizing for inclusion in the project cost allowances. A written record of this and other air usage in the preliminary feasibility study serves as a flag for designers who will follow the final feasibility study with the basic engineering stage.

Tailings Disposal

All process operations must address the subject of tailings. The tailings can range from the storage of concentration products to detoxified heaps. In the early stages of the project, the process engineer must collaborate with the geologist and mining engineer in establishing the most convenient areas for the eventual storage of waste products and attempt to visualize the eventual storage methodology. The early-stage site visit should yield a good view of the topography of the site, identify areas with sufficient suitable area for accommodating a tailings receptor, and determine the approximate size of a starter tailings facility.

Possible Design Configurations

Frequently there will exist, in close proximity to the mill operations, an area that requires minimal additional structure to be turned into a tailings repository. Beginning with a starter dike, usually of mine run materials, the project can then expand with a containment made wholly from cyclone tailings. The initial capital estimate will include the cost of the starter dikes and piping systems. Given the costs associated with closure, it is possible that lining the facility will provide insurance against a long-term problem of releasing contaminated water into the groundwater system.

Tailings facilities located in arid areas may be constrained in size by filtering the tailings prior to deposition in conjunction with placement in a lined facility.

In areas of high seismicity, the range of tailings dam construction techniques will be limited. The engineered design will probably demand dam construction methods that are associated with water retainment structures, structures that are maintained with low phreatic levels, and structures with a high tolerance for seismic and/or large precipitation events. Chile is one country that requires compaction of cycloned tailings in critical areas, and this has proven to be a very successful technique to improve dam stability where it is needed.

For example, Chile is a very progressive country, but it has more earthquakes than other mining countries. In 1965, the El Cobre dam failure killed 200 people in a downstream village. Consequently, the Chileans began requiring compaction in critical situations. In 1985, two major tailing dams above Santiago held firm during an 8.2 earthquake with the epicenter near Valparaiso. They later overtopped, yet did not fail (R.L. Bullock, personal communication).

At the intermediate feasibility (or prefeasibility) stage of project development, the preliminary tailings design will usually be assigned to a geotechnical consultant. The consultant will be responsible for the preliminary design, engineering, capital cost estimate, and definition of the permitting requirements for the proposed system.

Operations

The intermediate feasibility study should include an operations plan for the proposed tailings repository. Detailed piping and filling schedules should be prepared so that adequate labor and equipment resources can be assigned to the project. Well-designed tailings facilities with high-quality construction materials and operating systems can often be operated with minimal staff. The philosophy of tailings structure additions and the selection of either in-house or contractors to do the work should be identified and planned. Tailings dam costs, particularly in the early years of a project, are almost always underestimated, as are the costs of future remediation of the impoundment and/or heaps and dumps.

Maintenance

Responsibility for the routine maintenance of the tailings impoundment should be assigned early. Not all tailings impoundments are maintained by the process plant. Frequently, facility maintenance is jointly shared between mine and process plant personnel.

Management

The management plan for the facility should also be outlined during the intermediate feasibility stage. The plant will include the accommodations for stormwater runoff and run-on, water ponding levels, freeboard requirements, tailings distribution plans, and instrumentation and monitoring. The facility must have an independent engineering control system installed to warn of possible failures, minimize erosion, accommodate future remediation efforts, and minimize risk. From an environmental point of view, it is usually advisable to pump the excessive mine-water flow to a separate impoundment for decanting and a clear water discharge.

FINAL FEASIBILITY STUDY

By now the reader should have reached the logical conclusion that little remains to be done to complete the final feasibility study for the project. In many ways, this is a proper and planned conclusion. Those things that are critical to project success and which may have been minimized or overlooked in the preliminary feasibility or intermediate feasibility (or prefeasibility) stages of the project will now be identified. The final feasibility study is almost always used to justify capital spending for the project, whether this comes from internally generated funds or from the equity and/or bank lending sources. The study will be complete. The risks for all process variables identified and compensated for and all necessary technical information for guiding the design of the project should be accounted for and placed in the design basis report (see Chapter 12). The project team will be in place. In the case of the mill, this usually means a minimum of two persons to a maximum of four or five.

Project Organization

Final feasibility studies that do not identify the key process plant personnel by name and position are almost always doomed to poor design decisions, chaotic start-ups, and high costs for both capital and initial operations. The project organization is one of the preeminent requirements of the final feasibility study and one of most frequently overlooked. When third parties are evaluating the viability of a potential investment, the quality and tenure of key personnel are highly important. It is also imperative that these persons have long-term involvement with the project.

Description of the Project Organization

The process plant organization should include the process manager, metallurgical manager, maintenance manager, and instrumentation manager. The term *manager* is used as a catchall for identifying areas of responsibility, as the actual position titles will change with differing corporate cultures and the size of the operation. Figure 7.4 identifies a typical project organization for a proposed milling operation ranging in size from 1,000 to 5,000 stpd.

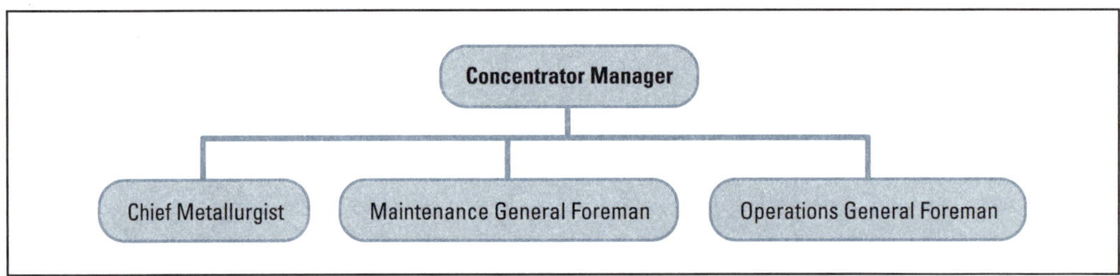

FIGURE 7.4 Small process plant organization chart

Large process operations ranging from 10,000 stpd to upward of 100,000 stpd will require a more compartmentalized organization and a larger project team during the completion of metallurgical testing, design of maintenance programs, development of operating/maintenance manuals, design of instrumentation programs, and the overall guidance of the engineering and construction manager. A typical project team for a large process plant may look like the group shown in the organization chart in Figure 7.5.

Qualifications of Assigned Personnel

All members of the process plant management and staff should have mutually inclusive skills inasmuch as any one of them could replace one another if need be. This demands meaningful experience, adaptation to new environments, good communication skills, and enthusiasm. Strong process plant management teams have been known to overcome a high level of problems associated with new plants, isolated environments, and untried designs. Cutting-edge technology need not necessarily be left to the adventurous and well financed, but if such a course is chosen, the project team will have to be first rate.

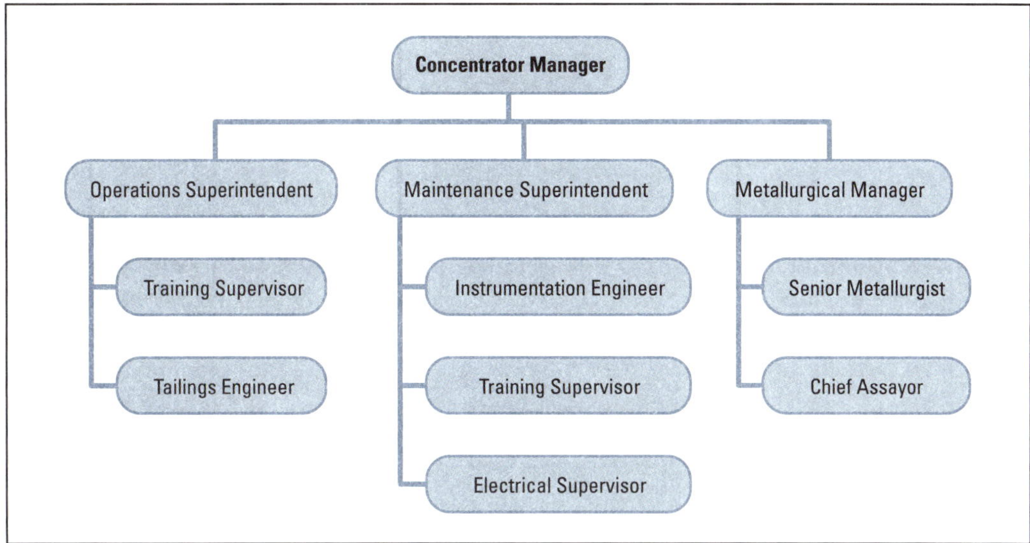

FIGURE 7.5 Large process plant organization chart

Geological Information

The geological information gathered since completion of the intermediate feasibility study should be complementary and confirming. If spectacular downgrades in the quality of the resource and outlined reserves have been found, the project team needs to regroup and slow down the finalization of the project. This is particularly true if geological data or geotechnical data result in the changing of mining plans, rates, or designs.

Project Location

The project location remains unchanged from that given in the intermediate feasibility study and may or may not include the discovery of new or satellite deposits.

Geological Description

In most cases, with the exception of new discoveries, the geological description as it relates to the process metallurgy will remain unchanged from the description given in the intermediate feasibility study. New discoveries will require complete descriptions and a plan for adapting to any unforeseen differences.

Exploration Program

Continuing exploration programs may result in increasing the project throughput or extending the mine life. Usually the effect on process operations is delayed into future years of operations and does not require immediate changes to plant sizing and/or operating philosophy. In any case, management is forced into accepting a certain level of resources and associated ore reserves and planning accordingly. Left to their own strategies, the geologists will never be completely without positive aspects for the project.

Sampling and Assaying

For the final feasibility study, the quality of samples and assays as it relates to metallurgical samples, bulk samples, and composites should be of the highest quality. All risk associated with assay quality and sample representation should be reduced to a low level.

Physical Characterization

The various ore types within the resource have been completely characterized and their geologic and mineralogical impact on processing clearly identified. All planned annual production from the project can be related to the specific characteristics of the ores to be processed and the metallurgical nuances required to achieve maximum mineral extraction and optimum final product quality.

Mining

Mining as relates to the final process design will have established a final mining method and a practical level of annual production. This includes the scheduled hours of the mining operation in contrast to the hours of operation for the concentrator, with properly allocated surge capacities between the two. The process department can then design a processing facility that takes advantage of the mine output and which is complementary to the mining process.

Mining–Processing Interface

The final feasibility study will illustrate the relationship between the mine and process organizations. Responsibilities for infrastructure, maintenance, crushing operations, tailings disposal, and other interfacial activities will be clearly defined and the staff and costs planned. In some cases, specialized maintenance and shop tradespersons will work under a mechanical/electrical department, and these relationships will also need to be defined.

Metallurgical Testing

Metallurgical testing leading up to the issuance of the final feasibility study should be in the final stages of development and should not be waiting on any critical data. All process variables associated with comminution, classification, screening, separation, dewatering, and tailings disposal should be clearly identified and accounted for in the basic engineering design.

Complete, confirming pilot-plant data or the equivalent laboratory-scale testing should be finalized and be the major source of support for the project. It is well to remember that negative feasibility studies are seldom written; incomplete and inaccurate studies are published all too frequently.

Historical Metallurgical Testing

All historical metallurgical laboratory bench-scale or pilot-plant test work as it relates to the following should be included in the final feasibility study:

- Sampling and assaying
- Comminution
- Screening and classification
- Extractive techniques
- Dewatering

- Concentrate grade
- Recovery
- Reagents and consumables

These assessments should be summarized and included in the final feasibility study. It is particularly important to point out the confirming nature of the data or, in the case of conflicting results, indicate the conditions and events that have led to change.

Contemporary Laboratory Testing

The volume of metallurgical test work completed on a bench scale since the completion of the intermediate feasibility study should be at a minimum. Because basic engineering work has been initiated in support of the final feasibility study, it does not bode well to have a variety of questionable results bombarding the designers. Certain confirming work may be required, particularly in the areas of dewatering and tailings disposal, but only from the standpoint of improving prior assumptions and not in reinventing the wheel. Laboratory tests are then essentially complete for the following:

- Sampling and assaying
- Comminution
- Screening and classification
- Extractive techniques
- Dewatering
- Concentrate grade
- Recovery
- Reagents and consumables

Contemporary Pilot-Plant Testing

Pilot-plant testing is usually completed as part of the final feasibility stage of a project. If the deposit is easily bulk sampled or if large-diameter core samples were available, the opportunity to include some pilot-plant work may have been practical at the intermediate feasibility level.

Pilot-plant testing is an extensive subject. For the purposes of this chapter, it is assumed that the reader is familiar with pilot-plant design and the difficulty of achieving representative results for some process variables when dealing with very low head-grade ores. In addition, certain low-pulp flow rates cannot be easily accommodated in pilot-plant pumping, piping, and launders. Regardless of how good it looks on paper, anything smaller than a 1-in. pipe or a 2-in. pump will probably not work or will cause so many process upsets that the results will be costly and yield them useless.

Pilot-plant tests can be very expensive, and oftentimes the data are no more revealing than that obtained from well-designed laboratory tests. New or novel approaches to separation or comminution may demand pilot-plant level work and, as has been mentioned previously, may mollify some corporate boards of directors. Successful pilot-plant testing may be the catalyst for generating project funding given practical solutions to mining and environmental concerns.

The pilot plant will offer very valuable data for estimating comminution power requirements, determining near-optimum process retention time, and confirming the choice of the reagent suite. Mineral recovery will also be determined to a level very similar to actual plant operations. The term *pilot plant* is used synonymously with *large column leach* tests and the dedication of certain mill circuits as test platforms.

Sampling and assaying. The sample provided for pilot-plant work must be meticulously assembled. The pilot plant may be used to test one or more ore types, and the samples selected need to very closely characterize each ore type. Given the expense of pilot-plant test work, data achieved from biased or nonrepresentative samples are "expensively worthless."

Comminution. One of the most important pieces of information that will be garnered from the pilot-plant run is an estimate of comminution power requirements. The pilot-plant data will confirm laboratory-scale estimates of crushability and grindability and serve to provide an accurate foundation for making equipment selections. In ores that have straightforward metallurgy, very similar ore types, and no suspect characteristics, the only data required from the pilot plant may be that associated with comminution system design.

In the latter instance, the preliminary selection of a vendor for the comminution section of the project will enable the pilot-plant work to be accomplished in the vendor's testing facility. This is usually less expensive than erecting one's own plant or contracting the testing at an industrial facility. A certain level of technical paranoia is evident in the industry, with testing on the same project being spread to several testing venues. The only thing accomplished is to confuse the report reader and offer several often-opposing answers to the same question. In most cases the "victim," at least financially, is the project.

Screening and classification. Continuous pilot-plant operations and very large-scale leaching tests will yield confirming data for liberation estimates and provide a platform for design of the grinding circuits. Ores that did not exhibit any troubling physical problems in laboratory testing may turn out to have serious pulp rheology problems when large, more reliable samples are tested. The pilot plant may also yield data useful in choosing between cyclone classification and the more traditional mechanical classifiers.

The ores may also exhibit peculiar screening problems, such as high clay content and/or sizing phenomena, that generate platey or other nonconforming particle shapes.

Extractive techniques. The pilot plant is not usually used to select between possible separation techniques. It does provide the opportunity for inserting reinforcing process steps such as gravity concentration ahead of leaching in the case of a gold ore.

The pilot plant will confirm selection of the process technique and provide information that is invaluable during start-up of actual operations.

Dewatering. Testing for the determination of the characteristics of process products for thickening and filtration will have been, up to this point, limited to very small samples from laboratory tests. This is particularly true for concentrates. The pilot plant is expected to provide data that are useful for plant scale-up based on large-sized samples, which may be collected under a variety of operating conditions.

Concentrate grade. The use of pilot-scale operations to verify concentrate-grade expectations is frequently troublesome, particularly in the case of low-grade sulfide ores. Large bench-scale tests yield small amounts of concentrates, and pilot plants are difficult to operate with the limited pulp flows associated with attempting to clean the concentrates to marketable levels.

Frother addition, in many cases, also severely limits process control. Pilot plants treating low-grade iron ores, phosphates, lead, zinc, coal, and industrial minerals usually retrieve sufficient weight into the concentrate to enable a good estimate of cleaning and regrinding requirements.

Estimates of concentrate grade from pilot-plant results utilizing accumulated weights of rougher concentrates and then performing large-scale laboratory cleaning tests are acceptable when corroborated by initial bench-scale work.

Recovery. Tailings assays gained under steady-state operating conditions will yield mineral recoveries, which, for the sample tested, are an accurate estimate of future plant operations. The pilot plant can be expected to provide recovery data over a wide range of throughput rates, reagent conditions, and grinds.

Reagents and consumables. Reagent addition rates, which are typically lower than laboratory bench-scale tests, will be expected from the pilot-plant data. The effects of recirculated water will often enhance recoveries and provide a clearer look at what future plant performance will be.

Metallurgical and Process Criteria

Metallurgical and process criteria will provide the basis for the basic engineering supporting the final feasibility study. The results should not differ dramatically from the intermediate feasibility data and need to be available to the designers and cost estimators early. Significant changes to any of the following variables can be extremely costly to project:

- Recovery
- Concentrate grade
- Abrasion index
- Crushability
- Grindability
- Magnetic susceptibility
- Reagent consumption

Flow Sheet and Material Balance

An important piece of mineral processing mythology states, "Late changes to the flow sheet and material balance are as knife thrusts to the pocketbook." If there is any controversy attached to the flow sheet and material balance, it needs to be resolved prior to beginning basic engineering work. Exorbitant feasibility fees are usually associated with a fickle client or one who does not understand the final feasibility process.

Flow Sheet and Material Balance Development

The flow sheet and material balance, if not the same as anticipated in the intermediate feasibility study, should be finalized well before launching into basic engineering, design, and cost estimating. The example included in the "Intermediate Feasibility Study" section of this chapter holds true for the final feasibility study and can be copied whole (see Figure 7.3).

Process and Operations Description

The process and operations description should describe in detail the following items:

- Unit operations
- Maintenance program

- Process control program
- Scheduling program

A comprehensive discussion of the preceding subjects in the final feasibility study carries a high level of confidence over to the management reader or the third-party due-diligence effort. An intimate knowledge of the process along with planning for maintenance, process control, and operations/maintenance scheduling are critical to the successful start-up and commissioning of a plant of any size or complexity.

As with the intermediate feasibility study, the process description will describe the flow sheet, the size and capacity of all major pieces of equipment, and the expected production or metallurgical results from each process step. Where possible, the results achieved should be referenced to the appropriate sections of the metallurgical test work.

Process Plant Infrastructure

At the final feasibility stage of the project, the process engineer will have identified and sized the concentrator infrastructure in sufficient detail to provide a basis for making a detailed cost estimate for each. Included in typical mill infrastructure are the following:

- Assay and metallurgical laboratory
- Compressed air system
- Instrument air system
- Dust collection system
- Process water supply
- Communications system
- Process control instrumentation
- Maintenance shops and equipment
- Mill transport and service vehicles

The costs for the preceding systems and equipment may be quite definitive, as in the case of vehicles, reliable and design-based systems, or the product of historical data and factoring for plants of the suggested size. What continues to be important is that all areas be addressed and not necessarily the ultimate accuracy of the estimate. The allowances will remain unchanged until definitive estimates are made in the final design following approval of the final feasibility study.

Electrical

Process plant electrical and the electrical personnel to support administrative and maintenance operations are usually included in, or as an adjunct to, the process plant capital. The final feasibility study will include an evaluation of power draw from the process operations, and the power associated with infrastructure (i.e., water wells, shops, offices, etc.) will be allowed.

Water

The process water system will be detailed with the freshwater supply characterized and the internal water conservation and recycle streams identified. The water system must be designed for the initial stages of start-up when little or no recycle water from tailings is available. In arid areas, specific allowances for evaporation must be addressed, along with the location and drawdown capability of the water resource.

Transportation

The process plant will depend on surface and/or air delivery of heavy maintenance and operating supplies. The cost structure for highway, railroad, and air freight must be evaluated and specific routings assigned to individual supplies. In cases of foreign or offshore locations, the extra burden of ocean freight, duties, and taxes must be adequately accounted for either in the process plant costs or in general project account.

Communications

The process plant communications system must be separated from the project communications system if necessary. Frequently, the plant will incorporate the project-wide radio communications system, a system of cellular phone applications, and an internal plant intercom system.

Compressed Air

Compressed air requirements for process blowdown and instrument air must be identified and allowed. Projects anticipating wet, sticky ores should be advised to apply a generous safety factor to compressed air requirements in that "there is never enough air" when constant air usage to free chutes, screens, and bins is required.

Instrument air requirements need to be identified and given a preliminary sizing for inclusion in the project cost allowances. A written record of this and other air usage in the preliminary feasibility study serves as a flag for designers who will follow the final feasibility study with the basic engineering stage.

Tailings Disposal

All process operations must address the subject of tailings. The tailings can range from the storage of concentration products to detoxified heaps. In the early stages of the project, the process engineer must collaborate with the geologist and mining engineer in establishing the most convenient areas for the eventual storage of waste products and attempt to visualize the eventual storage methodology. The early-stage site visit should yield a good view of the topography of the site, identify areas with sufficient suitable area for accommodating a tailings receptor, and determine the approximate size of a starter tailings facility. By this point in time, the proposed tailings area should have had condemnation drilling to prove that the tailings impoundment will not interfere with mining in later years.

Possible Design Configurations

Frequently there will exist, in close proximity to the mill operations, an area that requires minimal additional structure to be turned into a tailings repository. Beginning with a starter dike, usually of mine run materials, the project can then expand with a containment made wholly from cyclone tailings. The initial capital estimate will include the cost of the starter dikes and piping systems. Given the costs associated with closure, it is possible that lining the facility will provide insurance against a long-term problem of releasing contaminated water into the groundwater system.

Tailings facilities located in arid areas may be constrained in size by filtering the tailings prior to deposition in conjunction with placement in a lined facility.

In areas of high seismicity, the range of tailings dam construction techniques will be limited. The engineered design will probably demand dam construction methods that are associated with water retainment structures, structures that are maintained with low phreatic levels, and structures with a high tolerance for seismic and/or large precipitation events.

At the final feasibility stage of project development, the tailings impoundment design will have been assigned to a geotechnical engineering company. The engineering company will be responsible for the preliminary design, engineering, capital cost estimate, and definition of the permitting requirements for the proposed system.

Operations

The final feasibility study will include an operations plan for the proposed tailings repository. Detailed piping and filling schedules should be prepared so that adequate labor and equipment resources can be assigned to the project. Well-designed tailings facilities with high-quality construction materials and operating systems can often be operated with minimal staff. The philosophy of tailings structure additions and the selection of either in-house or contractors to do the work should be identified and planned. Tailings dam costs, particularly in the early years of a project, are almost always underestimated, as are the costs of future remediation of the impoundment and/or heaps and dumps.

Some well-established mining companies might believe that they have enough expertise within their organizations to design their own tailings impoundments. But an important point to keep in mind is that the designer must have considerable geotechnical knowledge in dam construction. Worldwide, one or two tailings dam failures occur almost every year. This should be testimony enough to motivate all mining companies to leave dam design to the experts. The most recent lawsuit in Brazil will result in a multi-billion-dollar settlement, which validates the seriousness of this problem (Els 2016).

Maintenance

Responsibility for the routine maintenance of the tailings impoundment will be assigned in the final feasibility study. Not all tailings impoundments are maintained by the process plant. Frequently, facility maintenance is jointly shared between mine and process plant personnel.

Management

The management plan for the facility will be outlined during the final feasibility stage. The plant will include the accommodations for stormwater runoff and run-on beyond the property, water ponding levels, freeboard requirements, tailings distribution plans, and instrumentation and monitoring. The facility must have an independent engineering control system installed to warn of possible failures, minimize erosion, accommodate future remediation efforts, and minimize risk.

REFERENCES

Els, F. 2016. 2016. BHP, Vale hit with $44 billion lawsuit over deadly spill. www.mining.com/bhp-vale-hit-44-billion-brazil-lawsuit/?utm_source=digest-en-mining-160503&utm_medium=email&utm_campaign=digest (accessed May 3, 2016).

Mular, A., and Bhappu, R. 1980. *Mineral Processing Plant Design*, 2nd ed. Littleton, CO: SME. pp. 4–10.

Taggart, A. 1976. *Handbook of Mineral Dressing: Ores and Industrial Minerals*. Wiley Engineering Handbook Series. New York: Wiley.

CHAPTER 8

Market Analysis for Mineral Property Feasibility Studies

Roderick G. Eggert

The commercial success of a mine depends not only on the quality of the ore body, mine design, and how the ore body is managed during mining, but also on the market environment in which the mine operates.

A market analysis aims to understand the market environment and is essential in answering questions such as: To what degree are market prices outside the control of any single company? Or, alternatively, are one or more firms, either producers or users, sufficiently large that they have some degree of control or influence over prices? Do decisions to develop a mine influence the behavior of other mine operators? If a metal concentrate or some other form of semiprocessed material is produced, are there downstream processors available to take this intermediate product and transform it into something that manufacturers would buy? Or does developing a mine require a smelting and refining capacity to be built too? Who are the likely customers, and do they require a standardized product capable of being supplied by many alternative suppliers; or do they require specialized products? What long-term price is appropriate in evaluating the potential profitability of developing a mine?

This chapter offers a five-step framework for answering these and related questions that influence the decision about whether to develop a mine (Figure 8.1). The steps are not strictly sequential and linear. Rather, they overlap to some degree, and sometimes inferences drawn during one step require returning to and revising inferences drawn at an earlier step. The analysis becomes increasingly nuanced and complex as it proceeds from one step to the next. The goal is an appropriate balance between clarifying simplicity, highlighting key features of a market, and realistic complexity, which will facilitate an expert understanding of a market.

STEP 1: DEFINING MARKETS AND COMPETITORS

The first step in market analysis is defining the market and one's competitors. Basic information needed to define a market and carry out the analysis is summarized in Table 8.1.

A market represents the interactions of buyers and sellers that determine what is produced and sold, in what quantities, where, and at what price. Competitors are the participants in a market and include any entity whose actions influence the actions of others. Who competes with whom, in turn, is defined both (a) geographically and (b) in terms of the nature of the product sold and bought. For both market dimensions, the concept of substitution is critical.

The geographic extent of a market reflects how far a product can be shipped and be a substitute in the eyes of customers for the product sold by another supplier in a different location.

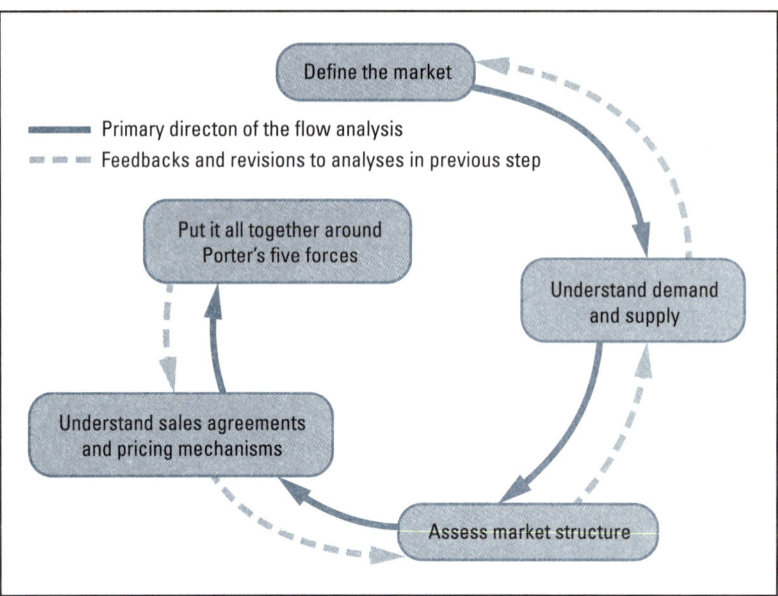

FIGURE 8.1 Five-step method for analyzing mineral and metal markets

Some markets—including those for most major and precious metals such as refined copper, lead, nickel, zinc, gold, and platinum—are essentially single, global markets because transportation costs represent a small portion of the delivered costs of a product, no matter how far the point of use is from the location of production. In these cases, there is a single global price for the product. All producers anywhere in the world are, in effect, competitors. Other markets are regional—like those for coal, iron ore, metal concentrates, and bulk industrial minerals—because transportation costs become a significant (and usually increasing) fraction of delivered costs the farther a product is shipped. At some point, transportation costs limit the geographic extent of a market. In such cases, customers in different locations pay different prices. Not all suppliers compete in all locations. The degree of competition among suppliers may be higher or lower in one region than another.

The product dimension of a market is determined by all products that are close substitutes for one another, again in the eyes of customers. Some products are homogeneous, such as most refined metals, and traded in standardized shapes, sizes, weights, and purity levels; one lot of metal is a perfect substitute for a lot of metal from another supplier. Other products, such as coal, iron ore, and metal concentrates, are differentiated in that the product sold by one supplier may be slightly different from that of another supplier in terms of grade, impurities, and other physical or chemical characteristics. Iron ore, for example, comes as lump ore, fines, or pellets, and different iron ores have different iron contents and impurities. When similar products are differentiated, a product from one supplier is not a perfect substitute for that of another supplier, and the nature and degree of product differentiation becomes an important characteristic of the market.

The output of step 1 in market analysis is a succinct statement defining the market and identifying the competitors relevant for the mineral property under study.

TABLE 8.1 Basic information for market analysis

Markets, production, prices	**Marketable form of product:** concentrate, metal, other; specifications, regulations, restrictions **Market location and alternatives:** likely purchasers, quality requirements, typical purchase agreements, geographic extent of the market **Intermediate products and end uses:** data and information on the materials, components and assemblies, and final products important for the mine output; possible future substitutions **Current and possible new future production:** data and information on existing production capacity, as well as potential new production (idled capacity, mines in construction, known but undeveloped deposits, exploration projects); role of by-product and co-product production; role of scrap recovery **Prices:** price data, trends, pricing mechanisms, and transparency
Inputs	**Land, water, and mineral rights:** ownership, terms, costs **Labor:** availability, rates, housing, transport **Transportation:** property access, product transportation **Utilities:** availability of electric power, natural gas, grid versus on-site generation of electricity, rates **Equipment and spare parts:** types, sources, costs
Government	Land access, preproduction rules and procedures, environmental and worker health and safety, tax and fiscal systems, mine closure, and rehabilitation

Adapted from Gentry and O'Neil 1992.

STEP 2: UNDERSTANDING DEMAND AND SUPPLY

Once the geographic extent of a market is defined, and all products that are close substitutes and the competitors in this market are identified, an in-depth evaluation of mineral or metal use (demand) and production (supply) can be performed.

Demand

Demand represents the perspective of users. The demand for minerals and metals is derived from the properties they provide to materials that, in turn, make up final products. For example, copper is demanded most importantly for its electrical conductivity; aluminum for its low density, high strength, and easy formability; and zinc for the corrosion resistance it provides for certain types of steel. Antimony is demanded primarily for its flame-retardant properties; cobalt for its corrosion and abrasive resistance, high-temperature strength, and magnetism; and neodymium for its magnetic properties. Each element has its own set of properties that can be used in materials, components, and final products.

A starting point for demand analysis is collecting and understanding information on mineral or metal usage. The point in the supply chain at which usage is measured varies from situation to situation. For ores and concentrates, statistics reflect usage by metallurgical operations, often smelters and refiners. For most metals, statistics reflect the use of refined metals by producers of intermediate products (ingots, alloys, shapes, wire, etc.). Finally, end-use statistics reflect the final products and sectors in which minerals and metals are embodied. For example, nickel use in the United States in 2015 can be described by *type of intermediate product* (45% stainless and alloy steels, 43% nonferrous alloys and superalloys, 7% electroplating, and 5% others) and *end use* (34% transportation and defense, 20% fabricated metal products, 14% chemical and petroleum industries, 13% electrical equipment, 5% construction, 5% household appliances, 5% industrial machinery, and 4% others) (Kuck 2016). Statistics like these should be used as a point of entry into understanding the driving forces behind the demand for a mineral or metal. More formally, think of demand as the relationship between (a) the quantity of mineral or metal demanded and (b) the determinants of this quantity. The exact determinants of demand vary from case to case; the following is a list of the more important determinants:

- **Price.** Normally it is expected that the higher the price, the smaller the quantity users will demand, and vice versa.
- **Prices of substitutes and complements.** A substitute is another mineral, metal, or material that provides properties similar to those of the material under study. For example, aluminum is a potential substitute for copper in applications requiring electrical conductivity. Aluminum and steel are potential substitutes in the outer body panels of automobiles and light trucks. Usually, the higher the price of a substitute, the higher the quantity demanded of the material under study.
- **Level of aggregate or sectoral activity.** Ordinarily, the higher the level of activity in the economy or sector, the higher the quantity demanded. The level of steel production directly influences the demand for iron ore. The level of construction activity influences the demand for copper wires and brass (copper/zinc) pipes and other plumbing fixtures. For the economy as a whole, the *level of gross domestic product*[*] or *industrial production*[†] influences demand for minerals and metals.
- **Technology.** Both product and process technologies influence demand. Consider battery (product) technologies. The lead-acid battery is the mainstay of automotive starting-lighting-ignition batteries, and this technology is the basis for most lead demand today. Nickel-metal-hydride rechargeable batteries, initially developed in the 1960s and 1970s and then commercialized in the 1990s, represented an important new use for rare-earth misch metal (a mixture of cerium, lanthanum, neodymium, and praseodymium). In the 2000s, the development of lithium-ion batteries has led to increased demand for lithium but reduced demand for nickel-metal-hydride batteries and their constituent raw materials because lithium-ion batteries have replaced nickel-metal hydride batteries in some applications.

  As for process technologies, an improvement in manufacturing efficiency reduces wastes and the amount of material demanded by the manufacturing process. Indium provides a good example here. In the late 1990s and early 2000s, demand for indium increased along with demand for flat-panel displays in television sets and computer monitors. Indium-tin oxide thin films are used in these displays. Early in this period, only about one-third of the indium that manufacturers purchased ended up on the thin films; the rest was wasted in the sputtering process (a little like spray painting). The increase in demand for indium led to higher prices, which in turn led to efforts to become more efficient in the use of indium in the manufacturing process. Today, something like two-thirds of the indium purchased finds its way onto the thin films because of recycling of indium wastes and improvements in the sputtering process.

---

* Gross domestic product, often referred to as GDP, is an estimate of the value of final goods and services produced within the geographic boundaries of a country during the specified quarter or year.

† Industrial production, usually stated as an index, measures the amount of production in manufacturing, mining (including oil and natural gas), and electrical and natural gas utilities.

- **Government policy.** Policies can serve to increase or reduce demand for a mineral or metal. Government policies to reduce air pollution encouraged or required use of catalytic converters with internal combustion engines to reduce emission of toxic gases. Catalytic converters use platinum, palladium, and rhodium, and so these policies stimulated demand for these materials. Conversely, government policies discourage or prohibit use of lead in paints, motor fuels, and many other applications—other than in automotive batteries—because lead (primarily lead oxide) is poisonous to human beings and many animals.
- **Expectations.** Demand is influenced by expectations. If users or investors, for example, expect prices to rise in the future, they may increase purchases of a metal today to avoid paying a higher price in the future, which increases current demand compared to what it would have been otherwise (and vice versa).

Identifying and understanding the specific determinants of demand for the mineral or metal the mine is producing is important in assessing how demand may change in the future, after a mine comes into production.

Supply

Supply represents the perspective of producers. A starting point for supply analysis is collecting and understanding information on mineral or metal production. A basic data set should include existing production capacity and costs by facility and company. An expanded data set should include information on projects in exploration and under development, which have the potential to come into production in the future, as well as mines and processing facilities that are scheduled or expected to close. For many, if not most minerals and metals, building these data sets is difficult to do by oneself. For precious or base metals, what is needed is a worldwide data set, which illustrates where the cost of production will rank in a cost seriatim. Firms would like their costs to lie in the lower quartile or so of the cost seriatim to be viable during downturns in mineral and metal markets. Much of the data are not publicly or freely available. Assembling this information requires detailed knowledge of the industry. As a result, many analysts rely on information developed by consulting firms that specialize in assembling and organizing data on operating mines, as well as exploration and mine-development projects.

The information in the basic and expanded data sets should be used as a point of entry into understanding the driving forces behind the supply of a mineral or metal. Think of supply as the relationship between (a) the quantity of mineral or metal supplied and (b) the determinants of this quantity. The following determinants are among the most important:

- **Prices.** Normally, the higher the price a producer receives for supplying a product, the greater the quantity it would like to supply, and vice versa.
- **Costs.** Usually, the higher the costs that a producer incurs, the lower the quantity it would like to supply, and vice versa. Costs come in different varieties. Capital costs make operations possible and include the costs of land, buildings, equipment, and other related items both prior to mining and during mining to refurbish or replace facilities that wear out. Operating costs are incurred during mining and processing, are a function of the rate of production, and include items such as labor, explosives, fuel, electricity, chemical reagents, and spare parts.

- **Technology.** The nature of processes and skills used to discover, evaluate, mine, and process a mineral resource influences what is possible in a technical sense and significantly influences costs of production. A technical innovation or improvement in technology reduces production costs compared to what otherwise would have been the case.
- **Government policy.** Policies can encourage, discourage, or be neutral with respect to mineral exploration, mine development, and mining. Policies influence whether a mine is developed and, if so, how it is operated. Policies influence the allocation of risks. Policies can be thought of as reducing or increasing production costs relative to a world with no policies or in which policies are neutral with respect to mining. Relevant policies range from those designed for the economy as a whole (such as basic commercial laws and rules, business taxation), to those governing land use and the environment (whether or not mining is involved), and finally to those designed specifically for mining (such as mining royalties and mine reclamation).
- **Wars, strikes, civil disturbances, natural catastrophes, and so forth.** In the short term, metal supply sometimes is constrained by wars, strikes, civil disturbances, and other disruptions to existing operations.
- **Expectations.** Finally, expectations influence supply. In the short term, if sellers expect prices to rise in the future, they may withhold supplies in the present to take advantage of higher future prices, and vice versa. In the long term, if sellers expect prices to rise in the future, they may invest in additional production capacity that will take advantage of the higher future prices; if sellers expect prices to fall, they may decide not to refurbish depreciating existing capacity, leading to lower production capacity in the future.

Identifying and understanding the specific determinants of a particular mineral or metal supply is important in assessing how supply may change in the future, after a mineral property being evaluated comes into production.

For many minerals and metals, joint production and recycling are important parts of supply, in addition to supplies from single-product mines. When this occurs, joint production and recycling should be included in the analysis of supply. Joint production occurs when multiple products are produced at a single operation. There are three types of joint products: main products, by-products, and co-products. A main product is so important to the commercial viability of a mine that the mine is designed, optimized, and operated around this product and its prices and markets. A by-product, recovered along with a main product, is so unimportant to the overall commercial viability of the mine that its prices and markets have, at most, a small degree of influence on mine design and the level of ore output. Whether a by-product is recovered depends on whether the by-product's price is sufficient to cover only the additional costs of recovering the by-product rather than leaving it in a waste stream. All shared costs, in effect, are borne by the main product. Co-products represent an intermediate case. Each of the two or more co-products significantly influences a mine's commercial viability, and prices for all co-products jointly determine the level of mining. A key economic idea when joint production occurs is a sharing of costs by two or more minerals or metals. Sharing of costs, in turn, means that by-product or co-product output of a mineral or metal typically has lower production costs than the same mineral or metal produced as a main product or single product.

Molybdenum, whose principal use is as an alloying element in various types of iron and steel, is recovered as a by-product or co-product at copper mines and also as a main product or single product at molybdenum mines. By-product and co-product molybdenum, representing more than half of world molybdenum production, typically has considerably lower costs per unit of molybdenum because of the sharing of costs between copper and molybdenum. The costs of mine design, drilling and blasting, ore haulage, and milling are shared.

A number of specialty metals are produced solely or predominantly as by-products. For example, germanium and indium are recovered almost entirely as by-products of zinc metallurgical operations. When a material is produced solely as a by-product, its availability and supply are determined not just by conditions in the by-product market but also market developments for the associated main products. Consider copper and tellurium. Almost all tellurium produced each year is as a minor by-product of the smelting and refining of copper ores. Since about the early 1990s, an increasing fraction of refined copper has been produced through an alternative technology, solvent extraction and electrowinning, a process that does not allow for tellurium recovery. A technological development in copper has influenced the availability of tellurium, although to date, the demand for tellurium has not been large enough for this development to significantly constrain tellurium availability.

Recycling, known as secondary production, has two very different feedstocks: manufacturing wastes and end-of-life products. The availability of metal from the recycling of manufacturing wastes depends substantially on the efficiency (or rather, inefficiency) of manufacturing. Cast metal production—pouring of liquid metal into forms—typically generates little waste for recycling; probably 90% or more of the metal fed into the process makes its way into cast metal. In contrast, wrought metal production—generating metal shapes out of sheet metal—is typically less efficient and generates significant volumes of trimmings available for recycling. The costs of recycling manufacturing wastes are usually low relative to metal production from mines because recycling avoids all costs of mining and initial processing. As a result, most manufacturing wastes generated in a given time period are recycled reasonably soon after they are generated. Moreover, whether recycling of manufacturing wastes occurs is usually not sensitive to price changes for the relevant metal, given the low costs of recycling.

The characteristics of metal supply from recycling of end-of-life products are quite different than recycling of manufacturing wastes: Not all metal available for recycling is recycled, and whether recycling occurs depends importantly on price. The quantities available for recycling depend significantly on past metal use in products and the lifetime of these products. The larger the level of past use and the shorter the product lifetime, the greater the quantities available for potential recycling relative to current demand. Whether these quantities actually are recycled depends on a metal's price relative to recycling costs, which in turn are influenced by the (a) the ease or difficulty of collecting, sorting, and transporting products to recycling and processing facilities and (b) the material complexity of the products. Collection, sorting, and transporting products typically have lower costs in urban than rural areas, because of greater population densities. The simpler the product, the lower the likely costs of recycling. Government policy sometimes plays a critical role in whether recycling occurs, for example, regulations requiring recycling of lead-acid batteries. Joint production also is an important issue in recycling. Recyclers are not interested so much in recycling a particular metal as they are in maximizing profits from all recyclable materials in a junked product. For fluorescent lamps, glass is the major material, and a well-established market and acceptable price for recycled glass would encourage recycling of rare-earth phosphor materials, which by themselves are not sufficiently valuable to justify recycling of fluorescent lamps.

Demand and Supply Determine Price and Quantity

In well-functioning markets, demand and supply—or more precisely, buyers and sellers informed by the underlying determinants of their behavior—interact to determine price and the quantity produced and consumed. Over time, prices and quantities change in response to changes in the underlying determinants of demand and supply. Exactly how and to what extent prices and quantities change depend on the time period of adjustment and the constraints on adjustment in different situations. The concepts of short run and long run help clarify the nature of change.

The *short run* is an adjustment period in which something is fixed or unchangeable. For a producer, typically production capacity is fixed in the short run. For a consumer, the plant and equipment using a mineral or metal as an input are fixed (e.g., a steel mill for iron ore, a wire mill for copper metal). So both producers and users are constrained in the short run to adjusting the rate at which they use their capacity that produces or uses a mineral or metal.

One consequence of these short-run constraints is that mineral and metal prices tend to be volatile from one year to the next. Metal prices may double, triple, or more over two or three years. Metal users, say automobile manufacturers, are constrained by their existing car designs and associated metal requirements. Metal producers are also constrained. Even if demand for automobiles increases by more than expected because of stronger than expected macroeconomic growth, metal producers cannot expand output beyond capacity in the short run. As a result, the way markets adjust to an unexpected increase in demand is through higher prices. The major episodes of booming commodity prices over the last century are all associated with periods of significant demand growth that were not anticipated in sufficient time to invest in the new capacity necessary to meet these demands without price increases (see Radetzki 2008).

When prices fall, they can fall significantly in the short run because it makes commercial sense for producers to continue to produce as long as they receive revenues that cover all their operating costs, even if revenues are insufficient to cover total costs, including repayment of capital and the minimum-acceptable profit required by the asset owners. By continuing to operate in these circumstances, producers minimize losses. They earn revenues that cover all of their avoidable, variable costs and at least some of the additional costs not directly a function of operations, such as repayment of capital expenses and the minimum profit demanded by investors, which leaves them better off than shutting down temporarily. Eventually, however, capital costs must be repaid and asset owners require profits, which leads us to the long run.

The *long run* is an adjustment period in which all factors of production are variable. Asset owners can decide to permanently shut a facility. Or they can invest in new production capacity in response to the unexpectedly large-demand growth referred to earlier when describing short-run price volatility. This new capacity can be expansions of existing mines or development of known but undeveloped deposits. With even longer time horizons, firms can invest in mineral exploration and technological innovation with an eye toward expanding output and lowering production costs.

The concept of long run is essential in thinking about what price to use in evaluating the commercial viability of developing a mineral deposit into a mine—a process that takes 5 to 10 years or more even after exploration has discovered and feasibility analysis has proven an ore reserve. One possibility is to project volatile (short-run) metal prices out into the future. More typically, given the difficulty of projecting the level and timing of future price volatility with any degree of confidence, analysts estimate a long-run price based on the total costs of production of existing mines and the likely costs of production at mines anticipated to come into production over the life of the mineral deposit under consideration. Existing and possible new operations, and their associated output rates, are arranged from low cost to high cost. The estimate of a long-run price is the total cost at the operation that just satisfies expected total demand in the future.

The outputs of step 2 are succinct statements and data summaries of

- End-use markets and determinants of changes over time in demand;
- Existing producers, possible new producers, production costs, and determinants of changes over time in supply (especially important here is whether production costs would be relatively low or high compared to existing and potential new operations);
- The magnitude and timing of historical, year-to-year price volatility; and
- An estimate of long-run price against which to compare estimated production costs at the mineral property being evaluated.

STEP 3: ASSESSING MARKET STRUCTURE

Step 2 treated production and use of minerals and metals separately. Step 3 brings them together. It focuses on the structure of production and use and how different market structures result in different types of behavior and strategies among market participants.

Among the many characteristics of market structure, three are key:

1. **Industry concentration, determined by the number and size distribution of sellers and buyers.** The larger the number of sellers and buyers, and the more nearly equal they are in size, the smaller the likelihood is that any single seller or buyer has a significant effect on price—and the less concentrated the industry. The smaller the number of firms, and the less equal they are in size, the more concentrated the industry.
2. **The degree to which products are differentiated or not.** Product differentiation occurs when similar products have differences from one supplier to another. As noted in the "Step 1" section, refined metals, which are greater than 99% pure metal, are nearly identical from one supplier to another; they are perfect substitutes from the perspective of buyers. Mineral ores and concentrates, in contrast, vary from one supplier to another in terms of grade, impurities, and other characteristics; even though there is some substitutability from one mineral ore or concentrate to another in the eyes of downstream processors, different ores and concentrates are not perfect substitutes. A low-grade concentrate will require more concentrate to yield the same amount of metal compared to a high-grade concentrate. A concentrate with hard-to-process impurities will be less desirable than a concentrate with easy-to-remove impurities. These differences in concentrate grade and impurities will be reflected in lower prices, or higher processing fees, for the lower-quality concentrates. More generally, the greater the degree of product differentiation, the lower the intensity of rivalry in that sector.
3. **The presence or absence of barriers to entry and exit.** Entry barriers are features of a market that make it difficult for new firms to enter the market, shielding existing producers (often referred to as incumbents) from competition. Barriers to entry are discussed in greater detail in the "Step 5" section. The larger the entry barriers, the slower

the speed and degree of entry into markets in which incumbents are earning profits that attract entry and the more sustainable the profitability of incumbents will be.

Exit barriers are market features that discourage firms from exiting a market even when they are losing money and would like to exit, making competition more intense and profitability of other firms lower than otherwise. Important forms of exit barriers include high costs of exit (such as agreements with labor to make large severance payments to workers), assets with low resale or salvage value, and government requirements that can be deferred by continuing to operate (e.g., final costs of mine closure). The larger the exit barriers, the greater the propensity for persistent excess production capacity during economic downturns and the lower the likely profits of operating in this sector.

These structural characteristics of markets define a number of idealized types of markets that, while not fitting any specific market exactly, focus our attention on important tendencies in the way actual markets work.

Perfect Competition and Monopoly

In perfect competition, there are (a) many buyers and sellers, (b) homogeneous, identical products, and (c) no barriers to entry and exit. These structural characteristics lead to the following:

- **Price-taking behavior.** The large number of buyers and sellers means that no single seller or buyer has appreciable influence by itself over price. All market participants are "price takers" in the sense that they simply accept the price the market determines through the independent interactions of all sellers and buyers.
- **Perfect substitutability.** Given that all producers supply homogeneous and identical products, all products are perfect substitutes for one another in the eyes of buyers, which intensifies the degree of competition.
- **Profits that are competed away quickly.** Profit accruing to an incumbent firm encourages entry by firms seeking to earn profits themselves. If entry is easy, entry occurs quickly, expanding supply, driving down prices, and reducing or even eliminating any profits beyond the minimum-acceptable profits required by investors and asset owners.

In a monopoly, the structural characteristics are polar opposites. There is (a) one supplier, (b) a unique product with no close substitutes, and (c) significant entry barriers, all of which together lead to the following:

- **Price-searching behavior.** A monopolist searches for its optimal price, which typically will be higher than if the market for the same product were perfectly competitive. A monopolist is not completely free to choose any price. It is constrained by market demand and its own production costs. A monopolist typically follows a trial-and-error process of determining a price that is higher than would exist in a perfectly competitive market but not so high as to encourage significant substitution away from its product or to encourage significant entry over the longer term.*

* A market with one buyer is known as monopsony. A monopsonist is similar to a monopolist except that it is a price searcher with respect to the price it pays for inputs, such as labor, raw materials, or transportation services. A monopsonist is able to pay a lower price than would exist in a perfectly competitive market.

- **Imperfect substitutability.** A product that is unique will have no substitutes, and thus users will have little flexibility to avoid paying the price a monopolist charges. In practice, almost every product has some type of substitute, and thus the issue is the relative availability of substitutes.
- **Profits that are sustainable.** Significant entry barriers make it difficult for entrants to replicate the activities of the profitable monopolist. For a significant period of time, a monopolist is shielded from the profit-reducing effect of entry.

Perfect competition and monopoly are simplified versions of reality. Few if any actual markets conform precisely to their structural requirements. Nevertheless, these market models focus our attention on three important attributes of more realistic, complicated markets: (a) the degree to which an individual firm controls the price at which it sells or buys its product, (b) the degree to which products from different suppliers are substitutes for one another, and (c) the relative ease or difficulty of entry and exit and the resulting likelihood of sustainable profitability.

A Dominant Firm with a Competitive Fringe

Some markets have characteristics that draw from both perfect competition and monopoly. One firm can be sufficiently large that its investment and production decisions influence price—it is the dominant firm. A dominant firm is a price searcher like a monopolist. Other firms make up the competitive fringe—firms sufficiently small that their individual investment and production decisions have no influence on price. Firms in the competitive fringe are price takers like firms in perfect competition.

A dominant firm, like a monopolist, is constrained by the response of consumers if the dominant firm charges a higher-than-competitive price. A dominant firm does not want to charge such a high price that buyers substitute away from the product it sells. Unlike a monopolist, a dominant firm is also constrained by the response of the competitive fringe if it charges a higher-than-competitive price. High prices encourage entry into the sector by firms in the competitive fringe or others, weakening the dominance of the dominant firm. A dominant firm does not want to charge such a high price that new entrants expand supply, drive down price, and weaken the dominance of the dominant firm.

Oligopoly

An oligopoly exists when there is a small number of relatively large sellers.* Each seller is sufficiently large that its individual investment and production decisions influence price. Moreover, when making these decisions, an oligopolist needs to consider the reactions of other oligopolists to its decisions. If one seller lowers its price to gain a larger market share, it will not expand its market share if the other oligopolists also lower their prices. If one seller makes a credible commitment to expand production capacity, it may discourage other oligopolists from expanding their production capacities. There is a mutual interdependence among oligopolists. Company behavior in oligopolistic industries is a little like behavior in a chess match.

* A market with a small number of relatively large buyers is an oligopsony.

Examples and Implications for Strategy

As noted earlier, no industry exactly fits into these market models. But the models can be considered first approximations of the structure of specific mineral and metal markets and the behavior of firms in these markets. These market models offer insights into business strategies appropriate in these sectors.

Markets for most major metals—copper, lead, nickel, zinc—as well as gold come close to the requirements of perfect competition. Most if not all firms are price takers, and products are essentially identical from all sellers. Business strategy focuses primarily on costs, that is, being a low-cost producer. Striving for low costs informs decisions at existing operations, as well as for acquisitions, mineral exploration, and development of new mines. In these markets, one new mine by itself will have little or no influence on price, and therefore, firms do not need to worry that their individual investment decisions will influence price or, more specifically, that building a new mine will expand supply to such a degree that price will fall.

Where the markets for major metals, and other minerals and metals, deviate from perfect competition is in the area of barriers to entry and exit. There are entry barriers that slow the entry of new production into metal markets even when existing firms would like to expand production and new firms would like to enter. There are exit barriers that slow the exit of operations that have become unprofitable. Barriers to entry and exit are discussed in more detail in the "Step 5" section.

Turning to the other market structures, there are few, if any, examples of monopoly. But there are a number of examples of the price-searching behavior of the dominant-firm and oligopoly models. In seaborne iron ore, Rio Tinto, BHP Billiton, and Vale each has the ability to influence iron-ore prices. Moreover, iron ores are not identical to one another. Physical size and shape include lump ores, fines, and pellets. Different ores have different ore grades and different impurities. As a result, iron ores are not perfect substitutes for one another from the perspective of iron and steel mills. These product differences are reflected in the desirability of different ore types and the resulting differences in ore prices from one type of ore to another.

The potash market is oligopolistic. The firms PotashCorp, Mosaic, Agrium, Uralkali, and Belaruskali represent two-thirds or more of world production. Moreover, at times these companies cooperate with one another. PotashCorp, Mosaic, and Agrium export their Canadian potash through the jointly owned company Canpotex. For many years, Russian and Belorussian potash was jointly marketed by Uralkali and Belaruskali, although this cooperative arrangement broke down in 2013. As with iron ore, potash comes in a variety of product grades, each with its own price.

Business strategy for dominant firms and oligopolists extends beyond striving to be a low-cost producer. At a minimum, strategy involves price-searching behavior, recognizing that one's investment and output decisions influence price and, in turn, the behavior of users and other producers. Restricting output to increase price encourages users to economize on their use of your product and other producers to enter the market—both of which erode one's market power. A dominant firm or oligopolist may undertake actions to deter entry into the sector, such as holding excess production capacity or holding exploration licenses for more land than it intends to actively explore to keep the land out of the hands of rivals. For more on business strategy in markets that are not perfectly competitive, see Besanko et al. (2016).

The first output for step 3 is a succinct statement summarizing the market structure for the minerals or metals the property under study would produce:

- What is the number and size distribution of sellers and buyers, the degree of product differentiation, and the presence or absence of barriers to entry and exit?
- Which market model most closely fits the minerals and metals this particular property would produce?
- In what ways does the model not fit this particular mine's minerals and metals well?

The second output for step 3 is a more-focused, preliminary evaluation of this mine's role in the market or markets for products this particular mine would sell:

- Would this mine be a price taker or price searcher? If a price taker, would this mine be vulnerable to actions by dominant incumbent firms (e.g., use of low-cost excess capacity to drive down prices and hurt this mine's profitability)? If a price searcher, would bringing this property into production have an effect on prices? Could production or capacity at this mine strengthen its market power?
- Would a standardized product be produced with many potential buyers? Or would the mine be a supplier of a specialized product, requiring close coordination with buyers and their special needs?
- How easy or difficult is it for other firms to enter and exit the market? If entry is easy, then any profitability will be temporary, unless this mine can consistently reduce costs relative to mineral or metal prices. If exit is difficult, this mine's profitability might become at risk because firms that should leave the sector do not.

Answers to these questions are important inputs to overall evaluation of the market in step 5. But first, prices and mechanisms through which prices are determined will be considered in the following section, "Step 4."

STEP 4: UNDERSTANDING SALES AGREEMENTS AND PRICING MECHANISMS

Up until now, this chapter has referred somewhat casually to a mineral's or metal's "price" as if there were only one price. Different types of prices have been ignored and, moreover, the mechanisms through which they are formed have been disregarded. Additionally, the chapter has ignored the practical issue of understanding sales agreements. Step 4 considers both sales agreements and pricing.

Sales Agreements

Sales agreements have three basic elements: price, quantity, and a variety of other terms. Among the important other terms are dates and locations of delivery; product specifications and quality parameters; mechanisms for dispute resolution; allowable damages should one party not live up to its obligations; and who is responsible for absorbing the costs of new, unanticipated government impositions. The Australian resources-law association AMPLA publishes model agreements for mineral sales (www.ampla.org).

Types of Prices

The important types of prices include

- **Spot or cash prices** for transactions with immediate delivery;
- **Forward prices**, which are determined today for a delivery in the future;

- **Futures prices**, which is a more-organized, standardized version of a forward price, made possible through pricing on a commodity exchange, which is discussed in the following "Pricing Mechanisms" subsection; and
- **Options on futures**, which is the price for a financial instrument that gives the holder the option but not the obligation to purchase a futures contract on or before a specified future date.

Pricing Mechanisms

Actual prices of any of these types are determined through a variety of mechanisms. Radetzki (2008) and Humphreys (2011) identify five principal mechanisms.

The first is commodity exchanges—institutions that serve as intermediaries bringing together buyers and sellers anonymously to determine price. Examples include the London Metal Exchange and the New York Mercantile Exchange. Trading occurs on a regular basis, usually daily, for a standardized set of products. Each institution defines types of contracts it will trade (e.g., spot, futures, options on futures) and standardized quantities, grades, delivery dates, currency of price quotations, and other specifications for the metal being sold or bought. Commodity-exchange pricing exists for many of the major metals (e.g., copper, lead, zinc, nickel, tin) and precious metals (e.g., gold, platinum, silver). But not all minerals and metals have commodity-exchange pricing (e.g., lithium, bauxite, gallium, indium, tellurium). Prices are typically determined by open outcry or some other means allowing agents for sellers and buyers to make their intentions known. Commodity exchanges facilitate both investing and hedging through the use of futures and options-on-futures contracts.*

The second mechanism is over-the-counter (OTC) trading, such as the London Bullion Market Association (LBMA). Although similar to pricing on a commodity exchange, OTC price formation is not anonymous. Rather, sellers and buyers negotiate directly and determine price without using intermediaries. Organizations such as LBMA provide standardized terms and a mechanism through which sellers and buyers interact.

The third is bilateral negotiations. Two parties negotiate prices, quantities, delivery dates, and other terms of a sales agreement.

The fourth mechanism is a price dictated by producers (producer pricing). Producers announce that they are offering material at a certain price, and customers choose whether to buy at that price.

The fifth mechanism is a price dictated by users (user-dictated pricing). Similar to producer pricing, in this case users announce they are prepared to buy material at a certain price, and sellers choose whether to sell at that price. For both producer- and user-dictated prices, some degree of pricing power must be in the hands of the producer or user; otherwise their counterparties would negotiate a more favorable price. Producer- and user-dictated prices are less common today than in the past.

Of the five mechanisms, bilateral negotiations and commodity exchanges are the most common. In fact, many sales agreements arranged through bilateral negotiations involve specific quantities to be delivered over the life of the contract (often monthly for a period of

---

* *Investing* here refers to using trading on commodity exchanges to take advantage of price volatility to make profits; in other words, investors earn profits by taking advantage of price risks. *Hedgers*, in contrast, seek to reduce and manage price risks through the use of commodity-exchange trading.

a year), while setting price on the basis of the commodity-exchange price in the month of actual delivery.

Transparency

A price is transparent if it is publicly available for all to see, including actual and potential sellers and buyers, as well as market analysts, commentators, investors, and speculators. Commodity exchange and OTC prices are the most transparent and, when they exist and are accepted by market participants, tend to be the basis for transaction prices even for transactions not conducted through commodity exchanges or over the counter. Prices determined through the other three mechanisms are less transparent. Negotiated prices often are kept confidential, although at times parties to a transaction may announce a price as part of efforts to influence prices in other transactions. Producer- and user-dictated prices, although typically announced, are not always the same as transaction prices because sellers or buyers may have to offer price premiums or discounts to achieve their desired quantities sold or bought.

For minerals and metals not priced on commodity exchanges or over the counter, prices are often published in the trade press based on information reporters obtain from parties involved in recent transactions. These prices can be considered generally indicative of current market conditions (i.e., useful first approximations) but have to be used with caution, as they may reflect an unrepresentative sample of recent transactions.

The outputs of step 4 are statements summarizing the types of sales agreements, pricing mechanisms, and the degree of transparency for pricing in the sectors relevant for the property under evaluation.

STEP 5: PUTTING IT ALL TOGETHER USING PORTER'S FIVE FORCES

The outputs of the first four steps serve as inputs into the overall market evaluation. Of the different ways one might organize this evaluation, a method Michael Porter proposes is especially informative (Porter 1980, 1998, 2008). Porter argues that competition comes in many forms, which can be grouped into five categories, the Porter five forces. Porter emphasizes that competition is a threat to profitability: the higher the degree of competition, the lower the potential profitability, and vice versa.

The goal of a five-forces analysis is to (a) identify the major threats to profitability in both the short run and the long run and then (b) develop business strategies that minimize the threats of the strongest forces, target areas where competitive forces are weak, and in so doing, enhance profitability.

A Porter analysis begins with a market definition, using the output of step 1 described earlier in this chapter. Once a market is defined, then the analysis proceeds to evaluating the relative strengths of the five forces: internal rivalry, threat of entry, threat of substitution, bargaining power of input suppliers, and bargaining power of buyers.

Internal Rivalry

The first competitive force is rivalry among existing firms in the sector, or internal rivalry. The important determinants of the internal rivalry are number and size distribution of existing firms, degree of product differentiation, and exit barriers, which are all concepts introduced in steps 1–4. Each of these will be considered in turn.

The larger the number of firms and the more nearly equal in size they are, the lower the likelihood that any single firm by itself will influence price. The Herfindahl–Hirschman Index (HHI) is a measure of market concentration, which incorporates both the number of firms and their size distribution. The HHI is the sum of the squared market shares of all producers in a market, with market shares expressed in the range of 0% to 100% of the market. HHI scores, in turn, can range from 0 to 10,000. Markets with HHIs of less than 1,500 are considered un-concentrated, and firms can be considered price takers. Markets with HHIs between 1,500 and 2,500 are moderately concentrated and with HHIs of greater than 2,500 are concentrated; the larger firms in sectors here are likely to have some degree of pricing power and are price searchers. See the website of the U.S. Department of Justice for more on HHI and how it is used in evaluating whether to permit company mergers (www.justice.gov/atr/herfindahl-hirschman-index).

Estimating a useful HHI depends critically on defining a market correctly, which the "Step 1" section focused on. Consider iron ore. Calculating HHI using all producers in the world and their shares of world production is inappropriate. Small Chinese producers of low-grade iron ore are not competitive in North American steel markets, and North American producers of iron ore generally are not competitive in Asian steel markets; their ores are not of sufficiently high grade and quality to offset the disadvantages of distance and high transportation costs. An appropriate HHI for iron ore depends on which region of the world the iron ore will be sold and should include only those iron-ore producers that can viably sell in this region. Moreover, calculating an appropriate HHI for iron ore requires deciding whether lump ore, pellets, and fines are sufficiently close substitutes to be included in the same market. Most analysts include all three types of ferrous material in a single market for purposes of estimating HHI.

The degree of product differentiation influences internal rivalry in the following way. With no product differentiation, all producers supply exactly the same product, which thus are perfect substitutes for one another. Markets for refined metals come to mind here. Any product differentiation reduces, to some degree at least, the intensity of internal rivalry. A higher-grade iron ore is more valuable than a lower-grade ore. A copper concentrate with easy-to-process impurities is more valuable than a copper concentrate with difficult-to-process impurities such as arsenic.

Exit barriers make rivalry more intense than would be the case if there were no exit barriers, as discussed earlier.

Threat of Entry

The second competitive force is the threat of entry, which is a function of entry barriers, a concept introduced in step 3. The higher the threat of entry, the lower the potential profitability, and vice versa. The idea here is that the threat of entry discourages existing firms from taking advantage of short-term ability to raise prices or in some other way take advantage of market power over buyers. Even if a firm has monopoly or oligopoly power today, it might choose not to use this power to avoid encouraging entry into the sector. Moreover, regardless of whether incumbent firms benefit from monopoly or oligopoly power, the higher the threat of entry, the more temporary any profitability in the short term will be. So threat of entry is a threat to profitability.

The threat of entry is determined by the extent to which entry barriers exist. As Besanko et al. (2016) discuss, there are two types of entry barriers: structural and strategic. Structural entry barriers represent cost advantages enjoyed by incumbents that entrants find difficult to overcome. Strategic entry barriers represent actions incumbents undertake to deter entry.

In mineral and metal production, among the potential structural entry barriers are those that follow:

- **Availability of low-cost, well-located, high-quality mineral resources.** Mineral deposits are heterogeneous. Relatively few are large, high grade, located close to the earth's surface, and with impurities that are easy to separate and remove from the desired mineral or metal. Some deposits with these natural (favorable) characteristics are nevertheless at a disadvantage because of remote locations. Thus owners of desirable deposits can earn profits that are sustainable over extended periods of time because of the difficulty potential entrants face in finding or acquiring similar deposits.
- **Scale and scope economies.** Scale economies exist when average costs per unit of output decrease as the scale or size of an operation increases. Often, scale economies exist but only up to certain scale or annual output rate. In these circumstances, there is a minimum efficient scale of operation, the scale beyond which average costs no longer decrease as scale increases further. When there is a minimum efficient scale of operation, there is a barrier to entering at an operating scale lower than this minimum because an entrant would have higher costs than larger operations. Scale economies often exist for bulk commodities (e.g., bauxite, coal, iron ore) and base metals (e.g., copper, lead, nickel, zinc), and thus it is rare for small mines to come into production. Conversely, gold mining often occurs at small scales without suffering from cost disadvantages, suggesting that scale economies are less prevalent in this sector.

  Scope economies are similar, except that they reflect falling costs per unit of output as the number of products recovered and sold from a mine increases. Sharing of costs is the key idea here. As discussed earlier in the "Step 2" section, more than half of the world's molybdenum is recovered as a by-product or co-product at copper mines, while the remainder comes from single-product molybdenum mines. By-product and co-product molybdenum typically have considerably lower costs per unit of molybdenum because of the sharing of costs between copper and molybdenum. The costs of mine design, drilling and blasting, ore haulage, and milling are shared. The general point is that the presence of significant by-product or co-product production serves as an entry barrier for higher-cost, single-product output of a mineral or metal.
- **Lack of access to proprietary technology.** If access to proprietary technology is key to profitability, then lack of access to this technology is a barrier to entry.
- **Government restrictions.** Government restrictions can make it difficult for entry to occur, or at least delay the timing of entry. For mining, lack of access to potentially mineralized lands, permitting and other preproduction government approvals, and fiscal and tax rules all have the potential to discourage or slow entry.

The degree to which any of these structural entry barriers actually discourage or slow entry into an otherwise profitable industry varies across minerals and metals and, in the case of government restrictions, from one political jurisdiction to another.

As for strategic entry barriers, an incumbent can hold excess production capacity to deter entry. Excess production capacity discourages entry if an incumbent can credibly signal the ability to bring unused, low-cost capacity into production should entry occur, making the entrant's higher-cost capacity unprofitable. It costs money for an incumbent to hold excess capacity. Thus, whether it makes sense to hold this unused capacity depends on the trade-off between (a) the costs of holding excess capacity and (b) the additional profits the incumbent earns compared to what it would earn post-entry should entry occur.

Threat of Substitution

The third force is the threat of substitution. The key idea here is that substitute materials, technologies, or systems are threats to profitability. Substitutes represent competition from the demand side of a market.

Substitutes come in different forms, which can be evaluated stage by stage over a material's life cycle. Consider a producer of iron ore in the form of fines. In the most direct sense, lump ore and pellets are substitutes for fines. Moving down the supply chain, iron and steel scrap is a substitute for all forms of iron ore in iron and steel making. Most five-forces analyses of iron ore, however, probably would include lump ore, fines, pellets, and scrap metal as sufficiently close substitutes that they would be included earlier in evaluating internal rivalry.

So for purposes of most five-forces analyses in mining, threat of substitution begins with substitutes for metal or intermediate product derived from the mineral resource. In this sense, the most obvious form of substitution is *material for material*: glass or plastics instead of aluminum in beverage containers, fiber-optic materials for copper in certain communications applications, and aluminum and composite materials for steel in the outer-body panels of motor vehicles, for example.

Further down the supply chain, a second form of substitution is *technological*, in which a new processing technique or new product alters demand for the material under study. Aluminum cans now contain less aluminum than they did when aluminum cans were first used in the 1970s because of improvements in the rolling of aluminum sheet. Regarding new products, the demand for samarium and cobalt for use in samarium-cobalt (Sm-Co) permanent magnets is lower today than one might have imagined in the late 1970s because of the development of more-powerful neodymium-iron-boron magnets in the 1980s, which replaced Sm-Co magnets is some uses.

Finally, a third form of substitution is also technological but in a broader *systems* sense. One system for delivering a desired set of material properties replaces another system. Consider lighting and incandescent, fluorescent, and light-emitting-diode (LED) systems. Incandescent lightbulbs rely on heating a wire filament, usually tungsten, to a high temperature to create light. Fluorescent lights, both compact and linear tubes, rely on a different technological system to create light, using electricity to excite mercury vapor, which in turn causes phosphor materials to light up. The phosphors in fluorescent lamps contain yttrium and the lanthanide elements lanthanum, cerium, europium, and terbium. LED bulbs use semiconductor materials to create light. Although they contain yttrium, europium, and terbium like fluorescent lamps, LEDs use much less per lumen of light generated. As incandescent bulbs are being replaced by fluorescent and increasingly by LED lights, the relative demands for the constituent raw materials (tungsten, yttrium, and the lanthanide elements) are changing considerably. For example, the substitutions embodied in the switch from fluorescent to LED lighting are significantly reducing demand for the lanthanide phosphor elements.

Bargaining Power of Input Suppliers and Buyers

The fourth and fifth forces involve bargaining power—of input suppliers (the fourth force) and buyers (the fifth force). The key ideas are the same in both cases: a powerful input supplier or buyer can negotiate more favorable terms than otherwise, reducing the profits of the firm buying the inputs or selling output. To link this analysis with step 3 and our evaluation of market structure, these forces contemplate situations in which an input supplier has monopoly or oligopoly power (the fourth force) or a buyer has monopsony or oligopsony power (the fifth force).

For mineral development and mining, important input suppliers to consider are providers of labor, equipment and spare parts, explosives, chemical reagents, transportation services, electricity, and fuel. In addition, local communities and local, regional, and national governments should be viewed as input suppliers. Without their approvals, formal or otherwise, mining cannot occur. Without the goodwill of communities in which mining occurs, mining will be more difficult and costly than it would be if goodwill did not exist.

The degree to which input suppliers are threats to profitability varies considerably from one location to another and from one input market to another. Labor may be strongly unionized in one location and not in another. Transportation services may be subject to strong competition in one location and not in another. Local communities may support mining in some locations but not in others, and so on.

The evaluation of buyer power is more straightforward because the number of relevant markets is smaller than for inputs. A mine has a limited number of products it sells, perhaps only one. Producers of many major and precious metals, such as copper or gold, produce a standardized product with many potential buyers and prices for almost all transactions determined on commodity exchanges. In these cases, there is no special threat to profitability because of the buying power of one or a small number of large buyers. In other cases, a mine produces a concentrate for which the number of potential buyers or processors is small, and the threat of buyer power can be larger. Concentrate producers sometimes have fewer buyers or processors to consider, and thus are at greater risk being taken advantage of by powerful buyers, because (a) it is costly to ship concentrate long distances, and thus a mine is limited to relatively nearby processors; or (b) the mine's concentrate is of a particular mineralogy or contains impurities that only some metallurgical facilities are capable of processing.

Many of the smaller mineral and metal markets, such as lithium and the various rare earths, do not have a single standardized product at all. Lithium is used in carbonate, chloride, hydroxide, metal, and other forms, and different users have different purity requirements. For rare earths, some users require mixed rare-earth oxides while others require separated oxides of specific elements; many applications require purities of only 99.9%, while others require 99.999%, and even at the same purity level, different customers have different tolerances for specific impurities. In these cases, a buyer can develop a strong bargaining position relative to its supplier if the supplier has invested in difficult-to-redeploy assets to satisfy the needs of the buyer.

Factors that influence the bargaining power of input suppliers and buyers include the following:

- **Relative industry concentration.** In an un-concentrated market, with many small sellers and buyers, neither suppliers nor buyers will tend to have an advantage in bargaining power. But if one side of a transaction is more concentrated than the other, then firms in the concentrated side will tend to have an advantage in bargaining power.

- **Purchase volumes.** The larger the purchase, whether of inputs or by buyers, the larger the potential bargaining power.
- **Availability of substitutes and switching costs.** The greater the availability of substitutes and the lower the switching costs, the larger the potential bargaining power. This statement is true whether considering substitute inputs or substitute products.
- **Threat of forward or backward integration.** If an input supplier can threaten credibly to integrate forward, then a purchaser of inputs may choose to pay a higher price rather than risk facing a new competitor. If a buyer can threaten to credibly integrate backward, then a seller may choose to accept a lower price rather than risk facing a new competitor.

Thinking About Government in the Five Forces

Porter only includes government in his five forces when discussing threat of entry. But clearly, government policies can influence other competitive forces. In internal rivalry, government tax policies can influence behavior of existing rivals. In threat of substitution, government policies influence what materials are used in specific applications. For example, government banning of lead from paints, gasoline, and most other products other than batteries significantly influenced the demand for lead. Local building ordinances in some places prohibit use of aluminum in wiring of homes. Government is a supplier of inputs in the sense that government approvals are one type of input necessary for mining to occur. In some cases, governments are important buyers of minerals or metals.

Summing Up the Five Forces

Porter's five forces provide a framework for incorporating a significant amount of detail about a market. It is important, however, not to be overwhelmed and confused by the detail and complexity of a market. In other words, detailed analysis needs to lead to an overall evaluation of the major threats to profitability to a firm operating in a particular line of business. Table 8.2

TABLE 8.2 Market summary evaluation template using Porter's five forces

<table>
<tr><td colspan="3">Market Definition
Geography (world, regional, local?)
Nature of the product (homogeneous or differentiated?)</td></tr>
<tr><td>Threats to Profitability</td><td>Short Term</td><td>Long Term</td></tr>
<tr><td>Internal rivalry</td><td></td><td></td></tr>
<tr><td>Threat of entry</td><td></td><td></td></tr>
<tr><td>Threat of substitution</td><td></td><td></td></tr>
<tr><td>Bargaining power of input suppliers</td><td></td><td></td></tr>
<tr><td>Bargaining power of buyers</td><td></td><td></td></tr>
<tr><td colspan="3">Implications for Strategy
▪ Cost leadership or differentiation?
▪ Degree of focus (or not)?
▲ Geographic?
▲ Which segments of supply chain?</td></tr>
</table>

Adapted from Besanko et al. 2016

offers one possible template for organizing an overall market evaluation using Porter's five forces. It starts with a restatement of the market definition, including both the geographic and product dimensions of a market. The template then provides spaces for succinct statements about the degree to which each force is a threat to profitability—first in the short term (one to a few years, up to about a decade) and then in the long term (a decade and beyond).

This overall evaluation of threats to profitability and the relative strengths of competition in each Porter force leads naturally to a consideration of business strategy, that is, how to operate in this particular market to minimize the major threats to profitability and target areas where competitive forces are weak. Detailed discussion of strategy is beyond the scope of this chapter. But it is useful to link the five-forces analysis with the three generic strategies that Porter (1980) identifies: cost leadership, differentiation, and focus. *Cost leadership* means striving to be a low-cost producer. It is the obvious starting point for strategy when internal rivalry is intense and your firm has no influence over the price at which it sells its product.

Differentiation refers to producing a superior product for which a premium price can be charged; producing a superior product usually involves incurring higher costs than other firms in the sector. Differentiation is not an option when producing a standardized product, such as refined metals. But it is an option in markets with significant product differentiation.

Focus emphasizes the strategic decision firms must make about where to operate (geographic focus) and whether to be active in more than one stage in the production process (mining, metal production, intermediate products, etc.).

CONCLUSIONS

Deciding whether to develop a mineral deposit into a mine depends not just on the characteristics of the deposit but also on the market environment for the products the mine would sell. This chapter suggests a five-step framework for market analysis: define the market and competitors, understand demand and supply, characterize market structure, understand sales agreements and pricing mechanisms, and put it all together using Porter's five forces. As noted at the beginning, the steps are not strictly sequential and linear. They overlap to some degree, and inferences drawn during one step may require reconsideration of inferences drawn at an earlier step. The analysis becomes increasingly nuanced and complex as it proceeds. The framework seeks to balance clarifying simplicity with realistic complexity.

SELECTED READING

For readers interested in more-detailed treatments of mineral and metal markets, Tilton and Guzmán (2016) provide an introduction to the economic behavior of mineral and metal markets aimed at students and professionals in the mining sector. Radetzki and Wårell (2016) contains a broader examination of primary commodity markets. Maxwell (2013) is an edited series of papers by experts in market analysis, finance and investment decision making, and public policy. Humphreys (2015) provides historical commentary on how mineral and metal markets evolved and changed during the first decade and a half of the 2000s. All of these books are accessible to interested professionals, regardless of whether they have formal education in economics.

Besanko et al. (2016) provide an introduction on how to use microeconomic principles to inform business strategy. Porter first presented his five-forces model in Porter (1980), and he provided updates and commentaries in Porter (1998, 2008). For those interested in marketing industrial minerals, Chapter 2 provides advice on quality problems for industrial minerals, and Kogel et al. (2006) provides considerable general information on markets for industrial minerals and rocks.

ACKNOWLEDGMENTS

The author is grateful to his colleagues and graduate students in mineral and energy economics at the Colorado School of Mines, especially John Tilton, who provided comments on an earlier version of this chapter. The author also thanks colleagues and students at the Curtin University of Technology, Australia, and the University of Chile, where he has had the privilege and pleasure of teaching short, intensive courses on analyzing mineral and metal markets. Very little of this material is original, and the author is thus indebted to their many suggestions and contributions. Any errors or flawed thinking, of course, are the author's alone.

REFERENCES

Besanko, D., Dranove, D., Shanley, M., and Schaefer, S. 2016. *Economics of Strategy*, 7th ed. Hoboken, NJ: John Wiley and Sons.

Gentry, D.W., and O'Neil, T.J. 1992. Mine feasibility studies. In *SME Mining Engineering Handbook*, 2nd ed. Edited by H.L. Hartmann. Littleton, CO: SME.

Humphreys, D. 2011. Pricing and trading in minerals and metals. In *SME Mining Engineering Handbook*, 3rd ed. Edited by P. Darling. Littleton, CO: SME.

Humphreys, D. 2015. *The Remaking of the Mining Industry*. Houndsmill, Hampshire, UK: Palgrave Macmillan.

Kogel, J.E., Trivedi, N.C., Barker, J.M., and Krukowski, S.T. 2006. *Industrial Minerals and Rocks: Commodities, Markets and Uses*, 7th ed. Littleton, CO: SME.

Kuck, P.H. 2016. Nickel. In *Mineral Commodity Summaries 2016*. Reston, VA: U.S. Geological Survey. doi:10.3133/70140094.

Maxwell, P., ed. 2013. *Mineral Economics: Australian and Global Perspectives*, 2nd ed. Monograph 29. Carlton, Australia: Australasian Institute for Mining and Metallurgy.

Porter, M.E. 1980. *Competitive Strategy: Techniques for Analyzing Industries and Competitors*. New York: Free Press.

Porter, M.E. 1998. *On Competition*. Boston: Harvard Business School Press.

Porter, M.E. 2008. The five competitive forces that shape strategy. *Harvard Business Review* 86(1):78–93, 137.

Radetzki, M. 2008. *A Handbook of Primary Commodities in the Global Economy*. Cambridge, UK: Cambridge University Press.

Radetzki, M., and Wårell, L. 2016. *A Handbook of Primary Commodities in the Global Economy*, 2nd ed. Cambridge: Cambridge University Press.

Tilton, J.E., and Guzmán, J.I. 2016. *Mineral Economics and Policy*. New York: RFF Press and Routledge.

CHAPTER 9

Environmental Considerations During Feasibility Stages

Scott Mernitz

INITIAL CONCEPTS FOR NON-ENVIRONMENTAL SPECIALISTS

The issues related to the environment were often those most neglected in mining and energy project development in the middle and later decades of the 1900s. Yet these issues are usually critical to project success. The environmental laws of the United States—beginning with water pollution initiatives in the 1950s through the National Environmental Policy Act (NEPA) and the clean air, clean water, hazardous wastes, and abandoned waste sites cleanup initiatives in the 1970s through 1990s—recognized this growing concern. Other countries around the world were sometimes leading or often actively following these efforts in the United States.

Prudent environmental planning for mining impacts, impact mitigation, and reclamation and closure can usually save hundreds of thousands or even millions of dollars on the bottom line throughout project life. Such planning and financial management during operations can result in more attractive mining company annual reports, with the added benefit to shareholders of progress on a "green," sustainable, profitable set of projects. Figure 9.1, which depicts a prospective site for a mining project from the mining company's perspective, shows where environmental planning and permitting fits in.

Thus it makes sense for mining company staff members, bankers and other investors, and students to view environmental planning and potential impacts from a project in a certain way. Following is advice from the perspective of mining company staff when contemplating a project.

Initially, it is helpful to view those potential mining impacts in terms of management of *physical resources* during mine development, such as

- **Air**, including noise, dust, and chemical concentrations in the air affecting quality of life, which may affect human health and environmental risks;
- **Water, both surface and groundwater**, and the **sediments** in streams and lakes, and overall water management plans for the project;
- **Soils and underlying rock**, the surficial and bedrock geology; and
- **Land use** in terms of present and future (post-mining) land use, and reclamation of the land, including the preceding components.

Look also to the *biological resources*, including

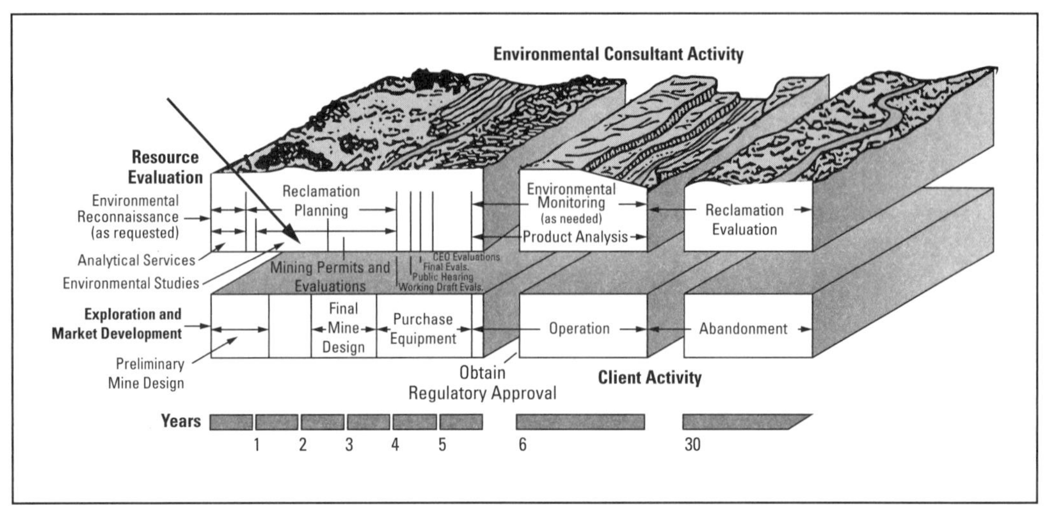

Source: CDM 1980

FIGURE 9.1 Environmental planning and permitting (large arrow) as it fits in the sequence of mining development on a prospective site

- The biological and physical components of area **soils**, such as weathered vegetation, microbes, and other soil organisms, as well as fertility to be conserved for future reclamation and geotechnical characteristics for construction;
- **Vegetation and wildlife (hereafter *flora* and *fauna*)**, including all plants and animals cataloged in the area, especially those that may be threatened or endangered or otherwise sensitive species, and even insects (in a rare case, such as in California); and
- **Aquatic ecology**, organisms in the surface water, and sediments in the water (lakes, ponds, streams, wetlands, and even those in groundwater and moist areas, such as in Australia).

And finally, consider the *human resources* that could be affected:

- The local and regional people, or more scientifically, the population demographics and its **socioeconomics**, including age and sex structure, employment, income, and related demographic issues
- The population's **sociocultural** characteristics, such as social groups, history, religion, origin, traditions, and practices, tribes, views of development, governance, and similar issues
- The **historical and archaeological (cultural) resources** in the project area of potential significance, which may require special treatment during development (and even sometimes paleontological or fossil resources, which may exist in bedrock or sediments)
- **Traditional and artisanal land uses** and relationships to the current land uses and planned mining development
- Modern, human-made arrangements, such as transportation, energy, and other **infrastructure** issues

- The **visual and aesthetic** resources and how project development could affect views, vistas, air quality, and haze, smoke, noise, and similar human perceptions

Thus a project needs *environmental baseline studies* in many or all of the preceding areas, which are highlighted in bold, in order to document conditions that existed before the mine is developed. These studies will help to prove, during and after the mining project has run its course, that the development of these natural resources has only had certain, documented before-and-after impacts. It is important that the proposed operation is not saddled with others' problems as future impacts are predicted or tallied. And further, *environmental background levels* of metals in water and soil in a mineralized area, for example, may already be higher than regulatory action levels for the state, province, or country, and baseline information is needed to substantiate this point before development begins.

Mitigation, defined as things an operator can do to minimize or eliminate an impact by changing some part of the project or the way it operates, is used to offset impacts. For example, a company may *mitigate* the visual impacts of a large volume of waste rock by backfilling it into an opening or pit, and then perform *reclamation and closure* on that pit by proper reapplication of topsoil, replanting of vegetative cover (if that is to be the future land use), and monitoring of both surface and subsurface conditions. Of course, the costs of all this activity are balanced by the benefits it brings in both tangible and intangible ways to the mining project bottom line. And *environmental permit requirements* as negotiated will, of course, dictate many expenditures of capital or operating costs for environmental matters.

Using the preceding basic environmental principles, the next sections examine the three stages of project development and how environmental considerations are best managed. Some of the preceding terms are further defined, and advice is offered on how certain issues can be handled.

PRELIMINARY FEASIBILITY

Types of Planning Studies

Various terms are used for the environmental and other studies undertaken by company staff members or consultants during the preliminary feasibility stage. The following is a list of some of the more common ones and their differences and similarities:

- **Scoping study.** This is a quick look, usually in the space of a few weeks on a desktop basis, at the key environmental issues the project may face and some qualitative estimates of risk.
- **Fatal flaw analysis or environmental reconnaissance.** Similar to the scoping study but with a focus on "project stopper" issues, this analysis may discover a fatal flaw in the project planning. A two-day to one-week site visit is usually made, which includes some local meetings and telephone calls; and a verbal report is usually given immediately after to client management. The analysis discusses in a preliminary sense why, how, and when issues may arise, and what could be done to mitigate them in project planning.
- **Permit planning.** A longer study, usually taking two or three months, is necessary to assess the environmental permits for air, water, land use, and other matters required for the project and to develop a rough, coordinated permit and public participation schedule. Regulatory agencies may be briefly contacted.

- **Environmental and permitting risk and liability analysis.** As part of one of the preceding studies, management may ask for an environmental and permitting risk analysis, listing technical risk areas in permitting, scheduling, and expenditures, and projecting quantitative levels of risk, cost, and delay in terms of percent, money, and months.

One or more of these studies may be performed in the preconstruction or early project planning stages to assess feasibility.

Global Issues and Perceptions

From the start, all modern mining projects, especially in this second decade of the 21st century, must keep in mind the global environmental issues of sustainability, maintenance of biodiversity in the project setting, and similar matters. These issues will probably be superseded in 10 years by another set. Chapter 10 addresses the issues of sustainability and the social license to operate in more detail.

For example, maintaining *biodiversity*—that is, attempting to maintain the diversity of flora and fauna during and after development in a project setting—is a relatively simple concept using baseline and impact data. If a sensitive plant or animal species is eliminated or driven away because of mining activity, biodiversity is affected. Mitigation may require special baseline studies, relocation of species, scheduling of operations or disturbance during certain seasons of the year, purchase of new compensatory habitat, or similar.

Operations staff should address biodiversity and other global environmental issues, which may be gaining current prominence during several phases: (1) as a project is initially planned, (2) as expansions or partial closures are considered, (3) as acquisitions are reviewed, (4) as agencies and nongovernmental organizations (NGOs) are contacted for their views, and (5) as reclamation and closure are contemplated.

Lenders and Lender Liability

The fine points of project financing are discussed elsewhere in this handbook. From an environmental perspective, however, lenders are often keenly interested in the risks related to a project going forward because some issues often become frustrating "project stoppers" or "delayers" that are difficult to solve. Environmental due diligence analyses frequently conclude with a section on risks. As a potential problem is assessed, mine project staff members and investors have found the following definitions of *risk* useful in their environmental decision-making:

- **Low risk.** The problem is unlikely to occur, and if it does, the affected systems are reinforced by adequate contingencies, engineering, and detailed cost information. The severity of impacts to operations is predicted to be negligible or easily remedied.
- **Medium risk.** Average risks exist, and there is good potential for the problem to need to be addressed at some point. These risks are typical for the industry but require continuing scrutiny, because they may cause impacts to cash flow significant enough to occasion a notable divergence from economic models and unexpected additional work and expense for staff.
- **High risk.** Risks that are highly likely to occur are considered unacceptable; a significant problem or divergence from plan is likely. The impacts on service to financial commitments and project development will require significant additional work. These risks are frequently referred to as "fatal flaws."

An environmental due diligence analysis will, therefore, often rate the notable risks in environmental quality sectors, so that lenders can be advised, and lenders can in turn query project staff. Operations staff will, of course, seek to identify and head off such risks from rising above the "low or negligible" level.

Regarding other guidance on risks, as of 2016, more than 80 global banks and lending institutions had adopted a set of guidelines called the *Equator Principles* (Equator Principles Association 2013). These principles are based on the World Bank and its International Finance Corporation (IFC) guidelines, regarding attention to and compliance with environmental and social matters, specifically for review of project loans. Chapter 10 addresses the Equator Principles and their importance for environmental and social aspects of project finance.

Bonding

A *reclamation and closure bond* is required on nearly all mining projects now permitted in the United States and is increasingly required on international projects with any potential liabilities and NGO attention. The bond, or *surety*, is a financial assurance that reclamation and closure required under permits by the regulatory agencies and other voluntary commitments by the operator will be met, notwithstanding the financial viability of the project and its sponsor at closure.

A few other terms also apply to bonding documents:

- The bond *permittee* receives the permit from the regulatory agency based on the posting of the bond.
- *Surety* may refer to the bonding documents themselves or to the insurance company as the *obligor or principal* backing the bond.
- The *obligee* is the party (usually a state or federal agency for mining projects on U.S. public lands) to whom the bond amount is paid, if the operating mining company forfeits its obligation to reclaim and close under normal business conditions.
- *Forfeiture* is that act of nonperformance, whereby the bond may be "called" by the obligee.
- *Collateral* are those monies, written instruments, equipment, or property put forward by the mining company to guarantee the bond, usually constituting 0.5% up to 15%–20% (and even more, in the 50%, 75%, or 100% range in extreme cases) of the total bond amount.
- *Accruals* are those company funds, on paper or in actual cash accounts, accrued during operations, often on a per-unit-of-production basis, to be used for later reclamation and closure.

Bonds are generally of four major types.

1. A *self-bond* is sometimes allowed to be put forth by a large, profitable mining company with a long history to guarantee the bond using the company's good name and credit performance.
2. An *irrevocable letter of credit* (which is sometimes known by its abbreviation, ILOC) may be obtained by a similar, well-funded company after payments or deposits in a bank, to be then guaranteed by the bank to back the bond.

3. A full *cash bond* may be required upfront by the agencies from a mid-level or junior mining company for a specific reclamation or closure item, usually in the few millions or tens of millions of dollars.
4. A broker may secure for the mining company a *bond policy*, from a surety or insurance company, to guarantee the reclamation and closure work, with payment of a premium and broker fee by the mining company.

The fourth type is perhaps the most common. However, beware that the surety will attempt to protect itself strongly in the event that the bond is called and may forestall payment through various administrative and legal proceedings over many months or years. Increasingly, bonding companies are cautious and aware of risks as metals prices decrease; production costs increase; reclamation and closure estimates escalate; more mining companies have financial difficulties; more bonds are called; and bankruptcy, especially among small- and medium-sized mining companies, is more prevalent than in better times.

The bond obligee in the United States is typically one agency or party. However, joint federal/state agency bonds have been discussed and administered in the United States, especially if the agencies have a written memorandum of understanding (often just called an MOU) or similar, and have common legal, administrative, and financial requirements. In such a case, one agency may be the obligee and the other, whose public lands or public welfare responsibilities may be affected, stands by to see that it is satisfied with final progress. Table 9.1 shows some U.S. mining project bonding examples in a time of active mining activity during the 1990s. These data were publicly available in environmental impact statements (EISs) or similar documents and from agency interviews during this period. One could compile a similar list for any type of project and locality.

As recently as 2015, one Montana metal mine renegotiated its bonds on two properties with federal and state agencies as surety bonds with two different insurance companies—each in the tens of millions of dollars and totaling around US$40–50 million. Bond amounts and details are sometimes posted on mining company websites under the "Environmental" or similar heading to document environmental responsibility for shareholders, prospective investors, and other interested parties.

In contrast, coal mining companies were under increasing stress in 2016 as climate change advocates and the U.S. presidential administration disfavored emissions from coal-fired power plants and favored solar and wind energy projects. It was reported (Gruver 2016) that more coal companies were facing bankruptcy as a result, and many were self-bonded, as explained earlier. The news story noted that the three biggest bankrupt coal companies had US$2.3 billion in IOUs (self-bonds) across five states regarding their reclamation and closure funding, creating concern among many parties.

It follows that the act of bonding and bond negotiations is critical in the early project stages for a mine and/or mill operator. Calculations of bond requirements are typically performed by both the operator (internally) and the government agency, then compared and negotiated. Work estimates for a third-party contractor—coming in from outside and bringing its own equipment and personnel—are typically used by the government, usually creating higher estimates than those for the mining company's ongoing, operational cleanup. Also, bond amounts may be expected (and requested by the obligee) to increase during project life to deal with increased reclamation grading, top-soiling of waste areas with planting requirements,

TABLE 9.1 U.S. bond amounts and obligees*

Year of Study or Public Document Consulted	Operating or Project Site Location	Operator	Bond Amount, US$	Bond Obligee	Comments
1994	Lamefoot Gold Mine near Republic, Washington	Echo Bay Mines	$83 million (initially; reduced by 2002)	Washington Department of Ecology	BLM was originally listed on bond rider.
1995	Zortman and Landusky Gold Mines, Zortman, Montana	Pegasus Gold	$10 million (Zortman) $15 million (Landusky)	Montana Department of Environmental Quality (MDEQ; formerly Montana Department of State Lands)	"Joint bonding" was in effect to the extent that release of reclamation bond on Bureau of Land Management (BLM) surface is approved by MDEQ, but subject to concurrence by BLM on completion of successful reclamation.
1998	Florida Canyon Gold Mine near Winnemucca, Nevada	Pegasus Gold	$17 million	U.S. Department of the Interior BLM, Winnemucca	Reclamation and closure plan's primary focus is protection of waters of State of Nevada; Nevada Division of Environmental Protection issued six permits for operations.
1998	Black Pine gold mine near Burley, Idaho	Pegasus Gold	$3 million	U.S. Forest Service (USFS)	
1998	Atlanta Gold Project, Atlanta, Idaho	Twin Gold Corp.	$1.4 million (projected estimate)	USFS	Cyanidation bond
			$0.1 million (projected estimate)	Idaho Department of Environmental Quality	
1999	Tonkin Springs gold mine near Austin, Nevada	Gold Capital Corp.	$2 million	BLM	
1999	Yankee Gulch Sodium Minerals Project near Meeker, Colorado	American Soda	Total unknown	U.S. Environmental Protection Agency (EPA), BLM, Colorado Division of Mining and Geology (CDMG)	Joint bonding among these agencies for: EPA-in-hole closure of product wells; BLM-well and core hole closures, surface reclamation, federal royalty obligations; and CDMG-surface disturbance
2001	Stillwater (Nye) and East Boulder Platinum Group Metals (PGM) Mines near Nye and Big Timber, Montana	Stillwater Mining Company	$14.8 million (East Boulder operations-pending) $8.7 million (Nye operations)	MDEQ	Federal agency, USFS, is interested in joint bonding and would have same requirements as state, but financial and administrative agreements are not yet complete.

* Precious metals and industrial minerals project data compiled from Behre Dolbear public reports, agency interviews, and other publicly available information.

backfilling issues in surface pits and underground workings, changes in water quality, regulatory changes, regulatory agency administrative fees, and the like.

Corporate Commitment and Quality

A key issue regarding environmental protection for a project is public commitment by the corporation. Shareholders, NGOs, and others reviewing a project and a mining company's reputation often look to a signed statement by the chief executive—a *corporate environmental policy*—as a commitment to a policy of sound environmental management while ensuring corporate economic health. This could be described as a corporate commitment to *environmental quality*. Related policies may exist regarding employee health and safety, community participation, and similar sustainability-type issues (see Chapter 10). Reviewers of a particular mining project also evaluate actual on-the-ground progress and actions backing the policy statement and will survey opinions from regulatory agencies, local groups, and other stakeholders.

The mining firm may also be concerned about the technical and environmental performance of its products in the international marketplace or *product quality*. As goods travel farther and farther abroad, vendors and consumers alike want the assurance that, for example, the vermiculite used in potting plants or in home insulation is free of asbestos-type fibers from a health and safety standpoint. They want to know that other industrial minerals and precious metals they purchase will perform as well as those they have purchased in the past. International standards help to ensure this product quality.

An interesting point of trivia regarding minerals quality and performance especially concerns coal and industrial minerals, notwithstanding the varying quality of oil and gas resources. Coal especially is not considered *fungible* or interchangeable as an energy mineral; that is, coal resources vary in moisture and ash content, heating value, and mineral matter and elemental content dependent on their nature and on their processing before going to market. And different grades of coal have different values in the market. Industrial minerals have similar quality variations. Gold, silver, copper, and several other metals, however, have distinct value characteristics and are fungible in the market. An ounce of pure silver, after refinement, has the value and qualities of an ounce of pure silver worldwide given published exchange global market prices. Iron ore after processing, oil, and others, including industrial minerals, are often chemically and physically variable, with variable prices because they are nationally, regionally, and globally market driven. The strict definition of *mineral* relates to a homogeneous element with a constant chemical composition and formula.

It follows that salable minerals and mineral processing wastes are variable, and firms are increasingly looking to organizations, such as the International Organization for Standardization (ISO) in Europe, for certification of product quality of salable minerals on world markets (ISO 9000 program and related standards). Regarding globally recognized standards for environmental management systems and mining wastes management, the ISO 14000 family of standards for environmental management has been substantially refined with related topics since it was first published in November 2001. ISO also created a standard for environmental due diligence, which was augmented by ISO 14001 (*Environmental Management Systems*) and ISO 14010 (*Environmental Auditing*), both commonly used in Australia and internationally in the early 2000s. Note that ISO 14010 has been superseded by ISO 19011 (*Guidelines for Quality and Environmental Management Systems Auditing*), which encompasses a number of audit-related functions.

As to global reach, because each country administers its own ISO certificates, a tally is difficult. One ISO informant estimated that by the end of 2002, more than 560,000 firms worldwide enrolled in the ISO 9000 program and about 49,000 in the ISO 14000 program. As of 2015, ISO 14001 claimed growth to more than 300,000 certifications for companies in 171 countries. The ISO website (www.iso.org) provides many details.

Environmental management programs usually specify *environmental audits, reporting,* and *corrective actions* regarding company facilities. In the United States, ASTM International (formerly American Society for Testing and Materials) offers ASTM E1527-13 and ASTM E1528-14e1 (as updated) as guides for environmental audits or site assessments, especially for mining and mining-related projects of small to medium size. These projects are sometimes of the "commercial real estate" type, such as industrial minerals companies or drilling companies under review by investment banks. These site assessments can give an initial view of potential risks to investment at a small property, and in advance of an aggressive scoping study on a larger property.

The full technical audit of permitting activities with risk and liability analysis, which was noted earlier, is a more comprehensive review of such topics than the site assessment and can be very detailed and instructive, involving several weeks.

Environmental due diligence reviewers of mining projects, be they government or private consultants, look for these audit documents prepared under recognized standards and evidence that their existence is backed by actions. These actions may take the form of company responses to customer complaints, remediation expenditures for waste emissions problems, attention to community concerns, and similar, risk-mitigating actions.

Large mining companies often have their own internal and external environmental auditing programs, the latter employing independent contractors and reported annually on their company websites under "environmental" or "sustainability." Such companies may have developed their own product quality and environmental management systems outside of the ISO and follow them with various levels of attention, corporate commitment, and funding. Systems like these are desirable in the global mining marketplace for substantiating consistent product quality, assessing mining wastes management, and reporting on environmental responsibility.

Scoping Out the Project

Those planning a mining project are advised to keep several environmentally related issues in mind during the preliminary feasibility study. These are

- Environmental baseline conditions and projected impacts to geology, soils, water, and air that could be project stoppers (e.g., air issues might have been first priority, but now water management in quantity and quality is often predominant);
- The community setting and the sociocultural climate, how it is formally governed, and existing informal groups, tribes, factions, political action organizations, and others who may react to the project;
- The risk of obtaining permits in a reasonable time, agency perceptions, politics (local, state, regional, national); and
- Other issues in the regional, national, and global setting, such as international environmental protection initiatives, World Bank and IFC guidance, and recent sustainability initiatives from both mining and environmental perspectives.

Conclusions during the preliminary feasibility stage can thus be generated based on this first pass at environmental issues and risks analysis.

INTERMEDIATE FEASIBILITY

After favorable conclusions during the preliminary feasibility stage are gained from environmental studies and in all other areas addressed in this handbook, the project is carried on to the next step of intermediate feasibility (or prefeasibility). For the environmental staff, this involves attention to several issues during the intermediate stage.

Plan Specifications

For each of the environmental disciplines discussed at the beginning of the chapter (see the "Initial Concepts for Non-Environmental Specialists" section), it follows that the baseline, existing conditions—however affected by other previous human activities—should be clearly sampled, monitored, and described during the intermediate feasibility stage. A detailed plan of study for each relevant environmental discipline is thus developed. Once again, this plan has the benefit to the mining company of establishing its responsibility for future actions against the baseline of any prior contamination or alteration. This prior disturbance could be by another mining company, artisanal miners (historic or current), a power plant, transmission lines or other rights of way, water dams, or other impoundments of various sorts, logging enterprises, fishing or aquaculture, agriculture, or numerous other activities.

The concept of *adequate data* is often debated with regulatory agencies and NGOs. Some U.S. agencies have published *data adequacy standards* describing in detail the types of data, data quality, and length of monitoring period required. Mine environmental staff should try to confirm data adequacy with the agencies in writing, in a *plan of study*, so that surprises do not come a year or two down the road. The company needs to negotiate out of improper demands to conduct university, NGO, or agency "research projects" with the company's time and money.

Usually, contractors who are specialists in the environmental disciplines set up the monitoring stations, maintain them, conduct the field studies, and prepare the *environmental baseline report*. Sometimes, special studies are needed for issues such as sensitive plant and animal species (flora and fauna), hydrogeology and groundwater impacts modeling, or soil and rock geochemistry. Often, one primary contractor has most of the discipline experts to conduct the studies, but usually at least a few are subcontracted out by the primary contractor, or the mining company contracts directly through its environmental staff and budgets the planning process. A checklist of environmental disciplines for baseline studies may include the following topics:

- Air quality
- Noise
- Climate and meteorology
- Soils and reclamation
- Geology
- Flora or vegetation and wetlands (including sensitive species)
- Fauna or wildlife (including sensitive species)
- Aquatic ecology (including sensitive species)

- Recreation
- Land use
- Energy
- Transportation
- Visual resources
- Groundwater
- Surface water and stream sediments in drainages
- Geochemistry of ore and waste rock
- Cultural resources, that is, archaeology and history (and sometimes paleontology, such as notable fossils in the geologic setting)
- Socioeconomics
- Special areas of critical environmental concern
- Use of hazardous materials

Some U.S. states have their own environmental quality or policy acts and their own EIS requirements for projects affecting state and sometimes private lands, often jointly with a federal government agency if federal public lands are involved. Check out all the angles, and attempt to prove compliance and attention to issues through receipt of written agency concurrence. Pay particular attention to EIS "scoping" type sessions where issues are raised by the general public, NGOs, and other stakeholders and are formally addressed. Written responses by the agencies will establish for the record that some studies are irrelevant or non-applicable to the project setting, so that time and money are not wasted on them and later questions can be quickly resolved.

Cost Estimates and Starting Work

Update meetings with the agencies will confirm that environmental baseline studies for various disciplines are necessary for the permits and an environmental impact assessment (EIA; if the latter is a necessary requirement in the country of operations). The company should use the list of environmental disciplines in the introduction to this chapter and form agreements with the agencies as to which disciplines will be applicable. The preliminary scoping comments should be reviewed and NGO requests for level of detail in various studies should be carefully analyzed. University commenters, researchers, educated retirees living in the project vicinity, biologist consultants, hydrologists with public agencies, and numerous others may have suggested what the mining industry often calls "research projects" to further the science and understanding of mining impacts, irrespective of whether such projects or studies are required by the regulations. The company may wish to volunteer to do such further studies for goodwill.

Much of this work, of course, depends on the regulatory interpretations of key regulatory agency staff. For example, a pro-environmental protection staffer, prompted by NGOs and with close communication ties to them, can delay acceptance of mining company study reports, permit applications, and mitigation plans. Or a balanced agency staffer can expedite the schedule with skill while staying within the usually balanced purview of the mining and environmental laws, regulations, and guidelines, gaining concessions from both camps. In another situation, a pro-mining staffer can raise NGO attention and may cause the mining project to become a target of such regional or national groups if perceptions exist that there is a lack of study and regulation.

With these issues in mind, it is important for the company to negotiate a thorough and reasonable set of baseline studies, line up contractors as necessary, and get on with the one year or more of field sampling and analysis. It is consequential that these studies are mostly completed during intermediate feasibility for several reasons related to project design and engineering.

Of the 100 units of costs the company may spend on baseline studies, the breakdown of discipline-by-discipline costs and major issues may be something like the following for a project in a temperate, semi-humid, mid-latitude location. It can be tailored to the company's setting:

- **Air quality and meteorology** (15 units)

  This includes four seasons of monitoring, stations and equipment, quarterly and annual reports, monitoring prior to and during operations of process equipment and facilities, addressing priority pollutants, "major" air pollutant source requirements, regional haze, acid rain allegations, and similar.
- **Surface and groundwater** (20 units)

  Again, this includes four seasons of stream and well monitoring, springs and seeps, nearby well users and effects, mine pit lake water quality modeling and groundwater effects, ponds and flow effects on wildlife and fish, and similar. An aquifer drawdown study is useful for mine planning.
- **Aquatic ecology** (5 units)

  Baseline studies are made, especially of fisheries in lakes and streams and aquatic insects, amphibians, and vegetation that sustains them; the health of aquatic sediments (any acid drainage effects from past operations or mineralization); and similar.
- **Cultural resources** (10 units)

  Costs usually include studies of prehistoric (archaeological) and historic resources of potential or identified significance from previous investigations in the mine study area. Additional field transects, literature surveys, and other research or cultural setting reports are often required.
- **Geology, soils, and geochemistry** (10 units)

  Reclamation and closure planning, geochemistry of ore and waste rock leachable characteristics are commonly required in detail here. Paleontology (fossils of significance) in the rocks, which could be disturbed, may be included.
- **Socioeconomics and sustainability** (10 units)

  A formal social impact assessment document may be prepared to accompany the EIS or EIA regarding controversial mining projects.
- **Flora and fauna** (15 units)

  This includes sensitive species of plants and animals as listed by various agencies and tribes. This might even include sensitive insects and their habitats, as in a State of California project.

- **The remainder** (15 units)
  - Recreation
  - Land use
  - Energy
  - Transportation
  - Visual resources—the U.S. Department of the Interior Bureau of Land Management (BLM) and U.S. Forest Service (USFS) have specific visual resource management systems with key viewpoints, color, form, and similar characteristics considered for planned built structures, which these document planned construction and assess impacts, often with computerized visual simulations.
- **Special areas of critical environmental concern**
- **Use of hazardous materials**

Permit Application Specifications

Clues to environmental baseline study specifications are often given in the permit requirements of the regulatory agencies in each state, province, department, or country. The requirement for a particular map or data table will often be filled by the baseline study report and its contents.

However, permits require additional, special submittals to comply with agency legal and regulatory requirements. The regulatory framework in the country of proposed operation is best initially researched by an experienced environmental regulatory and/or permitting specialist or an environmental lawyer. A combination of the two researchers is often optimum, as the science and engineering perspective merged with the legal perspective can give a keen view toward future liabilities. This review can also help to confirm that environmental baseline studies and permit applications are properly compiled the first time around, if at all possible. Some universities and agencies have an online legal and regulatory database for quick research. Many satisfactory permit lists are available to use as guides, and the researcher may also

- Consult any recent EIS for a similar project in a U.S. state or an EIA in an international setting; or
- Consult various legal summaries written by law firms experienced in these matters in the world's mining districts (e.g., see Rocky Mountain Mineral Law Foundation 1996).

The usual round of agency requests for additional data, explanation, justification, maps, and presentations often follow the initial permit submittal. However, patience and perseverance usually pay off. Anticipation of agency "pet" topics and emphasis on these issues, without too much commitment of time and money on the part of the company, can often pay dividends.

It follows that expenditure of time and money on particular topics in the environmental baseline studies should emphasize particular disciplines based on project design and the full range of alternatives envisioned by the in-house mine staff and key impacts. The level of controversy of particular impacts should be anticipated. Conflict resolution should receive attention. Coordination of interdisciplinary analysis, so that the water specialist is talking to the land use specialist and is talking to the reclamation planner, for example, is the responsibility of the mining company project manager or the contractor hired to manage and present this work.

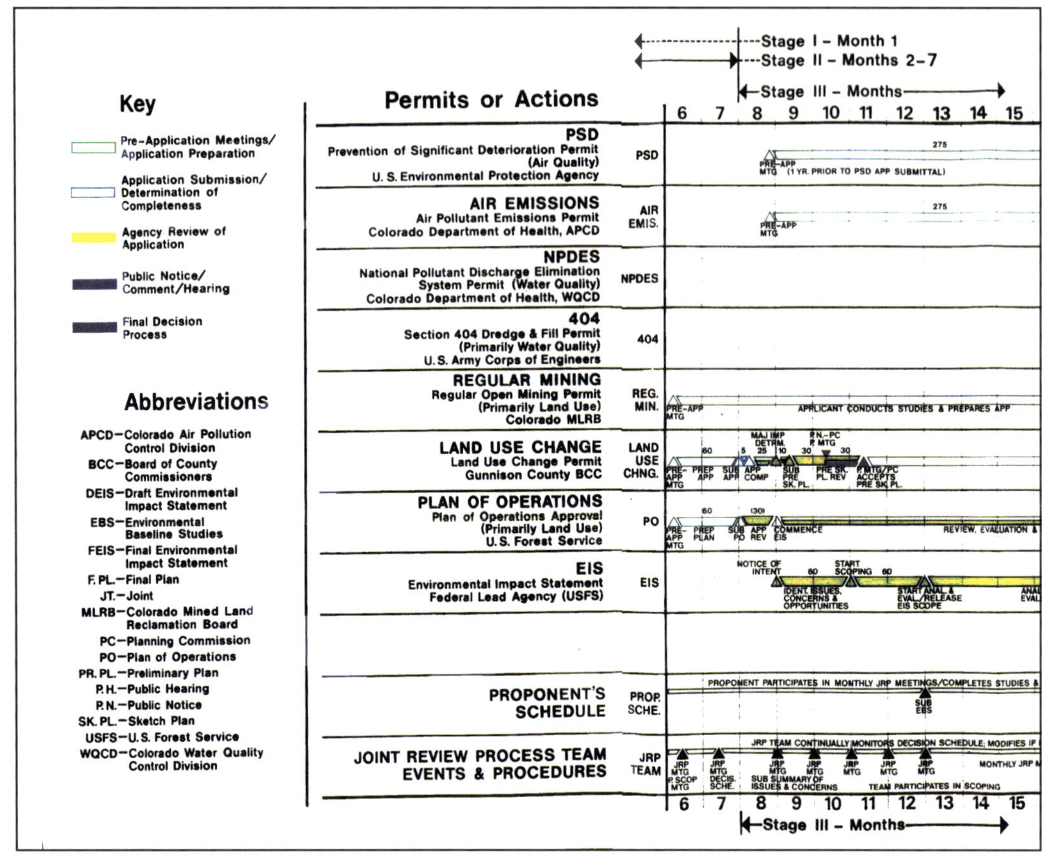

Source: CDNR 1980 (Figure continues)

FIGURE 9.2 Permits decision schedule for a metals project

Critical Path Planning: The Permit List and Schedule

The planning and scheduling of baseline studies and permit processing must take into account critical path items. For environmental matters, project success will hinge on these items. These are also items that affect construction and operations issues in many cases and may directly affect project profitability because of a late start or, for example, a redefined access road alignment.

Everyone has a favorite schedule format. Some prefer hand-drawn horizontal bar charts; others, bullet lists with dates; others, tables with many columns; and others, one of the many computerized project management and planning/critical path software programs. These latter are visually impressive, attractive, and in color, and can often be easily updated and distributed to the team. However, the initial input data, designation of early starts and late finishes, changing critical paths, and the computer program's perception of when the project will now end, can be debatable and exasperating.

Figure 9.2 is an example of an early manually developed permit schedule from a 1980s project-planning effort in Colorado, partially funded by federal agencies (CDNR 1980), and illustrates a happy medium.

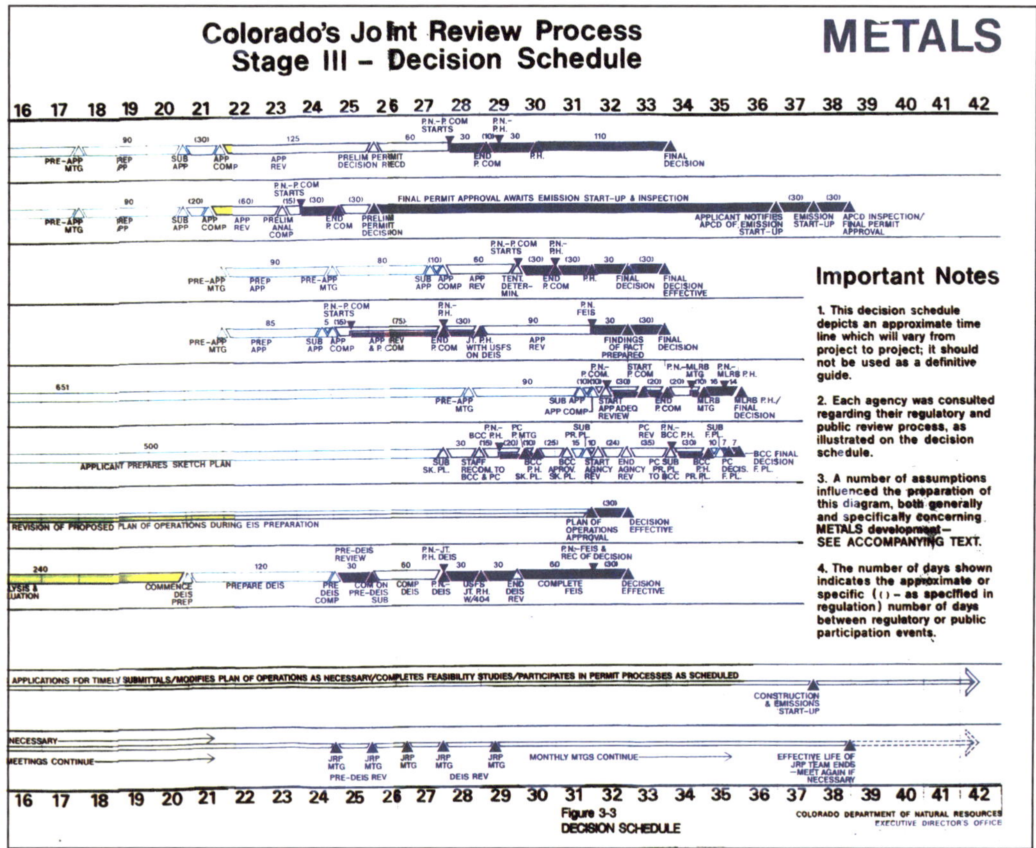

FIGURE 9.2 (Continued)

The schedule shown in Figure 9.2, originally and effectively presented in color in a large foldout format, shows the relevant major permits for a metal mine project in one of the western U.S. states in a simple, straightforward fashion. Coordination of data collection, permit application preparation, agency reviews, public hearings, and similar permit process features continue to appear relevant years after the initial presentation. Mining specialists who are comfortable with computers but not highly computer literate may prefer the manual schedule as the most efficient planning tool. The main point is for mining specialists to carefully prepare, update, and use any type of schedules they are comfortable with to keep on track and to keep the environmental team focused on the construction start.

Pros and Cons of Using a Contractor

The pros and cons of the preparation of requests for contractors' proposals for environmental work are well known to many mine operations offices. Such a process involves entertaining bids and then choosing and managing a contractor's technical effort and budget, versus performing these tasks with in-house staff. Use of a contractor for such tasks as environmental permitting, EIS preparation, field studies, agency strategy meetings and negotiations, and other helpful matters related to environmental and engineering progress is influenced by the size and capability of in-house staff, control, confidentiality desired, and budgets.

Often the magnitude of the EIS (U.S. terminology) or the EIA (international terminology) suggests using an impartial, professional third-party contractor with an interdisciplinary environmental expert team. This contractor may work under the direct supervision of the regulatory agency but receive project engineering data, alternatives analyses, baseline studies, and special impact studies from the mining company. The company usually pays all the bills plus the bill for documented time by the federal agency staff on the EIS effort.

The debates continue within the mining industry on the value of contractors or consultants. However, with staff cutbacks, the mining company is often forced to contract for services if it wants to make decent progress. Larger environmental and engineering consulting firms are sometimes called on to do the "heavy lifting" on several tasks. These can involve initial data review, field research and inspection, baseline report preparation, telephone and direct interviews, library and Internet research, mapping, engineering analysis, data and report compilation, quality control, presentations to agencies and the public, tabulation and analysis of public issues, and assistance in strategy and negotiations, all in support of mining company clients and agency reviewers. Specialty firms may perform geotechnical or geochemical work, tailings dam engineering, and similar, as previously noted.

From the author's perspective as a consultant to the mining industry, financial institutions, and governments, the consulting company desires to be seen as a trusted, knowledgeable, discreet advisor and colleague to help clients through difficult problems, often providing another valuable perspective to justify its fees. The contractor often reviews the voluminous work of others and advises on risks to investment at various project stages.

Mineralized Areas and Background Concentrations

During intermediate feasibility studies, as regulatory agencies and the public become aware of the project plans, the issues of mineralized soils and water and chemical analyses of background concentrations of metals in the mine project area often come out in full force. Unless the project is truly a "greenfields" discovery in a heretofore unmined area, some effects from past mineralization will be evident in soil, sediments, and surface and groundwater samples as baseline data are gathered. Most mines are in historic mining districts. Past mining will have disturbed and unearthed soil and rock, exposing it to air and water and causing concerns over such matters as acid rock drainage, blowing dust, effects on a fishery, or similar.

It is, therefore, important that baseline environmental data and information recognize the effects of past mining and other human-caused land disturbance on project areas and buffer zones before the company begins its work on-site. Baseline conditions should be documented, and written acknowledgment and acceptance of these conditions should be acquired from the regulatory agency before site development begins, and if at all feasible, before exploration.

Artisanal Miners

Artisanal miners is a curious term and encompasses all of the historical activities of local and in-migrant artisans, who, by mostly hand methods, have mined the surface or near-surface exposures of the ore zones of interest. Disturbance can be minor, such as a few rock chip sites or trenches; or major, such as hillsides that have slumped because of hydraulic mining, large hand-dug pits with many ladders to access individual claims, heavily worked placers, and contamination from hand processing using various reagents. Many Latin American case studies address these topics.

Artisanal miners have, of course, affected baseline conditions, and their effects must be documented by the mining proponent before permitting large-scale commercial operations. Issues of land and mineral title, displacement of local livelihoods, and security for the new operation will often arise as new larger-scale mining opportunities are explored and developed (e.g., see Mernitz 2005). Further, the mining company is sometimes not native to the country under development, causing cultural resistance. Specific community relations plans, social (and economic) impact assessments, training programs, local hiring practices, and special management efforts are required to effectively deal with a history of artisanal mining in a project area.

Land Positions and Ownership Claims

It will be well known whether the minerals being prospected for and mined are of the hard rock "locatable" type on federal lands (as they are known in the United States) or otherwise. Under U.S. laws, unpatented and patented mining claims for the underground minerals, combined with or even lacking surface ownership, give certain development rights and uses of the surface lands to the mining company, with environmental implications. This topic is complicated and the subject of numerous handbooks and articles (e.g., Holme Roberts and Owen LLP 2001). Federal agencies, such as the USFS and BLM regulate such public lands and minerals in much of the western United States.

In the United States, if the ownership of surface lands and subsurface minerals is private, then uses and impacts are somewhat unrestricted. To state it simply, however, such uses cannot affect the public health and welfare adversely. If the land is federal, state, or local "public" land—that is, surface access is open to the public, or federal or state agency surface or subsurface mineral ownership or "rights" exist—then an extensive permitting process and environmental impact analysis is often required. This is also often the case in other countries, which have varying minerals concession arrangements.

The company lawyer or permitting specialist should be consulted for the details. The environmental protection measures to address impacts to all of the environmental disciplines named earlier are the linchpins of this environmental considerations review. Whether the minerals are "locatable," "leasable" (see next section), or otherwise under the country's mineral concession requirements, these classifications will, in good part, determine the levels of environmental study, impacts, and mitigation required by the regulatory agencies.

Surface Owners' Rights and Mineral Leases

In the United States, energy minerals—oil and gas, coal, coal-bed methane (natural gas contained in the coal seams) and shale gas—and oil and gas extraction using directional drilling and hydraulic fracturing ("fracking") are termed *leasable*. The term refers to minerals that are subject to leases by federal agencies if in the federal mineral estate. This situation is also usually the case for industrial minerals, such as sand and gravel, specialty clays, potash and phosphate, lime, and the like, if on federal- or state-leased land in the United States. The lease will have different restrictions, stipulations, and emphases (with environmental implications) for these leasable minerals compared to the locatable minerals previously discussed because of resource determinations and markets, mine life, mine types, chemicals used, locations of the mines and process plants related to geology, and similar factors.

Means of extraction, surface and underground impacts, workforce, transport, types of process plants, and other factors are often quite different between leasable and locatable minerals.

These differences will become apparent on first research and meetings with the agencies regarding their permit requirements. Leasable minerals often have fewer adverse impacts to the environment than locatable hard rock minerals during their development because of geochemistry and other factors. However, recent conflicts have occurred throughout the United States as subsurface lessees and surface residential owners have disputed the impacts of leasable mineral extraction in expanding communities, in terms of such factors as light, noise, odors, transport, health effects, setback distances, and regulatory authority.

International Mining Projects, Land, Title, and Environmental Regulations

International mining projects are often developed in concessions, blocks, mineralized districts, or sections of surface lands that are known to contain mineral potential in either surficial or bedrock geology, sometimes both. Leasing or granting of concessions precedes mining project development. The government or its international contractors may have conducted remote sensing, field exploration and analysis, or similar geological, geophysical, and geochemical work. Alternately, one company or another may have performed the initial or subsequent exploration to prove up the reserves or resources. Artisanal miners may have defined the surface and shallow soils and bedrock mineralization and grades.

Land access, title, environmental restrictions, royalties to the government, and other factors may affect environmental permitting and controversy. The mine plan may affect existing residences and land uses. Local citizens may be organized into community groups, tribes, factions, native or worker associations, or others opposing or favoring the mine. Welcoming or conflicting local groups may also have rights in the land, such as the *ejidos* in Mexico. (These are communal farmsteads, which could involve several families affected by nearby mining in terms of agricultural production, irrigation, employment, transport, or access.) In recent decades, mining title opposition and environmental and social impacts have continued to stall projects in Peru, Argentina, Romania, the United States, and many other countries.

It follows that early work with the federal and provincial environmental agencies, and with federal, state, and local political and financial decision makers will pave the way for project success. Prudent actions by the mining company's environmental staff at this stage may include

- Study of laws and regulations;
- Retention of local lawyers and environmental study contractors;
- Community relations and sustainability efforts, and hiring of local staff or contractor specialists for same;
- Provision for improvements to local infrastructure; and
- Further understanding of sustainability issues before major project decisions.

In an interesting analysis, countries have been ranked for risks to mineral investment from the late 1990s through recent years by consulting firm Behre Dolbear (2014, 2016). These listings and risk rankings will aid the prospective investor and environmental specialist in the initial assessment of the regulatory climate—along with many other aspects of the investment climate—in the mining countries of the world.

Similarly, for several years, the Fraser Institute in Canada ranked the policy and mineral potential and overall investment attractiveness index for major mining regions around the world (*Mining Journal* 2003). The Heritage Foundation's *Index of Economic Freedom* (Heritage

Foundation 2017) is another such ranking and description, which addresses such mining investment issues for nearly every country. Law Business Research Limited (http://lbresearch.com) provides a good businessperson's snapshot of the means for "Getting the Deal Through" for 28 countries, including such newer targets as Kazakhstan.

Designing for Reclamation and Closure

It follows that designing the project with several issues in mind will alleviate many environmental headaches during construction, operations, and closure. It will be prudent to

- Commission a set of thorough environmental baseline studies to later assess project impacts against this baseline;
- Plan and budget to effectively monitor, report, and mitigate impacts during operations and inform shareholders or the public about the company's successes;
- Plan to use operations equipment, budgets, and staff to reclaim and close, as project areas are retired;
- Set aside real monies for closure in the form of bonds, cash, or other instruments in case of adversity; and
- Develop a written, funded closure plan designed to leave the site with an effective, sustainable future land use.

In this manner, the company's "legacy site" will be a well-founded one and not a set of problems for future governments, agencies, and generations. Project sustainability in community, ecology, economy, and governance will be the result.

Developing In-House Environmental Impact Analyses

For the mining company with an active environmental staff and available budget, it is prudent to anticipate what topics of focus the government regulatory agency or a third-party contractor will develop in the EIA for the project. A full EIA or an executive summary with a key-point bullet list could be developed in-house by a mining or energy company to plan for conflicts or data gaps and mitigate them early in the process.

Although this environmental impact document is known by various acronyms (EIS, EA [environmental assessment], or similar, given the language), it essentially contains the same subjects in varying levels of detail. Often, detailed appendixes, including the environmental monitoring and mitigation plan (EMMP; see the following section), are included. Appendix 9A shows a generic, detailed, annotated EIA outline including appendixes developed by the author for an energy project in Africa. This outline gives some definition to the magnitude of the effort, recommended page lengths, and types of graphics and details. Many examples are available in the literature and for advance planning, it would make sense to acquire one that details a recent project in the country where the firm plans to develop and meet with regulators.

Developing In-House EMMPs

The EMMP is frequently required for major projects by various governments. The United States is learning that gaps in recommended and committed mitigation often lead to later project conflicts, if such mitigation is not well-specified and funded, and U.S. regulatory agencies are increasingly specifying EMMP-type documents to accompany project permits. The

EMMP is usually very detailed, rigorous, and often specifies responsibilities and costs in addition to

- All of the environmental disciplines of concern;
- Environmental monitoring and reporting requirements with maps and specifications;
- Desired data quality;
- Inspections, compliance, and corrective actions documentation; and
- Commitments from the natural resources developer regarding funding, staff, and legally binding mitigation measures that are present to address potential problems.

Appendix 9C, prepared by a Canadian mining company and their contractor for a gold project in South America, is an example of a table of contents for a major project EMMP.

Permit Application Details

It is at this stage in intermediate feasibility that the details of the permit application requirements gained from the environmental baseline and project engineering studies are refined. The permit schedule and coordinating data efforts, meetings, and public disclosures are similarly refined and scheduled (see Figure 9.2).

FINAL FEASIBILITY

The exciting stage of final feasibility is achieved with much anticipation and a clear view of favorable project engineering, economics, public acceptance, and environmental suitability. A balance of economic development plans and environmental sensitivity seems to exist. Staff can now move to the next steps of refining detailed environmental field studies, continuing permitting activities, and public outreach.

Refine Impact Predictions In-House

As the baseline studies are underway during the intermediate and final feasibility stages, it is time once again for the environmental manager and staff to revisit the initial scoping studies and fatal flaw analyses developed during preliminary feasibility study. The company should assess whether expected issues at that time—many months and perhaps a few years ago—have come into prominence. The following are important questions to ask: Have the project plans been affected and changed? Have project development costs been impacted substantially because an issue was considered a medium or high risk? Will that risk likely be effectively mitigated? Is the project plan better as a result in terms of chances of success?

The impact predictions for each environmental discipline can at this time (during final feasibility) be further refined. What will the EIS or EIA focus on, and how will mitigation affect the government agency (decision maker or stakeholder) and be addressed in the final decision documents (permits)? Are proper funds and staffing being allocated by the mining project managers to environmental issues of highest risk? Has proper planning been done?

Reassess Project Opponents

The NGOs who are watching the project have probably refined their tactics by this time as well. If "stopping the project flat" was their initial goal, they have not succeeded because of the

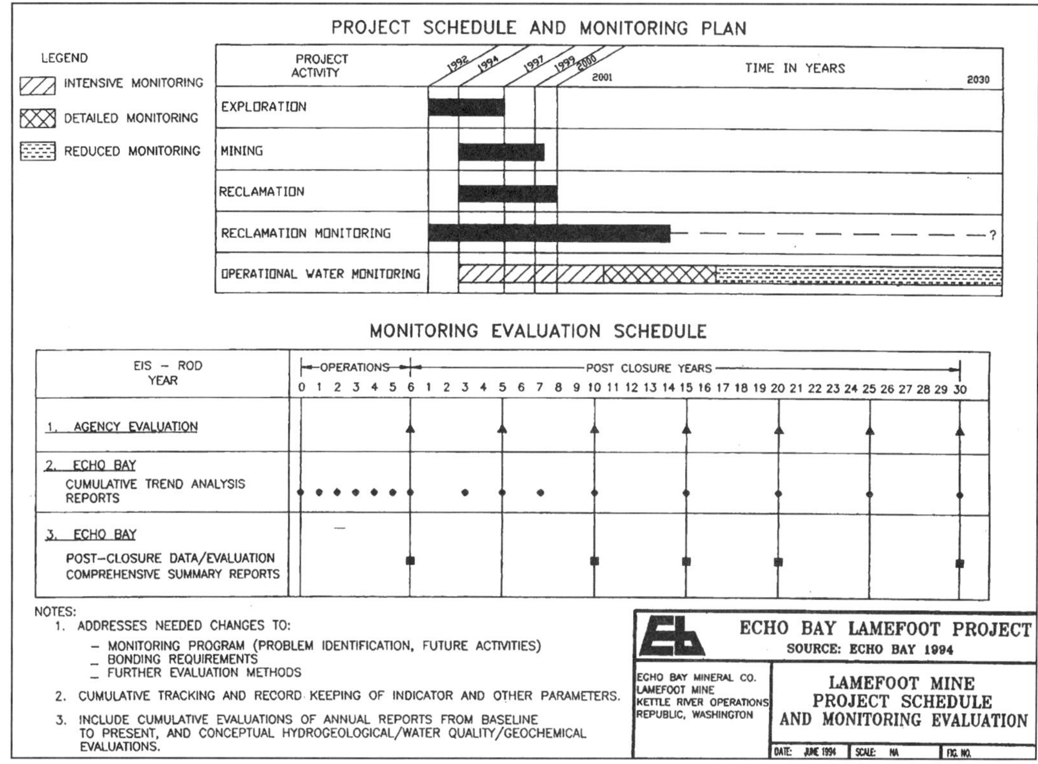

Source: BLM 1994

FIGURE 9.3 Lamefoot environmental impact statement compliance schedule

good efforts of the mining company's environmental, community outreach, engineering, and mining staff to balance profitable development and environmental protection.

So now perhaps the NGO tactics are "better project, more mitigation, and undermine profitability." Compromises can perhaps be reached here, but beware once again of the costly "research projects" that are not required by law and regulation and will delay development just to delay it. Negotiate hard and well, remind agencies of their mission and responsibilities, get legal and permitting advice, be vigilant regarding instances of improper influence by NGOs on agency staff, and carry on with open communications with the communities, NGOs, and agencies.

Or better yet, the NGOs have recognized this project as one that is carefully planned, recognizes environmental impact issues properly, and is a "good project" in comparison to the many they review. They will focus on targets elsewhere.

Refine the EMMP

The EMMP can be refined at this time to reflect critical issues, budget plans, expected staffing, negotiated reporting, corrective action requests from the agencies, and related matters.

For example, a negotiated schedule, over a 30-year period, was developed to address NGO and agency concerns (Figure 9.3) about an underground gold mine in Washington State. Because estimated impacts of the higher-elevation mine to groundwater and nearby surface water (of a lower-elevation, large, multipurpose recreational, residential, water-supply lake) continued to be debated among the parties, it was agreed to "let the actual data speak" as groundwater monitoring and field observations over the years reported actual, rather than predicted, effects. Monitoring and reporting schedules and adjustments to mitigation could then be negotiated quarterly, semiannually, or otherwise as the EIS had specified. Mining companies are continually pushing for "good science" to prevail, rather than the wish lists of project opponents who may desire to render the project uneconomic.

Refine Conceptual Reclamation and Closure Plans and Costs

At this stage, the conceptual reclamation plan can be refined in terms of project details. It is expected and recommended that this plan is fairly well defined in terms of funding during the intermediate feasibility study. The mine and plant footprint, facilities, topsoil stockpiles, equipment to be on-site, project development, production schedules, and similar details are all relevant.

For example, around 2005 one base metals company in Mexico had plans to discuss details of hectares of area, vegetative species, equipment, movement of materials, and similar matters in a reclamation and closure plan just as metals production was beginning. A series of outlined standard operating procedures (SOPs; in Spanish, *procedimientos de operaciónes estándar* [PEOs]) were developed to guide operations and reclamation personnel as production was occurring. In this way, the footprint at the end of mine life would be efficient for final closure in terms of soil, vegetation, water management, waste rock and tailings, and other materials that would remain on-site after demolition, decommissioning, and decontamination of facility sites.

Prepare First Permit Applications

The first permit applications—those which are expected to be the most difficult and have the longest lead times—should be prepared at this time using the latest project data. If some applications were submitted earlier, updates or amendments reflecting the latest project plans should be carefully prepared and highlighted for the agencies and public commenters.

Updating permit details is an important procedural point, as later administrative or legal actions by opponents may cite outdated or incomplete project information, which can prompt different baseline studies and impact analyses than those currently reported in project documents. Full disclosure of all anticipated project features and impacts, including later mineral processing technologies, is the most conservative approach.

Again, the company should closely consult the in-house permitting schedule to confirm that the permit durations, coordination of data and public forums, and other issues are current. Recent experiences of other mining firms with permit lead times, schedules, and success for comparable projects in the specific country, state, or area need close attention so that pitfalls, project design issues, and other delays can be avoided. Local media announcements and NGO websites should be closely followed.

Proceed with Construction and Operating Permits

Other construction and operating permits can continue to be processed at this time. Close coordination among engineering, geology, maintenance, health and safety, and other mine and plant staff will be necessary, and this coordination task often falls to the environmental specialist as water management and discharges, air emissions, soil and rock management and dust, wildlife habitat or tree removal, or other environmental issues are involved. Agencies must be notified and timely monitoring reports submitted. Agency coordination demands mining company time because local, state, provincial, and federal agencies may not talk and trade information for optimal permit administration. To promote minimal surprises and maximum scheduling and financial success, the company should be vigilant of the many cross-discipline issues as the mining project develops.

Prepare Reports to Aid Agencies

As the permitting begins and continues, it is often important to tailor internal environmental reports for agency and public consumption. These reports may have been prepared by technical mine staff or specialized consultants to address a particular issue. The issues addressed are often the subject of one or several permits and will be highlighted in the EIS. As an example, for a revived metals project in the Rocky Mountains, the following reports were scoped:

- An internal plan of operations in the model of that to be presented to the USFS, because the project was located on National Forest lands on patented and unpatented mining claims
- Sensitive plant and animal species surveys
- A water management plan for surface and groundwater quality and quantity, and process and stormwater segregation and management, and any mine and plant discharges, plus a conceptual project water balance
- Acid rock drainage potential, ore and waste rock geochemistry, and mining waste management plans
- A conceptual reclamation and closure plan with SOP guidelines for different facilities
- An environmental protection plan, as required under state mining reclamation regulations
- An alternatives analysis report to aid the EIS writers in assessing the engineering and economic feasibility of the alternatives internally considered in mining company planning
- A geotechnical mill tailings plan and mine backfill report
- Socioeconomics and community affairs mitigation plan
- An impact mitigation plan, with recommended and committed measures, with costs, implementation, monitoring, and reporting addressed
- Discussion of the water treatment plant operation, costs, reporting, and bonding
- Reclamation performance bonding cost estimates

Another project in South America presented the following items in the public EIA:

- A valued environmental components analysis
- An environmental management plan (EMP; in the past, the EMP was often less detailed than the previously described EMMP, with broader policy statements and discussion of recommended and committed mitigation measures, some to be negotiated with the agencies)
- Monthly capital expenditures during construction
- Project capital costs

- Chronology of local community meetings and individuals attending
- Environmental management system handbook
- Six appendixes regarding acid rock drainage test results

This list is exhaustive. The company may wish to do only a few of the preceding, but consider all angles and have information ready if a key issue could become controversial.

Analyze Bonding Details and Negotiate with Agencies

Early in the final feasibility stage, it is timely to begin a conceptual reclamation and closure plan and to discuss financial surety internally. Surveys of agency requirements and other bonding amounts for projects in the vicinity are appropriate (Table 9.1). The means for providing financial surety by the mining company should be developed. In the United States, bonding amounts have quickly risen from the hundreds of thousands of dollars in the 1990s to the millions, tens of millions, and a few in the hundreds of millions of dollars in the 2000s. Requirements in other countries vary widely in amounts, administration by the agency, and types of bonding instruments required.

As internal bonding amounts are calculated, the company should prepare for agency negotiations. The agency will likely have its own calculations based on a third-party contractor performing all work, no consideration of operations cost savings, sizeable engineering contingencies, add-on fees, very conservative estimates of monitoring station maintenance, sampling, analysis, reporting costs, and similar. Again, it is important to fall back to previously negotiated issues and concerns, meeting notes, and letters of agreement. As project plans change during operations, production increases, more facilities are designed and used, and project attention grows, the company should be prepared for amended bonding amounts every few years. Some successful mining projects, with more than 10 years of project life, have continued with profitability through several bonding changes.

Implement Public Outreach

During the life of the project, the mine staff and its contractors should continue to implement the public and community outreach program through final feasibility and beyond. Often a community affairs or public relations specialist is on staff by this stage and continuing to prepare press releases, visible at the legislature, testifying at administrative hearings, dealing with the media, and distributing the balanced mining economics and environmental protection message to the stakeholders and interested public. Website contents may be especially useful here to answer initial questions about project plans and expected impacts and to provide telephone or online contact information for queries.

Prepare the Environmental Sections of the Final Feasibility Study

The final feasibility study is in preparation at this time to document to a fairly precise level (±10%–15%) all project parameters in terms of facilities, predicted operations, costs, disturbed areas, environmental performance, and several other matters. Contents of the final feasibility study are described in detail in other chapters. For medium and small mining companies, this document is usually prepared by an independent third-party consultant who is well versed in mine development matters and mining economics. If a public company, major shareholders will closely review this document to anticipate project profitability. Environmental issues are given a long look during this phase.

As a result, for environmental and regulatory matters, a concise chapter in the final feasibility study will usually summarize the status of permits and approvals, discuss any notable risks or liabilities, and note mitigation measures or expenditures to minimize risks. Results of any agency, local landowners, community group, and/or NGO interviews will be documented to discuss the perceptions of these stakeholders as to expected project performance and the issues they will closely monitor.

Prepare for Due Diligence

As the final feasibility study is released and lending institutions are considering a bank facility on the project, due diligence of the project should be contemplated. Another consultant (different than the consultant preparing the final feasibility study) is often contracted to send a due diligence team to the site to verify the final feasibility study findings. Because this topic is one of the primary focuses of this handbook, additional details of the general process are not discussed further here.

However, once again the environmental aspects of the due diligence are a key check on the project's operational and financial success. If any critical environmental, regulatory, sustainability, health and safety, community, or similar risk or liability has been missed, the due diligence consultant should catch it here. Under this system of checks and balances, millions of dollars of the investment can be saved in terms of schedule and operational problems if remedied at this time. The bank, investor, mining company, or government may then go forward with some "insurance" and "assurance" that their money is safely invested.

Project staff can prepare for the environmental portion of the due diligence by maintaining organized records as follows:

- Agency files by chronological date on each permit, with phone records, amendments, maps, bond documents, and other paperwork
- An updated summary permit list and schedule, with agency, permit, dates of submittals, status, agency contacts and coordinates (telephone and email), and any relevant comments
- Meeting records, including presentation materials, speakers, attendees, questions and answers, and commitment statements made by the company
- NGO correspondence, media statements, and so forth, and company responses
- Community group meeting transcripts and comments, with documented follow-up by the mine staff
- Sustainability commitments and funding by the company (e.g., water systems, health clinics, housing, transportation, and other facilities established or to be improved by the mine, which may survive its closure)
- Reclamation and closure plans
- Similar documentations relating not just to prospective risks and *liabilities*, but also to environmental *assets* this project may have in its unique setting, given its careful planning

Consider International Environmental Guidance Standards

Especially in international settings, it is important for the mine environmental staff to be aware of these standards; assess the local, regional, and national situations; and attempt to attain the funds, staffing, and backing of management to comply as feasible. With respect to global standards, the author's work on a project in Saudi Arabia in 2005 focused on the following guidance documents, among others:

1. The World Bank's OP 4.01 (1999a) guidelines regarding category A, B, and C projects for EAs. Category A are those large projects with major, significant, and potentially irreversible impacts, requiring a large-volume EA (an EIS- or EIA-type document, as previously discussed) with detailed analysis; while those in Category B have lesser, predicted impacts and require smaller-volume, less-detailed analysis (as an EA under NEPA regulations).
2. The *Pollution Prevention and Abatement Handbook* (World Bank 1999b), an instructive volume regarding impacts and mitigation measures in many natural resource development sectors. Specific chapters in the handbook address various mining and processing sectors and their unique impacts.
3. The *Environmental, Health and Safety Guidelines* (*EHS Guidelines*; IFC 2007a), regarding assessment of impacts from underground mining and milling, address recommended water and air standards for such projects, and many other issues.

As a more comprehensive update to the *EHS Guidelines*, the *IFC Environmental, Health and Safety Guidelines for Mining* (IFC 2007b) addresses and recommends risk, health and safety, standards, monitoring and reporting, and best practices issues from 1990 to present. Several of the preceding documents have also been updated, and the websites should be consulted for details.

As previously addressed in this chapter, it is important to recall that the Equator Principles address the commitment of more than 80 global banks and investment firms to require a project-specific EMMP to accompany the EA, for mining projects to which they lend money. The mining company's borrower must covenant (formally agree in writing) to comply with the mitigation, reporting, and communication's responsibilities, corrective action measures, and other commitments specified in the EMMP.

Conclusions on the Final Feasibility Study

Now everything is in place to move the project forward. The project appears to be proceeding, and numerous obstacles, environmental and otherwise, have been addressed and handled. As the project moves into the construction phase and then into operations, the company's environmental staff must continue to be diligent, vigilant, and progressive. Few obstacles are ever fully "conquered," and issues often return, months or years later, to be addressed by the mining company. Successful projects continually anticipate issues and monitor potential effects on environmental performance, sometimes with an internal annual audit (or an audit performed by an outside party) to take a fresh look at various matters. The mine's management should be encouraged to conduct such an exercise. Audits specified by the bank, reports by the bank's independent engineer, and international cyanide code management audits are other examples of such checks on performance to help minimize catastrophic, or even minor, incidents.

During such audits, confirm that the proper tools, facilities, staff, plans, budgets, and other items are in place to move properly forward to another year of mine profitability and ventual effective reclamation and closure. It is time to again consult the usual players and gauge the environmental performance of the project according to the following mining company staff and stakeholders:

- Mine operations and infrastructure
- Mine engineering
- Mine health and safety
- Process plant
- Regulatory agencies at all government levels
- NGOs and other stakeholder organized groups
- Local community residents
- Any others affected either positively or adversely by the mining operation

Through conduct of such a careful and thoughtful program, the environmental considerations due diligence will be a positive contributor to a successful project, balancing environmental and economic priorities to help ensure long-term sustainability.

REFERENCES

ASTM E 1527-13. 2013. *Standard Practice for Environmental Site Assessments: Phase I Environmental Site Assessment Process*. West Conshohocken, PA: ASTM International.

ASTM E 1528-14e1. 2014. *Standard Practice for Limited Environmental Due Diligence: Transaction Screen Process.* West Conshohocken, PA: ASTM International.

Behre Dolbear. 2014. Where to Invest 2014. www.dolbear.com/resources/where-to-invest-2014/.

Behre Dolbear. 2016. *2014 Ranking of Countries for Mining Investment: "Where Not to Invest."* www.dolbear.com/wp-content/uploads/2016/04/2014-Where-to-Invest.pdf.

BLM (U.S. Bureau of Land Management). 1994. Lamefoot Mine (Republic, Washington) Final Environmental Impact Statement. Spokane, WA: BLM. Prepared with the assistance of Woodward-Clyde.

Cambior Inc. 2002. Rosebel gold mine (Suriname) environmental impact assessment. Quebec: Cambior and Rescan Environmental Services.

CDM (Camp Dresser and McKee). 1980. Environmental planning marketing brochure. Cambridge, MA: CDM.

CDNR (Colorado Department of Natural Resources). 1980. Colorado's joint review process for major energy and mineral resources development projects. Stage III—Decision Schedule—Metals. Denver, CO: CDNR, Executive Director's Office.

Equator Principles Association. 2013. *Equator Principles III.* www.equator-principles.com/index.php/equator-principles-3.

Gruver, M. 2016. Taxpayers likely to pay as mines close. *Denver Post.* May 20, p. 14a. www.pressreader.com/usa/the-denver-post/20160520/textview (accessed Feb. 19, 2017).

Heritage Foundation. 2017. *Index of Economic Freedom.* Washington, DC: Heritage Foundation. www.heritage.org/index.

Holme Roberts and Owen LLP. 2001. Millsites and mining claims: Recent developments in land tenure select documents. Presented at the International Standards Organization (ISO) Mining Law roundtable, Denver, CO, April 3.

IFC (International Finance Corporation). 2007a. *Environmental, Health, and Safety Guidelines.* www.ifc.org/ehsguidelines.

IFC (International Finance Corporation). 2007b. *Environmental, Health and Safety Guidelines for Mining.* www.ifc.org/wps/wcm/connect/1f4dc28048855af4879cd76a6515bb18/Final++Mining.pdf?MOD=AJPERES.

ISO 9000. 2015. *Quality Management Systems: Fundamentals and Vocabulary.* Geneva: International Organization for Standardization. www.iso.org.

ISO 14001. 2015. *Environmental Management Systems: Requirements with Guidance for Use.* Geneva: International Organization for Standardization. www.iso.org.

ISO 14010. 1996. *Guidelines for Environmental Auditing: General Principles.* Geneva: International Organization for Standardization.

ISO 19011. 2002. *Guidelines for Quality and Environmental Management Systems Auditing.* Geneva: International Organization for Standardization. www.iso.org.

Mernitz, S. 2005. Environmental geology and sustainability. *The Professional Geologist* 42(2):41–43.

Mining Journal. 2003. Fraser Institute ranks mineral attractiveness. *Mining Journal* (Jan. 31): 69.

Rocky Mountain Mineral Law Foundation, ed. 1996. Title XV: Environmental regulation of the mining industry. In *The American Law of Mining*, 2nd ed. New York: Matthew Bender. Chapters 165–180.

World Bank. 1999a (revised 2013). OP 4.01: Environmental assessment. In *Operational Manual.* https://policies.worldbank.org/sites/ppf3/PPFDocuments/Forms/DispPage.aspx?docid=1565&ver=current (accessed Feb. 19, 2017).

World Bank. 1999b. *Pollution Prevention and Abatement Handbook, 1998: Toward Cleaner Production.* doi:10.1596/0-8213-3638-x.

APPENDIX 9A

Suggested Environmental Impact Assessment Contents Using Typical International Practices

The analysis in the foregoing initial environmental evaluation (IEE) has suggested that a more detailed analysis—an environmental impact assessment (EIA) under the country's regulations—be performed for the subject project. With the unique project development issues in mind, and considering the scrutiny of this EIA that is likely from the international government, financial, industrial, and nongovernmental organization (NGO) community, the following annotated outline is proposed as a basis for EIA preparation.

SECTION I—EXECUTIVE SUMMARY [~10 PAGES]

i. Summary of the nature of this EIA and previous IEE, and the IEE → EIA process
ii. Principal government authorities
iii. Project description and location (the Proposed Action by the Project Proponent)
iv. Other alternatives considered in the impact analysis, and those eliminated
v. Key impacts and impact comparisons among alternatives
vi. Mitigation and the environmental management and monitoring plan and responsibilities
vii. Recommended actions

SECTION II—POLICY, LEGAL, AND ADMINISTRATIVE FRAMEWORK [~5 PAGES]

i. List of key government laws and regulations
ii. Relevant government agency guidelines for EIA preparation
iii. List of other relevant legal and policy guidelines regarding wastes, pollution, effluent limits and similar matters
iv. List of relevant global environmental policies

[Refer back to each of these in later EIAs as they are relevant.]

SECTION III—PROJECT DESCRIPTION [~50 PAGES]

Prepare a concise description of the Proposed Action, addressing the project's geographic, ecological, social, and temporal context, including the following:

i. Location, including precise boundary coordinates for each permit area, off-site water or other well locations, and neighboring leases and adjacent controlling interests (tables and maps, some color)

ii. Phases of development and operations, and the production schedule
iii. Related facilities descriptions, with diagrams and artist's conceptions as available
iv. Production plan, technology and techniques, including use of computer monitoring, remote imaging, and related innovative facilities as they may be relevant to impact analysis during construction, operations, and closure
v. Related facilities and investments, including pipelines, vessels, roads, water supply, housing, and similar
vi. Environmental and engineering control facilities planned, including backup power generation, water treatment, fire suppression, emergency facilities, and similar issues, and any engineering risk analysis performed
vii. Hazardous and solid materials to be used and wastes generated
viii. Employment, work schedules, and related human resources matters
ix. Other local economic plans including equipment and supplies purchase, taxes and royalties, and related matters

SECTION IV—ENVIRONMENTAL AND SOCIOECONOMIC SETTING [~75 PAGES]

Describe and map the dimensions of the varying study areas by discipline in each section and address any changes anticipated in these study areas before the project begins, including current and proposed development by others (e.g., on nearby leases). Address data for nearby locations and discuss their relevance. Present a summary of environmental baseline conditions by discipline, with liberal use of maps, tables, and diagrams, some in color, for each discipline regarding the following:

i. Physical conditions including topography, geology, soils, geomorphology
ii. Climate and meteorology
iii. Water and sediment quality, including surface and groundwater quality and quantity for onshore facility areas
iv. Air quality and noise
v. Biological conditions, including flora and fauna and sensitive species
vi. Fish and fisheries, both commercial and subsistence
vii. Comments on aquatic and marine ecology in general, addressing water quality and marine life in areas of proposed development and infrastructure, including coastal support facilities
viii. Socioeconomic and cultural attributes, with particular emphasis on indigenous peoples;
ix. Demographics or basic population characteristics
x. Employment, income, and other economic factors
xi. Social organization and culture, with emphasis on both the potential work force and those peoples directly or indirectly affected, positively or negatively, by potential development
xii. Community facilities, services, and education
xiii. Current land uses and special reserve areas

xiv. Impacts of past mining exploration and development on the affected peoples
xv. Community expectations and level of public participation

[Appendix 9B presents additional socioeconomic details that could be provided in this EIA or in a separate social impact assessment (SIA).]

SECTION V—ENVIRONMENTAL MANAGEMENT PLAN [~40 PAGES]

In advance of the presentation of analysis of predicted environmental impacts, the Proponent's committed environmental management, monitoring, and mitigation plan is summarized here, to address the following issues:

i. Water quality control and water management
ii. Process fluids management
iii. Mitigation of disturbance
iv. Minimization of impacts on subsistence activities
v. Air emissions controls
vi. Hazardous and solid waste management
vii. Noise abatement
viii. Monitoring, reporting, and corrective actions, and responsibility therefore
ix. Reclamation, decommissioning, and closure plans, and any bonding for such purposes

Again, liberal use of charts, maps, and diagrams is suggested to allow the document to be understood by the lay public.

[See Appendix 9B for an example outline for a more detailed environmental monitoring and mitigation plan (EMMP).]

SECTION VI—ENVIRONMENTAL AND SOCIOECONOMIC IMPACTS AND MITIGATION [~100 PAGES]

With Section V in mind regarding committed mitigation and monitoring, this "Impacts" section will present predicted impacts of the Proposed Action on the same outline for those topics in Section IV for the physical, biological, and human environment, to discuss the following:

i. Impact assessment methodology, including impact indicators by discipline and measures of significance
ii. Methods for estimating positive or negative impacts
iii. Recommended and committed mitigation measures and any predicted residual, negative impacts
iv. Opportunities for environmental enhancement, sustainable development, cleaner production processes, waste minimization, maintenance of biodiversity, and pollution prevention
v. Identification of remaining data gaps and uncertainties
vi. Other reasonable foreseeable development and the cumulative impacts of those projects with the Proposed Project
vii. Documentation of topics that do not warrant further attention

SECTION VII—ANALYSIS OF ALTERNATIVES [~20 PAGES]

A full and fair range of alternatives should be considered to satisfy government, global industry, and NGO reviewers. The Proponent may wish to present some of its internal alternatives analysis, showing alternatives that are feasible from an engineering and economic perspective, and others that were eliminated with reasons for elimination. Each alternative not previously presented in Section III should be briefly discussed (~2 pages each) with mention of its positive and negative attributes. Details are to be presented in a comparison table. The range of alternatives considered in the EIA might include the following:

i. Proposed action by the project proponent (summarize details briefly again here; Alternative 1)
ii. No Action (Alternative 2)
iii. Reduced development scenario over proposed action, still economically viable (Alternative 3)
iv. Increased development scenario over proposed action, still economically viable (Alternative 4)
v. Increased development scenario with additional mitigation, still economically viable (Alternative 5)

A key purpose of the EIA analysis is to enhance the design of a project through consideration of alternatives. Environmental costs and benefits of each alternative are presented in quantitative or qualitative terms in a comparison table. This is the heart of the EIA document. The basis for the selected alternative should be stated, or, if the agencies prefer, the EIA is used as a decision document and the agencies choose their Agency Preferred Alternative in some sort of decision record.

SECTION VIII—PUBLIC CONSULTATION [~10 PAGES]

In contrast to addressing such issues in the project description or socioeconomic sections, public consultation could be addressed in a separate section given the level of controversy concerning the Project and its impacts on the host country. This section will include the following:

i. Summary of public meetings and contacts, surveys, newsletters, press releases and media coverage, and other community relations efforts
ii. Specific efforts to reach indigenous peoples and their spokespersons, and results
iii. Government agency coordination
iv. Lists of meeting attendees, and comment letters received
v. Lists of EIA scoping issues that resulted from this public consultation
vi. Tracking of these scoping issues through the foregoing EIA document
vii. Planned future efforts to ensure the Proponent's accountability to public concerns
viii. (Optional) Lists of EIA preparers and their qualifications

APPENDIXES

Additional, detailed material that is critical to the EIA analysis could be placed in appendixes to the EIA. However, to reference such a file, material available in a reading room or local library—for readers with special interests—is a better option. For ease of access, handling, and readability by the lay public, an EIA document totaling about 300 pages is recommended.

APPENDIX 9B

Socioeconomics Detailed Outline

This outline presents additional socioeconomic details that could be provided in Section IV of Appendix 9A or in a separate social impact assessment (SIA).

1. SOCIOECONOMIC ENVIRONMENT AND EFFECTS

1.1 Demographic Profile and Effects

1.1.1 Summary
- Construction phase
- Operational phase

1.1.2 Existing Demographic Profile
- Population size and distribution
- Population growth
- Age structure
- Sex structure
- Ethnicity
- Length of residency
- Education status
- Health

1.1.3 Existing Industries and Labor Force
- Labor force characteristics

1.1.4 Projected Population Growth

1.1.5 Demographic Impact
- Construction phase
- Operational phase
- Combined effects

1.2 Workforce

1.2.1 Summary

1.2.2 Construction Phase

1.2.3 Operational Phase

1.3 Housing and Accommodation

1.3.1 Summary

1.3.2 Construction Phase

1.3.3 Temporary Accommodation

- 1.3.4 Permanent Accommodation
- 1.3.5 Construction Phase Effects
- 1.3.6 Operational Phase Effects

1.4 Government and Community Facilities and Services

- 1.4.1 Summary
- 1.4.2 Health and Medical Facilities
- 1.4.3 Emergency Services Facilities
- 1.4.4 Education Facilities
- 1.4.5 Cultural and Recreational Facilities
- 1.4.6 Community and Welfare Services

1.5 Land Tenure and Use

- 1.5.1 Summary
- 1.5.2 Land and Ocean Resources Tenure
- 1.5.3 Existing Natural Resource Use Site
  - Surrounding areas
- 1.5.4 Planning Controls and Legal Requirements

1.6 Visual Effects

- 1.6.1 Summary
- 1.6.2 Primary Visual Catchment
- 1.6.3 Methodology
- 1.6.4 Visual Character of the Development
- 1.6.5 Visual Interactions
- 1.6.6 Mitigation Measures

1.7 Cultural Heritage

- 1.7.1 Summary
- 1.7.2 Aim
- 1.7.3 Survey Methodology
- 1.7.4 Survey Results

1.8 Economic Impacts

- 1.8.1 Economic Model
- 1.8.2 Construction Phase
- 1.8.3 Operational Phase
- 1.8.4 Decommissioning Phase

1.9 Occupational Health and Public Health

APPENDIX 9C

Mining Project Environmental Management Plan Example

This table of contents is reproduced from Cambior Inc. 2002. At the time it was created, it was a very detailed environmental management plan (EMP) and similar to the environmental monitoring and mitigation plans (EMMPs) described earlier.

TABLE OF CONTENTS

SECTION VII: ENVIRONMENTAL MANAGEMENT PLAN

Cambior Inc. – Rosebel Gold Project - xxiv - Rescan™ Environmental Services Ltd.

TABLE OF CONTENTS

TABLE OF CONTENTS

TABLE OF CONTENTS

TABLE OF CONTENTS

CHAPTER 10

Sustainability and the Social License to Operate

Scott Mernitz

From the start, all modern mining project planners, especially in this second decade of the 21st century, must keep in mind the social, regulatory, and political setting in which they operate or plan to operate. The issues of social sustainability, governance, and the overall social license to operate have become watchwords for mining executives. This has occurred for good reason. Opponents of mining are many, and, often woefully misinformed and easily persuaded by frightening stories rarely based in modern science or fact.

The term *social license to operate (SLO)* can be confusing when used in international discussions. It is not an actual license or permit but rather a combination of the social and political agreements that must be gained with the local and regional populace and the local, state, and federal governments to proceed without bureaucratic delays, roadblocks, sabotage, demonstrations, sometimes violent riots, and other disruptions to a functioning commercial mining operation.

SLO often arises because locally, regionally, or globally connected nongovernmental organizations (NGOs), such as environmental protection groups, community activist groups, and others, look to any mining project announcements and pick targets for critical attack. These attacks often come if indigenous (native) peoples, relocation of communities, changes in infrastructure (e.g., potential effects on drinking or irrigation water supply), changes to agricultural practices, or other controversial aspects of the human environment are not, in the opinion of the NGOs or other groups, appropriately considered and addressed with concrete mitigation by the formal permitting process. The use of the World Wide Web is skillful, extensive, and sophisticated and aptly manipulated by NGOs to gain financial and political support for anti-mining views. Currently, this global tool—the Internet—and its practiced use has sometimes even exceeded the mining industry's ability to counter attacks from activist groups.

In the last 20 years, many mining projects in the world have struggled with SLO. For many recent projects, all permits for construction and even operations may have been approved, based on proper company applications, by the respective governments and their regulatory agencies. These have likely included permits for comprehensive environmental impact and mitigation (the environmental impact assessment, or EIA), water quality, mining and reclamation, air emissions, local land use, and so forth. Yet social acceptance of the project in the locality, region, and larger governed areas may not have the political weight to carry the project forward. Lack of SLO delays or stops projects. Conflict resolution is the key, and it is difficult in some situations. Projects have stopped or failed when SLO collides with mining operations.

SUSTAINABILITY

A key measure in attempting to gain project success is to successfully demonstrate *sustainability* to project opponents. This demonstration must continue prior to, during, and following project planning, development and construction, operations, and postclosure. Based on this author's experience, there are five factors that comprise the sustainability picture:

1. A mining or energy project must have *financial sustainability* to succeed for a minimum project life, say five years or more, and in turn to have a positive impact on its environmental, social, and political surroundings.
2. A project's next priority should be the state of *environmental sustainability*, addressing protection of the natural and physical environment and its ecosystems with a strong view to long-term productivity (including the concept of biodiversity maintenance).
3. In combination with its preceding financial position, a project should promote *economic sustainability*, that is, primary (basic extraction and processing), secondary (suppliers of equipment and hard goods), and tertiary (suppliers of soft goods and related services) activity and economic growth and reduction of poverty.
4. *Social sustainability* must be a key component of minerals project development, including numerous opportunities for fulfilling human potential (and, one could say, maintenance of human biodiversity).
5. Appropriate *governance* structures, both official and unofficial, formal and informal, corporate and local, must exist and be developed to make the project work and contribute favorably, in all aspects of sustainability, to its setting. For the company, the principles of accountability, transparency, and fairness in returns to the local community and infrastructure are increasingly expected to provide for future generations while addressing the needs of the current generation. For the government, similar characteristics of openness, efficiency, lack of corruption, and timely processing of reviews and permits are necessary.

RECENT MINING INDUSTRY EFFORTS

Many reviews of sustainability efforts by the mining industry have been conducted by favorable commentators in the past two decades in an attempt to improve the industry's position, both with the public perception of mining's value to society and with documentation of its truly good works on the ground as it attempts to profitably extract mineral resources for the human good.

An excellent two-part series in *The Professional Geologist* (Johnson 2007a, 2007b) presents an example of such an analysis. The articles include a useful discussion of sustainability definitions, a status report, and examples of project progress, limitations, and a rating scale. Johnson proposes a three-P's (people, planet, and profit) sustainability index with a triangular radar plot.

Johnson (2007a, 2007b) also raises such issues as socially responsible investment funds and attempts by the mining industry to attract their investments. Organizations such as the International Council on Mining and Metals and the Global Reporting Initiative as well as measures such as corporate social responsibility ratings have arisen during the past 15–20 years as the mining industry strives for sustainability. And the need to convince the public that such is not "greenwash" is an ever-present battle in the mining media. Mining companies that emphasize "green" gold or diamonds (as opposed to "blood diamonds") are noted in publicity venues such as magazines, films, and TV advertisements, some even prevalent in primetime media.

There have been many other efforts by mining sustainability organizations to enlist mining company members and solicit enrollment fees, perform independent annual inspections and report on results, and generally provide "good news" for the mining company to report to their stockholders and the global community. Among these are the Mining, Minerals and Sustainable Development project, the establishment of a sustainability index for mining companies, a discussion of the five steps to sustainable natural resources development, the Extractive Industries Transparency Initiative (EITI) to promote transparency in minerals projects in developing countries, various writings on governance, and other works. These examples are not given citations here but rather show the wide range of efforts of this type and suggest some of the key words for website searches by the interested reader.

The author worked on one challenging and interesting project in the Democratic Republic of the Congo regarding the EITI and the planned percentage of project development costs to be devoted to social and community improvements. Again, a web search on "sustainability" will yield much related reading.

MINING FINANCE EFFORTS AND THE ENVIRONMENT

Perhaps the most diverse and far-reaching of these sustainability efforts have been those of the World Bank and its mining finance efforts through its International Finance Corporation (IFC) with its project-recommended performance standards and through other World Bank entities. The World Bank has encouraged large- to medium-sized global banks and financial institutions to embrace and follow its sustainability principles in lending decisions to energy and minerals projects—and the banks have come on strongly. Currently, 91 financial institutions have adopted a set of guidelines called the *Equator Principles* (Equator Principles Association 2013). See the Equator Principles website (www.equator-principles.com) for many of the details and for citations on much of the discussion that follows.

The Equator Principles are well known to many in the mining and environmental industries. Originally developed by fewer than 10 major banks in 2003, with technical advice from this author and many in the industry, the list has steadily grown to cover many of the countries of the world and countless financing arrangements. As a voluntary measure of goodwill and actual and potential sustainability practices during the life of the loan, mining companies and banks often contract a third-party consultant to oversee a mining project and analyze compliance with the Equator Principles.

Exhibit II of *Equator Principles III* provides a checklist of issues to be addressed by the mining project, as applicable, to provide evidence that the project is suitable for a loan by the Equator Principles Financial Institution (EPFI) as reported (Equator Principles Association 2013). A reviewer will check for the following aspects and issues to be adequately addressed in project documents and commitments by the operator.

a. assessment of the baseline environmental and social conditions

b. consideration of feasible environmentally and socially preferable alternatives

c. requirements under host country laws and regulations, applicable international treaties and agreements d. protection and conservation of biodiversity (including endangered species and sensitive ecosystems in modified, natural and Critical Habitats) and identification of legally protected areas

e. sustainable management and use of renewable natural resources (including sustainable resource management through appropriate independent certification systems)

f. use and management of dangerous substances

g. major hazards assessment and management

h. efficient production, delivery and use of energy

i. pollution prevention and waste minimisation, pollution controls (liquid effluents and air emissions), and solid and chemical waste management

j. viability of Project operations in view of reasonably foreseeable changing weather patterns/climatic conditions, together with adaptation opportunities

k. cumulative impacts of existing Projects, the proposed Project, and anticipated future Projects

l. respect of human rights by acting with due diligence to prevent, mitigate and manage adverse human rights impacts

m. labour issues (including the four core labour standards), and occupational health and safety

n. consultation and participation of affected parties in the design, review and implementation of the Project

o. socio-economic impacts

p. impacts on Affected Communities, and disadvantaged or vulnerable groups

q. gender and disproportionate gender impacts

r. land acquisition and involuntary resettlement

s. impacts on indigenous peoples, and their unique cultural systems and values

t. protection of cultural property and heritage

u. protection of community health, safety and security (including risks, impacts and management of Project's use of security personnel)

v. fire prevention and life safety

As background, starting in 2006 the Equator Principles have been steadily refined, with new best recommended practices evident on their website in 2017 as *Equator Principles III*, effective in 2013. Mining project staff members will find the following of interest:

- Project financings with total project capital costs of US$10 million or more (versus US$50 million previously) must have the principles applied.
- A social and environmental assessment (SEA) is required relevant to the level of expected impacts. This combines the EIA and the social impact assessment reports (see aspects and issues in the preceding Equator Principles Exhibit II list).
- IFC performance standards (IFC 2012) and industry-specific *Environmental, Health and Safety Guidelines* (*EHS Guidelines*; IFC 2007a) apply to projects in countries without well-developed environmental regulatory systems.

- For category A and B projects (major to moderate potential impacts), a brief action plan (based on the environmental monitoring or management plan) is required, outlining mitigation commitments looking forward, funding, monitoring, reporting, and corrective actions.
- More formal consultations with affected parties, disclosure, and grievance mechanisms are specified.
- Independent consultant review of the reports and process is also specified, plus independent expert monitoring and reporting over the life of the loan.

To give an example of how the IFC performance standards (IFC 2012) and *EHS Guidelines* (IFC 2007a) could be used to evaluate a mining activity, a project that includes an open pit copper mine, process plant, and nearby smelter is considered. Nearly all of the eight specific IFC performance standards would apply, such as the SEA requirement, labor and working conditions, and biodiversity conservation. Regarding the industry-specific *EHS Guidelines*, those on mining (reformulated and expanded in 2007) and base metal smelting and refining would apply (IFC 2007b, 2007c). Other infrastructure guidelines, such as those regarding waste management facilities, may also apply. The general EHS and mining guidelines are quite comprehensive for all types of mining and milling potential impacts. Here the IFC website (and World Bank website) should be consulted for citations and details.

In the *Equator Principles III* (2013), the following new points and trends in environmental and sustainability reviews of projects are given attention:

- The Equator Principles are to be applied to both project-related corporate loans and bridge loans.
- It is no longer adequate for the EPFI to only report the annual number of transactions, but rather it must report the actual number of projects closed; names and details as to country, category, sector, and region are to be specified.
- A general summary of the SEA for each project must be online for reviewers, and it must include greenhouse gas (GHG) emission levels for large emitters.
- Previously, only social risks and free prior and informed *consultation* for stakeholders needed to be included in the analysis. Now human rights due diligence and free prior and informed *consent* is to be noted.
- Overall due diligence to anticipated climate change effects and GHG project reporting must be specified.
- Social risks and impacts analysis must be included.
- The glossary of terms (found in the *Equator Principles III* Exhibit I on their website) refines several definitions. Of special note is the definition of designated countries and the standards that apply to those with robust environmental programs.

Mining companies, banks, investors, and government entities can work with consultants familiar with the Equator Principles to meet compliance on their global projects.

CONFLICT RESOLUTION

In view of the many conflicts regarding mining projects' approval, which have been publicized in recent years, a few comments on conflict resolution may be useful. For example, one mining project on the mountainous border between Chile and Peru has been in development for more than 10 years. The lengthy delay has created escalating costs, estimated initially at US$100 million to upward of US$8 billion. The project is facing a US$16 million fine from one government and a claim of US$140 million in reparations costs to potentially affected parties. Another mining project in Alaska, to which this author has contributed, has similarly seen ascending estimated costs, ownership changes, and extensive NGO activity, halting progress and creating litigation over five to six years. An open pit copper mine project in California with a Canadian sponsor went to NAFTA (North American Free Trade Agreement) court in Washington, D.C., seeking US$50 million in damages and claiming unfair treatment by California and U.S. regulators because of inability to earn potential revenues from the mine.

Many of these conflicts, of course, relate to sustainability—of the existing environmental conditions (e.g., a fishery and fish habitat), economics (local employment), sociocultural conditions (way of life for indigenous peoples), finance (mining company right to mine and profitability), and governance (state and federal regulatory approvals and local rights to favor or protest the project). And these conflicts generally lead to litigation, administrative law judgments, or binding arbitration. But what is the alternative in such cases?

One suggestion is environmental mediation. In such cases, the conflicting parties would agree by memorandum of understanding (which is politically, if not exactly, legally binding) to share information, participate in negotiation sessions with an impartial third-party mediator, retrench and renegotiate, and attempt to reach a compromise settlement. The parties would contribute to a common fund to compensate the mediator and pay expenses for the meetings, field trips, exhibits to view during negotiations, specific additional research agreed on by all parties, bringing in other experts as needed on conflict topics, and similar needs to clarify and educate.

Mediation and a mediated settlement are not binding. However, the objective would be for all parties to ratify an agreement of what could be legally binding stipulations and conduct. It is clear that many millions of dollars could be saved if successful.

The author (Mernitz 1980) has explored this topic in detail and used mediation informally in many of his mining and environmental consulting experiences. Some of the key factors in environmental dispute resolution that may help to make a project "mediable" are

- Favorable regional and local physical and cultural characteristics and a history of local environmental conflicts for learning purposes;
- Manageable, stable (local, regional, state, or federal) levels of government;
- Some atmosphere of existing compromises (i.e., no powerful "no project under any circumstance" forces but rather a sense of "some project" rather than the "company planned project" being worthy of negotiation);
- Negotiable ancillary (side) issues that may promote compromise;
- A keen and focused multidisciplinary analysis of the parties (stakeholders) to the conflict, regarding such issues as economic self-interest, social self-interest, personal value judgments, and concern for natural systems; and
- A state governor role that can be explored and worked to bring parties to the table and promote political compromise.

It follows that past mining disputes in a mining company's region of interest can be explored and assessed to learn of effective approaches and compromises that may make for a successful project, whether in the South American Andes, hills of Romania, outback of Australia, or dry plains and fault block mountains of Nevada.

INDUSTRY SUSTAINABILITY ADVICE EFFORTS

In the lessons taught to his students by one of the primary authors and editors of this due diligence text, Bullock references a few of the commentators (Bennett, Joyce, and Thomson; see Bullock's discussion and citations in Chapter 17) in the early days of sustainability thinking. Several of their ideas and concepts will be instructive if one is interested in further reading.

In Chapter 17, Bullock also presents a lengthy treatise on social and political risk from the perspective of a long-time mining engineer with much global experience. That chapter will also be useful for the environmental specialist with a keen interest in presenting sustainability advice to a client or company.

The Society for Mining, Metallurgy & Exploration (SME) offers a notable collection of writings by its members on sustainability, social license to operate, community and social issues in mining, and related topics (see the "Store" tab at www.smenet.org). Further, for the reader with an avid interest in environmental issues and SLO, updated discussions are available in Parts 16 (Environmental Issues) and 17 (Community and Social Issues) of the venerable *SME Mining Engineering Handbook* (Darling 2011).

In closing, it remains that sustainability and SLO are important parts of the mining company staff, banker and investor, government, and student analyses of a mining project—from various engineering, geologic, environmental, economic, social, and operational aspects—as the project moves along during its life.

REFERENCES

Darling, P., ed. 2011. *SME Mining Engineering Handbook*, 3rd ed. Englewood, CO: SME.

Equator Principles Association. 2013. *Equator Principles III*. www.equator-principles.com/resources/equator_principles_III.pdf (accessed March 30, 2017).

IFC (International Finance Corporation). 2007a. *Environmental, Health, and Safety Guidelines*. www.ifc.org/ehsguidelines.

IFC (International Finance Corporation). 2007b. *Environmental, Health, and Safety Guidelines: Base Metal Smelting and Refining*. www.ifc.org/wps/wcm/connect/4365de0048855b9e8984db6a6515bb18/Final%2B-%2BSmelting%2Band%2BRefining.pdf?MOD=AJPERES&id=1323152449229 (accessed March 30, 2017).

IFC (International Finance Corporation). 2007c. *Environmental, Health and Safety Guidelines for Mining*. www.ifc.org/wps/wcm/connect/1f4dc28048855af4879cd76a6515bb18/Final++Mining.pdf?MOD=AJPERES.

IFC (International Finance Corporation). 2012. *Performance Standards on Environmental and Social Sustainability*. Washington, DC: IFC. www.ifc.org/wps/wcm/connect/115482804a0255db96fbffd1a5d13d27/PS_English_2012_Full-Document.pdf?MOD=AJPERES (accessed March 30, 2017).

Johnson, G.W. 2007a. Defining sustainability and evolving metrics in the mining and metals sector, Part 1. *The Professional Geologist* 44(3):45–50.

Johnson, G.W. 2007b. Defining sustainability and evolving metrics in the mining and metals sector, Part 2. *The Professional Geologist* 44(4):40–46.

Mernitz, S. 1980. *Mediation of Environmental Disputes: A Sourcebook*. New York: Praeger.

CHAPTER 11

Phased Approach to Mineral Property Feasibility Study and Economic Analysis

Richard L. Bullock

Many approaches can be used to perform a mineral property feasibility study. Several of them have been briefly discussed in Chapter 1, and a few of those and others are discussed in further detail in this chapter. All of them involve studying various aspects of the potential mineral operation, such as the mineral resource and determining the reserve estimate (if there is one); determining a mining method based on the measured and indicated resource in the final feasibility study; preparing the mineral extraction flow sheet; performing a market analysis; determining the infrastructure needs; quantifying the environmental and socioeconomic impacts and mitigation required; estimating the cost of all of the above; and then performing an economic analysis of the assumed revenues versus the costs to determine whether the project meets the company objectives. The next objective, of course, is to determine the optimum method of designing and developing the mineral property to yield the greatest economic reward that can be achieved with that particular property, at that point in time, in whatever environment or location where it is found.

However, one consideration is most important: The prudent company will want to minimize the money and time spent on the property until it is reasonably confident that the mineral resource is indeed an ore reserve. This means that the company will want to limit its look at all of the aspects of the potential mineral operation listed previously and only spend money where it is absolutely necessary to document the factual data and determine economic viability. Furthermore, the early look at the potential operation may not necessarily optimize the potential operation. In fact, it would probably be a miracle if it did. What that first look should do is put together a semi-engineered, logical approach for the entire operation, and test this method for economic viability. It can be difficult to determine how deep to go with each aspect of the project team's study during each phase of the investigation. Just how much analysis should be done on each of these elements—the mineral resource, the mining method, the metallurgy, the infrastructure, and so on—and all of the hundreds of items that make up these various aspects of the mineral evaluation study? Obviously, one needs a very systematic approach to solve this problem. Keep in mind that at the end of the final feasibility study, the company will want an optimal configuration of mine, process plant, waste disposal facility, and infrastructure, not just a mine or plant that will produce a product under less-than-optimal conditions.

THE CLASSIC ENGINEERED APPROACH

The authors reviewed below take the approach sited in the previous paragraph: that is, all recommend a phased approach to mineral property feasibility. Lee (1984) is quoted by Hustrulid and Kuchta (2006) as taking a three-phased approach:

> ***Stage 1: Conceptual* [pre-feasibility or scoping] *study***
> *A conceptual (or preliminary valuation) study represents the transformation of a project idea into a broad investment proposition, by using comparative methods of scope definition and cost estimating techniques to identify a potential investment opportunity. Capital and operating costs are usually approximate ratio estimates using historical data. It is intended primarily to highlight the principal investment aspects of a possible mining proposition. The preparation of such a study is normally the work of one or two engineers. The findings are reported as a* [scoping study] *preliminary valuation.*
>
> ***Stage 2: Preliminary or pre-feasibility study***
> *A preliminary study is an intermediate-level exercise, normally not suitable for an investment decision. It has the objectives of determining whether the project concept justifies a detailed analysis by a feasibility study, and whether any aspects of the project are critical to its viability and necessitate in-depth investigation through functional or support studies.*
>
> *A preliminary study should be viewed as an intermediate stage between a relatively inexpensive conceptual study and a relatively expensive feasibility study. Some are done by a two or three man team who have access to consultants in various fields; others may be multi-group efforts.*
>
> ***Stage 3: Feasibility study***
> *The feasibility study provides a definitive technical, environmental and commercial base for an investment decision. It uses iterative processes to optimize all critical elements of the project. It identifies the production capacity, technology, investment and production costs, sales revenues, and return on investment. Normally it defines the scope of work unequivocally, and serves as a base-line document for advancement of the project through subsequent phases.*

With more than 25 years of experience conducting feasibility studies, evaluating mineral properties, and performing due diligence reviews, the chapter author has found the same problems duplicated repeatedly throughout the industry using the above-mentioned classic engineered approach.

In Chapter 2, the reporting standards for various countries are given. These must be followed no matter which system of feasibility studies described in the following text are used. This handbook was authored by those who primarily performed their work under the rules of the U.S. Securities and Exchange Commission (SEC 1992) or the Canadian Securities Administrators (CSA; NI 43-101). The rules you will follow while performing feasibility studies of due diligence reports depend on where the company financial equity listings are exchanged. A good guideline for the United States is *The SME Guide for Reporting Exploration Information, Mineral Resources, and Mineral Reserves* (SME 2017).

Conceptual or Scoping Study

A conceptual or scoping study can be extremely misleading. Nearly any and all exploration projects that are only slightly submarginal can be shown to be worthy of further development based on casual educated guesses and optimistic simplified, or even biased, evaluations. In this author's opinion, back-of-the-envelope approaches to mine feasibility study need to stay on the backs of envelopes and out of formal, official-looking reports. This type of report can be at its worst when done by the exploration firm or group itself to try to sell the project to someone else or just raise financing. However, when an independent third party does the conceptual or scoping study, it can and is used as a useful tool for the company to determine whether they wish to go to the next phase of feasibility study or what it might be worth on the open market. Also, this approach might be very appropriate for looking for commodity targets for the exploration group, but not for further in-house decisions to move the project to the next level, based on the exploration group's mining and milling judgment. This is not to say that conceptual unclassified screening studies do not have their place in justifying other types of work, but care and caution are needed not to dignify the conceptual study beyond the engineering basis that it really has. In fact, some countries' security exchange agencies, such as the CSA, do allow and specify such a preliminary study, which they call a *preliminary economic assessment* (PEA). But it is not a class of feasibility study, given that it is completed without a substantial engineering basis. A further difference is the inclusion of inferred resource materials, which is not allowed in feasibility studies of the CSA or in those of the SEC. It was offered as a way for the exploration groups to show the economic potential of the inferred resource material.

The CSA justifies the use of the PEA in lieu of a preliminary feasibility or scoping feasibility study, as explained by Gosson (2011/2012):

> *The practical reality in the mining business is that investors want to have an understanding of the economic potential of a mineral project, even at a preliminary stage of assessment. At this early stage, a significant portion of the Mineral Resources may be in the Inferred category—particularly for deeper deposits only amenable to exploitation using underground mining methods. To restrict the economic analyses to just the Measured and Indicated Mineral Resource categories could result in a meaningless and potentially misleading result. Canadian Securities regulators recognized this investor need for information and provided a carve-out for studies at the level of Preliminary Assessments. Mining companies may disclose the results of economic analyses that include Inferred Mineral Resources at an early stage of the mineral project. "Early stage" was defined as being prior to the completion of a Preliminary Feasibility Study. There were certain conditions to the carve-out:*
>
> - *Results of the Preliminary Assessment must be material to the company.*
> - *Cautionary language must be included with the disclosure of the Preliminary Assessment.*
> - *The disclosure must include the basis of the Preliminary Assessment and the assumptions used.*
> - *The disclosure would require the filing of a technical report supporting the Preliminary Assessment.*

In addition, S. Vézina (2013) further clarifies:

> *To ensure that a preliminary assessment is not equated with a pre-feasibility study, CSA staff recommend that issuers ensure the results of their assessment include the cautionary language required by section 3.4 of Regulation 43-101, indicating that the economic viability of the mineral resources has not been demonstrated. Also, it would be prudent to include a detailed description of the risks associated with the project in the assessment so that the public is able to understand the importance and limits of its results.*
>
> *CSA staff consider that a preliminary economic assessment is, by definition, a study other than a pre-feasibility or feasibility study. Two parallel studies done at, or nearly at, the same time are not in substance separate studies, but components of the same study. Thus, the staff indicate that a preliminary economic assessment done concurrently with a pre-feasibility or feasibility study will likely be treated as a pre-feasibility or feasibility study if it:*
>
> - *has the effect of incorporating inferred mineral resources into the pre-feasibility or feasibility study;*
> - *updates a pre-feasibility or feasibility study to include more optimistic or even more aggressive assumptions and parameters than the initial study;*
> - *is essentially a pre-feasibility or feasibility study in all respects but name.*

The interpretation of this explanation is that they are re-enforcing the separation of inferred resources. Inferred mineral resources must be upgraded to indicated or measured resources first. There is no doubt that the PEA is useful in documenting potentially viable resources, but it does allow for some misunderstandings when a company presents the PEA along with a prefeasibility study. The investor must remember that the accuracy of the economics at all levels of feasibility studies depends on how much good engineering directed at this specific project has been done. If none has been completed, then one should not expect the accuracy of the project's related costs and economics to be accurate, and thus very misleading. Typically found in many projects, only 1% or 2% of the total engineering on the project will have been completed. For a small project, this may amount to 1,400–2,800 worker-hours of engineering work. But for a large project, this 1% to 2% may come to 12,500–25,000 worker-hours of engineering. With this amount of engineering, experience, and cost on similar projects, the accuracy of the scoping study feasibility may be nearer to ±45%. Other authors claim that an accuracy of 30% is achievable for a prefeasibility or scoping study (White 1997). To obtain this accuracy, it is this author's opinion that the real amount of engineering should be 6%–8% of the total project hours. This is discussed more in later sections. Unless the project is being developed in an old district where a mine or plant has recently been built and this is simply a look-alike, it is unlikely that this accuracy can be achieved.

Prefeasibility Study

The problems that have been found with many prefeasibility studies that followed a conceptual or scoping study as previously outlined is that many times this phase simply follows the path set by the scoping study. There is reluctance to go back and justify the processes that were chosen for the mining method, processing method, infrastructure needed, waste disposal method, overall size of the operation, and so forth. That is, the project is not usually optimized at a time when the project team is small, has plenty of time to think in terms of multiple methods, and the projected construction schedule is not yet pushing the project.

Another problem that has been observed is that some of the elements or activities of the prefeasibility study, once applied, will be taken too far, while others may not be taken far enough. Thus, because some of the elements of the study will go too far, the client will invariably believe and proclaim to others in management and the investors that "the study is really more than a preliminary feasibility study." This will probably not be true, but it will give members of management (and possible financiers) some unjustified overconfidence in the project.

Another very critical failure of the prefeasibility study in this system of feasibility progression is that the preliminary study is the last moment when you must find the "fatal flaw" of the project if it has one. It may be found in the conceptual or scoping study, but it may not. One definitely does not want to wait until a very large engineering group has been assembled to work on the (final) feasibility study. By this time, the momentum to the project is well in force and it will cost a lot of money just to stop it. If the project is being done by a company listed on the Canadian or U.S. stock exchanges, then inferred mineral resources may not be used for mine planning purposes, unless it is a very small portion of interburden located between measured or indicated resource material that must also be mined.

Feasibility Study

When using the classic approach, by the time one starts the feasibility study, the project direction of each element has usually been set. For all aspects of the project to proceed immediately at the same pace from this point, there is little opportunity to stop and look at the many interrelated operating variables that should have been examined at an earlier stage. Therefore, it is likely that a nonoptimized design will emerge from this type of study. In this author's opinion, the mining industry is full of nonoptimized mines and plants that have been built because the optimization studies did not take place at the proper time (i.e., during the intermediate feasibility study). Sometimes it is realized toward the end of such a study that certain aspects have not been optimized, and subsequently, major last-minute adjustments are attempted. Such actions are not usually based on the same amount of engineering analysis that went into the original planning. The inaccuracy of such last-minute changes, and the ripple effect to all other aspects of the project (particularly the environmental and regulatory engineering), damage the credibility of the entire project.

Gentry and O'Neil (1992) discuss the number of analyses done through various stages of the project, based on increased amounts of data, and consequently, how each analysis will take increasing amounts of time (and expense) to perform. What is implied in their discussion is preparation of a prefeasibility study, followed by an intermediate feasibility study, which would contain the following information described by Gocht et al. (1988; as cited in Gentry and O'Neil 1992):

1. **Project description:** geographic area, existing access routes, topography, climate, project history, concessionary terms, schedule for development of mine and any processing facilities.
2. **Geology:** regional geology, detailed description of the project area, preliminary reserve calculations, plans for detailed target evaluation.
3. **Mining:** geometry of the ore body, proposed mining plan (and alternatives), required plant and equipment.

4. **Processing:** technical descriptions of the ore and concentrate, processing facilities.
5. **Other operating needs:** availability of energy, water, spare parts, and equipment (diesel oil, explosives, replacement parts, etc.).
6. **Transportation:** description of the additional, necessary transportation facilities (roads, air strips, bridges, harbors, rail lines).
7. **Towns and related facilities:** housing for workers, schools for children of workers, medical facilities, company offices.
8. **Labor requirements:** estimates of work force broken down according to qualifications (skills) and local availability.
9. **Environmental protection:** plans to reduce or minimize environmental damage, description of relevant environmental legislation.
10. **Legal considerations:** review of mining laws, taxation, foreign-investment regulations, and political risk.
11. **Economic analysis:** cost estimates for plant and equipment, infrastructure, materials, labor, other factors; market analysis, including production, consumption, and price formation for the relevant minerals; revenue forecasts based on expected production and mineral prices; cash flow and net present value analysis; sensitivity analysis.

If Gocht et al. (1988) would have done their writing in today's social awareness environment, they would probably have included "social issues" of mine development that must be considered.

Following the positive results of an intermediate feasibility study as previously outlined, Gentry and O'Neil (1992) suggest a "comprehensive feasibility study," which leads to what some call a "bankable" document. They do not give the details of what should be studied, but they do cite Taylor (1977) in stating what essential functions the final report must fulfill:

1. Provide a comprehensive framework of established and detailed facts concerning the mineral project.
2. Present an appropriate scheme of exploitation complete with plans, designs, equipment lists, and so forth, in sufficient detail for accurate cost estimation and associated economic results.
3. Indicate the most likely profitability on investment in the project, assuming the project is equipped and operated as specified in the report.
4. Provide an assessment of pertinent legal factors, financing alternatives, fiscal regimes, environmental regulations, and risk and sensitivity analysis on important technical, economic, political, and financial variables affecting the project.
5. Present all information in a manner intelligible to the owner and suitable for presentation to prospective partners or to sources of finance. The document must be based on the judgment of a financial institution as "bankable."

Others have also documented a particular approach to mineral property feasibility studies, and one has even included five stages, beginning in the exploration phase even before the so-called engineering prefeasibility stage (Stone 1997). This raises the issue of exploration targets and decisions regarding what commodities a company wants to try to develop. Although this author agrees that this must be done, the activities during exploration cover areas that are not discussed in this chapter.

What is recommended in this chapter is a three-phased approach to mineral property feasibility, as previously introduced by Gentry and O'Neil (1992) using works by Gocht et al. (1988) and Taylor (1977). This method is considerably different than what is suggested by these authors and will be explained by listing the details that are contained in each phase of the approach. It is the sheer magnitude of details enumerated and the description of detail that is contained in the iteration of each phase that makes this method unique from any of the preceding descriptions. To this author's knowledge, this amount of detail of the tasks required in a mineral property feasibility study have not been documented elsewhere. The original method was developed by the Mine Evaluation and Development Group of a company that does not wish to be identified. The two principals who developed the system during the 1976–1977 time period were W.J. Bulick and G.D. Mittelstadt (personal communication). Many parts of the system have been modified considerably by the author since that time.

The need for such an approach became evident when the company realized there was no efficient way to manage 8–12 concurrent mineral project feasibility studies without a formal procedure. The studies involved four different mineral commodities, located in five countries, all having different starting dates and mine lives. The projects were being studied by different project teams, and the company believed it was absolutely necessary to develop, in writing, the detailed procedures for each team to follow. It was only by formalizing the feasibility study process that management could ensure the following:

1. Items would not be left out.
2. Activities would be studied in equal depth.

This made certain that the results would be on an equivalent or comparable basis for the quality of study thus completed. Although other companies may not have 8–12 projects going at one time under the conditions previously described, the established procedure will serve any user of the system very well and yield project results that are comparable for financial decisions.

THE THREE-PHASED APPROACH TO MINE FEASIBILITY STUDIES

While looking at the long and detailed lists of items that need to be studied in the different phases described in this section, the reader may believe there are too many activities to accomplish and that the time and expense to accomplish them is too great. Some may choose to combine many of the activities of the preliminary feasibility study with the intermediate feasibility study. This may be possible, and it is discussed in this chapter. A company must be careful that it does not dilute the preliminary feasibility study so that it does not retain the validity from which a confident financial decision can be made. Some may also believe that items can be eliminated or that the study of certain items is not applicable. But a great amount of caution should be exercised before eliminating any study aspect unless the company has so much experience and data on that particular aspect that the item is simply not necessary.

Phase One: The Preliminary Feasibility Study

The objectives of each phase of all mineral property feasibilities should be the same, and the following objectives of the preliminary feasibility study were given in the beginning of Chapter 1:

- To develop the value of the mineral property to the company by determining the optimum method of either exploiting it, selling it, or holding it; and
- To reach that decision point as early as possible, while spending the least amount of money.

These are still the primary objectives. However, more specific to the preliminary feasibility study, the object is to consider those logical mining and processing methods and the other project elements in sufficient detail to be able to determine that they will work together to meet the company's objectives (which are usually financial) and to estimate the capital and operating cost commensurate with the engineering that has been expended.

This study will be primarily based on information supplied through exploration. The exploration department will record the results of their work in a formal report for use in project evaluation. The exploration report should contain the following information with appropriate maps and cross sections:

- Property location and access
- Description of surface features
- Description of regional, local, and mineral-deposit geology
- Review of exploration activities
- Tabulation of potential ore reserves and resource material
- Explanation of resource calculation method, including information on geostatistics applied
- Description of companies' land and water position
- Ownership and royalty conditions
- History of property
- Rock-quality-designation values, as a minimum, and any rock mass classification work that has been done
- Results of any special studies or examinations the exploration department has performed (metallurgical tests, geotechnical work, etc.)
- Report on any special problems or confrontations with local populace of the area
- Any other pertinent data, such as attitude of local populace toward mining, special environmental problems, availability of water and hydrologic conditions in general, infrastructure requirements, and so on

Ideally, several mining and processing alternatives will be examined as a screening process. Obviously, these cannot be in-depth studies, but most mining engineers with many years of industrial experience can determine very quickly what mining methods will be applicable and can then place costs on the alternatives for this application. Likewise, an industry-experienced mineral processor can determine the candidate process flow sheets and can place costs on these options.

At the same time, all the other elements of the project must be considered, and these must be studied in adequate detail to discover any fatal flaws or problems that need engineering mitigation. Certainly, environmental and socioeconomic issues need to be studied and scoped to the extent that any existing or expected problems will be detected. All of these items can then be examined for future cost and prospective work plans.

Costs and expenditure schedules will be based on industry-factored historical experience. Major capital costs can be based on telephone quotes from suppliers or "canned" commercial programs built for this type of application. Usually, no fieldwork or metallurgical testing will be conducted unless a definite metallurgical problem with the resource has been recognized. If the problem is suspected to be a fatal flaw, by all means it should be studied. By the end of the preliminary feasibility study, the completed engineering on the project should be 6%–8%. The probable error (PE) of cost estimates should be about ±30%–45%. Contingencies of 20% (for surface facilities) to 35% (for mine facilities) for capital costs will apply. For a small project, this may amount to 8,400–11,200 worker-hours of engineering-type work. But for a large, multibillion-dollar, multisite project, this may amount to 74,000–100,000 worker-hours.

Major Activities

A condensed version of the° major activities of the preliminary feasibility study are shown in Figure 11.1, with more emphasis on the work that must be done in the geology, environmental, and geotechnical areas. A detailed description of each of these activities, and more, is provided in Appendix 11A at the end of this chapter. Note that the appendixes at the end of this chapter are itemized by a method of cost control called the work breakdown structure, or WBS. A detailed discussion of the WBS is included in Chapter 13.

Results of this study will be adequate for comparative screening of those mining or processing alternatives, and an economic analysis will determine whether to proceed with or reject the project. A primary objective of the study is to plan and estimate costs for a further predevelopment program if warranted. This will require approximately 6%–8% of the entire amount of engineering and study for the entire project to be completed. The PE of cost estimates ("accuracy") should be about ±20%–30%. Contingencies of 20%–30% for capital costs will apply and an economic analysis will be performed. The preliminary feasibility report should be fully documented. Presentations to management will be made and, depending on the results of the economic analysis, approval to proceed to the next step or to shelve the project follows.

Phase Two: The Intermediate Feasibility (or Prefeasibility) Study

Based on the results of the preliminary feasibility study showing a project that has the potential to achieve the desired company goals, the intermediate feasibility study should be initiated. The basic objective previously stated for feasibility studies has not changed for the intermediate feasibility study; however, the specific object for this study is much different than the preliminary feasibility study. Now that it has been shown that the mineral resource being examined has potential economic viability by using at least one mining or processing system, the objective must now focus of examining methods to *optimize* these systems, while at the same time taking an *in-depth look* at all of the project parameters briefly studied in the preliminary feasibility study.

At this time, accurate topography maps specific to the area must be generated if they are not already available. Any shortcomings in the land and water status that may have been discovered in the preliminary feasibility study *must be corrected* at this point.

Mine design will be based on information from the early exploration (delineation) drilling program plus any additional exploration sampling that has taken place between the two phases.

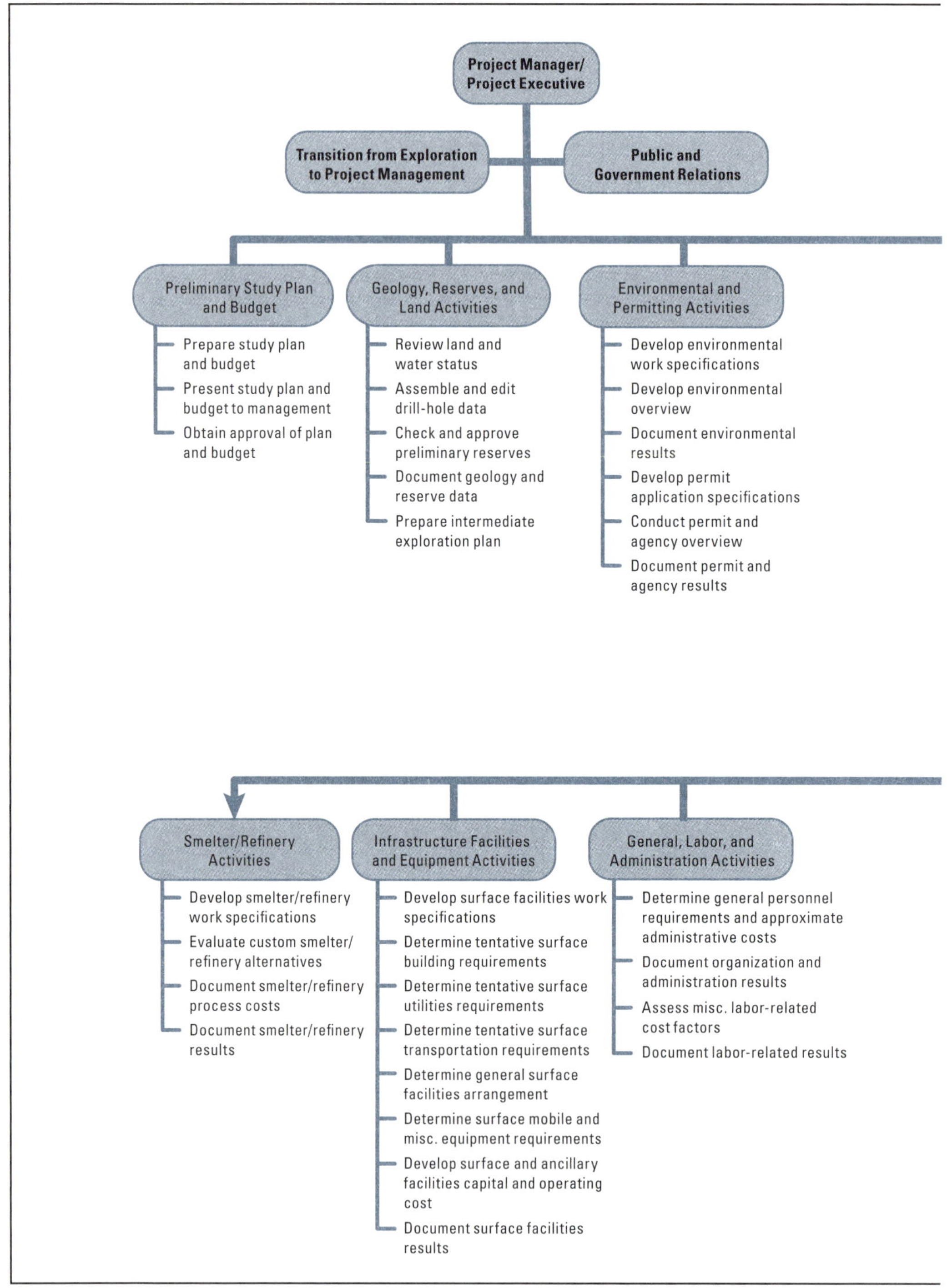

FIGURE 11.1 Preliminary feasibility study activities

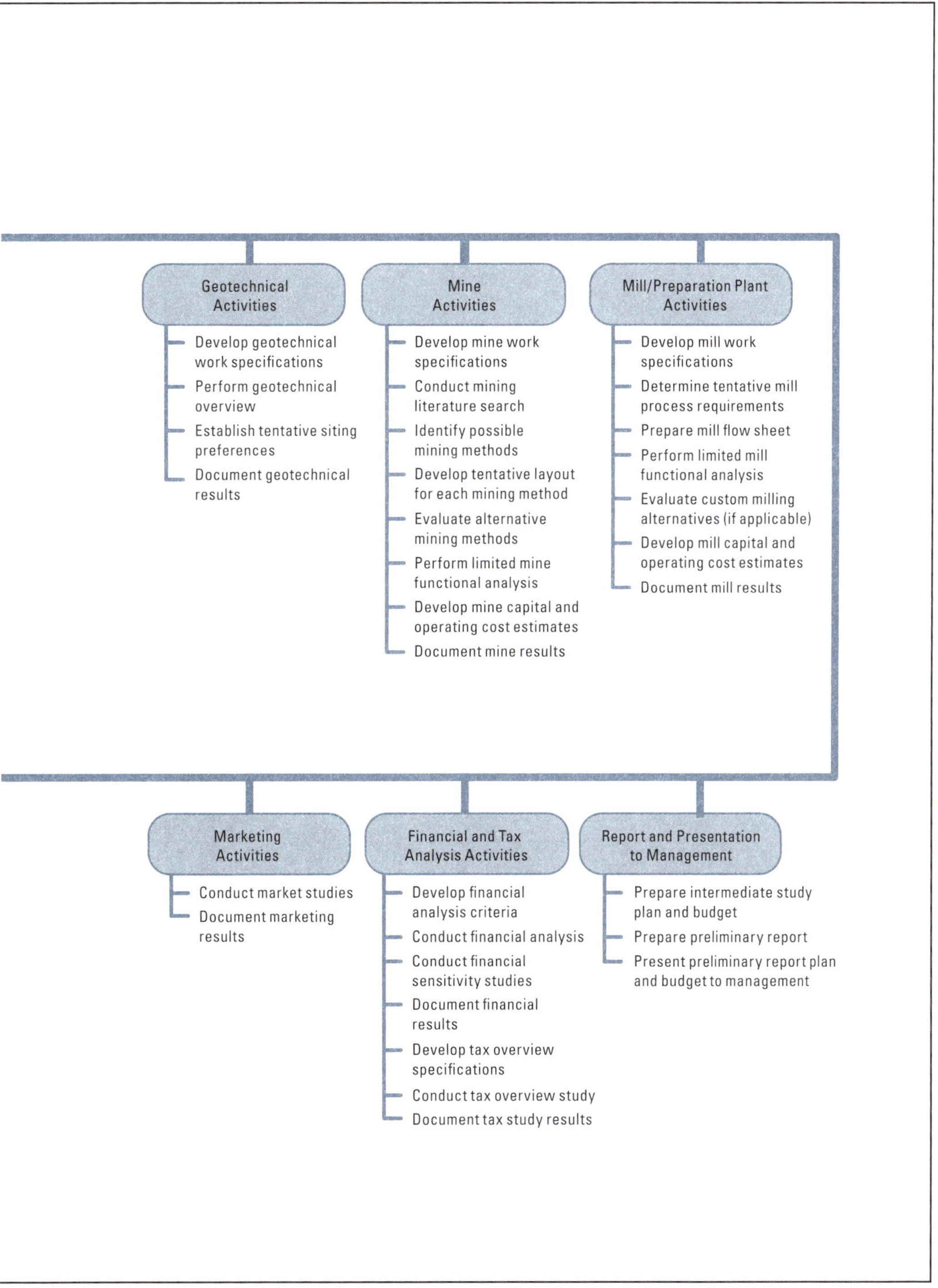
Geotechnical Activities
Develop geotechnical work specifications
Perform geotechnical overview
Establish tentative siting preferences
Document geotechnical results
Mine Activities
Develop mine work specifications
Conduct mining literature search
Identify possible mining methods
Develop tentative layout for each mining method
Evaluate alternative mining methods
Perform limited mine functional analysis
Develop mine capital and operating cost estimates
Document mine results
Mill/Preparation Plant Activities
Develop mill work specifications
Determine tentative mill process requirements
Prepare mill flow sheet
Perform limited mill functional analysis
Evaluate custom milling alternatives (if applicable)
Develop mill capital and operating cost estimates
Document mill results
Marketing Activities
Conduct market studies
Document marketing results
Financial and Tax Analysis Activities
Develop financial analysis criteria
Conduct financial analysis
Conduct financial sensitivity studies
Document financial results
Develop tax overview specifications
Conduct tax overview study
Document tax study results
Report and Presentation to Management
Prepare intermediate study plan and budget
Prepare preliminary report
Present preliminary report plan and budget to management

In some cases, bulk sampling may be required. Consequently, if permits can be obtained, a *test mine* may be justified *after this phase* of the study. Further exploration drilling may be needed and should be initiated. If further drilling or trenching is to be done during this phase, permits and contracts must be prepared for the permits and drilling contractors to be acquired. The sampling program, now under the control of the project team, must complete the following:

- Prepare a sample flowchart.
- Prepare a "chain of custody" procedure (if not already in place).
- Procure and analyze the new samples.

The new geology and mineral information must be fed back into the database and evaluated. After rebuilding and analyzing the new database and documenting the current reserves and resources, new reserve and resource maps can be constructed for mine planning.

Given the shape and character of the ore reserve identified to this point, the *mine planning* will begin. While those methods considered in the preliminary feasibility study may be reexamined, other methods should also be considered, since the ore-body shape, size, character, and grade may have changed. The methods described in Chapters 5 and 6 should be followed. But this time, after a rough screening of multiple mining methods, two or three of the most probable mining methods (or variations of the mining method) that are considered safe and environmentally permissible, and which will probably yield the lowest cost (or greatest recovery), should be carried through the study until an economic comparison of the methods can be made.

Likewise, with the latest mineralogical data and mining methods, several *mineral processing* and waste disposal alternatives should be considered. Those that seem likely to yield the best economics should be carried through the study until a true economic comparison can be made between the methods. The methods employed in Chapter 7 should be used.

Facilities *siting* and *geotechnical investigations* will need to be conducted. Contract preparation to cover the scope of work for these activities must be done if competent staff members are not available within the company. These activities include the following:

- Contractor bidding lists must be prepared.
- Requests for proposals (RFPs) must be issued.
- Bid evaluation criteria must be written.
- The bid evaluation must be performed on all submitted bids and the contractor's bids ranked.

Contracts are then awarded and negotiations on details of the contract are completed. The time that it will take to perform all of the tasks for each contract should not be underestimated. It may take several months to get the contractor on board even after the contract is let. But with contractors committed, work can now be started.

Environmental baseline studies will be initiated, impact assessments will be conducted, and some long lead-time permit applications may be started. Refer to Chapter 9 for specific guidance in the task that must be done. Again, contract preparation to cover the scope of work for these environmental activities must be done if competent staff employees are not available within the company. Bidding lists must be prepared, RFPs issued, and bid evaluation criteria written and the bid evaluation administered. Finally, environmental contracts are then awarded when negotiated details of the contract are completed. With contractors on board, the environmental work can now be started. The baseline studies will take time, but they should be completed by the time that the intermediate feasibility study is completed, which will allow for this information to be submitted along with the intermediate mining and process planning to be presented to the agencies.

Results of this intermediate feasibility study will be adequate for determining economic feasibility and defining additional predevelopment and/or metallurgical testing requirements. In many cases, the benefits and the requirements for *a test mine or bulk sampling* will be fully recognized and will be defined at this point. In most cases, specific permitting will be required, and this will take time to receive such permits. The subject of a test mine is discussed later in this chapter.

The *costs estimates* for the (two or three) alternatives developed during this phase should be based on detailed functional analysis of each operation of the mining and processing methods of operation, on suppliers' written quotes, and bench-scale metallurgical testing (see Chapters 14–16). By the end of the intermediate feasibility study, the completed engineering on the project should be 15%–20%. For a small project, this may amount to 21,000–28,000 worker-hours of engineering-type work. But for a large, multibillion-dollar, multisite project, this may amount to 185,000–250,000 worker-hours. The PE of cost estimates should be about ±15%–20%. Contingencies of 15%–20% will apply. *Economic analysis* will be performed on the favorable sets of alternatives selected from above. The intermediate feasibility study is fully documented. Presentations to management will be made and, depending on the results of the economic analysis, approval to proceed to the next step or shelve the project follows.

Major Activities

A condensed version of the major activities involved in an intermediate feasibility study is shown in Figure 11.2. A detailed description of each of these approximately 150 major activities is found in Appendix 11B at the end of this chapter.

Why a Test Mine May Be Needed

On every mineral project, there may be some aspect about the potential project that may require the need for a test mine. This could be either from a geologic point of view, from mining continuity or a ground support viewpoint, from a metallurgical recovery position, or from some other potential engineering problem.

In all of these cases, the test mine will drastically lower the investment risk for the full project. The downside of a test mine is that it may reveal a fatal flaw that will disqualify the project for development at that point in time. But it will do so at a much lower cost than the full capital cost of the entire project.

From a geologic point of view, the test mine will allow the project team to

- Examine a much larger sample of the mineralized vein or area, thus being able to
  - Better identify the model of mineralization;
  - Better determine where further exploration is needed;
  - Better predict mineralization continuity; and
  - More accurately estimate reserve and resource size and grade.

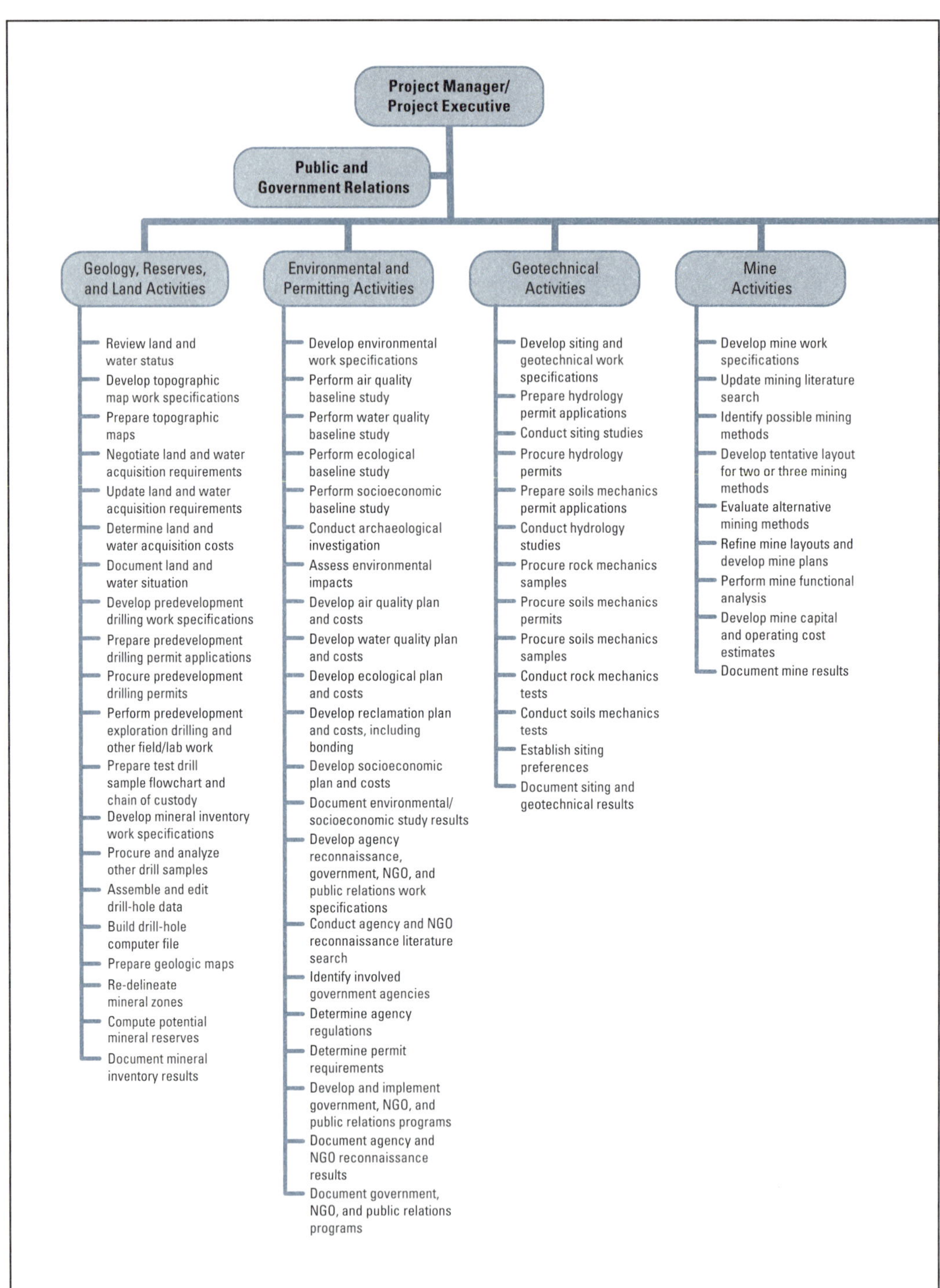

FIGURE 11.2 Intermediate feasibility study activities (Figure continues)

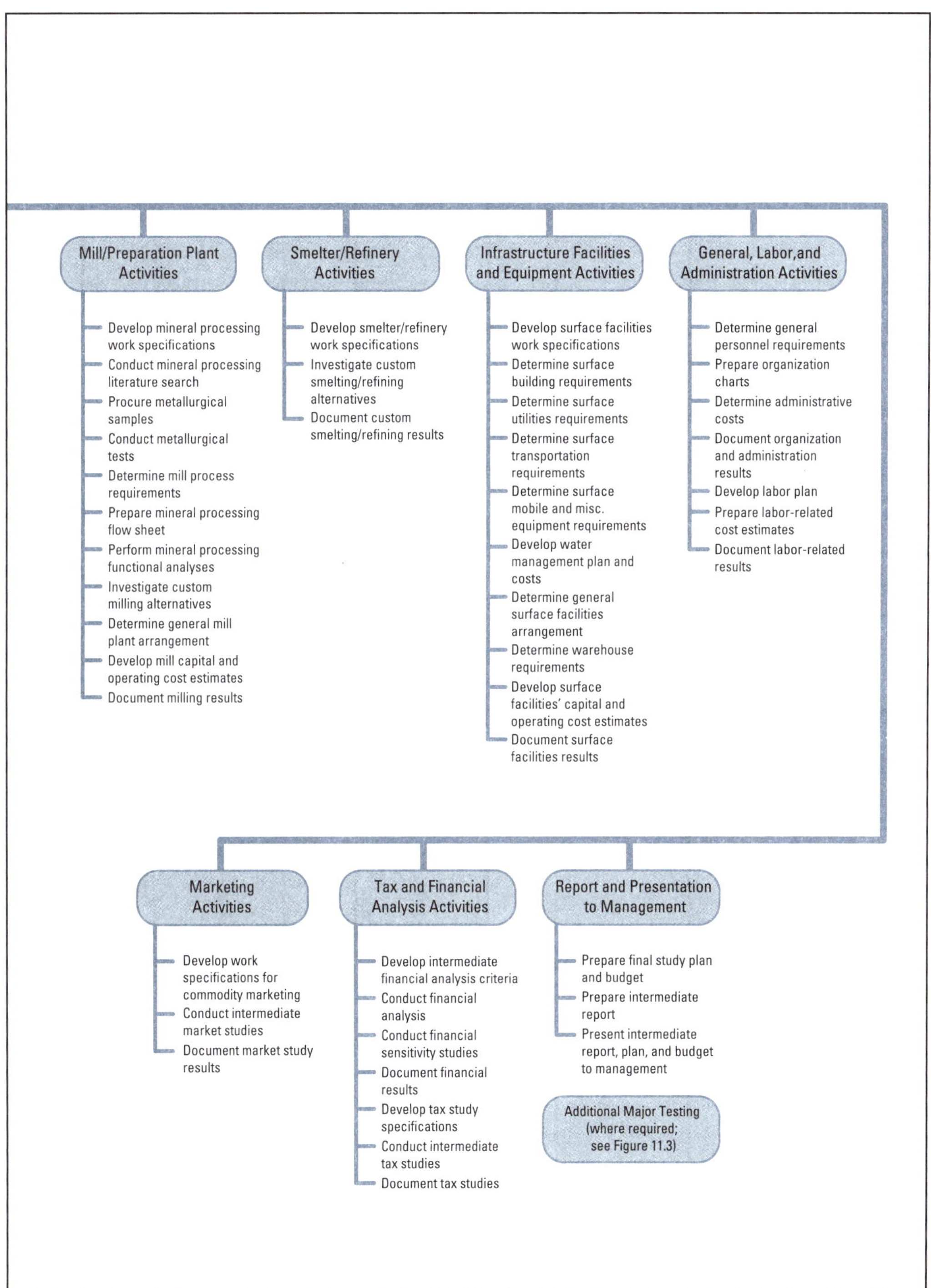

FIGURE 11.2 (Continued)

(Figure continues)

From the mining aspects, the test mine will allow the project team to

- Verify the expected ore continuity, thus eliminating surprises in mining methods;
- Accurately assess the rock strength, which will allow a prudent planning and sizing of the commercial mine opening;
- Verify mining efficiency, productivity, and dilution as it relates to drilling, blasting, and materials handling, thus making the economic analysis more accurate;
- Determine the nature of mine water inflows from more reliable water studies, which will allow for an adequate water-handling procedure to be installed before problems are encountered;
- Better quantify the mine ventilation friction factors and requirements; and
- Confirm the character of the waste product and how it will be handled in the commercial operation.

From the metallurgical aspects, the test mine will allow the project team to

- Verify and optimize the metallurgical flow sheet with a pilot-plant process that is continuous and not done with a series of bench tests done in batches;
- Determine the size and type of equipment that will be optimal for the metallurgical recovery;
- Determine what type and amount of reagents will lead to the best recoveries and concentrate grades;
- Determine the required amount of water and how to achieve a water balance;
- More accurately predict the concentrate grade, moisture content, and impurities; and
- Obtain a much more accurate assessment of the work index from a bulk sample than small samples.

From an environmental point of view, the test mine will allow the project team to

- Demonstrate the company's ability to control the operation in such a manner that it will not do harm to the environment;
- Allow a complete study of the characterization of the different waste products and determine any future problems; and
- Allow the study of the difficulties with settling solids in the discharged water or by removing unwanted elements, and determine what might be necessary to mitigate any future problems (if water discharge is involved).

From other points of view of project management, the test mine will allow the project team to

- Improve the ability to make more accurate cost estimates, given that the project team will have a better knowledge of the abrasiveness of the rock and of the ground/stope or slope control of the pit walls. This could actually lower the cost estimate, because less contingency may be used;
- Improve the labor estimates, as the team will have a better understanding of the productivity of each unit operation after test mining; and

- Predict a more accurate schedule, because the team will have a better understanding of unit productivity.

From the company's point of view, a test mine will

- Lower the overall risk of the project in virtually every aspect;
- Provide early access to develop the commercial underground mine (if that is what is needed), thus shortening the overall schedule from the end of the final feasibility study to the end of construction;
- Enable the openings to be utilized as part of the commercial mine operation if the operation goes forward; and
- Provide a training facility prior to the commercial mine start-up.

Given that test mines are expensive, they must be justified to management based on the latest economics from the last phase of study that was prepared. Usually, the level of feasibility will need to at least be the intermediate feasibility phase to provide sufficient confidence in the economic analysis and to justify such an expense. The activities for a test mine as well as any additional exploration drilling are found in Figure 11.3, and details are provided in Appendix 11C.

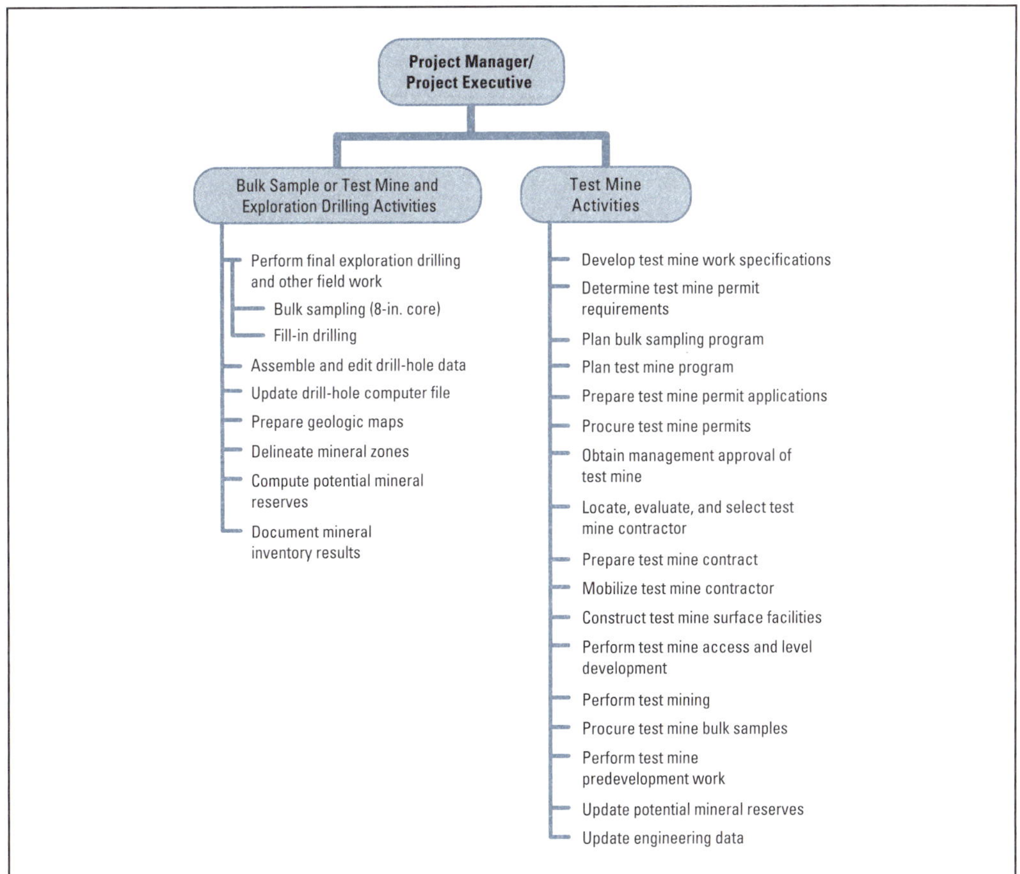

FIGURE 11.3 Activities required for additional testing

Phase Three: The Final Feasibility Study

The final feasibility study should be initiated when results of the intermediate feasibility study show that the project still has the potential to achieve the desired company goals. The objective, as it was in the first two phases, is still to determine the potential value of the property to its owners either by determining the optimum method of developing it, selling it, or putting it on the shelf. However, more specific to the final feasibility study, the objective now becomes one of minor refinements to all of the details of the intermediate study that yielded the results that met the company objectives, thus optimizing the return on future investment. The final feasibility study will be in preparation during the permitting time. Assuming that the project will still show favorable results at the end of this phase, it is the design parameters that are set in the final feasibility study that will feed into the design basis report (DBR). The DBR is what guides the project into the design and construction phase, and finally into operations.

Test mining with bulk sampling and pilot-plant testing may have been completed, but if not completed earlier, it may now become a part of the final feasibility study. Mine and process facilities will be further studied, and the best alternative that was developed in the intermediate study will be optimized. Using the latest exploration and metallurgical test data, probably from the test mine bulk sample, the reserves will be updated and the metallurgical flow sheet will be optimized. Final environmental impacts will be determined. (Follow the guidelines supplied in Chapters 9 and 10.) Applications for construction and operating permits will usually be made early in this phase of study (subject to later modification). Mine and process operating cost estimates will again be made by performing functional analyses. Capital cost will be refined by again soliciting written quotes from vendors. By the end of the final feasibility study, the completed engineering should be 20%–30%. For a small project, this may amount to 28,000–42,000 worker-hours of engineering-type work. But for a large, multibillion dollar, multisite project, this may amount to 185,000–250,000 worker-hours The PE of cost estimates should be about ±10%–15%. Contingencies of 10% will apply to most *engineered structures*. Other less well-defined aspects of the project (e.g., mine development) should have contingencies of at least 15%. An economic analysis will be performed. The final feasibility report is fully documented. Presentations to management will be made and, depending on the results of the economic analysis, approval to proceed to the design and construction phase of project development will be made with budget approval, or a decision will be made to shelve the project.

Major Activities

A condensed version of the major activities of the final feasibility study is shown in Figure 11.4. A detailed description of the 100+ activities is provided in Appendix 11D at the end of this chapter.

The approved project final feasibility study will be presented in sufficient detail for a design basis report or DBR (sometimes called a *design basis memorandum* or *design basis document*) to be produced. The DBR is the document that will guide the project through the next step: designing the project based on all of the preceding studies. The purposes and content of the DBR is discussed in considerable detail in Chapter 12. With a well-documented DBR, some additional configuration work may be required, but modifications to the basic plan will be minimal during the design phase, thus minimizing the engineering design cost.

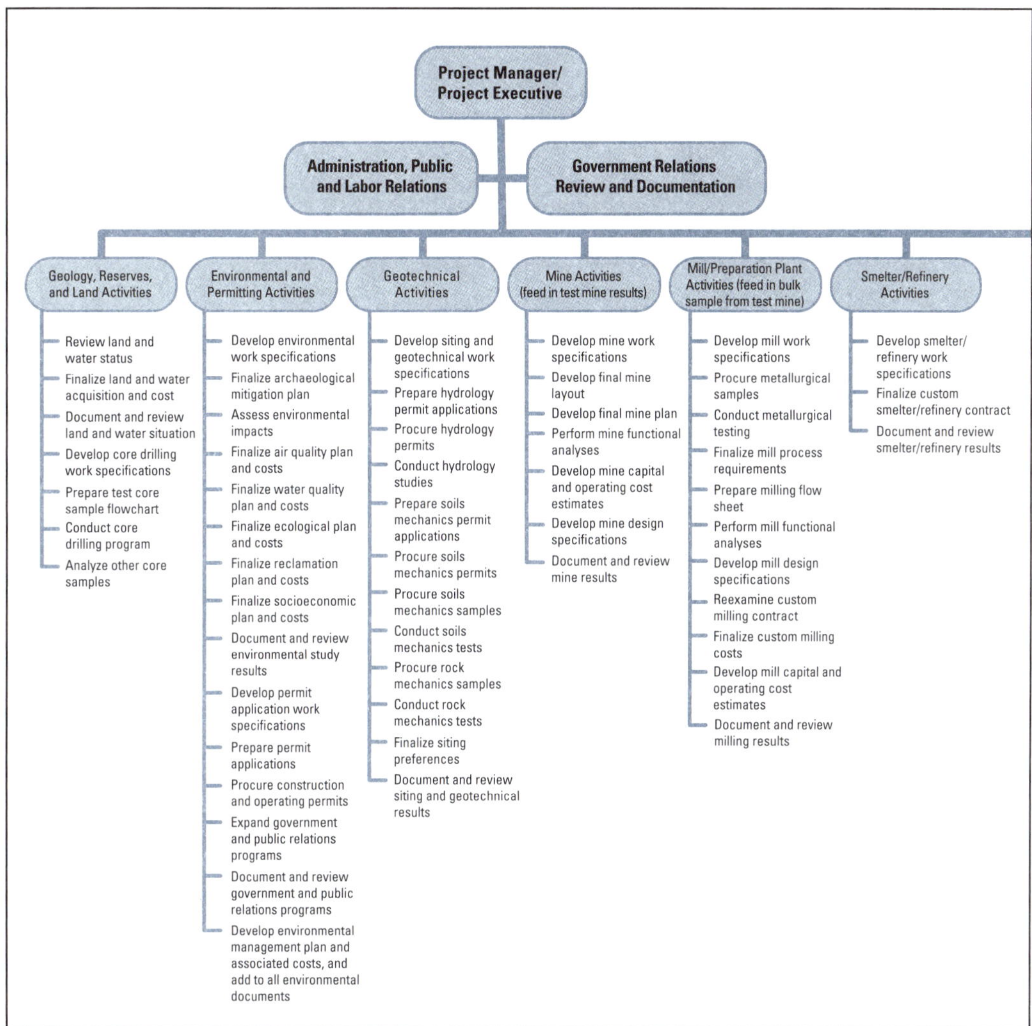

FIGURE 11.4 Final feasibility study activities (Figure continues)

Project approval and appropriation of funds for design and construction will normally occur after the final feasibility phase. Using the DBR results will enable bids to be solicited and final design and construction to begin.

Combining the Classical Approach with the Three-Phased Approach

Because there are good reasons, as outlined at the beginning of this chapter, to sometimes use the conceptual or scoping study approach, if the decision is to move the project to the next level of study, then one should still convert to the three-phased approach outlined previously. In this case, one should scrutinize the conceptual or scoping study that was done and compare it to the details of phase one: the preliminary feasibility study. Whereas in the scoping study only 1% to 2% of the engineering may have completed, the preliminary feasibility study (phase one) as outlined previously would have completed as much as 6%–8% of the engineering. This means that if the plan is to now proceed to the intermediate feasibility (or

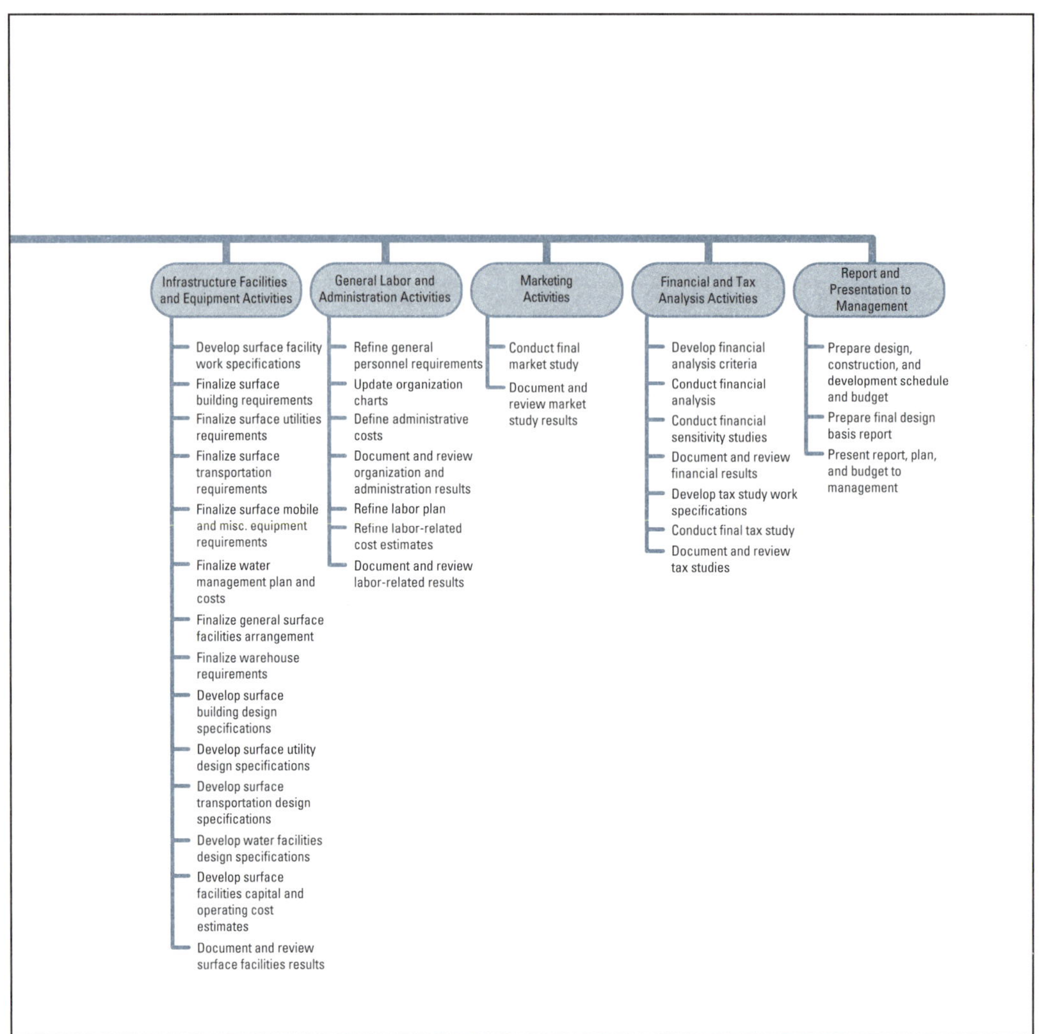

FIGURE 11.4 (Continued)

prefeasibility) study, one must plan on doing the extra work in phase two: the intermediate feasibility (or prefeasibility) study. By the time the end of the intermediate feasibility study has been reached, 15%–20% of the engineering work should have been completed before going into the final feasibility study. When the final feasibility study is completed, 20%–30% of the total engineering for the entire project should be completed.

Feasibility Timing and Schedule

The time it takes between the discovery of a resource that may be a potential ore body and the time that the ore body is brought into production can vary significantly. Obviously, if you have an extremely high-grade ore body, it may take less time to identify enough ore to justify the start of mining. Likewise, if you can make money no matter how you mine it, then the company may not want to spend a lot of time optimizing the mining and milling methods (though this could prove, and has proved, to be a mistake in past situations). On the other hand, very large, marginal mineral resources may take many years to define and to optimize every aspect of the study to make the resource into a viable reserve.

Technological changes over time may also be a factor that allows the property to finally be developed after many years of study. Another factor is the changes that are continuously occurring for environmental permitting. Permitting can vary considerably. Overall, it usually takes from two to six years to complete the mineral property feasibility evaluation study. This subject is addressed in Chapter 13.

REFERENCES

Gentry, D.W., and O'Neil, T.J. 1992. Mine feasibility studies. In *Mining Engineering Handbook*, Vol. 1. Edited by H.L. Hartman. Littleton, CO: SME. p. 393.

Gocht, W.R., Zantop, H., and Eggert, R.G. 1988. *International Mineral Economics*. New York: Springer Verlag.

Gosson, G. 2011/2012. The inclusion of inferred mineral resources in economic analyses. December 2011/January 2012. Canadian Institute of Mining, Metallurgy and Petroleum. http://web.cim.org/standards/MenuPage.cfm?sections=185,331&menu=352 (accessed March 21, 2017).

Hustrulid, W., and Kuchta, M. 2006. *Open Pit Mine Planning and Design*, Vol. I, Fundamentals, 2nd ed. London: Taylor and Francis.

Lee, T.D. 1984. Planning and mine feasibility study—An owner's perspective. In *Proceedings of the 1984 NWMA Short Course "Mine Feasibility—Concept to Completion."* Compiled by G.E. McKelvey. Spokane, WA: Northwest Mining Association.

NI 43-101. *Standards of Disclosure for Mineral Projects*. Canadian National Instrument 43-101. Ontario Securities Commission. www.osc.gov.on.ca/documents/en/Securities-Category4/ni_20160509_43-101_mineral-projects.pdf.

SEC (U.S. Securities and Exchange Commission). 1992. *Industry Guide 7: Description of Property by Issuers Engaged or to Be Engaged in Significant Mining Operations*. sec.gov/about/forms/industryguides.pdf (accessed Jan. 17, 2017).

SME (Society for Mining, Metallurgy & Exploration). 2017. *The SME Guide for Reporting Exploration Information, Mineral Resources, and Mineral Reserves*. Englewood, CO: SME. www.smenet.org/docs/publications/2014 SME_Guide_Reporting_%20June_10_2014.pdf

Stone, I. 1997. Orebody definition and optimisation (Coal). In *MINDEV 97, The International Conference on Mine Project Development*. Victoria, Australia: Australasian Institute of Mining and Metallurgy. pp. 39–45.

Taylor, H.K. 1977. Mine valuation and feasibility studies. In *Mineral Industry Costs*. Compiled by J.R. Hoskins and W.R. Green. Spokane, WA: Northwest Mining Association. pp. 1–17.

Vézina, S. 2013. Canada: Clarifying the preliminary economic assessment under regulation 43-101. www.mondaq.com/canada/x/226622/Mining/Clarifying+The+Preliminary+Economic+Assessment+Under+Regulation+43101 (accessed March 21, 2017).

White, M.E. 1997. Feasibility studies—Scope and accuracy. In *Mindev 97, The International Conference on Mine Project Development*. Victoria, Australia: Australasian Institute of Mining and Metallurgy. pp. 27–34.

APPENDIX 11A

Phase I: Preliminary Feasibility Study Complete Activity Definitions

Activity No. (from WBS) **Activity Title and Description**

10100 Public Relations

- Do in-house determination of public relations' perceived responsibility and identify company official to serve as spokesperson for the project.
- Inform state government officials if the mineral resource is a new major discovery, prior to official public announcement of the discovery. Then prepare and disseminate initial press release announcing discovery.
- Identify statewide media editors and set up proper liaison and method of briefings.
- Identify concerns of local, regional, and state population, and prepare proactive response demonstrating how each concern will be mitigated. As more data are received, update these proactive responses to the public.

10101 Review Exploration Report

Review report prepared at end of exploration phase. Report should contain information on

- Mineral deposit,
- Property location and access,
- Area surface features,
- Exploration activities completed and planned,
- Geology (regional, local, and deposit),
- Potential ore reserves,
- Company's land and water position,
- Property ownership and royalties,
- Property history,
- Special studies performed or environmental problems noted, and
- General data.

Review should include trip to project site to familiarize team members with site and area.

10102 Prepare Preliminary Study Plan and Budget

Prepare preliminary feasibility study schedule, with labor and cost budgets necessary to complete preliminary study. Prepare schedule to show activities and time for remainder of project phases:

- Intermediate and final studies
- Design, construct, and develop (through start of production)

10103 Present Preliminary Study Plan and Budget to Management

Present schedule, plan, and budget for review.

10104 Obtain Approval of Preliminary Plan and Budget

Obtain approval from appropriate levels of management to proceed with preliminary feasibility study outlined in schedule, plan, and budget.

10201 Review Land and Water Status

Review land ownership and water rights, control, royalty, and lease situation developed during exploration phase. Project team members should review land status with personnel in the company's land office or its land agent.

10301 Assemble and Edit Drill-Hole Data

Assemble drill-hole data pertaining to deposit. Audit data for correctness and completeness.

10302 Check and Approve Preliminary Reserves

Check and modify or approve preliminary reserves calculated by exploration group.

10303 Document and Review Geology and Reserve Data

Write report documenting drill-hole and reserve data. Report should contain appropriate tables, maps, sections, and written information concerning mineral inventory and reserve data, regional and local geology, and other pertinent information. Review assembled information with appropriate levels of management. Write report in style and format suitable as a chapter in the preliminary feasibility study report.

10304 Prepare Intermediate Exploration Plan and Budget

Develop scope of work, schedule, and budget for exploration drilling program for next phases of work.

10401 Develop Environmental Work Specifications

Define scope of work for an environmental overview of project and surrounding area.

10402 Develop Environmental Overview

Develop general environmental plan for protecting quality of water, land, ecology, cultural resources, and socioeconomics of project area during construction and operation. Determine costs, if applicable, to prevent or mitigate environmental damages and return area to near original condition at project end. Costs should have accuracy of +30%.

10403 Document and Review Environmental Results

Write report documenting environmental overview. Review results of study with appropriate levels of management and other personnel. Write report in style and format suitable as chapter in preliminary feasibility report.

10501 Develop Geotechnical Work Specifications

Develop scope of work necessary for siting, soils mechanics, rock mechanics, and hydrology studies.

10502 Perform Geotechnical Overview

Examine drill cores or send cores for testing, if necessary, to determine unusual characteristics that may impact mining costs. Evaluate potential problems and associated costs. Perform field reconnaissance, with appropriate lab and field tests if necessary, to determine soils and surface hydrology conditions in and around potential mine, mill, tailings, and surface facility sites. Evaluate potential problems and associated costs.

10503 Establish Tentative Siting Preferences

Select tentative mine, mill, tailings disposal, and surface facilities sites based on preliminary evaluation of costs, soils mechanics, surface hydrology, and general environmental conditions. Costs should have accuracy of +30%.

10504 Document and Review Geotechnical Results

Write report documenting geotechnical and siting overview. Review results with appropriate levels of management and other personnel. Write report in style and format suitable as a chapter in the preliminary feasibility study report.

10601 Develop Permit Application Specifications

Develop scope of work necessary to determine governmental agencies involved and permits required for every stage of project through design/construct and into operation.

10602 Conduct Preliminary Permit and Agency Overview

Conduct literature search and telephone conversations to determine permits required to develop, construct, and operate project. Determine local, state, and federal agencies involved. Evaluate the time and cost of permits and bonds needed. These costs should have an accuracy of ±30%.

10603 Document and Review Permit and Agency Results

Write report documenting results of permit and agency overview study. Review with appropriate levels of management and other personnel. Write report in style and format suitable as a chapter in the preliminary feasibility study report.

10701 Develop Mine Work Specifications

Develop scope of work necessary for conducting mining evaluation studies.

10702 Conduct Mining Literature Search

Review available literature involving mining methods and schemes for deposits of similar nature. This activity could involve visiting similar operations to gather data pertaining to mining methods, equipment, personnel, and costs.

10703 Identify Possible Mining Methods

Through literature search, personal knowledge, mine visitations, and discussions with other people, identify technically feasible mining methods applicable to this type of deposit.

10704 Develop Tentative Layout for Each Mining Method

Lay out preliminary mine plan for each technically feasible mining method considered.

10705 Evaluate Alternative Mining Methods

Perform rough capital and operating cost calculations for each technically feasible method selected. Evaluate several production rates. Perform quick discounted cash-flow analysis and rank methods in order of economic preference. Eliminate alternatives with little or no chance of economic success. Determine production rates that will satisfy market conditions and give best economic rate of return.

10706 Perform Limited Mine Functional Analysis

Evaluate operational cycles and requirements for labor, equipment, and supply for each mining function and for each alternative selected in Activity 10705 based on mine engineer's experience. The functions include drilling, blasting, loading, hauling, scaling, bolting, ground control, hoisting, primary crushing (if underground), maintenance, supply/debris handling, pumping, and other support services. Prepare cost and operational cycles for each function.

10707 Develop Mine Capital and Operating Cost Estimates

Estimate capital requirements necessary to bring mine on stream. Estimate operating costs required to produce ore. Estimate costs for two to four mining methods and production rates selected for study. Costs should have accuracy of ±35% to 45% (depending on method of functional analysis and geologic definition). List costs in format for financial analysis.

10708 Document and Review Mine Results

Write report documenting mine study work performed. Review results with appropriate levels of management and other personnel. Write report in style and format suitable as a chapter in the preliminary feasibility study report.

10801 Develop Mill Work Specifications

Develop scope of work necessary for conducting milling evaluation studies.

10802 Determine Tentative Mill Process Requirements

Use literature search, company personnel experience and discussions with others to determine feasible process methods. Determine processing requirements for each alternative. Activity may require input from bench tests performed during exploration phase or during preliminary study phase.

10803 Prepare Mill Flow Sheet

Prepare flow sheet for each process alternative. Flow sheet should contain sufficient detail to allow selection and sizing of equipment. Perform capital and operating cost calculations for each technically feasible method. Perform quick discounted cash-flow analysis and rank methods in order of economic importance. Eliminate alternatives with little or no chance of economic success.

10804 Perform Limited Mill Functional Analysis

Evaluate operational cycles and requirements for labor, equipment, and supply for each milling function and for each alternative selected in Activity 10803 based on the metallurgical engineer's experience. The functions include stockpiling/reclaiming, crushing, grinding, screening, concentrating, classifying, clarifying, tailings/waste disposal, concentrate handling at the mill site, maintenance, supply/debris handling, and other support services. Prepare cost and operational schedules for each.

10805 Evaluate Custom Milling Alternatives

Investigate opportunities for selling run-of-mine (ROM) material. Determine sales price and charges associated with selling ROM (if a reasonable alternative). Investigate opportunities for tolling ROM material. Determine custom mill capacity, timing, and costs. Costs should include capital and operating expenses for everything associated with tolling. These include transportation of product to custom mill, losses/deducts for processing, tolling charge, sampling methods, transportation of concentrate, and personnel requirements.

10806 Develop Mill Capital and Operating Cost Estimates

Determine capital and operating cost estimates for all milling operation alternatives and different production rates. Costs should have accuracy of ±30%. Put costs in format suitable for financial analysis.

10807 Document and Review Mill Results

Write report documenting mill study work performed. Review results with appropriate levels of management and other personnel. Write report in style and format suitable as a chapter in the preliminary feasibility study report.

10901 Develop Smelter/Refinery Work Specifications

Determine general requirements for smelting and/or refining mill concentrates. Define and document the scope of work for the custom refining study.

10902 Evaluate Custom Smelter/Refinery Alternatives

Do preliminary investigation of opportunities for custom smelting and/or refining project concentrate. Determine custom refining capacity, timing, and costs.

Costs should include capital and operating estimates for everything associated with custom refining. These include transportation of concentrate to refinery, transportation losses (where applicable), smelting/refining charge (consider deducts and/or credits), transportation of refined product, and personnel requirements. Costs should have accuracy of ±25%.

10903 Document and Review Smelter/Refinery Results

Write report documenting refinery work performed, including cost. Review results of studies with appropriate levels of management and other personnel. Write report in style and format suitable as a chapter in the preliminary feasibility study report.

11001 Develop Surface Facilities Work Specifications

Develop scope of work and schedule necessary for study of project surface facilities not connected with mine and mill studies.

11002 Determine Tentative Surface Building Requirements

Make preliminary estimate of surface buildings required for project operation. Include buildings that serve support function for mine and mill but not buildings directly related to mining and milling activities. Types of buildings include ambulance garage, administration/office, dry/changehouse, guardhouse, security fence, surface shops, and port facilities (if applicable). Include concentrate storage and loading facilities capital and operating costs at shipping docks. These can run into millions of dollars if you have to supply them.

Determine operating and maintenance personnel and equipment requirements (includes shop, office, and dry equipment).

11003 Determine Tentative Surface Utilities Requirements

Make preliminary estimate of utilities required for project construction, development, and operation. Study does not include power distribution within facilities included in mining and milling studies. Utilities include electric power (including internal switching and transformers), fuel for buildings and fuel storage for operating equipment, communications (radio, voice/data telephone system, and GPS), potable water, water for dust control, water and system for fire protection, sewage system, and garbage/trash/solid waste removal and disposal area and system.

Determine operating and maintenance personnel and equipment requirements.

11004 Determine Tentative Surface Transportation Requirements

Make preliminary estimate of transportation needs for moving equipment, supplies, material, and mine/mill product into and out of project area during project construction, development, and operation. Study should include alternative transportation method such as truck, rail, ship/barge (if feasible), and air haulage (if feasible), combinations of above, and personnel transportation.

Study should determine access road, personnel, and equipment requirements.

11005 Determine General Surface Facilities Arrangement

Make preliminary estimate of requirements for

- Internal road for surface facility, plant, and disposal area(s) (does not include haulage roads for open pit mine but does include equipment for maintaining auxiliary roads);
- Parking areas;
- Construction laydown area(s); and
- Storage area(s).

Prepare preliminary plot plan showing arrangement of all surface facilities, including the listed items, water facilities, mine, mill, and tailings facilities. Determine maintenance material and supply requirements for these areas.

11006 Determine Surface Mobile and Miscellaneous Equipment Requirements

Make preliminary estimate of equipment requirements not covered under other activities. This includes equipment for ambulance(s); road and yard area maintenance; supervisor pickups/car(s); maintenance personnel pickups and trucks; loader(s)/ backhoe(s); forklift(s); crane(s)/cherry picker(s); portable welder(s), compressor(s), generator(s), and light set(s); small rear-dump truck(s); and crawler tractor(s) with dozer. Determine operating and maintenance personnel and equipment requirements.

11007 Develop Surface and Ancillary Facilities Capital and Operating Cost

Determine capital requirements necessary to build surface and ancillary facilities. Determine operating costs associated with surface facilities, including personnel, supplies, office, and safety equipment. List the costs in spreadsheet format suitable for financial analysis. Costs should have accuracy of ±30%.

11008 Document and Review Surface Facilities Results

Write report documenting results of surface facilities studies. Review results with appropriate levels of management and other personnel. Write report in format and style suitable as a chapter in the preliminary feasibility study report.

11101 Determine General Personnel Requirements

Determine approximate administrative and management personnel requirements, and operating, maintenance, support, and supervisory personnel requirements developed in preceding activities. Split requirements into salaried exempt, salaried nonexempt, and hourly classifications. Develop labor buildup schedules for each classification.

11102 Determine Approximate Administrative Costs

Determine salaries and wages of personnel identified in Activity 11101. Determine payroll burden associated with salaries and wages. Determine cost, type, and quantity of office equipment and supplies required for all offices, including administration, mine, mill, maintenance, and others. Prepare costs in form suitable for financial analysis. Costs should have accuracy of ±20%. Include costs for relocation and salaried personnel.

11103 Document and Review Organization and Administration Results

Write report documenting administrative costs and personnel requirements. Review results of studies with appropriate levels of management and other personnel. Write report in style and format suitable as a chapter in the preliminary feasibility study report.

11201 Assess Miscellaneous Labor-Related Cost Factors

Assess impact and cost of factors affecting labor recruiting, hiring, and retention. Factors include incentive system, labor setting, recruiting, training, retention, performance, and cost. Factors may also include employee housing and company-supplied transportation. List costs in spread format suitable for financial analysis. Costs should have accuracy of ±20%.

11202 Document and Review Labor-Related Results

Write report documenting labor-related studies. Review results of studies with appropriate levels of management and other personnel. Write report in style and format suitable as a chapter in the preliminary feasibility study report.

11301 Conduct Preliminary Market Studies

Perform market studies to determine selling price of salable products and probable product sales potential. Estimate price ranges for life of project (in terms of constant dollar, not inflation). If changes in product sales potential are identified for the future, they should be included as sensitivities. Prepare expected sales of products in format suitable for financial analysis.

11302 Document and Review Marketing Results

Write report documenting marketing studies. Review results of studies with appropriate levels of management and other personnel. Write report in style and format suitable as a chapter in the preliminary feasibility study report.

11401 Develop Preliminary Financial Analysis Criteria

Develop criteria for performing preliminary financial analysis. Criteria should include overall project schedule (includes intermediate and final evaluation, design, construction, development, and startup); mine and product estimated production; capital and operating costs estimates; royalties; escalation factors (only if this is not a constant dollar analysis); corporate overhead allocation; working capital; property acquisition costs; mill recovery; depreciation methods; depletion allowance; tax rates; weighted contingency for unforeseen factors if not included on every capital cost item; capitalization factor; salvage values; working capital; sensitivity analysis; and project alternative comparisons.

11402 Conduct Financial Analysis

Conduct financial analysis for total project using Apex or other suitable computer program. Print results of economic analysis.

11403 Conduct Financial Sensitivity Studies

Evaluate risk sensitivity of project to key factors such as operating costs, capital costs, reserves, grade, production sales, mill recovery, royalties, taxes, and other items with high degree of uncertainty. Conduct sensitivity analysis using computer program that will perform Monte Carlo simulation, which will assess combined risk sensitivities. Print results of the sensitivity analysis.

11404 Document and Review Financial Results

Write report documenting financial analysis and sensitivities. Review results of work with appropriate levels of management and other personnel. Write report in style and format suitable as a chapter in the preliminary feasibility study report.

11501 Develop Tax Overview Specifications

Develop scope of work and schedule necessary for conducting study of taxes applicable to project.

11502 Conduct Tax Overview Study

Make preliminary study of taxes applicable to project and their cost and impact on construction, development, and operation of project. Prepare tax rates in form suitable for financial analysis.

11503 Document and Review Tax Study Results

Write report documenting tax studies. Review results of studies with appropriate levels of management and other personnel. Write report in style and format suitable as a chapter in the preliminary feasibility study report.

11601 Prepare Intermediate Study Plan and Budget

Update intermediate feasibility study plan and schedule using formalized scheduling techniques. Update budget for intermediate study, including any additional exploration or sampling needed. Schedule and budget should conform to the activities outlined in Activity 10304 and used in financial analysis (11401).

11602 Prepare Preliminary Report

Prepare formal report detailing preliminary study work performed. Prepare report in style and format that is suitable for presentation to management, suitable for use by other project teams, and containing history and results of work performed.

11603 Present Preliminary Report Plan and Budget to Management

Present all data generated during preliminary study, plus plan and budget for intermediate study, to management for review. Present information in meeting(s) with hard copies of reports, schedules, and data. Distribute report at least 1 week prior to meeting to allow personnel time to read and review.

APPENDIX 11B

Phase II: Intermediate Feasibility Study Complete Activity Definitions

Activity No. (from WBS) **Activity Title and Description**

20100 Public Relations

- As soon as company management approves the intermediate feasibility study plan and budget, notify state government officials that the project will proceed to the next level of examination.
- Brief the media on the decision to proceed to the next level, making sure the media understand that the next level of decision making is many months away and that the final decision of whether or not to build a mine is probably years away.
- Set up in-house means to directly respond to questions from the public by disseminating information as it becomes available and presenting speeches at professional, civic, and labor organization meetings. This may also be done by mineral industry audiovisual presentations tied in with the local settings and issues.
- Conduct meetings to help determine the needs of the area, which will promote sustainable development and lead to a social license to operate.

20101 Get Management Approval of Intermediate Study Plan and Budget

Get approval from appropriate levels of management to proceed with intermediate feasibility study outlined in plan, schedule, and budget submitted to management for review at end of preliminary study.

20201 Review Land and Water Status

Review land and water ownership, control, royalty, and lease situation developed during exploration phase and updated during preliminary study (if any work done). Project team should review land status and water with personnel in company land office and other appropriate individuals with respect to site locations identified during preliminary study. All related water rights should be determined.

20202 Develop Topographic Map Work Specifications

Develop scope of work and schedule necessary for topographic mapping. Prepare request for proposal (RFP) to send to contractors capable of performing the work. RFP should include draft of proposed contract.

20205 Prepare Topographic Maps

Notify contractor to proceed with work required by contract. Obtain prepared maps and other data from contractor.

20206 Negotiate Land and Water Acquisition Requirements

The company land office should start negotiating options on land and water requirements identified in the preliminary study. Requirements should include alternatives, because generally the final mine, plant, tailings sites, and so forth are not yet identified at this stage of project evaluation. This probably means optioning some land not needed for final selected sites.

20207 Update Land and Water Acquisition Requirements

Land and water requirements should be updated as intermediate study progresses and project team learns more about requirements. This may allow land personnel to drop negotiations on certain pieces not needed for project. This updating will not normally select final land sites required for project development, construction, and operation.

20208 Determine Land and Water Acquisition Costs

Land personnel should determine approximate costs of buying and/or leasing land and acquiring water necessary to construct, develop, and operate project. Costs will become part of intermediate study financial analysis, so accuracy should be within ±15% to 20%. Put costs in format suitable for financial analysis.

20209 Document and Review Land and Water Situation

Write progress report documenting results and costs of land and water negotiations to date. Review results of negotiations with appropriate levels of management and other personnel. Prepare risk analysis and mitigations for land and water satisfactory procurement. Write report in style and format suitable as a chapter in the intermediate study feasibility report.

20301 Develop Predevelopment Drilling Work Specifications

Evaluate exploration report and data. Develop scope of work and schedule necessary to perform predevelopment drilling activities. This activity assumes no exploration drilling was performed during or after the preliminary study and additional drilling is required to bring the reserve accuracy to range to within ±15% to 20%.

20302 Prepare Predevelopment Drilling Permit Applications

Gather data required to prepare applications for permits to do predevelopment drilling field work. Prepare applications.

20303 Procure Predevelopment Drilling Permits

Submit completed applications for permit to appropriate governmental agencies. Wait for agencies to approve applications. Obtain approved permits.

20306 Perform Predevelopment Exploration Drilling and Other Field/Lab Work

Notify contractor(s) to proceed with work required by contract(s). Do other geologic field work as required. What is needed is sufficient drilling and field work to get reserve estimates within ±15%–20% accuracy range. Assay drill-hole samples and/or log drill holes and obtain rock quality designations (RQDs).

20307 Prepare Test Drill Sample Flowchart and Chain of Custody

Prepare flowchart (listing) of drill cores required for various tests (e.g., metallurgy, rock mechanics, porosity, permeability, density, and moisture). Chart should identify size and amount of cores required, location of procured cores (area of deposit), purpose of cores, place cores sent for testing, types of tests, method of assay, number of duplicate tests, number of blind tests, number of tests on standards, and core storage instructions. Establish a mandatory chain-of-custody protocol with proper check points and sign-offs.

20308 Develop Mineral Inventory Work Specifications

Develop scope of work and schedule necessary to determine deposit mineral inventory. Determine data requirements for computerized and/or hand-calculated mineral inventory system. In addition to mineral and geologic data, RQD must be developed. Inform data gatherers of required data format.

20309 Procure Other Drill Samples

Other core samples are necessary for porosity, permeability, Cerchar abrasivity, density, and moisture determinations. *Other* means samples needed for tests other than metallurgy and rock mechanics.

20310 Analyze Other Drill Samples

Send other core samples to appropriate lab or testing facility(ies). Analyze and test core samples for porosity, permeability, density, and moisture. Send test data to project team and other interested parties.

20311 Assemble and Edit Drill-Hole Data

Assemble drill-hole data pertaining to deposit. Edit data for correctness and completeness.

20312 Build Drill-Hole Computer File

Prepare mineral inventory data for entry in computer system. Build drill-hole files in computer. Types of data to include: identification; geologic parameters; RQD values for each interval of the mining horizon and 20 ft above and below; collar coordinates and elevation; assay values and intervals; hole depth, dip, and direction; and date hole completed.

20313 Prepare Geologic Maps

Prepare necessary drill-hole maps and cross sections, with computer and/or by hand, to help evaluate the mineral deposit.

20314 Re-Delineate Mineral Zones

Identify and re-delineate mineralized zones. Delineate by computer from drill-hole files and/or by hand.

20315 Compute Potential Mineral Reserves

Build computer block model and compute mineral reserves with various cutoff grades, mining heights, waste thicknesses, and so forth. Calculate reserves by hand if computer block model is not developed. Prepare a risk analysis and mitigation plan for the mineral reserve tabulated.

20316 Document and Review Mineral Inventory Results

Write report documenting results of Phase I predevelopment program. Report should contain data on geology, field work, and reserves. Review results with appropriate levels of management and other personnel. Write report in style and format suitable as a chapter in the intermediate feasibility study report.

20401 Develop Environmental Work Specifications

Develop scope of work and schedule necessary for environmental baseline studies, environmental impact analyses, and environmental control plans required for project analysis and costs. Work will serve as database for project permit applications.

20404 Perform Air Quality Baseline Study

Conduct literature search to determine amount and value of air quality and meteorological data available in project area(s). Establish project site monitoring stations to gather air quality, including radiological (if required) and meteorological baseline data. Gather data over required period of time.

20405 Perform Water Quality Baseline Study

Conduct literature search to determine amount and value of surface and groundwater quality data in project area(s). Gather and analyze samples of surface water in the area(s) over required period of time, generally on performing hydrology studies (Activity 20508).

20406 Perform Ecological Baseline Study

Conduct literature search to determine amount and value of ecological data available in project area(s). Gather and/or study samples of life forms (flora and fauna) in area(s) over required period of time.

20407 Perform Socioeconomic Baseline Study

Perform literature search to determine amount and quality of labor and housing available in general area. Conduct general area survey to verify and/or update statistics on amount and quality of labor and housing and other socioeconomic data. Study the sustained development options for the affected area.

Consider the need for a social license to operate in the area affected.

20408 Conduct Archaeological Investigation

Conduct project area search for cultural resources or items of historical significance. Concentrate search in areas of potential land disturbance caused by mine, plant, tailings, and surface facilities construction and development.

20409 Assess Environmental Impacts

Assess impact to baseline environment caused by construction, development, and operation of mine, plant, tailings disposal, and surface facilities.

20410 Develop Air Quality Plan and Costs

Develop plan, with one or more alternatives, to maintain air quality in and around project area. Evaluate effects caused by construction, development, and operation of mine, plant, tailings disposal, and surface facilities. Estimate capital and operating costs to implement the plans and put costs in format suitable for financial analysis. Cost accuracy should equal ±15%.

20411 Develop Water Quality Plan and Costs

Develop plan, with one or more alternatives, to maintain surface water quality in and around project area. Evaluate effects caused by construction, development, and operation of mine, plant, tailings disposal, and surface facilities. Estimate capital and operating costs to implement the plans and put costs in format suitable for financial analysis. Cost accuracy should equal ±15%.

20412 Develop Ecological Plan and Costs

Develop plan, with one or more alternatives, to maintain quality of ecology in and around project area. Evaluate effects caused by construction, development, and operation of mine, plant, tailings disposal, and surface facilities. Estimate capital and operating costs to implement the plans and put costs in format suitable for financial analysis. Cost accuracy should equal ±15%.

20413 Develop Reclamation Plan and Costs, Including Bonding

Develop plan, with one or more alternatives, to reclaim land disturbed by effects of construction, development, and operation of mine, plant, tailings disposal, and surface facilities. Estimate capital and operating costs to implement the plan and put costs in format suitable for financial analysis. Cost accuracy should equal ±15%.

20414 Develop Socioeconomic Plan and Costs

Develop plan, with one or more alternatives, to estimate requirements for community development plan may include requirements for employee housing; medical and dental facilities; schools; community utilities (power, water, sewer, etc.); community services; skills and occupational training other than for the operation; recreational activities; and potential sustained development infrastructure. Estimate capital and operating costs to implement the plans and put costs in format suitable for financial analysis. Cost accuracy should equal ±15%.

20415 Document and Review Environmental/ Socioeconomic Study Results

Write report documenting environmental and socioeconomic work performed. Review results of studies with appropriate levels of management and other personnel. Write report in style and format suitable as a chapter in the intermediate feasibility study report. Contractors should submit report in draft form for project team review before finalizing. Prepare risk analysis and mitigation plan for environmental and socioeconomic concerns.

20501 Develop Siting and Geotechnical Work Specifications

Develop scope of work and schedule necessary for siting, rock mechanics, soils mechanics and foundation, and hydrology studies.

20504 Prepare Hydrology Permit Applications

Prepare necessary permit application(s) to perform hydrology studies of groundwater and surface water quantities and qualities. Permits are needed primarily to drill test wells.

20505 Conduct Siting Studies

Conduct studies to determine suitable locations of all surface facilities for the mine, plant, roads, tailings disposal, and surface facilities. Identify several sites.

20506 Procure Hydrology Permits

Submit permit application(s) to appropriate governmental agency(ies). Wait for permit approval. Get approved permit(s).

20507 Prepare Soils Mechanics Permit Applications

Prepare necessary permit application(s) to conduct soils and foundation investigations. Permit(s) are likely needed for test pits and test borings.

20508 Conduct Hydrology Studies

Conduct studies to collect groundwater quantity and quality data. This usually requires drilling and pump-testing wells to determine amount and quality of water expected during mining and amount and quality of water required for makeup. Conduct studies to determine surface water flow patterns and amounts to expect during possible maximum flood and 100-year-flood periods. Send study data to project team in report form. Contractor should submit report in draft form for review by project team before finalizing.

20509 Procure Rock Mechanics Samples

Procure drill-core samples to use for rock mechanics tests. Drilling is usually performed as part of predevelopment drilling program.

20510 Procure Soils Mechanics Permits

Submit permit application(s) to appropriate governmental agency(ies). Wait for permit approval. Get approved permit(s).

20511 Procure Soils Mechanics Samples

Procure soils samples to use for soils mechanics tests and foundation analyses. Procure samples in areas preferred for plant and surface facilities construction and tailings disposal. Samples usually consist of test borings and test pits.

20512 Conduct Rock Mechanics Tests

Ship rock mechanics samples to testing laboratory. Conduct appropriate tests to determine strength of rock. Analyze test results to determine size of openings and pillars in underground mine or pit slope angles (stability) in open pit mine. Send test results in report form to project team. If rock appears to be applicable to mechanical excavation methods, send rock samples for Cerchar abrasivity tests. Develop rock mass rating designation and/or Barton's Q designation. Contractor should submit report in draft form for review by project team before finalizing.

20513 Conduct Soils Mechanics Tests

Ship soils mechanics samples to testing laboratory. Conduct appropriate tests to determine physical and chemical properties affecting building foundations and tailings disposal areas. For open pit mines, tests are needed to determine slope stability. Conduct appropriate field tests to determine above-soils properties and water flow characteristics. Analyze test results. Field tests could include location of suitable construction materials. Send test results in report form to project team. Contractor should submit report in draft form for review by project team before finalizing.

20514 Establish Siting Preferences

Rank sites selected in Activity 20505 in order of preference. Consider factors such as relationship to existing facilities, capital and operating costs, environment, land position, topography, accessibility, capacity, surface water flow patterns, soils mechanics data, and relationship to mineral deposit. Perform ranking using matrix evaluation procedures.

20515 Document and Review Siting and Geotechnical Results

Write report documenting siting and geotechnical work performed. Review results of studies with appropriate levels of management and other personnel. Write report in style and format suitable as a chapter in the intermediate feasibility study report. Contractor should submit report in draft form for review by project team before finalizing. Prepare a risk analysis and mitigation plan for the siting, geotechnical, and hydrological information used in the design.

20601 Develop Agency Reconnaissance, Government, NGO, and Public Relations Work Specification

Develop scope of work and schedule necessary for agency reconnaissance, government relations, nongovernmental organizations (NGOs), and public relations studies and/or work.

20604 Conduct Agency and NGO Reconnaissance Literature Search

Search literature to get overview of permits required by various government agencies, which may affect design, construction, development, or operation of project. Study area projects and identify active NGO groups in the area.

20605 Identify Involved Government Agencies

Conduct telephone conversation and personal visits to update data on government agencies having jurisdiction over design, construction, development, and operation of project. Determine all legal and political jurisdictions and all laws, regulations, and legislative codes at the federal, state, and local levels that are applicable to the process of mine/mill planning and operation. The following must also be identified:

- List all political jurisdictions in which the mining operation will exist.
- Obtain copies of all federal and state laws and codes relating to the state and country in which you want to construct a mining operation.
- Obtain a list of all mining permits required and a description of the regulatory processes involved in obtaining the permits.
- For properties within the United States, request determination from the district engineer of the U.S. Army Corps of Engineers (and other federal agencies if federal lands are involved) on its possible involvement and the necessity for a federal environmental impact statement (EIS) under the National Environmental Policy Act.
- If federal EIS is required, determine if state environmental impact report may be used as database or whether operative federal agency will require more, less, or other data than that collected for the state agency.
- Participate in any hearings on any federal statement or actions that occur independent of the state.
- Determine with the state environmental agency which state laws will be applied to the mine development under its jurisdiction. Obtain in writing the rationale for elimination of any potentially applicable laws.
- Inventory other state laws, permits, and permissions applicable to a mine in that state.
- Obtain copies of county zoning codes, sewerage codes, and information on which districts, special assessment districts, or other subjurisdictions of the county may be influenced by the mine.
- Determine each code and jurisdiction that may be encountered and the responsible local administrators. Obtain requirements to be fulfilled under each code.
- Obtain copies of all codes and rules applicable in the township or incorporated area, especially zoning and those related to utilities, waste disposal, and highways.

20606 Determine Agency Regulations

Conduct telephone conversations and personal visits to update data on government agency regulations affecting design, construction, development, and operation of project.

20607 Determine Agency Permit Requirements

Conduct telephone conversations and personal visits to update data on government agency permit requirements affecting design, construction, development, and operation of project.

20608 Develop and Implement Government, NGO, and Public Relations Programs

Develop program(s) to keep government, NGOs, and public informed of nature and status of project. Implement one or more of the programs.

20609 Document and Review Agency and NGO Reconnaissance Results

Write report documenting agency and NGOs reconnaissance work performed. Review results of studies with appropriate levels of management and other personnel. Write report in style and format suitable as a chapter in the intermediate feasibility study report. Contractor should submit report in draft form for review by project team before finalizing.

20610 Document and Review Government, NGO, and Public Relations Programs

Write report documenting government and public relations programs developed. Review results of studies with appropriate levels of management and other personnel. Write report in style and format suitable as a chapter in the intermediate feasibility report. Develop a risk analysis and mitigation plan for the government, NGO, and public relations programs.

20701 Develop Mine Work Specifications

Develop scope of work and schedule necessary for conducting mining evaluation studies.

20704 Update Mining Literature Search

Review company, contractor, and general available literature published about mining deposits of similar nature. Search should identify and benchmark mine data related to the mining method, unusual characteristics, types of equipment, potential problems, and so forth. Search may include visits to operating properties of similar nature.

20705 Identify Possible Mining Methods

Identify mining methods suitable for developing and producing deposits of this nature. Use combination of literature search and experience of company personnel and contractor personnel. List and define the potential methods. Consider range of production from the various methods.

20706 Develop Tentative Layout for Two or Three Mining Methods

Prepare tentative mine development and production layouts for each possible mining method identified. Prepare layouts with sufficient detail and accuracy to allow comparisons of capital and operating costs.

20707 Evaluate Alternative Mining Methods

Evaluate operational characteristics of two or three possible mining methods. Perform comparative capital and operating cost analyses, using present worth techniques. Identify other factors influencing selection of preferred mining method(s) such as environment, safety, hydrology, recoveries, rock mechanics, potential for mechanical excavation, dilution, and production limitations.

Rank alternatives in order of preference using matrix system and considering all listed factors and others, if applicable. Select two or three best methods for more detailed evaluations.

20708 Refine Mine Layouts and Develop Mine Plans

Prepare mine layouts for chosen alternatives in sufficient detail to allow development of mine plans. Five-year mine plan and life-of-mine plan will be based only on proven and probable reserves for all properties within the United States and Canada or where the ownership stock is listed in those two countries. Mine plans should have an accuracy of ±15% to 20%. Mine plans should include layouts and schedules for head frame, shaft, stations, preproduction development, underground service area development, production, various sizes of underground storage, and the cost benefit of each size and continuing production development.

20709 Perform Mine Functional Analyses

Calculate operational and development cycles, labor requirements, equipment requirements, and supply requirements for each mining function and for each alternative selected in Activity 20707 based on mine industry experience in these mining methods. The functions include drilling, blasting, loading, hauling, scaling, bolting, ground control, mine backfill (if needed) hoisting, primary crushing (if underground), maintenance, supply/debris handling, pumping, and other supports services. For each function, develop personnel productivity learning curves and prepare cost and operational schedules for each.

20710 Develop Mine Capital and Operating Cost Estimates

Develop cost estimates for total mining operation and alternatives. Costs should have accuracy of ±15% to 20%. Because costs will become part of intermediate financial analysis, put in format suitable for financial analysis.

20711 Document and Review Mine Results

Write report documenting mine study work performed. Review results of studies with appropriate levels of management and other personnel. Write report in style and format suitable as a chapter in the intermediate feasibility study report. Contractor should submit report in draft form for review by project team before finalizing. Prepare a risk analysis and mitigation plan on the proposed mining operation. At this point, try to locate and evaluate mining contractors and perform a cost comparison analysis.

20801 Develop Mineral Processing Work Specifications

Develop scope of work and schedule necessary for process, milling, waste rock storage, and tailings disposal studies.

20804 Conduct Mineral Processing Literature Search

Review company, contractor(s), and general available literature published concerning processing material of similar nature. Search should identify and benchmark some possible processing methods, types of equipment, potential problems, and so forth. Search may include visits to operating properties using processes of similar nature.

20805 Procure Metallurgical Samples

Procure drill-core samples to use for metallurgical testing. Drilling may be performed as part of predevelopment drilling program. Make sure that the samples represent the ore body. Large core samples for autogenous grinding tests may be needed.

20806 Conduct Metallurgical Tests

Ship metallurgical samples to testing laboratory. Conduct appropriate tests to determine comminuting characteristics (work index); separation and concentration characteristics of all types of ores; reagent consumption; heads, tails, and concentrate analyses; process flow sheet; environmentally harmful gaseous, liquid, and solid products produced; complete waste characterization of all waste products; long-term leachability of metal ions from tailings; and areas of uncertainty. Send test results to project team in report form. Contractor should submit report in draft form for review by project team before finalizing.

20807 Determine Mill Process Requirements

Evaluate metallurgical testing results. Use test results, literature search, company experience, and contractor experience to select two or three best process variation methods. Determine processing requirements for each alternative.

20808 Prepare Mineral Processing Flow Sheet

Prepare flow sheet for each process alternative. Flow sheet should contain sufficient detail to allow selection and sizing of equipment and show material balance.

20809 Perform Mineral Processing Functional Analyses

Calculate the operational cycles and labor, equipment, and supply requirements for each milling function and for each process alternative selected in Activity 20807 based on the various sizes of production from the mine and industry experience. The functions include stockpiling/reclaiming (size of storage), crushing, grinding, screening, concentrating including heap leaching (if applicable), classifying, clarifying, tailings disposal, concentrate handling, maintenance, supply/debris handling, and other support services. Prepare cost and operational schedules for each.

20810 Investigate Custom Milling Alternatives

Investigate opportunities for selling run-of-mine (ROM) material. Determine sale price and charges associated with selling ROM. Investigate opportunities for tolling mine-produced material. Determine custom mill capacity, timing, and costs. Costs should include capital and operating estimates for all associated tolling activities such as mill expansion costs to company (if any), transportation of product to custom mill, losses/deducts for processing, tolling charge, sampling methods, transportation of concentrate, and personnel requirements.

20811 Determine General Mill Plant Arrangement

Determine arrangement of mill facilities, including tailings, for each alternative. Prepare design basis and general arrangement drawings.

20812 Develop Mill Capital and Operating Cost Estimate

Develop cost estimates for total milling operation alternatives. Costs should have accuracy of ±15%. Because costs will become part of intermediate financial analysis, put costs in format suitable for financial analysis.

20813 Document and Review Milling Results

Write report documenting milling and metallurgical work performed. Review results of studies with appropriate levels of management and other personnel. Write report in style and format suitable as a chapter in the intermediate feasibility study report. Contractor should submit report in draft form for project team review before finalizing. Prepare risk analysis and mitigation plan for the metallurgical processing.

20901 Develop Smelter/Refinery Work Specifications

Develop scope of work and schedule necessary for custom smelting/refining studies.

20904 Investigate Custom Smelting/Refining Alternatives

Investigate opportunities for custom smelting and/or refining project concentrate. Determine custom refining capacity, timing, and costs. Costs should include capital and operating estimates for everything associated with custom refining such as transportation of concentrate to refinery, refining charge (consider deducts, penalties, and/or credits), transportation of refined product, and personnel requirements.

20905 Document and Review Cutstom Smelting/Refining Results

Write report documenting smelting/refinery work performed. Review results of studies with appropriate levels of management and other personnel. Write report in style and format suitable as a chapter in the intermediate feasibility study report. Contractor should submit report in draft form for project team review before finalizing. Prepare a risk analysis and mitigation plan of the smelting/refining process.

21001 Develop Surface Facilities Work Specifications

Develop scope of work and schedule necessary for study of project surface facilities not connected with mine and mill studies.

21004 Determine Surface Building Requirements

Determine surface buildings required for project operation. Include buildings serving support function for mine and mill but not buildings directly related to mining and milling activities. Types of buildings include ambulance garage, fire-fighting facility, administration/office, dry/changehouse, guardhouse, and surface shops. Determine operating and maintenance personnel and equipment requirements (including shop, office, and dry equipment). Prepare design basis and general arrangement drawings. Develop capital and operating cost estimates with ±15% accuracy to include in intermediate financial analysis. Put costs in form suitable for financial analysis.

21005 Determine Surface Utilities Requirements

Determine utilities required for project construction, development, and operation. Study does not include power distribution inside mine and mill battery limits distribution within facilities included in mining and milling studies. Utilities include electric power; fuel for buildings and fuel storage for operating equipment; communications (radio, telephone, and computer networks required); potable water; fire protection; sewage system; and garbage, trash, and solid waste removal and disposal. Determine operating and maintenance personnel and equipment requirements. Develop applicable piping and instrumentation diagram/drawing (P&ID) and line drawings. Develop capital and operating cost estimates with ±15% accuracy to include in intermediate financial analysis. Put costs in form suitable for financial analysis.

21006 Determine Surface Transportation Requirements

Determine transportation needs for moving equipment, supplies, material, and mine/mill product into and out of project area during project construction, development, and operation. Study should evaluate alternative transportation methods such as truck haulage; rail haulage (both off-site and in-plant); ship/barge haulage and port facilities (if needed); air haulage (if feasible); and combinations of these methods.

Study should determine access road requirements and costs, capital and operating costs, personnel and equipment requirements, and other costs associated with each feasible transportation system. Develop capital and operating costs with ±15% accuracy to include in intermediate financial analysis and put costs in form suitable for financial analysis.

21007 Determine Surface Mobile and Miscellaneous Equipment Requirements

Determine the surface mobile and miscellaneous equipment requirements not covered under other activities. This includes equipment for emergency medical and safety; road and yard area maintenance; supervisor pickups and car(s); maintenance personnel pickups and trucks; loader(s)/backhoe(s); forklift(s); crane(s)/ cherry picker(s); portable welder(s)/compressor(s)/ generator(s); small rear-dump truck(s); and crawler tractor(s) with dozer. Determine operating and maintenance personnel and equipment requirements. Develop capital and operating cost estimates with ±15% accuracy to include in intermediate financial analysis. Put costs in form suitable for financial analysis.

21008 Develop Water Management Plan and Costs

Determine requirements for total project water management system. This will require evaluating one or more alternatives for mine, mill, tailings disposal, potable, fire protection and other water usage requirements; mine dewatering and pumping requirements; project water balance (identify all water sources and losses); makeup water requirements and source (if water short); water treatment, disposal, or evaporation system (if excess water); potable water system (include source, treatment, storage, and distribution); fire protection system (include source, treatment, storage, and distribution); and dust control water requirements. Determine operating and maintenance personnel and equipment requirements, including buildings or structures to house facilities. Develop P&ID. Develop capital and operating cost estimates with ±15% accuracy to include in intermediate financial analysis. Put costs in form suitable for financial analysis.

21009 Determine General Surface Facilities Arrangement

Determine requirements for surface facility, plant, and disposal area(s) internal road (does not include haulage roads for open pit mine); parking areas; construction lay-down area(s); and storage area(s) (including mine waste).

Prepare plot plans showing alternatives for arrangement of all surface facilities including listed items, water facilities, mine, mill, and tailings facilities. Rank alternatives in preference order using matrix system to evaluate factors such as facility spacing and location, environment, accessibility, utilization, capital cost of each alternative, and operating costs (if different and applicable).

Choose best alternative. Prepare design basis and general arrangement drawings. Show the capital costs of areas selected from the preceding matrix analysis. Determine operating and maintenance equipment and personnel requirements for areas identified in Activity 21007, and maintenance material and supply requirements for areas listed in Activities 21008 and 21009.

21010 Determine Warehouse Requirements

Determine size of warehouse and storage yard facilities; amount of warehouse inventory; equipment (mobile and stationary) necessary to store warehoused items, load and unload supplies, and move supplies within confines of project area; and operating personnel requirements. Develop capital and operating costs with ±15% accuracy to include in intermediate financial analysis. Put costs in form suitable for financial analysis.

21011 Develop Surface Facilities' Capital and Operating Cost Estimates

Operating cost estimates should have accuracy of ±15%. Because costs will become part of intermediate financial analysis, put in format suitable for financial analysis.

21012 Document and Review Surface Facilities Results

Write report documenting surface facilities studies. Review results of studies with appropriate levels of management and other personnel. Write report in style and format suitable as a chapter in the intermediate feasibility study report. Contractor should submit report in draft form for review by project team before finalizing. Prepare risk analysis and mitigation plan for any and all surface facilities or utilities.

21101 Determine General Personnel Requirements

Determine administrative and management personnel requirements; and operating, maintenance, support, and supervisory personnel requirements developed in all the preceding project activities. Split requirements into salaried exempt, salaried nonexempt, and hourly classifications. Develop labor buildup schedules for each classification using results of learning curve analysis for each classification.

21102 Prepare Organization Charts

Prepare charts showing how each project alternative should be organized. Charts should show lines of authority and responsibility.

21103 Determine Administrative Costs

Determine salaries and wages of personnel identified in Activity 21101; payroll burden associated with salaries and wages; and cost, type, and quantity of office equipment and supplies required for all offices including administration, mine, mill, maintenance, and others. Prepare costs in form suitable for financial analysis. Costs should have accuracy of ±15%.

21104 Document and Review Organization and Administration Results

Write report documenting administrative costs and personnel requirements. Review results of studies with appropriate levels of management and other personnel. Write report in style and format suitable as a chapter in the intermediate feasibility study report.

21201 Develop Labor Plan

Develop plan, with alternatives, for attracting and keeping productive, qualified personnel. Items to investigate include recruiting, training, absentee and turnover projections, commuting/fly in–fly out work force, community development, salaries/wages, fringe benefits/payroll burden, incentive system, and union/nonunion considerations.

21202 Prepare Labor-Related Cost Estimates

Identify personnel and equipment requirements for plans developed. Prepare capital and operating cost estimates associated with plans developed. Costs should have accuracy of ±15%. Put costs in format suitable for financial analysis.

21203 Document and Review Labor-Related Results

Write report documenting labor-related studies. Review results of studies with appropriate levels of management and other personnel. Write report in style and format suitable as a chapter in the intermediate feasibility study report. Prepare a risk analysis and mitigation plan for all administrative and labor-related issues.

21300 Develop Work Specifications for Commodity Marketing

Develop scope of work and schedule necessary for analyzing the marketing of the commodities that are to be produced by the mine/mill facility. For most of the industry, this will be done by a market specialist and will need to be contracted. However, for some commodities or companies, these may be completed in-house.

21301 Conduct Intermediate Market Studies

Update market studies to determine product requirements, supply and demand forecast, selling price and marketing strategy of salable products, and position relative to competitors. Review metallurgical results of most recent testing against product sales specifications. Estimate price ranges for life of project. Prepare prices in format suitable for financial analysis.

21302 Document and Review Market Study Results

Write report documenting marketing studies listed in Activity 21301. Review results of studies with appropriate levels of management and other personnel. Write report in style and format suitable as a chapter in the intermediate feasibility study report. Prepare a risk analysis and mitigation plan for all aspects of marketing the product.

21401 Develop Intermediate Financial Analysis Criteria

Develop criteria for performing intermediate financial analysis. Criteria include overall project schedule (includes final evaluation, design, construction, development, and start-up), capital and operating costs, royalties; escalation factors (though the analysis will probably be done in constant dollars), tax rates, working capital, property acquisition costs, mine and mill recovery, revenues, depreciation methods and depletion allowance, allowance for unforeseen factors, capitalization factors, salvage values, corporate overhead allocation, sensitivity and risk analysis (see Activity 21403), and project alternative comparisons.

21402 Conduct Financial Analysis

Conduct analysis for total project using a suitable computer program. Print results of economic analysis.

21403 Conduct Financial Sensitivity Studies

Evaluate sensitivity/risk of various key factors. Also, quantify the degree of risk and perform Monte Carlo risk analysis on the collective factors: operating costs, capital costs, reserves, grade, mill recovery, royalties, taxes, and other items with high degree of uncertainty. Print results of sensitivity/risk analysis.

21404 Document and Review Financial Results

Write report documenting financial analysis and sensitivities. Review results of work with appropriate levels of management and other personnel. Write report in style and format suitable as a chapter in the intermediate feasibility study report. Prepare a risk analysis and mitigation plan for the techniques used in the financial analysis.

21501 Develop Tax Study Specifications

Develop scope of work and schedule necessary for conducting study of taxes applicable to project.

21502 Conduct Intermediate Tax Studies

Update tax studies to determine taxes applicable to project. Analyze taxes to understand how they affect construction, development, and operation of project. Prepare tax rates in format suitable for financial analysis.

21503 Document and Review Tax Studies

Write report documenting tax studies. Review results of studies with appropriate levels of management and other personnel. Write report in style and format suitable as a chapter in the intermediate feasibility study report. Prepare a risk analysis and mitigation plan related to the taxing of future property and production.

21601 Prepare Final Study Plan and Budget

Update final feasibility study plan and schedule using formalized scheduling techniques. Update budget for final feasibility study. Schedule and budget should conform to those used in financial analysis (Activity 21401).

21602 Prepare Intermediate Report

Prepare formal report detailing intermediate study work performed. Prepare report in style and format that

- Is suitable for presentation to management,
- Is suitable for use by other project teams,
- Contains history and results of work performed, and
- Has a composite table of all the risk factors analyzed in the report.

21603 Present Intermediate Report, Plan, and Budget to Management

Present all data generated during intermediate study, plus plan and budget for final study, to management for review. Present information in meeting(s) with hard copies of reports, schedules, and data. Distribute report at least one week prior to meeting to allow personnel time to read and review.

APPENDIX 11C

Additional Testing Activities (Such as a Bulk Sample or Test Mine)

Activity No. (from WBS) **Activity Title and Description**

30300 Determine Need for Bulk Sample Only or Full Test Mine

It must be determined whether the need for additional information can be obtained by bulk sampling methods or if a test mine must be developed to obtain other mining and metallurgical information.

30301 Perform Final Exploration Drilling and Other Field Work

Continue work outlined under predevelopment drilling contracts (Activity 20306). Do other geologic field work as required. This may indicate obtaining bulk samples by drilling 8-in. cores. Final exploration means sufficient drilling and field work to get reserve estimates within ±10% accuracy range. Assay drill-hole samples and/or log drill holes.

30302 Assemble and Edit Drill-Hole Data

Assemble drill-hole data pertaining to deposit. Edit data for correctness and completeness. Types of data included: identification; geologic parameters; collar coordinates; assay values and intervals; hole depth, dip, and direction; and date hole completed.

30303 Update Drill-Hole Computer File

Prepare mineral inventory data for entry in computer system. Add to existing data to update drill-hole files in computer.

30310 Prepare Geologic Maps

Prepare necessary drill-hole maps and cross sections to help evaluate the mineral deposit. Prepare maps with computer and/or by hand.

30311 Delineate Mineral Zones

Identify and delineate mineralized zones by computer from drill-hole files and/or by hand.

30312 Compute Potential Mineral Reserves

Build computer block model and compute mineral reserves with various cutoff grades, mining heights, waste thicknesses, and so forth. Calculate reserves by hand if block model not developed.

30313 Document and Review Mineral Inventory Results

Write report documenting results of predevelopment program. Report should contain data on geology, field work, and reserves. Review results with appropriate levels of management and other personnel. Write the study in a style and format suitable as a chapter in the phase of the feasibility study report that the test mine study occurred.

30710 Develop Test Mine Work Specifications

Develop scope of work and schedule necessary for designing, constructing, developing, and operating test mine (assuming that test mine is needed). Prepare request for proposal (RFP) to send to contractors capable of performing the work. RFP should contain draft of proposed contract.

30711 Determine Test Mine Permit Requirements

Identify permits required to design, construct, develop, and operate test mine. Refer to work performed under Activity 20607. Recheck with government agencies for new or different requirements.

30712 Plan Bulk Sampling Program

Calculate amount of bulk sample required for pilot-plant testing. Evaluate geologic and reserve data to choose test mine bulk sampling areas representative of deposit. Interface bulk sampling plan with test mining plan, Activity 30713. Modify bulk sampling plan and areas based on realistic mining plan, time schedule, and budget. Select pilot plant to run bulk sample(s). Plan metallurgical tests required. Determine requirements for sample(s) handling and transportation and sampled material disposal.

30713 Plan Test Mine Program

Develop a mine plan to select layout and development necessary for the following: metallurgical bulk sampling program (underground [UG] and open pit [OP]); predevelopment drilling program (UG); test mining program; rock mechanics tests (UG and OP); pillar, drift, and stope size evaluations (UG); drilling and blasting or mechanical excavation tests (UG and OP); ground support tests (UG); slope-stability tests (OP); and materials handling tests (UG and OP). Design required test mine surface facilities, access system and development (surface mine stripping or underground mine station[s], and level[s]) needs. Develop schedule and budget for test mine activities. Prepare construction contractor bid package.

30714 Prepare Test Mine Permit Applications

Prepare permit application(s) necessary to perform test mining program.

30715 Procure Test Mine Permits

Submit permit application(s) to appropriate government agency(ies). Wait for permit approval(s). Obtain approved permit(s).

30716 Obtain Management Approval of Test Mine

Obtain approval from appropriate levels of management to proceed with test mining as outlined in budget and schedule developed under Activity 30713.

30717 Locate, Evaluate, and Select Test Mine Contractor

Identify contractors capable of performing the work. Send each contractor copy of bid package developed in Activity 30713. Evaluate bids received, contractor's financial status (Dun & Bradstreet report if required) and other pertinent data. Select preferred contractor, preferably using matrix evaluation if low bid is not only selection criterion. Inform contractor(s) of its selection. Give notice to proceed. Revise, if necessary, scope of work and contract to reflect information contained in bids.

30718 Prepare Test Mine Contract

Write contract, with assistance from law office and controllers. Get necessary company approvals. Send contract to contractor for signature. Get approved contract from contractor and review for signature correctness. Some delay can occur if contractor wants to negotiate terms before approving.

30719 Mobilize Test Mine Contractor

Require time for contractor to arrive on-site and set up once notified to proceed.

30720 Construct Test Mine Surface Facilities

Build or erect surface facilities necessary for construction, development, and operation of test mine, including hoisting facilities, shaft collar, and headframe for an underground test mine.

30721 Perform Test Mine Access and Level Development

For underground test mine, sink shaft (or other method of access), excavate, and construct station(s) and perform necessary level development. For open-pit test mine, strip necessary overburden and waste material.

30722 Perform Test Mining

Conduct mining tests as outlined under Activity 30713.

30723 Procure Test Mine Bulk Samples

Procure test mine bulk sample(s) as planned under Activity 30712 in intermediate study. (In many cases, this activity and Activity 30724 occur between the intermediate and final feasibility study.)

30724 Perform Test Mine Predevelopment Work

Perform test mine drilling and other geologic work as outlined under Activity 30713.

30725 Update Potential Mineral Reserves

Update mineral reserve calculations using drilling, assay, and geologic data gathered during test mining (Activity 30724).

30726 Update Engineering Data

Update all previously acquired engineering data with the data gathered during test mine operation.

APPENDIX 11D

Phase III: Final Feasibility Study Complete Activity Definitions

Activity No. (from WBS) **Activity Title and Description**

30100 Public Relations

As soon as the company management approves the final feasibility study plan and budget, notify state government officials that the project will proceed to the next level of examination. Brief the media on the decision to proceed to the final feasibility study, making sure to emphasize that the final decision of whether or not to build a mine is probably many months or even years away. Continue the dissemination of information as it becomes available and continue presenting speeches at local and state meetings. Set up local town meeting where stakeholders can question firsthand what will take place if a mine is built. Address their remaining concerns and the company's plans to mitigate their concerns.

30101 Management Approval of Final Study Plan and Budget

Get approval from appropriate levels of management to proceed with final feasibility study outlined in plan, schedule, and budget submitted to management for review at end of intermediate study (Activity 21603).

30201 Review Land and Water Status

Review land and water ownership, control, royalty, and lease situation updated during intermediate study. Project team should review land status and water with personnel in company's land office, and other appropriate individuals with respect to site locations identified during intermediate study. All related water rights should be determined.

30202 Finalize Land and Water Acquisitions and Costs

Land and water personnel should determine final costs of buying and/or leasing land and acquiring water necessary to construct, develop, and operate project. Commitments for land are probably required at this time. Costs will become part of final study financial analysis so accuracy should be within ±10%. Put costs in format suitable for financial analysis.

30203 Document and Review Land and Water Situation

Write report documenting results and costs of land and water negotiations. Review results of negotiations with appropriate levels of management and other personnel. Complete all land maps. Write report in style and format suitable as a chapter in the final feasibility study report.

30304 Develop Core Drilling Work Specifications

Develop scope of work and schedule necessary for core drilling program to obtain samples for the following tests: metallurgical (may not need if test mine program planned), rock mechanics (may not need if test mine program planned), Cerchar abrasivity, density, porosity, permeability, and miscellaneous. This program may be accomplished with regular predevelopment drilling program or as separate program.

30307 Prepare Test Core Sample Flowchart

Prepare flowchart (listing) of drill cores required for various tests (e.g., metallurgy, rock mechanics, porosity, permeability, density, moisture). Chart should identify size and amount of cores required, from where cores procured (area of deposit), purpose of cores, where cores are sent for testing, types of test, and core storage instructions.

30308 Conduct Core Drilling Program

Perform the field core drilling program as planned and scheduled under Activity 30307.

30309 Analyze Other Core Samples

Send other core samples to appropriate lab or testing facility(ies). Analyze and test core samples for porosity, permeability, density, and moisture. Send test data to project team and other interested parties.

30401 Develop Environmental Work Specifications

Develop scope of work and schedule necessary for environmental impact analyses and environmental control plans required for project analysis and costs. Work will serve as base data for final feasibility costs and probable update data for permit applications. Prepare request for proposal (RFP) to send to contractors capable of performing the work. RFP should include draft of proposed contract.

30403 Finalize Archaeological Mitigation Plan

Reassess impact to archaeological site caused by construction, development, and operation of mine, plant, tailings disposal, and surface facilities. Redevelop the mitigations plan as required and review with environmental agency. This should include input from the final feasibility study for mine, mill, tailings, and surface facilities. Finalize the archaeological mitigation plan that was agreed upon with the environmental agency. Estimate capital and operating costs to implement the plans.

Complete design drawings. Put costs in format suitable for financial analysis. Cost accuracy should equal +10%.

30404 Assess Environmental Impacts

Reassess impact to baseline environment caused by construction, development, and operation of mine, plant, tailings disposal, and surface facilities. This should include input from the final feasibility study of mine, mill, tailings, and surface facilities.

30405 Finalize Air Quality Plan and Costs

Finalize chosen plan to maintain air quality in and around project area. Update effects caused by construction, development, and operation of mine, plant, tailings disposal, and surface facilities. Complete design basis drawings. Estimate capital and operating costs to implement the plans. Put costs in format suitable for financial analysis. Cost accuracy should equal +10%.

30406 Finalize Water Quality Plan and Costs

Finalize chosen plan to maintain surface water quality in and around project area. Update effects caused by construction, development, and operation of mine, plant, tailings disposal, and surface facilities. Estimate capital and operating costs to implement the plans. Complete design drawings. Put costs in format suitable for financial analysis. Cost accuracy should equal +10%.

30407 Finalize Ecological Plan and Costs

Finalize chosen plan to mitigate ecological disturbances caused by effects of construction, development, and operation of mine, plant, tailings disposal, and surface facilities. Estimate capital and operating costs to implement the plans. Put costs in format suitable for financial analysis. Cost accuracy should equal +10%.

30408 Finalize Reclamation Plan and Costs

Finalize chosen plan to reclaim land disturbed by effects of construction, development, and operation of mine, plant, tailings disposal, and surface facilities. Estimate capital and operating costs to implement the plans. Complete design basis drawings. Put costs in format suitable for financial analysis. Cost accuracy should equal +10%.

30409 Finalize Socioeconomic Plan and Costs

Finalize chosen plan to estimate requirements for community development. Plan should include requirements for the following: employee housing, medical and dental facilities, schools, community utilities, community services, and recreational activities. Estimate capital and operating costs to implement the plans. Put costs in format suitable for financial analysis. Cost accuracy should equal +10%.

30410 Document and Review Environmental Study Results

Write report documenting environmental work performed. Review results of studies with appropriate levels of management and other personnel. Write report in style and format suitable as a chapter in the final feasibility study report. Contractors should submit report in draft form for project team review before finalizing.

30501 Develop Siting and Geotechnical Work Specifications

Develop scope of work and schedule necessary to finalize siting, rock mechanics, soils mechanics and foundation, and hydrology studies. Prepare RFP to send to contractors capable of performing the work. RFP should contain draft of proposed contract.

30504 Prepare Hydrology Permit Applications

Prepare necessary permit applications to perform hydrology studies of groundwater and surface water quantities and qualities. Permits are needed primarily to drill test wells.

30505 Procure Hydrology Permits

Submit permit application(s) to appropriate governmental agency(ies). Wait for permit approval. Get approved permit(s).

30506 Conduct Hydrology Studies

Conduct studies to finalize groundwater quantity and quality data. This usually requires drilling and pump testing wells to determine amount and quality of water expected during mining and amount and quality of water required for makeup. Conduct studies to determine surface water flow patterns and amounts to expect during possible maximum flood and 100-year-flood periods. Send study data to project team in report form. Contractor should submit report in draft form for review by project team before finalizing.

30507 Prepare Soils Mechanics Permit Applications

Prepare necessary permit application(s) to finalize soils and foundations investigations. Permit(s) probably necessary to dig test pits and do test borings.

30508 Procure Soils Mechanics Permits

Submit permit application(s) to appropriate governmental agency(ies). Wait for permit approval. Get approved permit(s).

30509 Procure Soils Mechanics Samples

Procure soils samples to finalize soils mechanics tests and foundation analyses. Procure samples in areas preferred for plant and surface facilities construction and tailings disposal. Samples usually consist of test borings and test pits. Samples will serve as basis for buildings, dams, shafts, and other foundation design specifications.

30510 Conduct Soils Mechanics Tests

Ship soils mechanics samples to testing laboratory. Conduct appropriate tests to finalize physical and chemical properties affecting building foundations, tailings disposal areas, and shaft collars. For open pit mines, tests are needed to determine slope stability. Conduct appropriate field tests to determine above-soils properties and water flow characteristics. Analyze test results. Field tests could include location of suitable construction materials. Send test results in report form to project team. Contractor should submit report in draft form for review by project team before finalizing.

30511 Procure Rock Mechanics Samples

Procure drill-core samples to finalize rock mechanics properties. Drilling is part of core drilling program (Activity 30310).

30512 Conduct Rock Mechanics Tests

Ship rock mechanics samples to testing laboratory. Conduct appropriate tests to finalize strength of rock. Analyze test results to finalize size of openings and pillars in underground mine or pit slope angles (stability) in open pit mine. If applicable, test for the application of mechanical excavation. Send test results in report form to project team. Contractor should submit report in draft form for review by project team before finalizing.

30513 Finalize Siting Preferences

Determine the final location of all surface facilities. This includes shaft and other mine facilities, mill and processing facilities, tailings facilities and pipelines, and surface ancillary facilities such as roads, buildings, power lines, gas lines, storage areas, waste disposal areas, parking areas, and construction lay-down areas. If an open pit mine, this would include location of pit. Locate all sites within a few feet of their planned constructed location. Complete design basis drawings. Final facilities location selection should include factors such as the following: facility spacing and location, environment, accessibility, utilization, capital cost, and operating costs (if different and applicable).

30514 Document and Review Siting and Geotechnical Results

Write report documenting siting and geotechnical work performed. Review results of studies with appropriate levels of management and other personnel. Write report in style and format suitable as a chapter in the final feasibility study report. Contractor should submit report in draft form for review by project team before finalizing.

30601 Develop Permit Application Work Specifications

Develop scope of work and schedule necessary for preparation of permit applications. Prepare RFP to send to contractors capable of preparing permit applications. RFP should contain draft of proposed contract.

30604 Prepare Permit Applications

Procure necessary forms and formats for all permits required to construct and operate project. Complete all permit applications as required by local, state, and federal agencies. Submit applications to appropriate governmental agencies.

30605 Procure Construction and Operating Permits

Wait for various governmental agencies to approve permit applications. Get approved applications from agencies. This task could require some application rewriting or amending if one or more agencies need data not presented in original application.

30606 Expand Government and Public Relations Programs

Expand and update program(s) to keep government and public informed of nature and status of project.

30607 Document and Review Government and Public Relations Programs

Write report documenting status of government and public relations programs. Review results of programs with appropriate levels of management and other personnel. Write report in style and format suitable as a chapter in the final feasibility report.

30701 Develop Mine Work Specifications

Develop scope of work and schedule necessary for conducting final mine evaluation study. Prepare RFP to send to contractors capable of performing the work. RFP should contain draft of proposed contract.

30704 Develop Final Mine Layout

Prepare final mine development and production layouts. Prepare layouts with sufficient detail and accuracy to develop mine plans and allow estimating capital and operating costs, and development and operating schedules to accuracies of ±10% to 15%.

30705 Develop Final Mine Plan

Mine plans should, include layouts and schedules for headframe, shaft, stations, preproduction development, underground service area development, production, and continuing production development. Complete design basis drawings. Develop life-of mine production plans as well as detailed five-year mine production plans. Mine plans should have an accuracy of +15%.

30706 Perform Mine Functional Analyses

Calculate operational cycles, and labor, equipment, and supply requirements for each mining function. Refine the functions analysis made earlier to include drilling, blasting, loading, hauling, scaling, bolting, ground control, mine backfill (if needed), hoisting, primary crushing (if underground), maintenance, supply/debris handling, pumping, and other support services. For each function, develop personnel productivity learning curves and prepare cost and operational schedules for each.

Use personnel productivity learning curves developed in intermediate feasibility study for each function. Prepare cost and operational schedules for each.

30707 Develop Mine Capital and Operating Cost Estimates

Develop cost estimates for total mining operation. Costs should have accuracy of ±10% to 15% and will become part of final financial analysis. Put costs in format suitable for financial analysis.

30708 Develop Mine Design Specifications

Develop design specifications for competitive bidding of mine design, construction, and development work. Prepare bid packages. Design specifications for an underground mine should include systems for mine access, materials and personnel handling, ventilation, communications, electrical, mine dewatering, and fuel storage and handling; maintenance and warehousing facilities; explosives handling and storage facilities; crushing facilities; and sewage system.

Design specifications for an open pit mine normally include communications system(s); sewage system; electrical system; mine dewatering system; fuel storage and handling system; oil, lubrication, and antifreeze system; and explosives handling and storage facilities.

30709 Document and Review Mine Results

Write report documenting mine study work performed. Review results of studies with appropriate levels of management and other personnel. Write report in style and format suitable as a chapter in the final feasibility study report. Contractor should submit report in draft form for review by project team before finalizing.

Note: If the need for a test mine was not recognized at the end of the intermediate feasibility (or prefeasibility) study and is now necessary, see Appendix 11C for WBS Activities 30710–30726 that are now needed.

30801 Develop Mill Work Specifications

Develop scope of work and schedule necessary for final process, milling, and tailings disposal studies. Prepare RFP to send to contractors capable of performing the work. RFP should contain draft of proposed contract.

30804 Procure Metallurgical Samples

Procure drill-core samples to use for metallurgical testing. Drilling is usually performed as part of predevelopment drilling program. This activity may be unnecessary if bulk sample is obtained from test mine.

30805 Conduct Metallurgical Testing

Ship metallurgical samples to testing laboratory or pilot mill, if bulk sample for test mine is used. Conduct appropriate tests to determine final metallurgical data and design specifications for comminuting characteristics; separation and concentration characteristics of reagent consumption; heads, tails, and concentrate analyses; process flow sheet; environmentally harmful gaseous, liquid, and solid products produced; and areas of uncertainty.

Send test results to project team in report form. Contractor should submit report in draft form for review by project team before finalizing.

30806 Finalize Mill Process Requirements

Evaluate metallurgical testing results. Use test results, literature search, company experience, and contractor experience to determine the best process method. This method, and other technical data gathered, will serve as basis for mill design.

30807 Prepare Milling Flow Sheet

Prepare flow sheet for chosen process. Flow sheet should contain sufficient detail to allow selection and sizing of equipment. After the flow sheet and equipment sizes are finalized, develop final mill plant arrangement drawings.

30808 Perform Mill Functional Analyses

Calculate the operational cycles, and labor, equipment, and supply requirements for each milling function and the process method used in Activity 30807. The functions include the following: stockpiling/ reclaiming (size of storage), crushing, grinding, screening, concentrating including heap leaching (if applicable), classifying, clarifying, tailings disposal, concentrate handling, maintenance, supply/debris handling, and other support services. Prepare cost and operational schedules for each.

30809 Develop Mill Design Specifications

Develop specifications for competitive bidding of mill and tailings facilities design and construction. Prepare bid packages. Complete design basis drawings and basic engineering drawings.

30810 Reexamine Custom Milling Contract (assuming this is the option)

Start finalizing contract terms with custom mill suitable for processing mined material. May want to finalize terms and sign contract before getting management approval of project.

This activity assumes no processing facility was built as part of the project.

30811 Finalize Custom Milling Costs

Determine costs associated with custom milling of mined material. Finalize with custom mill quantities of material for processing, timing, and costs. Costs need to include capital (if required) and operating estimates for everything associated with tolling:

- Mill expansion costs to the minerals company (if any)
- Transportation of product to custom mill
- Losses/deductions for processing
- Tolling charge
- Sampling methods and costs
- Transportation of concentrate
- Personnel requirements

If selling run-of-mine material represents the chosen alternative, determine final sales price and charges. This alternative would replace custom milling or a project mill for financial evaluation purposes.

30812 Develop Mill Capital and Operating Cost Estimates

Develop estimates for total milling operation. Costs should have accuracy of +10%. Costs will become part of final financial analysis. Put costs in format suitable for financial analysis.

30813 Document and Review Milling Results

Write report documenting milling and metallurgical work performed. Review results of studies with appropriate levels of management and other personnel. Write report in style and format suitable as a chapter in the final feasibility study report. Contractor should submit report in draft form for project team review before finalizing.

30901 Develop Smelter/Refinery Work Specifications

Develop scope of work and schedule necessary for finalizing custom refining plans.

30902 Finalize Custom Smelter/Refinery Contract

Start finalizing contract terms with custom refinery(ies) suitable for processing mill concentrates. Finalize with custom refinery(ies) quantities of material for processing, timing, and costs. Costs need to include capital (if required) and operating estimates for everything associated with tolling such as transportation of concentrate to refinery, refining charge (consider deducts and/or credits), transportation of refined product, and personnel requirements. Costs should have accuracy of +10%.

30903 Document and Review Smelter/Refinery Results

Write report documenting refinery work performed. Review results of studies with appropriate levels of management and other personnel. Write report in style and format suitable as a chapter in the final feasibility study report.

31001 Develop Surface Facility Work Specifications

Develop scope of work and schedule necessary for final study of project surface facilities not connected with mine and mill studies. Prepare RFP to send to contractors capable of performing the work. RFP should contain draft of proposed contract.

31004 Finalize Surface Building Requirements

Finalize surface buildings required for project operation. These include those serving support function for mine and mill but not buildings directly related to mining and milling activities. Types of buildings include ambulance garage, administration/office, dry/changehouse, guardhouse, and surface shops.

Finalize operating and maintenance personnel and equipment requirements (including shop equipment, office, and dry equipment). Complete design basis drawings. Develop capital and operating cost estimates with +10% accuracy to include in final financial analysis. Put costs in form suitable for financial analysis.

31005 Finalize Surface Utilities Requirements

Finalize utilities required for project construction, development, and operation. Study does not include power distribution inside mine and mill battery limits (distribution within facilities included in mining and milling studies). Utilities should include electric power; fuel for buildings and fuel storage for operating equipment; communications (radio and telephone); potable water; fire protection, sewage system, and garbage/trash/solid waste removal and disposal. Finalize operating and maintenance personnel and equipment requirements. Complete design basis drawings. Develop capital and operating cost estimates with +10% accuracy to include in final financial analysis. Put costs in form suitable for financial analysis.

31006 Finalize Surface Transportation Requirements

Finalize transportation method chosen in intermediate study for moving equipment, supplies, material, and mine/mill product into and out of project area during project construction, development, and operation. Methods evaluated include haulage by truck, rail, ship/barge (if feasible), and air (if feasible), or combinations of these.

Finalize access road requirements and costs, capital and operating costs, personnel and equipment requirements, and other costs associated with transportation system. Develop capital and operating costs with +10% accuracy to include in final financial analysis. Put costs in form suitable for financial analysis.

31007 Finalize Surface Mobile and Miscellaneous Equipment Requirements

Finalize requirements not covered under other activities. This includes equipment for medical emergencies, road and yard area maintenance, supervisor pickups/car(s), maintenance personnel pickups and trucks, loader(s)/backhoe(s), forklift(s), crane(s)/ cherry picker(s), portable welder(s)/compressor(s)/ generator(s), small rear-dump truck(s), and crawler tractor(s) with dozer. Finalize operating and maintenance personnel and equipment requirements. Develop capital and operating cost estimates with +10% accuracy to include in final financial analysis. Put costs in form suitable for financial analysis.

31008 Finalize Water Management Plan and Costs

Finalize requirements for total project water management system. This includes the following: mine, mill, tailings disposal, potable, fire protection and other water usage requirements; mine dewatering and pumping requirements; project water balance (identify all water sources and losses); makeup water requirements and source (if water short); water treatment, disposal, or evaporation system (if excess water); potable water system (include source, treatment, storage, and distribution); fire protection system (include source, treatment, storage, and distribution). Finalize operating and maintenance personnel and equipment requirements, including buildings or structures to house the water-related facilities. Complete design basis drawings. Develop capital and operating cost estimates with +10% accuracy to include in final financial analysis. Put costs in form suitable for financial analysis.

31009 Finalize General Surface Facilities Arrangement

Finalize requirements for surface facility, plant, and disposal area(s); internal road (does not include haulage roads for open pit mine); parking areas; construction lay-down area(s); and storage area(s).

Prepare plot plans showing final arrangement of all surface facilities, including items in list, water facilities, mine, mill, and tailings facilities. Finalize capital costs of these surface facilities. Determine operating and maintenance equipment and personnel requirements under preceding surface activities (31004 to 31009). Finalize maintenance material and supply requirements for areas identified in list. Complete design basis drawings.

31010 Finalize Warehouse Requirements

Finalize size of warehouse and storage yard facilities; amount of warehouse inventory; equipment (mobile and stationary) necessary to store warehoused items, load and unload supplies, and move supplies within confines of project area; and operating personnel requirements. Complete design basis drawings. Finalize capital and operating costs with +10% accuracy to include in final financial analysis. Put costs in form suitable for financial analysis.

31011 Develop Surface Building Design Specifications

Develop design specifications for competitive bidding of design and construction. Buildings are identified under Activity 31004. Prepare bid package(s).

31012 Develop Surface Utility Design Specifications

Develop design specifications for competitive bidding of design and construction. Utilities are identified under Activity 31005. Prepare bid package(s).

31013 Develop Surface Transportation Design Specifications

Develop design specifications for competitive bidding of design and construction. Facilities are identified under Activity 31006. Prepare bid package(s).

31014 Develop Water Facilities Design Specifications

Develop specifications for competitive bidding of design and construction. Facilities are identified under Activity 31008. Prepare bid package(s).

31015 Develop Surface Facilities Capital and Operating Cost Estimates

Assemble capital and operating cost estimates for surface facilities. Costs should have accuracy of +10%. As costs will become part of final financial analysis, put costs in format suitable for financial analysis.

31016 Document and Review Surface Facilities Results

Write report documenting surface facilities studies. Review results of studies with appropriate levels of management and other personnel. Write report in style and format suitable as a chapter in the final feasibility study report. Contractor should submit report in draft form for review by project team before finalizing.

31017 Develop Environmental Monitoring Plan (EMP) for Proposed Operation

Now that all of the mine/plant facilities are defined, complete design basis drawings, specify the labor disciplines, the types and requirements of the monitoring program, the inspections to be required, and the method whereby corrective action and compliance will be achieved.

31018 Define Cost of EMP

All professional and staff personnel cost and their equipment must be included. Also include allowances for outside testing on a scheduled basis.

31019 Amend All Permit Applications to Include Aspects of EMP That Pertain to Various Permits

Return to the various permit applications and insert those actions and plans that the operating company will take to monitor and control all aspects of the operation to remain in compliance with various regulators' requirements.

31101 Refine General Personnel Requirements

Finalize administrative and management personnel requirements; and operating, maintenance, support, and supervisory personnel requirements developed in preceding activities. Split requirements into salaried exempt, salaried nonexempt, and hourly classifications. Finalize labor buildup schedules for each classification.

31102 Update Organization Charts

Finalize organization charts showing the project organization. Charts should show lines of authority and responsibility.

31103 Define Administrative Costs

Finalize salaries and wages of personnel identified in 31101; payroll burden associated with salaries and wages; and cost, type, and quantity of office equipment and supplies required for all offices including administration, mine, mill, and maintenance. Prepare costs in form suitable for financial analysis. Costs should have accuracy of +10%.

31104 Document and Review Organization and Administration Results

Write report documenting administrative costs and personnel requirements. Review results of studies with appropriate levels of management and other personnel. Write report in style and format suitable as a chapter in the final feasibility study report.

31201 Refine Labor Plan

Finalize plan for attracting and keeping productive, qualified personnel. Plan should include items such as recruiting, training, absentee and turnover projections, commuting (including fly in–fly out), community development, salaries/wages, fringe benefits/payroll burden, incentive system, and union/nonunion considerations.

31202 Refine Labor-Related Cost Estimates

Finalize personnel and equipment requirements and capital and operating cost estimates for the plan developed. Costs should have accuracy of +10% and be in format suitable for financial analysis.

31203 Document and Review Labor-Related Results

Write report documenting labor-related studies. Review results of studies with appropriate levels of management and other personnel. Write report in style and format suitable as a chapter in the final feasibility study report.

31301 Conduct Final Market Study

Update market studies to determine selling price of salable products. Check product specifications of final metallurgical test with required product specifications. Estimate price ranges for life of project and prepare prices in format suitable for financial analysis.

31302 Document and Review Market Study Results

Write report documenting marketing studies. Review results of studies with appropriate levels of management and other personnel. Write report in style and format suitable as a chapter in the final feasibility study report.

31401 Develop Financial Analysis Criteria

Develop criteria for performing final financial analysis. Criteria should include overall project schedule (design, construction, development, and start-up), ore production and final production schedule, capital and operating costs, royalties, escalation factors, tax rates, working capital, property acquisition costs, mill recovery, depreciation methods, depletion allowance, allowance for unforeseen, capitalization factors, sensitivity and risk analysis (see Activity 31403), salvage values, and corporate overhead allocation.

31402 Conduct Financial Analysis

Conduct financial analysis for total project using a suitable computer program. Print results of economic analysis.

31403 Conduct Financial Sensitivity Studies

Evaluate sensitivity/risk of various project key factors. Also, quantify the degree of risk and perform Monte Carlo risk analysis on the collective factors, including operating costs, capital costs, reserves, grade, mill recovery, royalties, taxes, and other items with high degree of uncertainty. Conduct sensitivity analysis using a suitable computer program. Print results of sensitivity analysis.

31404 Document and Review Financial Results

Write report documenting financial analysis and sensitivities. Review results of work with appropriate levels of management and other personnel. Write report in style and format suitable as a chapter in the final feasibility study report.

31501 Develop Tax Study Work Specifications

Develop scope of work and schedule necessary for finalizing study of taxes applicable to project.

31502 Conduct Final Tax Study

Update tax studies to finalize taxes applicable to project. Analyze taxes to understand how they affect construction, development, and operation of project. Prepare tax rates in format suitable for financial analysis.

31503 Document and Review Tax Studies

Write report documenting tax studies. Review results of studies with appropriate levels of management and other personnel. Write report in style and format suitable as a chapter in the final feasibility study report.

31601 Prepare Design, Construction, and Development Schedule and Budget

Update and expand plan and schedule using formalized scheduling techniques. Update budget for project design, construction, and development. Schedule and budget should conform to those used for final study financial analysis (Activity 31401).

31602 Prepare Final Design Basis Report

Prepare the final design basis report documenting all of the technical parameters in a single document. Prepare formal report detailing final study work performed. Prepare report in style and format that is

- Suitable for presentation to management,
- Suitable for use by other project teams, and
- Containing history and results of work performed.

31603 Present Report, Plan, and Budget to Management

Present all data generated during final study, plus plan and budget for design, construction, and development phase to management for review. Present information in meeting(s) with hard copies of reports, schedules, and data. Distribute report at least 1 week prior to meeting to allow personnel to read and review.

CHAPTER 12

Design Basis Report

Richard L. Bullock

The primary purpose of the design basis report (DBR)* is to be able to convey to any future design engineers a consolidated document where all the needed information is contained in a somewhat condensed version. But it can also serve to inform others, such as financial organizations, construction personnel, or persons who may be interested in joint venturing the project, about some of the details of the project that is contained in the final feasibility study. While much of the information is also in the final feasibility study, this document is written more for the purpose of documenting for management that the project is indeed both feasible and economically viable. In contrast, the design basis document is written to convey all of the technical information that will be needed by the architecture/engineering (A/E) design organization that has already been worked out by the owner's project feasibility team. It will contain all of the drawings that were prepared during the final feasibility study, plus any other drawings that are required to convey the needed technical information to the architect/engineer (A/E) on what the owner wants built. It is also the document that will be used for information for the final bids by the various A/E organizations.

In the introduction, the writer should define the purpose and the use of the DBR. At a minimum, the DBR serves several purposes:

- Defines the technical basis for project design and construction so that downstream basic, detailed engineering can proceed;
- Provides the basis for a coordinated review by the organizational entities involved, (i.e., the future operations group, the engineering group, management, and the future A/E);
- Provides documentation for the technical basis and facilities description for which the development of the final feasibility cost estimates were completed; and
- Conveys the construction and procurement philosophy of the company at that point in time to the future A/E.

The DBR is usually written in several volumes. The example given in this chapter shows the DBR written in five volumes, listed in the following sections. Rather than try to describe in specific detail what should be written under each section and subsection of each volume, a brief description will be given concerning the general content of that volume and then a generic outline is presented of items that need this specific coverage.

* This document is also sometimes called the design basis memorandum or project design basis. It can be referred to as the DBM, PDB, or, in this case, the DBR. In any situation, they are all the same thing.

The concept that is presented here has been used in the petrochemical industry for decades to help the transition from feasibility to building the plant, on budget and on time. Unfortunately, the mining industry personnel normally think that this is extra work and cost at a time when all they want is to get the plant built and start making money. But they are missing the point; they are letting the engineering, procurement, and construction management (EPCM) contractor decide what will go into that plant, with little guidance from the experience of the mining company operations group.

VOLUME I: MANAGEMENT SUMMARY

The management summary, prepared by the project executive or project manager as applicable, summarizes the project objectives, the assumptions that were made, the work that has been completed, the economic analysis and the associated risk, and the recommendations of the project team. Other items that should be covered if they have been studied by the project team are funding of the project; the business plan, with market and competition analysis and strategies; and any outstanding major issues involving government agencies related to utilities, transportation, land, royalties, or potential project partnerships. Finally, the conclusions and recommendations, with discussions on the reserves, the feasibility of the project, the market, the schedule of the design, construction, and start-up as planned in the feasibility study, any preappropriation work contemplated, and funding that is needed to complete the first volume. The following is an example outline of the information that is contained in Volume I.

Introduction and Summary

- Technical feasibility
  - Ore deposit
  - Facilities
  - Viability
- Economic feasibility
  - Investment and capital cost
  - Economic analysis
  - Sensitivities
  - Operating cost
  - Sales price and operating profit
  - Construction and life of mine
- Product market analysis
  - Competitive situation
  - Market development activities
  - Execution plan
  - Project execution responsibility
- Division of project execution (if applicable)
  - Line of reporting
  - Planned method of contracting
  - Construction plan
  - Construction labor buildup
- Business plan
  - Objectives
  - Demand for finished or refined product
- Product market analysis
  - Competitive situation
  - Market development activities
- Market strategy
- Outstanding issues
  - Country mining law and code
  - Potential project partner (if being considered)
  - Land and water purchase from the local government or other sources
- Purchase of private land and rights-of-way
  - Interface with country agencies
  - Reestablishment of contractual basis
  - Project mobilization
  - Construction and operating permits
  - Power supply agreements
  - Concentrate transport rail agreement concentrator water supply agreement

Conclusions and Recommendations

- Conclusions
- Feasibility
- Market
- Preappropriation work funding
- Recommendations

VOLUME II: PROJECT ECONOMICS

Project economics, prepared by the project executive or project manager, summarizes the capital and operating costs, project schedule, market forecasts, inflation projections (if constant dollar analysis was not used), and other factors that affect the total erected cost and project economics. The project risks, as they have been identified, and the measures needed to mitigate those risks should be documented. The following is an example outline of the information that is contained in Volume II.

Section 1: Overview
Project schedule
Capital cost estimate
Operating cost estimate
Marketing
Business climate and investment outlook
Economic analysis
Finance/funding

Section 2: Schedule
Project schedule
Schedule basis and assumptions

Section 3: Capital Cost Estimate
Capital cost estimate summary
Initial facilities
Deferred/replacement capital
Owner's costs
Facilities cost estimate basis
- Schedule basis
- Sources of cost information
- Escalation basis (if used)
- Direct materials basis
- Direct labor basis
- Construction indirect costs
- Contractor engineering cost basis
- Contractor's fee basis
- Project contingency cost basis

Section 4: Operating Cost Estimate
Summary
Operating cost estimate basis
- Production schedule
- Sources of cost information
- Escalation (if used)
- Operating labor
- Operating supplies
- Repair and maintenance material
- General and administrative cost
- Other costs
- Operating cost contingency

Projected first-year operating costs
- Production factors
- Project timing
- Business factors

Position of this property in world seriatim of industry
- Comparison with other mine candidates for development
- Comparison with existing and potential mine producers

Section 5: Marketing
Summary
Overview of the commodity market
Commodity demand
Commodity mine supply/demand balance
Commodity price
Market analysis
Commodity concentrate market
Commodity finished product market
By-product market
Marketing and business strategy

Section 6: Business Climate and Investment Outlook
Summary
Political outlook
Economic outlook
Investment climate
Microeconomic outlook

Section 7: Economic Analysis
Summary
Basis of analysis
- Capital costs
- Operating costs
- Working capital

Production data
- Revenue
- Escalation
- Tax

Economic results
Sensitivities
    Revenue and cost

Section 8: Financial Strategy
Financial strategy objectives
Financing plan

VOLUME III: TECHNICAL NARRATIVE

The technical narrative, prepared by the project team, describes the technical basis for the project and lists the design considerations and constraints. This is the technical meat of the project. This narrative must convey to the future A/E constructor exactly what it is that is to be built, exactly what it is to do, and precisely how it will accomplish what it is supposed to do after it is built. Nothing can be left out. For this reason, all of the drawings prepared during the final feasibility stage, plus whatever drawings are necessary to convey the message to the A/E, must be in the DBR. The better defined the project is in the DBR, the more accurate the cost will be to the bid estimates, and the fewer exceptions that will have to be negotiated. The following is an example outline of the Volume III contents.

Note: Those items denoted with an asterisk (*) in Sections 2 through 8 require written summaries, technical design bases for the item, design considerations and assumptions that were made for that item, technical system descriptions of the components of that system within a subcategory, environmental control systems, and finally, the equipment list for the subcategory. These items will not be repeatedly listed under each subcategory but still must be documented.

Section 1: Overview
Introduction
    Mine
    Primary crushing
    Concentrating/cleaning
    Waste disposal
    Off-sites
    Waste dump leaching (if applicable)
    Leachate recovery plant (if applicable)
    Plans for future expansion (if applicable)
Design basis
    Production rates
    Start-up scheduled
    Production buildup schedule
    Objectives of each operational function in design

Section 2: Mine and Primary Crushing
General summary
Geology, exploration, resource, and reserve description
Mining*
    Mining plan
Layout of mining facilities
    Description and site conditions
    Design considerations
    Plot plan
    Type of building and construction
Civil works related to mining
    Summary
    Site investigations
    Site preparations
    Miscellaneous civil works
Primary crushing and storage facilities*
Maintenance facilities*
Auxiliary mine buildings*
Utilities
    Summary
    Water supply*
    Fire protection*
    Power*
    Compressed air*
    Fuel oil and lubrication handling facility*
    Communications*
Industrial wastewater collection, treatment, and disposal*
Other environmental control systems
Warehousing and supplies handling*
Discussion of preengineering trade-off studies

Section 3: Ore Conveyance System

*Whatever system is to be used must be fully described. Whether a mine hoisting shaft, a slope conveyor system, overland conveyor system, slurry pipeline, truck or rail system, and so forth, the design basis must be given.**

Section 4: Concentrator or Process Cleaning

General summary (battery limits)
Layout and civil considerations
  Location map
  Plot plan
  Site considerations
Course product storage*
Communitions circuit(s)*
Mineral extraction circuit(s)*
Thickening, filtering, and drying*
Chemical storage, preparation, and distributions*
Sampling and process control*
Utilities and yard facilities*
Yard and plant piping*
Maintenance facility for process plant*
Warehousing and handling of supplies of processing plant*
Auxiliary processing buildings*
Environmental control systems
Any product expansion plans*

Section 5: Waste Disposal and Water Recovery/Treatment

General summary (battery limits)
Waste system pipeline*
Waste disposal area description
  Summary
  Site selection
  Regional topography and geology
Local site geology
  Hydrology (groundwater)
  Meteorology
  Hydrology (surface water)
  Seismicity
Operation of waste disposal area
  Summary
  General features
  Description of proposed deposition system
  Completion plans
  Seepage mitigation plans
  Dust control plans
Waste dam construction
Summary future work
  Waste dam design basis (specify waste compaction if required)
  Waste dam details
  Stability analysis
  Construction materials specification and placement procedures
  Quality control plans
  Staged dam construction sequence
  Equipment list
Reclaim water system*
Seepage water recovery*
Waste utilities and services*
Waste pond area civil works and buildings*
Maintenance of waste facilities*
Other environmental control systems

Section 6: Off-sites

General summary
  Facilities
  Product storage, transport, and shipping
  Freshwater supply
  Electric power supply
  Access roads
  Communications
  Fire protection
  Mine area drainage and waste treatment plant
  Solid waste collection and disposal plans
  Plant security
  Product transport, storage, and shipping*
  Water supply*
  Electrical power supply*
  Access roads*
  Communications system*
  Security facilities
  Mine area drainage treatment plant*
  Solid waste collection and disposal*
Other environmental control systems

Section 7: Dump or Pad Leaching (if applicable)

Summary
Design basis
  Metallurgical process
  Operating schedule
  Projected tonnages and analysis
  Process flow sheet and mass balance

Leaching parameters
- Reagent requirements
- Environmental requirements

Design considerations
- Process design support documents
- Factors considered
- Environmental considerations

System description
General description
- Leach area preparation
- Pregnant leach solution collection
- Leach solution distribution
- Raffinate and pregnant leach
- Solution pumping

Emergency discharge handling
- Electrical system description
- Plant heating, ventilation, and air conditioning
- Control and instrumentation
- Sampling and analytical control
- Environmental safeguard description

Equipment list
- Mechanical process equipment
- Electrical equipment

Solution diversion system*
Waste diversion system*

Section 8: Downstream Extraction
*Any and all downstream extraction processes, such as solvent extraction electrowinning or metals smelting and refining must be fully described.**

VOLUME IV: PROJECT EXECUTION PLAN

The project execution plan, prepared by the project team, defines the real and potential problems in the detailed engineering, procurement, and construction of the project. Furthermore, it goes on to describe the best plans to ensure that these problems are mitigated or at least minimized. The recommended contracting plans are spelled out, as are the plans for engineering and design, procurement, and construction. The following is an example outline of the information that is contained in Volume IV.

Section 1: Introduction
Objectives and purpose:
- Clearly convey to company management how the project will be executed.
- Clearly convey to future A/E and construction contractors how the project will be executed.
- Provide organizational structure and divisional responsibility for the project.
- Complete safe, operable mine/plant, on schedule and within budget.
- Complete mine/plant, meeting all country and government regulations.
- Identify major outstanding issues and action that must be addressed prior to execution.
- Define complete basis to enable project to mobilize and accelerate critical early activities to achieve earliest project completion.

Conclusions

Section 2: Background
Project history
Project general description
- Mine
- Process plant
- Infrastructure/off-sites
- Other facilities

Project milestones
Guidelines to use of country resources

Section 3: Project Environment Controls and Business Environment
- Concerns and interest of country government
- Environmental protection required and permits needed
- Water supply (construction)
- Water supply (operations)
- Waste disposal impoundments
- Roads, electrical power, and communication
- Concentrate transportation
- Mine
- Process plant

Land acquisition
- Country taxes

Labor market
- General
- Market mechanism
- Subcontracting labor supply
- Direct hire

Employment requirements
- Competition for resources during project period

Country economy
Resources of concern
Public relations
Company public relations plan
Project team public relations plan
Contractors and subcontractor's public relations

Section 4: Project Execution Organization

Overall project organization
Engineering, procurement, and construction coordination task force organization
Company project organization
Project executive's organization
Contractor(s) project organization expected
Deputy project director in home country
Deputy project director in country of project
Deputy project director of engineering
Deputy project director of procurement
Deputy project director of construction
Project control director
Finance director
Human resource director
Turnover and replacement organization

Section 5: Schedules and Labor Requirements

Schedules
- Mine engineering
- Mine procurement
- Mine construction
- Process plant engineering
- Process plant procurement
- Process plant construction
- Infrastructure/off-sites engineering
- Infrastructure/off-sites procurement
- Infrastructure/off-sites construction
- Labor distribution to all areas in all phases

Section 6: Project Engineering Execution Basis (assuming appropriation approval)

Objectives
Detailed work plan for contract engineering
- Methodology
- Execution
- Division of work
- Amount of work in the home country
- Amount of work in the foreign project country (if applicable)
- Engineering personnel orientation
- Orientation meetings
- Site orientation and description
- Mine description
- Process plant description
- Infrastructure/off-sites
- Other facilities
- Planned organization
- Project basic documentation and references
- Engineering documents

Standards and criteria to be used
Engineering quality control
Quality assurance achievement expected
- Purpose
- Scope
- Audit methodology expected
- Contractor
- Company

Section 7: Project Procurement Execution Basis

Procurement organization, functions, and responsibilities
- Scope and policy
- Organization
- Responsibilities and functions

Procurement procedures and documentation
- Procedures expected
- Documentation expected
- Country vendor survey information

Available materials in country or nearest available
Available fabrication facilities in country or nearest available
Available subcontracting services in country or nearest available

Section 8: Project Construction Execution Basis

Construction management
- Project organization
- Construction management procedures
- Reporting
- Construction expected
- Construction management interfaces project management

Preappropriation activities (as applicable)

Labor
- Craft supply plan
- Logistics of labor source
- Supervision required
- Safety organization
- Expatriate housing (if applicable)

Construction equipment, tools and consumables in general
- Major equipment
- Procurement sources

Transportation of equipment, tools and consumables
- Maintenance program expected
- Tools
- Consumables
- Fuel
- Aggregate
- Concrete

Construction facilities

VOLUME V: OPERATING PLAN

The operating plan, prepared by the companies' operations department, presents the strategy to minimize the impact for identified potential problems in start-up and continuing operations. While much of the companies' operating philosophy should already have be placed into the design as presented in the final feasibility study, the writers of the operating plan should again emphasize that philosophy. The company's attitude toward mechanization and automation, and what it is willing to pay for it should be expressed. The company's philosophy on maintenance and contracting should be explained up front. Such things as staff recruitment and training will be planned, scheduled, and budgeted. Learning-curve estimates will be applied toward the production buildup, so the estimated production will be met on schedule and project economics will be preserved. The following is an example outline of the information that is contained in Volume V.

Section 1: Introduction

Section 2: Owner

Local organization
- Location
- Description of staff
- Type of management

Management committees (if applicable)

Business interfaces
- Transportation company
- Utility companies
- Adjoining property agreements
- Operating consulting agencies (if applicable)
- Government regulatory, licensing, and permitting agencies

Section 3: Operating Departments

General
- Operating schedule
- Mine camp (if applicable)
- Personnel transportation
  - Staff
  - Hourly
- Food service (if applicable)
- Medical facilities

Functional department and interrelationships
- Mining department
  - Mine operation
  - Geological/surveying
  - Mine engineering
  - Mine and field maintenance
  - Electrical
  - Primary communition
  - Other functional operating group specific to this property
- Concentrator/cleaning plant department
  - Plant operation
  - Metallurgical/process engineering

Maintenance
Electrical
Process control and instrumentation
Central maintenance and fabrication department
Transportation department
Other operating department specific to this property

Section 4: Recruiting

Labor needs
Staff (by function)
Hourly (by function)
Staffing plan buildup estimate
Staff (by function)
Hourly (by function)
Availability of personnel labor pool
Staff (by function)
Hourly (by function)

Section 5: Training

Training objectives
Initial training
Ongoing training
Management and professional development
Training program
Job positions to be trained to match job descriptions
Training organizations
Use of outside institutions
Vendors training
Inside training
Initial training time estimated

Section 6: Start-Up

Basis of start-up philosophy
Who will participate
Who will be in charge of start-up plan
Mining department's plan
Concentrator/process plant's plan
Transportation facilities' plan (if applicable)
Organization for start-up
Mine
Concentrator/process plant
Transportation
Start-up assistance
Operational staff
A/E contractor
Vendors and other consultants (if applicable)
Time and budget estimate for start-up

Section 7: Infrastructure and Support Services (any auxiliary operations that support the main production operations)

Housing (if applicable)
Food service (if applicable)
Personnel transportation (if applicable)
Power system
Power generation
Acquired power
Water system
Tailings/waste disposal system

Section 8: Maintenance

Company philosophy and policies
Maintenance control programs
Work control system
Preventive maintenance
Maintenance planning and scheduling
Maintenance management reports
Job priorities
Downtime analysis philosophy
Backlog reporting system
Numbering control system
Warehouse and inventory control system
Other tasks
Maintenance work requirements
Collection and cataloging of equipment information
Develop equipment
Identification codes
Develop preventive maintenance schedules
Develop maintenance forms
Execute contracts for rebuild and repair components

Section 9: Road Maintenance

Description of road system to be maintained
Responsibilities for specific areas to be maintained

Section 10: Environmental

Company policy and objectives
Present conditions by areas
Source of pollution by areas
Assessment of hazards from preceding sources
Objectives of monitoring program
Monitoring program recommended responsibilities

Internal responsibilities
Consultant responsibilities

Section 11: Administration and Support System

Purchasing department
Controllers department
Financial control
Registration and depreciation of property and materials
Accounting system
Financial reporting
Cost accounting and cost distribution
Capital and expense budgets
System development
Marketing
Marketing philosophy of major products
Marketing philosophy of by-products

Section 12: Communications

Company philosophy
Organization
Central database system
Computer applications support
Ore resource management and information
Operation production modeling and automation
Personnel
Management
Capital/financial
Equipment
Communication systems
Mine systems support (including GPS)
Plant systems support
Office systems support

Section 13: Safety

Company philosophy
Organization
Training
Safety protective equipment policies
Fire protection
Mine fire protection
Plant fire protection
Other surface area protection
Interrelationships between operations and safety/health and first-aid clinics

Section 14: Security

Company philosophy (this section dependent on geographic and political location of operation)
Organization required
Internal organization
Contracted organization
Function of organization
Areas of security concern
Mine
Plant
Transportation of product
Other surface facilities

POTENTIAL APPENDIXES TO THE DESIGN BASIS REPORT

Some additions to the DBR will probably need to be included. The following is a list of likely add-ons:

- Organization charts
- Condensed job descriptions of all jobs
- Maintenance management control system description and forms
- Business control system description and forms
- Environmental monitoring programs details
- Personnel training module details

CONCLUSION

It is this DBR document that is used as the basis for the subsequent engineering design. Not only does it contain the technical data and information decided on by the company during the final feasibility study but also the project execution plan for contracting, building, and constructing the mines of the project. It also contains the operating plan, which will guide the engineers and builder to construct the mine and plant so the operating philosophy of the company can be quickly achieved and maintained. Taking the time and expense of putting together a DBR will go a long way in identifying exactly what the client expects from the EPCM contractor. It is the first step in mitigating massive cost overruns.

For very large projects, or very complex projects, the second step in mitigating large cost overruns is described in Chapter 17 (see the section "Part II: An Engineering Approach to Risk Appraisal and Adjustment"), but it *should be based on the DBR.*

Anyone involved in mining projects is well aware of the drastic project overruns occurring in today's world of project execution. The average project overrun between 1965 and 2001 was documented at approximately 26% (Bullock 2011). But since that time, it has gotten much worse. A worldwide survey by Deloitte (2012) found that *mining project overruns* were increasing exponentially across continents:

- South America, 60%
- North America, 51%
- Australia, 40%
- South Africa, 30%

Although there are many reasons why this is happening, one of the major reasons is that the risk of project execution has not been thoroughly or properly assessed. After completing the DBR, a project-interested team gathers and completes a *project risk appraisal and adjustment study*. When performed properly, it should vent out all of those areas that are most vulnerable to potential project execution risk, and mitigations are planned that will keep the project on schedule and within the "adjusted capital cost." The details of this appraisal are found in Chapter 17.

REFERENCES

Bullock, R. 2011. Accuracy of feasibility study evaluations would improve accountability. *Mining Engineering*, 63(4):78–85.

Deloitte. 2012. *Tracking the Trends 2013: The Top 10 Issues Mining Companies May Face in the Coming Year*. Available from https://cmo.deloitte.com/content/dam/Deloitte/ca/Documents/international-business/ca-en-ib-tracking-the-trends-2013.pdf.

CHAPTER 13

Project Management and Control at the Feasibility and Evaluation Level

Richard L. Bullock

PROJECT CONTROL

All projects at all levels need to be managed and controlled. No matter which phase they are in, projects at the feasibility level need management and control. This chapter is not meant to be an exhaustive dissertation on the subject of project management but rather a simple condensed reminder of the bare necessities that must be in place. In this chapter the following subjects will be briefly covered:

- Cost control through the work breakdown structure
- Time estimates that might be expected in mineral property feasibility and evaluation studies
- The importance of project scheduling
- The project team organization at different levels of the study and on different sizes and complexities of projects

THE WORK BREAKDOWN STRUCTURE

Within each of the three levels of the feasibility study, there are between 50 and 150 major activities. For each major activity, there may be 10–20 elements, or types of work in the study. A large mining company, trying to grow, or even holding its own with its depleting asset, may have as many as 8–12 projects going at any one time, at various levels of study. One can only imagine the complexity of accounting for everyone's time and expenses being charged to the ongoing work for all of these projects without using a numbering system for tracking the various activities. Billing and accounting are not the only reasons for the need of such a system. From the project management point of view, there is an even better reason to organize all the activities into a numbering system. This way, all of the activities can be handled and scheduled on a computer. This is no small task because many of the activities feed information to other activities before they can begin.

The numbering system is the functional breakdown for all of the elements of the project, which can follow each division and subdivision of the element activities to as low a level a breakdown as is desired within the project. This usually goes by the name of *work breakdown structure* (WBS). All major projects use such a system. As defined by the American Association

TABLE 13.1 Typical work breakdown structure numbering system

Numbering Sequence	Project Study Level
1XXXX	Preliminary feasibility study
2XXXX	Intermediate feasibility (or prefeasibility) study, which may include the test mine/bulk sample
3XXXX	Final feasibility study, including the design basis document
4XXXX	Engineering design, including all preconstruction activities
5XXXX	Construction/mine execution phase development
6XXXX	Mine/plant operations

Source: Bullock 2011

of Cost Engineers, it is "a product-oriented family tree division of hardware, software, facilities and other items which organizes, defines and displays all of the work to be performed in accomplishing the project objectives" (Humphreys 2004). All government projects must have a WBS established.

Hustrulid et al. (2006) state, "A WBS is a simple common-sense procedure which systematically reviews the full scope of a project (or study) and breaks it down into logical packages of work. The primary challenge is normally one of perspective. It is imperative that the entire project be visualized as a sum of many parts, any one of which could be designed, scheduled, constructed, and priced as a single mini-project." The WBS system becomes even more valuable if one is looking at many projects either over a period of years or all at one time. The advantage is that if the WBS is written using a generic approach, then all of the projects within one company can follow the same structure, thus ensuring comparable completeness for any given level of study.

The sequences and study levels shown in Table 13.1 will familiarize you with a typical WBS numbering system. Although the table shows a particular WBS used by a large mineral company, there is nothing special about it, except that it was written as a generic WBS that could be used on many mineral projects that were active at one time, and it worked. Each phase of the project is considered as part of the identification. Writing a WBS for each project would have been possible, but then the comparison between all of the projects would have been more difficult, and possibly less accurate.

Notice that screening or scoping projects is not included as a category, for it is only when the project passes some screening activity that it officially becomes a project. Within each project phase, a further breakdown of the numbering sequence identifies major areas of work. An example of how this might be broken down is shown in Table 13.2.

The feasibility study definitions of each activity serve as a checklist, and with time elements applied to each activity and subactivity, they form the basis for building a project schedule. Each project will have unique characteristics, which will require changes to the activities listed, but the general logic and activity identifications should apply to most mineral projects to be evaluated. The more consistent the approach, the more accurate will be the comparison in choosing between the various mineral projects.

The activities within each level of the feasibility study that are shown in the appendixes of Chapter 11 correspond to the activity numbering that relates to the preceding WBS. Using this numbering system, and applying time elements to each activity number, allows one to build a computerized schedule network. In fact, a new schedule is built at the end of each feasibility level, projecting the estimated time for the remaining feasibility studies, and the design, construction, and start-up of the operation.

TABLE 13.2 WBS example

Work Area Numbering Sequence	Areas of Work
XX100	Preparation for reviews and management approval
XX200	Land and water status and mapping
XX300	Geology and predevelopment bulk sampling
XX400	Environmental and socioeconomic work (including permitting)
XX500	Geotechnical, siting studies, and planning
XX600	Agency reconnaissance, government and public relations, and permitting
XX700	Mining, including a test mine
XX800	Mineral processing and metallurgy sampling and testing (upstream)
XX900	Smelting/refining (downstream)
X1000	Surface and ancillary infrastructure facilities
X1100	Personnel
X1200	Labor planning and relations
X1300	Market investigation and planning
X1400	Financial analysis (cost estimates are within the preceding elements)
X1500	Tax studies and analysis
X1600	Planning, budgeting, project accounting, and reporting

Source: Bullock 2011

Feasibility Timing and Schedule

The time it takes between the discovery that there is a resource that may be a potential ore body and the time that the ore body is brought into production can vary significantly. Obviously, if the company has an extremely high-grade ore body, it may take a lot less time to identify enough ore to start mining. Likewise, if the company can make money no matter how the ore body is mined, then the it may not want to spend a lot of time optimizing the mining and milling methods (though this could and has proved to be a mistake in past situations). Conversely, very large, marginal mineral resources may take many years to define and optimize in every aspect of the study to make the resource into a viable reserve.

Technological changes over time may also be a factor that allows the property to finally be developed after many years of study. Of course, the other factor is the ever-changing environmental permitting. Permitting can vary significantly: It can take as little as a year for a small punch coal mine in some Appalachian states, whereas the development of a world-class zinc/copper mineral resource has been stopped for more than 20 years in states such as Wisconsin, even though it can be demonstrated that the underground mine can be built and operated in a manner that would be completely environmentally acceptable anywhere else. Overall, it usually takes from two to six years just to complete the mineral property feasibility evaluation study, and the time frame heavily depends on the state or country in which the application for permits are being made.

The overall time is logically divided between the classical phases of mineral development:

- Preliminary exploration and discovery
- Land acquisition
- Exploration

- Feasibility studies and environmental permitting
- Final engineering
- Development and construction
- Start-up to full production

All of these activities vary greatly in length, depending on their complexity and the location of the deposit.

Nelson and Associates (1979) completed a study of the states of Wisconsin and Minnesota for the U.S. Bureau of Mines on the time frames of many of the preceding elements. Four *major* mining projects by different mining companies were studied. (Three were operating and one was being studied.) It is well known that there is a very strict system of environmental permitting for this area for new mine development. So it is not at all surprising that the Nelson study found environmental permitting required a long period of time. In fact, their findings concerning permitting time for a metal mine in Wisconsin was nearly 100% optimistic, even if the project had been built. Table 13.3 is a summary of the elements and time periods.

TABLE 13.3 Timetable summary for four Minnesota and Wisconsin properties

Elements	Years
Environmental monitoring	2.18
Environmental impact report evaluation	4.85
State permits	5.25
Local permits	3.25
Environmental impact report preparation	1.55
Wisconsin Department of Natural Resources (WDNR) for the Wisconsin Environmental Protection Agency	3.90
Federal environmental impact statement	3.05
Master hearings	1.16
WDNR permits	3.05

In reality, some of these activities can go on simultaneously or overlap. But even the most optimistic schedule to receive permits in Wisconsin for a metal mine would have been 10–12 years. In fact, more than 25 years passed, with three different companies applying, and a permit was never issued to open the mine. This was despite every reasonable requirement being met and no detrimental environmental degradation expected to occur for developing an underground mine. Thus the world-class Crandon mineral resource was never developed.

In addition, in the same report, other time estimates were created from these four major projects for mine development items that are not necessarily restricted to Wisconsin or Minnesota. Table 13.4 lists these items and projected time frames.

Assuming that there would be enough overlap of activities to reduce the time estimate by 25%, this would still leave 19.33 years to take a project from preliminary exploration to development, not including environmental permitting time. If one assumes that the ore body has already been drilled and the reserve documented, then one might consider only the time required for feasibility studies, final engineering, and construction, and this would equal only *13.85 years* (without the environmental permitting time).

TABLE 13.4 Mine development and actual time frames for four projects

Development Items	Years
Preexploration	2.80
Land acquisition	4.30
Exploration	4.83
Preliminary engineering (feasibility studies)	8.00
Final engineering	2.26*
Construction (schedule and execution)	3.59*
Total	25.78

*Based on three actual projects.

The obvious time estimate that looks much too long is eight years for preengineering or feasibility studies. However, remember that this information came from four different major projects, which were completed by four different major mining companies. Could this be true that it actually takes eight years to complete a mineral property feasibility study? Yes, it certainly can take that long for many projects, but not necessarily all of them. Table 13.5 shows the average times expected to complete project evaluations on 10 small-to-large projects. The information was compiled by a minerals company that wished to remain anonymous.

TABLE 13.5 Average project evaluation time frames

Project Evaluation Phase	Time Duration
Preliminary feasibility study	7.5 months (156 working days)
Intermediate feasibility (or prefeasibility) study	2 years, 8 months (666 working days)
Final feasibility study	2 years, 10 months (709 working days)
Total	6 years, 1.5 months (1,531 working days)

Source: Bullock 2011

This schedule, however, takes into account having to do absolutely everything that could be needed to be done, overlapping all activities possible, but bringing everything to a very high level of engineering standards. (Remember from Chapter 11 that 6%–8%, 15%–20%, and 20%–30% should be completed for the three levels of total project engineering, respectively.) Depending on the size, grade, location, ownership of the project, and how much financing the owner would need, these time estimates can radically change.

Metal or Industrial Minerals in the United States

To give some simple approximate guidelines, the following requirements might be considered, based on the author's personal experience and observations.

- Preproduction of two years or less is required if the following conditions are evident:
  - There are near-optimum physical conditions with a fairly shallow, concentrated ore body in competent ground.
  - The involved government levels are very "friendly to mining."
  - The mineral is not complex to mine or mill.
  - The mine is started as a small- to medium-size production.

    - No large-scale transportation facilities are needed.
    - No large infrastructure is needed.
    - The mine is in or near an existing mining district.
    - Market conditions are such that there is an assured market for this commodity.
    - Only short-term or no financing is needed by the developing company.
- Preproduction of two to five years is required for most of the small- to medium-size mines that are not bonanzas.

  Note: There is one fallacy to this measure in that a mine can start production, but if it is still in development, then the mine will carry that classification for tax purposes.
- Preproduction of five to seven years is usually required for large open pits, which need a lot of stripping, or large underground caving operations that will take a lot of development. Any very deep underground mine can fall into this category.
- Preproduction of seven years or more is required if there are difficult physical conditions, large technologic or environmental problems, and complex economic problems. Very deep South African mines take eight years or more just to develop.

Coal Mines in the United States

In the 1970s, punch coal mines in West Virginia and eastern Kentucky could take only 60–90 days to develop. However, development time for a deep mine with multiple seams, with the upper seams gone, could take two to three years.

Large surface mines in the West could take two to four years to set up for production buildup. Smaller surface mines in the West took one to two years to start production during the 1970s–1990s, but now the environmental permitting will take twice this long. In contrast to these depressing time estimates, Cusworth (1993) presented the estimates shown in Table 13.6 for Australia.

TABLE 13.6 Time estimates for 1993 Australian studies

Type of Study	Months
Scoping Study	
Establish data and basis of study	2
Study core period	3–4
Review and evaluation	2–3
Total	7–9
Prefeasibility Study	
Establish data and basis of study	2–3
Study core period	4–6
Review and evaluation	3–4
Total	9–13
Feasibility Study	
Establish data and basis of study	3–4
Study core period	6–9
Review and evaluation	3–4
Total	12–17

Source: Cusworth 1993

One could conclude from Cusworth that all projects in Australia in 1993 varied only from a total of 28 months to 39 months. Unfortunately, there are no details given to see what is actually covered during these periods. One would have to assume that much of the difference between the United States and Australia is the governmental red tape of the U.S. environmental agencies. But two other factors may play a significant role: (1) There were probably more virgin deposits being discovered in 1993 in Australia than the United States, which might have been of a higher grade; and (2) Australians tend to turn everything over to contractors, which can move things along faster with their larger staffs. Things have changed, however, even in Australia. The massive Sino Iron project took 12 years from the beginning of the prefeasibility study to the project start-up and was a cost and schedule "blowout" of more than $1 billion extra over estimates (Department of Mines and Petroleum and Department of State Development 2009; Klinger 2011; MacKinnon 2012; Zhu 2010; Jun 2014).

Scheduling of each element of the project must be done from the beginning. This is one of the important reasons to document in advance all of the activities of each phase of the levels of feasibility. Then estimated labor hours must be assigned to each of these project activities and subactivities. And the precedent level of each activity in relationship to all the related activities must be determined. Setting up and maintaining the schedule of even a medium-size project is a major task.

ORGANIZING THE PROJECT TEAM

A project team can be organized in many ways, depending on

- Phase or level of the feasibility study,
- Size and complexity of the project,
- Location of the project, and
- Size and experience levels within the parent company.

First, consider what talent is needed either part time or full time on a project feasibility team. Certainly, there must be people on the team who understand and can perform project management, costing, and scheduling for the project. But also needed is every technical discipline that must be considered by the activities of the evaluation. This means the fields of geology, geostatistics, mining, metallurgy, environmental consideration, hydrology, geomechanics, civil infrastructure, and economic evaluation must be represented. But there must also be people who can furnish legal, land, water, public relations, marketing, tax, and financial information.

So how do we get all of this talent assigned to even one project, let alone eight or more projects going at once? Depending on the size of the parent company, the company must either build the organization within the company structure or depend on the consulting industry to supply the needed talent.

Taking the in-house approach, the company must form a project management and development organization, where staff will be assigned to the project management nucleus of each project, and then technical specialists will be assigned from a technical support organization on an as-needed basis to perform the hundreds of technical activities that will be required. This is a typical functional/matrix-type organization. By approaching the problem in this fashion, and using proper staffing scheduling, many projects can be handled at one time. This approach works well on small- to medium-size projects up through the intermediate feasibility (or prefeasibility) phase of study. For large or megaprojects, it would probably only work through the preliminary feasibility phase.

Taking the consultant approach, the company should still form a project management organization to manage each project, but then contract either one large multidiscipline consultant organization or individual discipline consultants to perform the various technical tasks of each project. The consultant approach is not discussed in detail, because the architecture/engineering consultant basically supplies all of the organization. The project company, however, must always maintain a presence in the contracting organization to ensure that the best interest of the company is always primary. The size of the presence would probably represent about 10% of the size of the contracting staff.

It is difficult to generalize, but from the author's experience, if the company is running several small- to medium-size projects, and these projects are in the preliminary or even the intermediate feasibility phases, then it is easier to organize a core group, consisting of the project manager, project cost coordinator, and scheduling coordinator. To this then is added the individuals by assignment from the technical organization, who are experts in their particular fields. The technical organization is their functional home, but they will be assigned temporary duty to the individual projects. By assigning work in this manner, each discipline can probably handle several projects at one time with proper scheduling. This has worked very well when handling multiple projects within one mine evaluation and development group, when the project studies are in the early stages. It can usually work well up through the intermediate feasibility phase, particularly if the projects are in the same country where the home office is located and where the mine evaluation and development groups are located. However, when the projects are overseas, or if it is a rather large project, there is so much fieldwork required during the intermediate and final feasibility studies that it is usually best to relocate a dedicated project team to a location near the site.

Organizational Role

The responsibility for feasibility studies within a company should be assigned to the feasibility and planning group of the engineering or project management and development department group within that company. The role of this group in conducting this work is to perform evaluation studies of mineral deposits and mineral processing facilities for projects discovered, acquired, or located in all countries in which the minerals company has an entire or part ownership interest. This group will be responsible for the project management and technical applications used in taking the project from the end of exploration through to operational development.

Assignment of evaluation studies of all types to a central headquarters organization study team has the following objectives and advantages:

- Ensures that all projects are treated uniformly and objectively, thus making certain that all projects have an equivalent economic comparison
- Ensures a more complete and thorough study through the use of a larger central staff
- Ensures centralized project planning and scheduling
- Provides for a larger experience base when all candidate projects are considered

Evaluation studies of the various levels or possibly even acquisition or expansion and modification studies should be handled by the central project management group. (The three steps, definition, and extent of a feasibility study—preliminary feasibility, intermediate feasibility, and final feasibility—are discussed in Chapter 11.) Normally a mineral property or process that is internally discovered or developed will be initially evaluated by a screening study when it is transferred from the exploration group to the project development group. This is followed by a preliminary feasibility study. A similar procedure will be followed for significant expansion or modification of an existing operation. A project that passes the preliminary stage and receives management approval would then be the subject of an intermediate feasibility (or prefeasibility) study.

Acquisition studies are a separate problem and would be performed as requested or required by other organizational components of the company, and then typically a task force organization might be utilized. In these cases, operational organization may join with the project management group, filling in those specialty areas that they know best, such as operational analysis, equipment maintenance, human resources, legal issues, taxes, and so forth.

Early evaluation and feasibility studies would be carried out by the matrix study team within the feasibility and planning division. At such time as the magnitude and importance of a project justifies it, a separate project team organization would be established. This would normally occur at the end of either the preliminary or intermediate feasibility study phase when the cost, duration, level of staffing, or overall importance indicates that a separate project should be created. At this point, the team would also be transferred to a location close to the site of that project but still function under the project development group. In a situation where the company had an operation in that country, then the group would have to function under the management policies of the country manager.

Often, a company's overseas staff is composed solely of exploration personnel. Thus, in these cases, there is no evaluation or development staff located in that country. But in this case, a project management and technical staff will have to be put into place using company staff, consulting staff, and local staff who can be found. It is also worth noting that in some countries in the world, there is little or no technical mineral engineering base on which to draw for a staff, and one must be literally imported. In such cases, company management must take these factors into account.

Organizational Concepts in Evaluation Studies

The project development department is organized as a matrix with support in the various technical specialties provided by a central engineering group, as well as other corporate components (legal, public relations, environmental, etc.) and outside contractors. In this setting, evaluation studies for a particular property or project would be carried out by a small study team group within the feasibility and planning division, drawing on other corporate components for contributions in the individual study areas.

Because preliminary evaluation studies rarely require significant physical work at the project site but do require effective staff and engineering work, the principal location of the study team should in most cases remain in the company headquarters. Such site or country work as may be necessary can be conducted by extended visits or by contractors or local staff in each area or country where such local staff with the required skills is presently available. In this context, the guiding principle should be to not add permanent personnel in each country for evaluation studies unless the requirements for physical work at the site or other special circumstances require it. For this reason, it is considered likely that the initial personnel assigned to a specific project location would probably be geologic personnel, who are handling the second phase of sampling for the next part of the feasibility study. While they are now assigned to this particular project in the project development group, their functional home, to which they may eventually return, is still the geology division.

When on-site, project or country site-specific personnel are required, either for purposes of development geology or evaluation studies. These personnel will be appointed to their respective positions by their respective divisions. Initially, when development geology personnel are required, they would be assigned by that group and would use the existing exploration structure for support and coordination, which may include such items as office space, pay, accounting, and other administrative support, as well as logistical support. Also, any engineering or other personnel required for evaluation studies would be assigned in a similar manner. This could be a feasibility study team or a specific project management group.

At the conclusion of the preliminary or intermediate feasibility study phases and when it is determined that a separate project organization should be established, an independent project team would be organized. The project's functional and technical control would continue to be supervised by the project development department, but from a personnel relations point of view, the project staff would work through the local organization as determined by management and coordination with the affiliate or country manager. In such cases, particularly for large multisite projects, the project might very well have a project executive who also serves on the local country management committee or staff.

The Study Team Approach to Evaluation Studies

The matrix form of project management has been found to work very well on small- to medium-size projects during the first two phases of a feasibility study. While conducting a preliminary or intermediate feasibility study for a property, the study team that is formed should approach some of the major activities and use contributions by other company components in a particular manner. This is accomplished by usually three or four individuals dedicated to the project and the rest of the team, depending on their professional expertise, assigned on a temporary basis from other departments within the company. While small mining companies may not have the support organization that is assumed here, in these cases they will have to rely on consulting organizations to furnish the needed expertise. These projects may be located domestically, but more often than not in today's environment they will probably be overseas. When these projects are located overseas, the other companies' country organizations must play a significant role in carrying out the company policies. The matrix approach to project management is desired because at this point, it is unknown if the project will be proven to be economically or legally viable, and it may soon be canceled or put on the shelf. Building up a large engineering staff that may soon have to be laid off does not bode well for a company's reputation. This type of matrix organization was successfully used by this author in managing seven projects in both the preliminary and intermediate feasibility (or prefeasibility) stages. Two other larger projects were ongoing with full staffs in the intermediate feasibility stage.

The Transfer to Project Development from Exploration

At the outset it should be planned that there will be an active liaison between the international exploration group, the company geology department, and the feasibility and corporate planning group. This is to ensure orderly planning by project management, as well as the development and transfer of the necessary data for evaluation, and to ensure that projects scheduled for evaluation meet minimum criteria. It is expected that this liaison will be developed through periodic reviews and reports by the geologic group and occasional site visits by the project management group.

It is also anticipated that the decision concerning the timing of the transfer of a project to the project development department will be a joint decision with the exploration group and based on the technical evidence as well as the business objectives. The responsible country affiliate, if there is one, should continue to maintain mineral and land rights throughout the evaluation period and probably, for that matter, throughout the life of the property. Additions or modifications to mineral, land, or water rights during the period the property is under the project development department should be planned and managed by the feasibility and planning group and implemented through the land staff personnel attached to the project development group and the country affiliate.

It is anticipated that prior to making a decision to transfer a property, a scoping-level economic screening study will have been prepared by the feasibility and planning group as an aid in the decision-making process. This study will take into account all the major economic considerations for a property typical of that under study, including business risk assessment, rate of return, and any additional economic guidelines prescribed for that country.

The feasibility and planning group, in conjunction with other involved departments, should coordinate the transfer of a property from exploration to project development and should develop and implement such plans as will be required to

- Transfer all required technical and financial data, including
  - All mineral deposit information,
  - Property location and access,
  - Area surface features,
  - Exploration activities completed and planned for this property,
  - Geology (regional, local, and deposit),
  - Potential ore reserves,
  - The company's land and water position,
  - Property and water ownership and royalties,
  - Property history,
  - Special studies performed,
  - All environmental or social problems noted at that point, and
  - General data;
- Arrange for proper security or maintenance of any physical facilities;
- Ensure that all regulatory requirements are fulfilled and maintained; and
- Ensure that mineral, land, and water rights are maintained.

The Study Team

Once a property is transferred from the exploration group to the project development group, a study team will be formed in the feasibility study group to carry out the preliminary feasibility study. Depending on the size of the expected project, appropriate personnel in the feasibility section, on a full- or part-time basis, and including a study leader, a project management specialist, and an accountant, will comprise the study team.

Table 13.7 shows the various elements that need to be considered in *support* of evaluation studies and the sources of information and assistance in each study area that are expected to be available in most medium- to large-size mineral company organizations. In each case, the sources of information are listed in their approximate order or sequence of importance.

After forming a study team, the team, in cooperation and with contribution support from other company components will prepare a study work plan, schedule, and budget. This plan

TABLE 13.7 Elements in evaluations and staff support needed

Study Element or Area	Departmental Sources of Information, Assistance, and Support*
Land, water, minerals, and rights	Land staff, country affiliate, legal, local government
Taxation and accounting	Corporate planning, tax counsel, legal, local government
Government regulations: environmental, occupational, social, business	Environmental affairs, legal, local government, training and safety, study team, corporate planning, other producers
Outside share: joint venture, royalty	Corporate planning, country affiliate, exploration, legal, local government
Markets and prices	Corporate planning
Labor and compensation	Employee relations, exploration, contractor, study team, other producers
Transportation	Study team, corporate planning, company transportation, contractor, other producers
Communications	Study team, company telecommunications, engineering
Mineral resource: geology, cut-off, reserves	Exploration, geology, study team, engineering/geology economic study team, geology/geostatistics
Public affairs	Public relations
Mining methods	Study team, engineering, or contractor
Mine design	Study team, engineering, or contractor
Mineralogy	Geology
Process development	Study team, engineering, or contractor
Production rate	Study team, engineering, or contractor
Utilities	Study team, engineering, other producers
Infrastructure	Study team, exploration, competitors, or contractor
Operating supplies	Study team, engineering, or contractor
Capital plant and equipment	Study team, engineering, contractor, or other producers
Engineering and construction services	Study team, engineering, contractor, or other producers
Estimation of costs	Study team, engineering, contractor, or other producers
Economic modeling	Study team, engineering, corporate planning
Escalation factors	Corporate planning
Scheduling	Study team, corporate planning
Return and cost of capital rate or other country-specific economic guidelines	Corporate planning

*In all cases, where the expertise does not exist within the company, then a consultant firm will need to be contracted. (Allow time to write the request for proposal [RFP] and allow time for bidding and bid analysis, and to award the contract and mobilize the contractor. Each contract could take many weeks.)

will take into account any special features that pertain to the property in question, the need for further resource definition by the geology department staff and the need for assistance by engineering staff assigned to that particular country affiliate. Special attention will be given to identifying any necessary programs for additional land, mineral, or water right acquisitions.

After the study has begun, the study team, which includes personnel from other organizational components, will gather necessary site- and country-specific data by means of

- Site reconnaissance visits (usually short duration),
- Extended working task visits (the length of time it takes to complete the task),
- Locally available contract engineering and other contractors,
- Local legal counsel, and
- Local staff.

Geology Department

The geology department should plan to carry out its responsibilities and obligations through the use of locally available personnel and through the transfer of expatriate personnel to the project. Work assignments and personnel administration should come from the geology department to the study team. The geology department will have functional responsibility for the technical aspects of all drilling and sampling, geologic information, minerals database, and geostatistical and reserve calculations performed during the evaluation and feasibility phases.

Legal Department

Legal counsel in the company's legal department will furnish the study team with legal support. In addition to providing legal advice on a direct basis, other sources of this support could include utilization of company affiliate attorneys attached to the local organization and developing outside legal counsel in each country that can provide specific advice on individual questions. All legal advice from outside counsel and from company affiliate attorneys should be coordinated by legal counsel in the company office. Areas of consultation could include regulation interpretations of all types; questions of law, land, mineral and water rights; certain tax questions; and business practices. The primary source of tax information should be a tax specialist within the company or a hired tax consultant. The legal department should also assist directly in the preparation and translation of contracts for work to be done in each country.

Environmental Affairs Department

The environmental affairs department will advise the study team on the company's environmental policies and requirements and on the interpretation of country-specific environmental regulations. Other activities of the environmental affairs group, which will provide support to the study team, include

- Forecasting environmental regulatory trends, analyzing proposed or developing regulations, and predicting their effects on company projects and plans;
- Jointly with counsel, providing summaries of country-specific environmental regulations and requirements;
- Advising project personnel about technical environmental considerations for project development and cost-effective environmental conservation methods;

- Coordinating permitting activities and assisting in negotiations for major permits; and
- Advising project development staff about selection and supervision of contractors and consultants for environmental impact studies and for designs of environmental protection systems.

Corporate Planning Department

The corporate planning department should be expected to provide the following data for evaluation studies as it applies to each country:

- Tax estimates and calculations (unless the study teams has been advised to use a local tax consultant)
- Accounting (depreciation and depletion)
- Prices, markets, smelter schedules
- Product transportation cost estimates
- Forecasts for escalation if a constant dollar analysis is not used
- Distributed proceeds share (royalty and joint ventures)
- Guideline return rates as they are affected by business climate
- Land political risk
- Cooperation in the formulation of economic models

Engineering Department

The engineering department will provide extensive technical support in a large number of technical engineering disciplines and specialties. In addition, individual engineer personnel will be available for assignment on a matrix basis to augment the study team. In all probability, personnel experienced in the techniques of project management will also come from the engineering department and be assigned to the study teams.

Employee Relations, Training, and Safety Departments

It is expected that the employee relations department and the training and safety departments located in the company headquarters should be able to provide consultation and advice on occupational regulation and compensation and labor-related issues if it is a domestic project. However, if it is a foreign project, most of the country-specific data in this area will be developed by the study team, unless there is a country affiliate office where such information can be obtained.

Public Affairs Department

The public affairs person responsible for each country, assuming there is one, should be expected to provide contribution and advice as required. In the likely event that this is the company's first time in that country, local socioeconomic consultants can usually supply the needed advice.

Special Circumstances in Joint Ventures

In joint ventures, it is anticipated that liaison will be necessary with joint venture participants during the course of any evaluation studies. It is assumed that the operator of the joint venture will, in all cases, be responsible for preparing the initial evaluation studies and furnishing these with development recommendations to the other participants. If so, it follows that the participants will seek a liaison during the evaluation study to ensure that the results of the study meet everyone's needs. After all, most evaluation studies are of significant duration and cost, and it would be wise to ensure that the basis of the study is mutually acceptable to avoid unnecessary disagreement. The appropriate forum for such liaison is to set up a joint management committee, where each participant can have the needed representation to ensure that the project being studied will be engineered to their satisfaction.

The objectives of the joint management committee are to view the situation from the viewpoints of both operator and nonoperator. Ideally, the following would be the objectives sought:

- Acquire such technical data as are necessary to independently appraise the results of the study.
- Assess the completeness of the physical data and data pertaining to all other aspects of the study.
- Seek a consensus on the design basis.
- Attempt to ensure that sound methods and procedures are employed in the evaluation study.
- Provide timely advice by each of the companies' designated joint venture representatives in regard to the preceding objectives.

Significant Expansion and/or Modification by an Operating Unit

To ensure that all types of projects are treated uniformly and objectively, major expansions and modification by an operating unit should require similar economic evaluations as a property developed by exploration or acquired by acquisition. This should include modifications to existing operations, such as major operating changes as well as new products produced. This type of project will be different as much of the initial data will be developed by the local operating unit. To take advantage of this knowledge, it is anticipated that the project development study team will include local staff temporarily assigned to it and the study team will be provided assistance by the local unit. The project management criteria applicable to other types of projects should also apply to these projects.

REFERENCES

Bullock, R.L. 2011. Mineral property feasibility studies. In *SME Mining Engineering Handbook*, 3rd ed. Edited by P. Darling. Littleton, CO: SME. pp. 227–261.

Cusworth, N. 1993. Predevelopment expenditure. In *Cost Estimation Handbook for the Australian Mining Industry*. Edited by M. Noakes and T. Lanz. Victoria, Australia: Australasian Institute of Mining and Metallurgy. pp. 252–259.

Department of Mines and Petroleum and Department of State Development. 2009. Full steam ahead for Sino Iron. *Prospect: Western Australia's International Resources Development Magazine*, (March-May): 2–4. www.dmp.wa.gov.au/Documents/About-Us-Careers/March_May_2009.pdf (accessed November 2016).

Humphreys, K.K., ed. 2004. *Project and Cost Engineer's Handbook*, 4th ed. Boca Raton, FL: CRC Press. p. 314.

Hustrulid, W., Kuchta, M., and Martin, R.K. 2006. *Open Pit Mine Planning and Design*, 3rd ed. Vol. 1. Rotterdam, The Netherlands: CRC Press/Balkema. pp. 19–20.

Jun, P. 2014. Can Sino Iron dig out of its investment hole? *Caixin Online*, Jan. 16. http://english.caixin.com/2014-01-16/100629880.html?p2 (accessed February 2017).

Klinger, P. 2011. Iron ore project faces $900m blowout. *The West Australian,* July 16. https://au.news.yahoo.com/thewest/wa/a/9858958/iron-ore-project-faces-us900m-blowout/#page1 (accessed November 2016).
MacKinnon, M. 2012. CITIC's Australia Sino Iron mine cost surges to $8bn-report. Edited by E. Lane. *Reuters Africa,* July 21. http://af.reuters.com/article/commoditiesNews/idAFB2643320120721 (accessed November 2016).
Nelson and Associates. 1979. *Report on Task 8, Time and Cost Estimates.* Iron River, MI: Ecological Research Services.
Zhu, C. 2010. Sino-Australian ore mine digs a deeper hole. *Caixin Online,* April 21. http://english.caixin.com/2010-04-21/100137639.html (accessed November 2016).

CHAPTER 14

Introduction to Cost Estimating for Feasibility Projects

Richard L. Bullock

One of the primary tasks that has to be done when preparing an initial property feasibility study is to estimate the capital and operating costs to bring the property to full production. This act of preparing a cost estimate is the one common element that connects all of the various types of engineering evaluations given in Chapter 1. Mine/plant costs can be estimated in many ways, and some are more accurate than others. Some cost estimate methods require much more detailed engineering than others, and it takes a lot more time and money to prepare the estimate. Obviously, if you are in the preliminary feasibility phase, you would not be expected to spend a great amount of time and money preparing the cost estimate, compared to the final feasibility phase. This chapter will cover some of the cost estimating methods and tools and the accuracies expected for the level of engineering performed for each phase of the evaluation study. There are good discussions of mine cost estimating methods supplied by InfoMine USA in both the *SME Mining Engineering Handbook* (Darling 2011) and SME's *Underground Mining Methods: Engineering Fundamentals and International Case Studies* (Hustrulid and Bullock 2001). It is not within the scope of this handbook to describe in detail the cost estimating techniques for the dozen or so mining methods and the half a dozen or so processing methods for the intermediate feasibility (or prefeasibility) study or final feasibility study in a lot of detail. These capital cost estimates depend so much on the 15%–30% of the total engineering that goes into the design, and a large part will depend on material take-offs of the many drawings that are produced in the *intermediate and final feasibility phases* of the process. However, general information is presented on benchmarking of the cost that is developed.

NATURAL ELEMENTS THAT AFFECT COST

Every item that becomes a part of the engineering design affects the estimated cost—some items more than others. The characteristics of the natural resource will usually influence the cost estimate the most. Generally, the depth of the resource is most influential at open pit mines, where it will strongly affect the stripping ratio and thereby affect the preproduction stripping capital. It will further influence the haulage cost for the entire mining life.

For underground mines, the deeper the resource to be mined, the deeper the shaft or decline that must be driven for production, workers, and materials transportation. This results in greater capital cost for these facilities. But the production cost for the entire life of the mine will also be proportional to the depth. Amounts of groundwater that may need to be pumped from the mined area also affect cost. Conversely, the lack of water will be a concern. Both situations can run up the cost of producing the mineral product from the area.

The geometry of the deposit also affects cost. Thin or narrow resources usually result in a less productive mining system and higher operating cost per ton, and usually, for the same tonnage mined, a higher capital cost per ton of daily production. Complex mineralogy may lead to complex metallurgy and a costlier system to extract the material being produced.

Location of the resource is also a strong cost indicator: the more remote the location, the more infrastructure required. Likewise, the further away from populated areas, the greater the amount of training that will be required and the greater will be the cost to transport trained personnel in and out of the mining camp. Power and other utilities will typically cost more in remote areas. Extreme weather conditions usually run up the construction and operating costs or cause production to shut down for certain periods of time.

Although the preceeding items are parameters of the nature of the mineral resource, all of them can and should be determinable in the very early stages of the mineral property evaluation and feasibility study. The point is, how good a job must we do at measuring the effects of these parameters while determining how well we can engineer a mitigation to each of the problems and, indeed, how much will each mitigation cost?

Two other areas of concern are related to location of the resource and can cause variations in the cost estimate: (1) environmental laws and regulations that must be met within the design of the future property, and (2) the socioeconomic improvements that must be met in the local communities. The cost estimator must be fully aware of the steps that will be taken to mitigate the real or perceived environmental and socioeconomic problems, just as much as he or she must be aware of every other aspect of the plant design criteria.

TYPES OF COST ESTIMATE AND WHEN TO USE THEM

There is a basic rule of cost estimating: the more engineering hours that are devoted to the planning and design of the project, the more accurate it is, and so should be the estimate of the capital and operating costs. Thus it goes without saying, a "screening" or "first pass" study cost estimate, based on very preliminary concepts of what the project might evolve into, may find the probable error (PE) somewhere between ±30% and 50%. That is because probably no real engineering specific to the particular property with the unique conditions of the deposit have been considered in the cost estimate. There is one notable exception to this error statistic: If a company has already built one or two mines in the same district, with the same mining conditions expected in the next mine development, then the screening cost estimates based on previous experience and engineering performance for the previous project may be accurate. These situations are the exception in mine feasibility work and cost estimation.

Before further discussion, the difference between *accuracy* and *contingency* needs to be reviewed. *Accuracy* of individual estimates is a function of the quantity and quality of data used, the estimating techniques employed, the experience and skill of personnel making the estimate, and the amount of engineering that has been applied to the project plan. Overall accuracy of project costs relates to the accuracies of the individual estimated elements, plus the contingency selected.

The *contingency* can best be described as "an allowance for unforeseen expenditures." It is a function of the percentage of the total number of items that are explicit estimates and the quality of the individual estimates. Selection of a contingency is a matter of judgment and experience with similar estimates. The contingency is a means for adjusting the overall project estimate to provide an equal chance of overrun and underrun. AACE International (formerly the American Association of Cost Engineers) defines *contingency* as follows:

> *An amount added to an estimate to allow for items, conditions, or events for which the state, occurrence, and/or effect is uncertain and that experience shows will likely result, in aggregate, in additional costs* (AACE International 2005).

Following is a more recent and somewhat expanded definition:

> *A specific provision added to a base estimate to cover indefinable items that have historically, by actual experience, been required but cannot be specifically identified in advance. The undefined elements are fully expected to occur in the forecast value of cost and/or schedule, but the exact nature and timing of their occurrence is indeterminate. It is not intended to cover scope changes or project exclusions* (Hickson and Owen 2015).

Contingency and accuracy are independent of each other. It makes sense, however, to require smaller contingencies in conjunction with higher levels of accuracy (i.e., lower PEs) for all capital and operating cost estimates. Obviously, the accuracy of the estimate depends on the amount of engineering that has been completed on the project.

Based on the phased approach to performing mineral property evaluations, as described in Chapter 11, a three-phased approach is suggested. Different companies have different names for these phases of evaluation, and some companies choose to perform the evaluation in two phases. The nomenclature found here follows that which is described in Chapter 11 and is considered a formalized approach to improving the cost estimate by increasing the amount of engineering applied in each phase, while at the same time minimizing the investment at each level. The various industry codes (NI 43-101, JORC, etc.) do require a phased approach, but the U.S. Security and Exchange Commission's (SEC's) Industry Guide 7 currently does not but is also in the process of being replaced (JORC 2012; SEC 1992). Changes to Industry Guide 7 have been proposed, but no final action has been taken. Recently, *The SME Guide for Reporting Exploration Information, Mineral Resources, and Mineral Reserves* (SME 2017) has not only advocated the three-phased approach but has recommended some minimum engineering standards. The objective of this method is to identify the value of the mineral property with the planned development and operation. The value determined may not be sufficient to warrant further study or investment, or it may result in a worthy project to move forward to the next step.

A review of the PE and the amount of contingency recommended in Chapter 11 is summarized in the discussions that follow.

Preliminary Feasibility Study

The preliminary feasibility study is based primarily on site information supplied by exploration. Ideally, several alternatives will be examined for screening purposes. Costs and expenditure schedules may be based on experience, historical costs, published cost models and curves, or regression analysis formulas from historical data. Major costs may be based on telephone quotes from suppliers. Usually, no field work (other than a site visit) or metallurgical testing is conducted, unless it is recognized that there is a definite metallurgical problem with the resource. Results of this study will be adequate for comparing several alternatives and/or rejecting the project. Another objective of this study is to plan and estimate costs for the intermediate feasibility (or prefeasibility) study and any additional predevelopment exploration work, if the next phase is warranted.

Preliminary Cost Estimates

For the preliminary feasibility study, it is estimated that between 6% and 8% of the total engineering and other studies for the project should have been completed. This first assumes that there has been a definite resource identified and quantified. The cost estimate comes from engineering estimates of daily production output of a specific grade of product; site location characteristics; probable utilities required; probable infrastructure needed; applicable mining method, with development method and its related cost; applicable processing method; general sketches of all of the above; preliminary equipment list, with appropriate sizing; approximate staff and labor requirements for each operating function; approximate quantity of utilities needed based on probable motor sizes or quantities of substance moved (e.g., air and water); unit cost of utilities; approximate number and size of buildings, probable waste disposal system; approximate shipping distances; and the expected direct environmental volumes of material to be handled and bonding cost. All of this is based on benchmarked historical costs and on information supplied primarily from the exploration effort. Next, the administration cost with indirect burden is added, and this forms the basis of the preliminary cost estimate. If you have followed the detailed analysis of all the individual elements listed in Appendix 11A, then the cost estimate should have a PE of ±20%–30%, which equates roughly to the 8%–6%, respectively, of the engineering and other studies that need to be done.

The contingency that should be added to the cost estimate is required to provide equal chance of overrun or underrun. These errors occur primarily from lack of advanced engineering design where many other elements of cost would be recognized and added up. Thus, contingencies are not intended to cover extra costs when these costs occur because of the scope changes in the project, but rather they are for omissions in cost that the engineering staff and cost estimating staff could not anticipate within the present scope of work. For a mineral property preliminary feasibility study, the contingency that should be added to the capital cost is 20%–30%.

In the process of developing the preliminary feasibility study, more than one mining method and/or processing method may have been considered, but it should not be an exhaustive trade-off study. Here is where an experienced mining and metallurgical engineer's judgments are needed. It is best to take the alternative evaluation at least far enough that a cost per ton for each method can be developed. If some of the alternative methods do result in different recoveries and/or grades of products, then the preliminary study should examine the likely alternatives through the calculation of a rate of return on the investment. Next, the best alternative for the intermediate feasibility study base case should be presented, but with a more in-depth review of the other possibilities that will be completed at that time.

Following are three general types of cost estimating methods used in preliminary cost estimates:

1. Methods based on historical experience in planning, designing, and building other mine/plant facilities and benchmarking detailed data from the projects, to make estimated projection on future projects
2. Methods based on a physical measurement of some element of the proposed project design (limited in application)
3. Unit of capacity cost method (very limited in application)

HISTORICAL METHODS USED IN DEVELOPING BENCHMARK COST ESTIMATING SYSTEMS

Of course, the best method for using historical data to develop the cost for a new mine is to find a previously developed mine that is exactly like the one a company wishes to develop, but it is a rare event to find two mines that are nearly alike. When a company is building several mines within the same mining district over a period of time, then it may be possible. Such an example is when a series of mines were built by St. Joe Minerals Corporation in the Missouri Viburnum Trend, where the Brushy Creek mine was built as a mirror image of the Fletcher mine that had been built a few years before. Consequently, the cost estimates were very accurately projected by escalating the actual cost of the first mine. Surely there would have also been similar situations with some of the mines that were built in the Carlin Trend by Newmont Mining and Barrick Gold. But in these cases, escalation indexes would still have to be used.

One method for doing screening cost estimates on mining properties is to use formulas that have been developed by studying a series of mines, and then using the actual cost from these mines to develop cost formulas by regression analysis. The U.S. Bureau of Mines (USBM) did this on several occasions to develop cost estimating systems. The original cost estimating system (CES) was developed by the A.A. Mathews Company (which later became STRAMM Engineering) who was assisted by Behre Dolbear & Company. The STRAMM system, as it was known, while originally published as a sort of handbook, was then revised and updated to the USBM Mineral Availability System (Clements et al. 1975), and still later was adapted to be set up on a mainframe computer using a tape drive (Lemmons 1984). The chapter author used that computer system in 1984 to estimate the operating cost of 28 mines that he had visited in various countries and found that the cost estimates produced fairly accurate results. However, to get the accuracies that were achieved, the site visits allowed for experienced judgment, based on observation, to adjust productivity factors of many of the cost elements that the program used in estimating the overall cost.

Still later, USBM developed a CES that used the average cost of the mines studied, and then a regression analysis for estimating the capital and operating cost of each mining method and minerals processing method. For rough screening cost estimating, the formulas developed by USBM (Camm 1994a) were a useful tool. This author used the system on 30 to 40 mines with confidence of its accuracy. However, when the USBM was dissolved, all the information was given to the U.S. Geological Survey, and it chose not to support and update the program. In this author's opinion, the 1994 mining data are no longer applicable because of changes in equipment productivity and mine mechanization since that time.

Because the technique used by USBM was valid, information based on that technique has been included in this chapter for future reference should any organization want to update the system. The regression formulas developed by Camm (1994a) in Tables 14.1 and 14.2 illustrate the methodology that could be used by future cost researchers. Consequently, to use them, one has to be aware of the technological improvements that may have taken place since that time and adjust the costs accordingly. Therefore, costs must be escalated to today's dollars but also to current technology, for which there is no published factor. Thus, their use is not recommended for most modern mining applications but may be pertinent to some purposes of mining and milling where the technology has not changed since 1994. For example, in shrinkage stoping, where the drilling and blasting operations are still using handheld drills and loading explosives by hand, then the application might still be relevant. Likewise, in many

TABLE 14.1 CES mine and mill total cost equations

Cost Model	Capital Cost (1994 dollars)*	Operating Cost (1994 dollars)*
Open Pit Mine Models		
Small open pit	$160{,}000(X)^{0.515}$	$71.0(X)^{-0.414}$
Large open pit	$2{,}670(X)^{0.917}$	$5.14(X)^{-0.148}$
Underground Mining Methods		
Block caving	$64{,}800(X)^{0.759}$	$48.4(X)^{-0.217}$
Cut and fill	$1{,}250{,}00(X)^{0.461}$	$279.9(X)^{-0.294}$
Room and pillar	$97{,}600(X)^{0644}$	$35.5(X)^{-0.171}$
Shrinkage stope	$179{,}000(X)^{0.620}$	$74.9(X)^{-0.160}$
Sublevel longhole	$115{,}000(X)^{0.552}$	$41.9(X)^{-0.181}$
Vertical crater retreat	$45{,}200(X)^{0.747}$	$51.0(X)^{-0.206}$
Mill Models		
Autoclave-CIL-EW (carbon in leach–electrowinning)	$96{,}500(X)^{0.770}$	$78.1(X)^{-0.196}$
CIL-EW	$50{,}000(X)^{0.745}$	$84.2(X)^{-0.281}$
CIP-EW (carbon in pulp–electrowinning)	$372{,}000(X)^{0.540}$	$105(X)^{-0.303}$
CCD-MC (countercurrent decantation and Merrill Crowe)	$414{,}000(X)^{0.584}$	$128(X)^{-0.300}$
Flotation-roast-leach	$481{,}000(X)^{0.552}$	$101(X)^{-0.246}$
Flotation, one product	$92{,}600(X)^{0.702}$	$121(X)^{-0.335}$
Flotation, two products	$82{,}500(X)^{0.702}$	$149(X)^{-0.356}$
Flotation, three products	$83{,}600(X)^{0.708}$	$153(X)^{-0.344}$
Gravity	$67.8(X)^{0.529}$	$67.8(X)^{-0.364}$
Heap leach	$296{,}500(X)^{0.512}$	$31.5(X)^{-0.223}$
Solvent extraction	$14{,}600(X)^{0.596}$	$3.00(X)^{-0.145}$

Source: Camm 1994a
*X = Capacity in short tons per day.

countries, the practice of "slushing" with rope buckets is still in use, and, in this case, the formulas would probably be accurate.

These cost formulas were derived from the same type of regression analysis of collected data from components of labor, equipment, steel, lubrication, construction material, electricity, reagents, and sales tax. USBM also developed approximate costs for underground depth factors, which are to be added to the mining costs given in Table 14.1. USBM also developed a system for estimating capital costs of access road and power lines, based on the length of these facilities to be installed. Also given was the estimate for building a tailings pond, based on the acreage of the facility. All of these items are contained in Table 14.2. Therefore, by using the formula from Tables 14.1 and 14.2, one can calculate what the approximate capital and operating cost would have been in June 1994. Again, the same caution for changes in technology and cost escalation applies. USBM also took a similar approach to developing a CES for use on a personal computer, but in this case, the method goes much further in allowing the user to adjust various factors specific to the individual property that is being estimated.

The ongoing gathering of costs from previous projects was found to be important to others, and USBM was not the only group using historical costs to develop CESs. This historical cost data form the basis of many CESs, which developed as cost curves (Mular 1978; O'Hara 1979), cost formulas, cost models (Schumacher 1993), and complete computer software

TABLE 14.2 Underground depth factor, infrastructure, and tailings pond equations

Cost Model (1994 dollars)	Cost Equation (1994 dollars)
Underground Mining Depth Factor	
Capital cost	$+371+180(D)(X)^{0.404}$
Operating cost	+2,343/(X) +0.440(D)(X)+0.00163(D)
Access Road Capital Cost ($/mile)	
40 ft wide	76,000 × miles of road(R)
50 ft wide	112,000 × miles of road(R)
60 ft wide	148,000 × miles of road(R)
Power Line Capital Cost ($/mile)	
20 ft pole height	298,200 × miles of line(P)
30 ft pole height	304,400 × miles of line(P)
40 ft pole height	310,400 × miles of line(P)
Tailings Pond Capital Cost	
Tailings pond, $ + $/acre	146,000 + 1,783(A)
Dam, $/linear ft of dam around tailings pond	161(L)
Liner, $/acre	5(L)+35,790(A)

Source: Camm 1994a

X = Capacity of mine, stpd
D = Depth of shaft bottom of ore body, ft
R = Length of road to construct, mi
P = Length of power line to construct, mi
A = Area of tailings pond, acres
L = Length of impoundment dam to construct around tailings pond, ft

programs (Camm 1994b; Henry and Wagner 1995; Aventurine Engineering 1998a, 1998b; Scott 1988; Stebbins and Schumacher 2001). Most recently, cost data from Stebbins and Schumacher (2001), Stebbins (2011), and Stebbins and Leinart (2011) have been partially updated later in this chapter. This information was gathered, based on individual company/agency experience, across the industry within a country or from worldwide applications.

Historical Methods Used in Performing a Preliminary Cost Estimate

There are several options for performing a preliminary cost estimate, all of which are based on *benchmarked historical cost information*. No matter which of the benchmark methods used, the benchmark cost will have to be converted to the projected new cost to account for the differences between the two. However, in all cases of using historical data, the data must be adjusted to the present situation (time, place, and size) compared to the benchmarked data. One common application of these cost adjustment techniques in preliminary studies is for individual pieces of equipment, where one knows the actual cost of that equipment previously purchased at some other operation. This raises some important aspects of using and normalizing historical costs on the project. The information that you have gathered may have

1. A difference in time between your project from the one you may be comparing it to,
2. A difference in location of the project from the one you may be comparing it to, and
3. A difference in size or horsepower rating compared to the item that you are trying to estimate.

Therefore, you may have to adjust for all three factors. What is important is to use the best data available for developing the indexes used for the adjustments.

TABLE 14.3 Mining and milling cost index codes published by the U.S. Bureau of Labor Statistics

	USBLS	
Index Code	Name	Item Used in Mining and Milling
CEU1000000008	Average hourly earnings, mining	Natural resources and mining
CEU2000000001	All employee's construction	All construction labor (surface and mill)
CEU3000000008	Average hourly earnings, manufacturing	Manufacturing labor average hourly earnings
wpu112	Construction machinery and equipment	Machinery, heavy equipment, and repair parts
wpusi012011	Construction material	The composite average for construction material
wpu101	Iron and steel	Drill bits and steel, track, mill balls, rods, and liners
wpu0811	Softwood lumber–timber	Mine timber
wpu057	Fuels and related products, refined	Diesel, gasoline, fuel oil, propane, and lubricants
wpu067902	Explosives	Explosives and accessories
wpu07120105	Rubber and plastic products	Truck and off-highway tires
ndu482111482111A02	Railroads (RR)–Metallic ores*	Discontinued in 2005
ndu482111482111A03	RR–Coal*	Discontinued in 2005
ndu482111482111A04	RR–Nonmetallic ores*	Discontinued in 2005
pcu482111482111	RR–Composite index of above	Above three RR indexes aggregated into one index
wpu0543	Industrial power, 500 kW demand	Composite of industrial electric power
wpu03 thru 15	Industrial commodities	Miscellaneous commodities not broken out
wpu061	Industrial chemicals	Milling reagents and any chemical used in mining
wpu05	Coal for power*	Where coal may be used as either a fuel or power
wpu0531	Natural gas for power	Where natural gas may be used as a fuel or power

Data from USBLS 2015

* These data are not used in Table 14.4.

Difference in Time

Accounting for the difference in time can be done in two ways. The first method is to use a common source that is available to everyone, the U.S. Bureau of Labor Statistics (USBLS). But it takes experience and judgment in knowing which one of the thousands of government statistical indexes to use that is applicable to mine plant development of costs estimates. The second method is based on inflation cost and is discussed later.

To assist the estimator in using the USBLS data, Table 14.3 summarizes the current applicable listings and the code that can be entered onto the USBLS web page (USBLS 2015). These indexes can then be used to adjust the change over time due to cost escalation. A word of caution: The government modifies the codes fairly regularly and changes the numbers and format. They have been changed many times in the last 10 years. So it may take some investigation or even communication to those officials within the USBLS to get help in finding the correct number in the index to use for the future. The indexes can be found on the Internet at www.bls.gov/. When the USBLS home page comes up, follow these steps to access the indexes:

1. Select the Data Tools tab.
2. Select the Series Reports item from the left side. When the Series Report page comes up, type the required index code(s) from Table 14.3 into the dialog box, and select Next.

3. When the Databases, Tables & Calculators by Subject page appears, select the time period that you need. For time frame, select All Time Periods; for output type, choose "HTML table"; and for annual averages, check the box for "include annual averages."
4. Select Retrieve Data. The report that is generated should have all of the codes selected for the years requested.

If you experience difficulties using the site, send an email to ppi-info@bls.gov, or go into the detail of any group and find the agency supervisor and send your question to that individual. The people there are very good at directing your question to the correct group, even if you initially send it to the wrong person.

Table 14.4 is an updated version of the indexes from 1970 through June 2015 that the USBM maintained and published at one time. These indexes can be used directly to scale a particular item or commodity from one time to another, or they can be combined to make a mining composite index. Of course, they should be updated to the current year that you wish to use them. The composite mining index would be formed by using the percentage of change for each of the various indexes as listed in Table 14.4 applied for that particular mining method. Thus, you would apply the rate of change between the two differences in time, multiplied by the percentage of that commodity that a particular mining or milling system uses.

Using Composite Indexes and Factor Weights

Cost data are often obtained for entire segments of a mine operation or capital cost for that particular type of operation. Mine operating cost is a composite cost that consists of wages, equipment, fuel and other supplies, electric power, and transportation. To update this composite cost, a composite index is necessary if a detailed breakout of the individual costs is not available. A composite index is created by means of factor weights. *Factor weights* are a set of decimals, summing to 1, that show the relative importance (weighting) of each individual type of cost in the composite cost. Factor weights are derived from detailed cost figures from mines that are similar to mines for which these factor weights are to be used. The detailed costs of those mines are divided into categories that are representative of the 15 different indexes, and the percentage that each category, when combined with the other index percentages, comprises for the total composite index from which operating and capital cost is calculated or scaled from one time period to another. To use the government cost indexes to construct a company's own set of composite indexes, one might follow the lead of what the USBM was using as the percentage proportions listed in Tables 14.5 and 14.6 for operating and capital costs for underground mines, and Tables 14.7 and 14.8 for operating and capital costs for surface mines, respectively. The percentages given in Tables 14.5 through 14.8 are the factor weights, but they should be expressed as a decimal when making the calculations. Factor weights have been derived individually for each commodity and for each type of operation. A few of these have been modified slightly by the author, based on his experience in cost estimating various types of mining methods.

The application of factor weights is shown in Table 14.9. A composite index (which is a weighted average of the individual cost indexes) is derived that can be used to update a mine operating cost from 1998 to 2002. (If this method was to be used in a foreign country, then a set of foreign indexes and local currency units would be applied.) The factor weights shown here were taken from Table 14.5. Ratios of factor weights have been derived individually for

TABLE 14.4 USBLS cost indexes applicable to mining and milling projects

Year	Mining Wage	Construction Wage	Manufacturing Wage	Equipment/ Machinery & Parts	Bits & Steel	Timber	Fuel & Lube	Explosives	Tires	Construction Material	Industrial Chemicals	Industrial Commodities	RR-LH* Transportation	Deep Sea Freight	Electricity	Natural Gas
Average 1970	3.77	4.74	3.36	33.70	33.95	36.60	13.26	35.69	38.80	39.1	28.61	35.22	26.7	—	22.53	7.92
Average 1980	8.97	9.37	7.27	84.00	89.97	104.80	88.58	84.00	91.93	92.5	91.69	87.80	75.9	—	77.80	63.12
1990	13.40	13.42	10.83	121.6	117.2	124.6	74.8	125.6	93.8	119.6	113.2	115.8	107.5	113.1	119.6	80.4
1995	14.78	14.73	12.34	136.7	122.8	178.5	60.8	144.2	93.0	138.8	128.4	125.5	95.4*	113.3	130.8	66.6
2000	16.55	17.48	14.32	148.2	116.6	178.6	91.3	136.9	89.0	144.1	129.1	134.8	102.6	155.8	131.5	155.5
2001	17.00	18.00	14.76	149.1	109.7	170.1	85.3	142.8	89.5	142.8	128.4	135.7	104.5	172.2	141.1	171.8
2002	17.19	18.52	15.29	151.1	114.1	170.8	79.5	147.8	90.5	144.0	127.3	132.4	106.6	185.8	139.9	122.5
2003	17.56	18.95	15.74	153.2	121.5	170.8	97.7	156.0	93.7	147.1	141.7	139.1	108.8	219.9	145.8	214.5
2004	18.07	19.23	16.15	158.5	162.4	209.8	119.9	168.9	98.3	161.5	162.8	147.6	113.4	225.9	147.2	245.9
2005	18.72	19.46	15.56	168.3	171.1	203.6	165.0	172.5	103.3	169.6	188.5	160.2	125.2	231.9	156.2	335.4
2006	19.90	20.02	16.80	175.4	186.5	189.4	193.2	179.9	106.7	180.2	212.4	168.8	135.9	233.3	172.8	280.3
2007	20.97	20.62	17.26	179.6	201.1	170.5	214.2	184.3	110.1	183.2	226.4	175.1	140.9	230.0	180.4	273.8
2008	22.50	21.25	17.75	185.3	246.4	156.3	271.7	201.2	120.8	196.4	274.6	192.3	157.3	258.3	189.1	344.0
2009	23.29	21.81	18.24	191.0	184.0	141.3	175.5	198.1	121.6	189.2	234.1	174.8	148.5	218.8	190.6	160.9
2010	23.82	22.22	18.61	191.4	223.5	160.8	225.2	188.7	129.7	195.5	269.2	187.0	156.2	244.8	193.1	185.8
2011	24.50	22.72	18.93	197.4	253.2	160.5	298.9	201.7	149.3	201.1	324.7	202.0	169.8	253.8	203.6	171.4
2012	25.79	23.22	19.08	205.4	240.7	171.7	306.5	207.6	155.8	205.8	306.9	202.1	177.4	249.9	209.6	117.9
2013	26.80	23.68	19.30	210.7	226.4	199.8	294.7	221.0	151.4	209.3	301.2	203.0	183.1	249.2	203.7	153.9
2014	26.92	24.67	19.56	214.3	232.1	205.6	278.0	220.9	147.7	214.5	288.9	204.1	186.5	N/A†	218.0	182.4
June 2015	26.69	24.20	19.84	217.0	195.4	192.9	175.9	227.2	143.8	213.5	241.4	188.9	179.8	N/A	223.6	106.3

Data from USBLS 2015

*This is a calculated value, based on when the USBLS changed the base of RR-LH (railroads–line haul) in December 2006 to 100.°

†N/A = not available

TABLE 14.5 Composite of underground mining operating cost

Cost Item	Block Caving, %	Cut-and-Fill, %*	Room-and-Pillar (CM), %†	Room-and-Pillar (Drill-and-Blast), %*	Shrinkage Stoping, %	Sublevel Caving, %	Sublevel Open Stoping, %	Vertical Crater Retreat, %
Labor and Administration	**56.6**	**60.3**	**47.0**	**51.0**	**63.1**	**46.5**	**54.4**	**52.5**
Operating labor	34.4	42.0	30.0	34.0	46.0	28.9	33.0	33.6
Administration	22.2	18.3	17.0	17.0	17.8	17.6	21.0	16.9
Supply	**21.0**	**24.0**	**30.0**	**23.8**	**22.9**	**25.4**	**23.4**	**21.3**
Steel drilling items	2.0	5.9	9.8	1.9	5.5	6.1	5.6	4.8
Steel pipe	1.0	0.3	6.6	0.9	0.8	0.6	1.1	0.9
Timber and lumber	0.0	6.6	2.0	0.0	2.6	0.0	0.7	0..6
Explosive	6.0	8.7	0.0	12.1	10.2	7.6	11.7	10.7
Construction material	10.0	0.0	0.0	0.0	0.0	4.0	0.0	0.0
Industrial material	0.3	0.1	0.0	2.6	0.0	0.4	0.0	0.0
Ventilation accessories	2.0	1.0	5.5	3.5	2.0	2.7	2.5	2.1
Electricity	3.9	1.4	6.1	2.8	1.8	4.0	1.8	2.0
Equipment	**18.5**	**15.7**	**23.0**	**22.1**	**14.0**	**28.1**	**22.2**	**26.2**
Equipment maintenance	8.8	12.4	12.4	9.4	12.1	19.4	17.3	21.9
Fuel	3.7	0.5	5.4	4.3	0.3	3.6	1.3	1.1
Lube	2.1	2.2	2.7	1.7	1.1	1.8	3.2	2.9
Tires	3.9	0.6	2.5	6.7	0.5	3.3	0.4	0.3
Total	**100**	**100**	**100**	**100**	**100**	**100**	**100**	**100**

Data from USBM 1994

*Data modified by author based on personal experience.

†Continuous mining in soft rock/coal.

each commodity and for each type of operation. A few of these have been modified slightly by the author, based on his experience in cost estimating various types of mining methods.

Mathematically:

$$\text{composite index} = + \text{ summation of } \{\text{index2014/index 2002}\} \times \text{commodity factor weight} \quad \textbf{(EQ 14.1)}$$

Or, for the example given in Table 14.9, the mine cost composite index for a room-and-pillar (R&P) operating cost inflated from 2002 to 2014 would be 1.688. So a benchmark R&P mine that had an operating cost of $17.50 per ton hoisted to inflate this direct mining cost for 2014, multiplied times the index, would bring it to a 2014 cost of $29.54 per ton.

Consequently, if this procedure were followed for all the types of mines shown in Tables 14.5 through 14.8 for all of the items for those years of interest, then the estimator would essentially have what the USBM at one time supplied to the industry for U.S. mining properties. This procedure would be repeated for all types of mineral processing.

A time-saving and probably cost-saving method is to utilize one of the commercially available service organizations that will supply not only cost information indexes for the United States but also mine equipment and mine cost model information for a fee, such as InfoMine USA (2016b). Scott Stebbins of Aventurine Engineering and Jennifer Leinart of InfoMine USA collaborated to provide cost models for the *SME Mining Engineering Handbook* (Stebbins 2011; Stebbins and Leinart 2011). As a service to the mining industry, they have updated the 10,000-t/d (metric tons per day) surface mine models and the block caving underground models for this chapter. This information is often used for intermediate-feasibility-level cost estimating.

TABLE 14.6 Composite of underground mining capital cost

Cost Item	Block Caving, %	Cut-and-Fill, %*	Room-and-Pillar (CM), %†	Room-and-Pillar (Drill-and-Blast), %*	Shrinkage Stoping, %	Sublevel Caving, %	Sublevel Open Stoping, %	Vertical Crater Retreat, %
Labor	**10.84**	**9.7**	**14.3**	**30.7**	**36.0**	**10.8**	**4.2**	**5.4**
Construction	10.84	9.7	14.3	30.7	36.0	10.8	4.2	5.4
Supply	**34.2**	**14.9**	**10.3**	**10.1**	**29.9**	**43.7**	**5.1**	**6.7**
Steel items and drill bits	2.9	1.8	4.4	5.9	3.8	12.8	0.7	0.9
Steel pipe	0.8	0.8	2.1		2.2	0.3	0.4	0.6
Timber and lumber		4.5	0.6		4.8		0.3	0.4
Explosives	12.1	5.7		4.2	14.1	4.2	2.7	4.5
Construction material	13.3					19.8		
Industrial material	0.2					0.2		
Ventilation pipe	2.8	2.1	1.8		4.9	1.5	0.9	1.2
Electricity	2.1	0.0	1.3		0.1	4.9	0.0	0.0
Equipment	**55.0**	**53.4**	**75.4**	**54.7**	**33.8**	**45.5**	**63.2**	**61.8**
Equipment and repair parts	49.4	74.6	73.8	56.6	33.4	44.2	90.6	87.8
Fuel	2.2	0.1	0.8	2.1	0.2	0.9	0.0	0.1
Lube	1.0	0.1	0.4	0.5	0.2	0.4	0.0	
Tires	2.3	0.2	0.4		0.3		0.1	

Data from USBM 1994

*Data modified by author based on personal experience.

†Continuous mining in soft rock/coal.

TABLE 14.7 Percentage distribution of surface mining operating cost*

Cost Item	Drill and Blast	Shovel and Trucks
Labor	**2.98**	**24.03**
Supply	**11.98**	**0.72**
Drill steel items and drill bits	2.75	
Explosives	9.14	
Electricity		0.72†
Equipment	**3.88**	**56.50**
Equipment and repair parts	3.02	36.91
Fuel and lube	0.86	15.05
Tires		4.54
Total	**100%**	

Data from USBM 1994

*Assumes 40,000 metric tons of ore, 2:1 stripping ratio.

†Omit and redistribute cost percentage for hydraulic shovel.

Difference in Time Index Application

No matter how the benchmarked information is obtained, whether it is from a commercial cost data supplier or a neighbor mine in the district, the cost data for time may need to be adjusted.

Having determined the cost or, by benchmarking, received the cost for a particular year, you must determine what it would be for the year needed for your estimate. Once you have a

TABLE 14.8 Percentage distribution of surface mining capital cost*

Cost Item	Drill and Blast	Shovel and Trucks
Labor	**0.37**	**2.39**
Supply	**1.48**	**0.09**
Steel items and drill bits	0.34	
Explosives	1.14	
Electricity		0.09
Equipment	**7.47**	**73.98**
Equipment and repair parts	7.36	69.38
Fuel and lube	0.11	2.38
Tires	<0.005	0.48
Transportation	<0.005	1.74
Surface Mine Infrastructure	**14.22**	
Mine buildings (10% operating equipment)	6.94	
Mine roads (7.5 km, 10 m wide)	5.75	
Electrical and communication systems	0.76	
Fuel station	0.77	
Total	**100%**	

Data from USBM 1994

*Assumes 40,000 t of ore, 2:1 stripping ratio. Includes preproduction stripping of 10,000,000 t.

TABLE 14.9 Indexing a composite cost for underground room-and-pillar operating cost (1998–2002)

Index Name	2014 Index	2002 Index	Ratio of Change	R&P Operating Cost Change Factor (Table 14.5)	Composition of Composite Index
Operating labor	26.85	17.17	1.564	0.510	0.796
Steel and drill bits	232.1	114.1	2.034	0.019	0.039
Steel pipe	232.1	114.1	2.034	0.009	0.018
Timber and lumber				0	0
Explosives	220.9	147.85	1.494	0.121	0.181
Construction material				0	0
Industrial material	204.1	132.4	1.542	0.026	0.040
Ventilation accessories	214.3	151.1	1.418	0.035	0.050
Electricity	218.0	141.1	1.545	0.028	0.030
Equipment and repair parts	214.3	151.1	1.418	0.094	0.107
Fuel	278.0	79.5	3.497	0.074	0.259
Lubrication	278.0	79.5	3.497	0.017	0.059
Tires	147.7	90.5	1.632	0.067	0.109
Transportation				0	0
R&P mine operating composite	For 2002 to 2014			1.00	1.688

cost index for the prior (past) year and a cost index for the present year, it is simple to find the present cost:

$$\text{(present \$ cost)/(past \$ cost)} = \text{(present cost of index)/(past cost of index)} \quad \textbf{(EQ 14.2)}$$

It was previously mentioned that there were two methods of escalating cost. The other way is to take the inflation rate for the item over the period of years and perform the following calculation:

$$\text{present \$ cost} = \text{(past \$ cost)} \times (1 + \text{average inflation for the item})^n \quad \textbf{(EQ 14.3)}$$

where n is the number of years that have elapsed.

Difference in Location Index Application

Accounting for the difference in location can be achieved by using cost indexes that are based on different country locations and, in the United States, for the individual cities, if there is a recognized difference in construction cost between specific areas (Grogan et al. 1998).

To use the cost index bases for a new location:

$$\text{(cost at new location)/(cost at past location)} = \text{(new location cost index)/(past location cost index)} \quad \textbf{(EQ 14.4)}$$

Foreign cost factors were published on 22 countries by USBM until 1994 costs (1992 cost basis). One final publication was published in 1995 that contained the foreign costs for 92 countries based on cost comparisons and exchange rates, but it was limited to 1992 data. Again, these valuable data were no longer published after the cost estimating information was turned over to the U.S. Geological Survey.

Another source of capital cost estimating assistance is to purchase Global Construction cost estimating books, which are published every year. Compass International publishes seven annual books containing the most accurate, current, and up-to-date cost information applicable to more than 120 countries (Compass International 2017). Table 14.10 shows an example of this source of cost indexes that may be obtained from the Internet. These books are devoted to general commercial construction, not mining projects or heavy mining equipment, and they are mostly tied to cities within a particular country. Several other sources are available for international construction cost information, although it is sometimes difficult to find information for many Central or Latin American countries.

Although not ideal for construction, if the location needed is not found in Table 14.10, then one might consider searching Numbeo.com's cost-of-living databases for other countries of the world (Numbeo.com 2017). This is not an ideal index for purchase of equipment, but it may be the best index available for the country needed.

Difference in Size/Capacity Index Application

If you are trying to adjust the cost for a particular piece of equipment or even an entire plant for order-of-magnitude estimation, the so-called six-tenths rule can be applied. Actually, the system originally developed by the chemical industry used a range of exponential values, not just six-tenths. Scaling to an exponential allows for a quick way to get order-of-magnitude cost estimates. Mular (1978) first started applying this method to mining and mineral processing

TABLE 14.10 International construction cost index

	Faithful Gould International Construction Index (Q2 2013) USA=100				
Country	**City**	**Exchange Rate**	**Currency**	**Index Range**	**Average Index**
Australia	Melbourne	0.9654	AUD	94.3–127.4	110.9
Austria	Vienna	0.7578	EUR	87.1–112.4	102.3
Brazil	Sao Paulo	2.0023	BRL	64.9–92.4	78.7
Canada	Toronto	1.0053	CAD	90.5–111.4	101.0
China	Shanghai	6.1697	CNY	56.1–72.1	64.1
Czech Republic	Prague	19.4453	CZK	57.4–77.6	67.5
Denmark	Copenhagen	5.6508	DKK	107.2–144.9	128.1
Finland	Helsinki	0.7578	EUR	104.2–139.9	126.1
France	Paris	0.7578	EUR	101.6–137.2	119.4
Germany	Frankfurt	0.7578	EUR	95–129.3	112.2
Greece	Athens	0.7578	EUR	72.6–97.7	85.2
India	Bangalore	53.6755	IRN	36.2–49	42.6
Ireland	Dublin	0.7578	EUR	81.8–100.3	91.1
Italy	Milan	0.7578	EUR	84.5–114.8	99.7
Malaysia	Kuala Lumpur	3.0425	MYR	39.1–55.5	47.3
Mexico	Mexico City	12.1380	MXN	71.2–96.3	83.8
Netherlands	Amsterdam	0.7578	EUR	80.5–109.5	95.0
New Zealand	Auckland	1.1657	NZD	99.5–141.5	120.5
Norway	Oslo	5.7519	NOK	121.2–163.9	142.6
Poland	Warsaw	3.1488	PLN	57.5–81.9	69.7
Portugal	Lisbon	0.7578	EUR	52.8–73.9	63.4
Russia	Moscow	31.1475	RUB	84.8–120.8	102.8
Spain	Madrid	0.7578	EUR	63.3–85.8	74.6
Sweden	Stockholm	6.4624	SEK	123.8–167.4	145.6
Switzerland	Zurich	0.9282	CHF	129.3–174.5	151.9
Thailand	Bangkok	29.2550	THB	60.5–86.2	73.4
United Kingdom	London	0.6422	GBP	96.5–118.3	107.4
United States	Chicago	1.0000	USD	90–110	100.0

Source: Wiggins 2013

systems and found that six-tenths was the most accurate exponential to use. As described by Mular for cost, original system/cost:

$$\text{new system} = (\text{capacity, original system}/\text{capacity, new system})^{0.6} \quad \textbf{(EQ 14.5)}$$

Adjustment factors will normally range between 0.1 and 0.9 for all industries. For the mining and mineral industry, the exponential adjustment factor will usually range between 0.6 and 0.7.

An exponential factor greater than 1.0 usually means that the plant being built requires significant infrastructure or environmental costs that the original plant(s) did not have. In fact, *if there are major differences in the technology between the two systems or a drastic change in the infrastructure, then the six-tenths factor can and should be scaled up or down.*

Example Combination of Time, Size, and Location Cost Adjustment

To demonstrate how the equations work in combination, suppose a new 170-ton truck cost \$2,500,000 in 2009 in the United States. What would it have cost to buy a 240-ton truck of the same brand priced in 2013 in Australia in their currency?

Step 1. Time

To get to 2013 dollars:

$$2009 \text{ cost} = \$2{,}500{,}000\ (1+ \text{average inflation rate})^{n}$$

$$\begin{aligned}\text{equipment inflation rate} + 2.5\% \text{ per year} &= \$2{,}500{,}000\ (1.025)4\\ &= \$2{,}500{,}000 \times 1.104\\ &= \$2{,}760{,}000\end{aligned}$$

Or using Table 14.4 for equipment in 2013:

$$\text{approximate cost from } 2009 = \$2{,}500{,}000 \times (210.7/191.0) = \$2{,}758{,}000$$

So the two methods of inflation are within 0.80% of each other, but the machinery index is probably more accurate for trucks. When rounded off to four significant numbers, it is \$2,800,000.

Step 2. Size

To adjust for size, an exponential factor is used. Since there have been considerable technological changes between 2009 and 2013 that need to be incorporated, an exponential factor of 0.72 is used instead of 0.60.

To adjust for size from 170 tons to 240 tons:

$$\begin{aligned}\text{cost} &= \$2{,}758{,}000\ (240/170)^{0.72}\\ &= \$2{,}758{,}000\ (1.412)^{0.72}\\ &= \$2{,}758\ {,}000 \times 1.282\\ &= \$3{,}535{,}000\end{aligned}$$

Step 3. Location

For estimating the cost of a 240-ton truck paid with Australian currency, the currency exchange rate must be applied to the cost from Step 2. The exchange rate between Australian currency and the U.S. dollar in 2013 was as follows:

From Table 14.10, the U.S. dollar is worth 0.9654 Australian dollar.

From Step 2:

$$\$3{,}535{,}000 \times 0.9654 = \$3{,}413{,}000$$

Although a cost index for world heavy equipment was not used in these calculations, such indexes are probably available from the heavy equipment manufacturers. In lieu of the proper type index, this example uses the construction index to illustrate the method. Then using the difference in the construction cost index of the United States in 2013 as 77.39 and Australia as 108.11, calculating the truck cost using the conversion index would be as follows:

$$\text{new truck cost in Australia, 2013} = \$3{,}413{,}000\ (110.9/100.0)$$
$$= \$3{,}413{,}000 \times 1.109$$
$$= \$3{,}785{,}000$$

So the 170-ton truck in 2009 that in the United States cost $2,500,000 would equate to a 240-ton truck purchased in 2013 in Australia at an estimated cost of $3,785,000.

The author acknowledges that the preceeding procedure is a stretch for this method, but it does illustrate the technique. For a piece of equipment, vendors will supply more accurate cost estimates as needed.

Measured Cost Information

For measured costs, some engineering must be performed on which the dollar value of the measured unit is applied. Most common are take-offs from drawings, or even sketches in the case of preliminary studies, that have been prepared. These usually cover material or units of work that must be undertaken. For example, if you had laid out a starter tailings dam to be built as an earth structure on a topographic map, then you could use this layout to determine the dimensions and calculate the earth volume.

Other measured costs may be related to time. For example, when you calculate the equipment ownership and operating (O&O) cost, you are estimating the O&O cost in dollars per hour. When the O&O costs are used in production estimate problems, a cost per ton is derived, but it is based on a measured cost per hour.

There are other measured cost methods, but they do not really relate to what might normally be used in a preliminary feasibility cost study.

Unit of Capacity Cost Method

The unit of capacity cost method simply multiplies capacity by a unit cost expressed as the installed capital cost per ton of annual capacity ($ per ton per year). For example, if the capital cost to develop a 35,000-t/d open pit copper mining/milling operation was $612,500,000, the unit capital cost works out to be $17,500 per annual capacity of copper. If you were then going to develop an open pit to produce 50,000 t of copper, based on the 35,000-t/d operation:

$$\text{capital cost} = \$17{,}500 \times 50{,}000 = \$875{,}000{,}000$$

Although this is a quick method of arriving at the capital cost of an open pit copper operation, *it certainly is not accurate enough to use for feasibility studies.* It might be useable for some macro-economic problem in a country for estimating the capital required to build several operations with many varying conditions, assuming that the average number would be typical. However, it assumes an average stripping ratio, an average pit wall slope, and an average grade for all mines.

This method works a little better for mineral processing plants, because most plants producing the same commodity typically use the same type of equipment. But even here, as an example, the "work-index" can vary greatly, and the cost estimate may be off considerably because one ore may require considerably more crushing and grinding equipment than the ore at another plant.

Use of Mining Cost Service*

Surface and Underground Cost Models

The cost models provided in Appendixes 14A through 14C at the end of this chapter can be used for rough preliminary studies estimates based on limited deposit information. The models are entirely theoretical and do not accurately replicate any existing mining operation.

The reader is cautioned against relying too heavily on these or any other models for making significant economic decisions. A cost model, no matter how carefully prepared, is only a model and should not be expected to represent projected costs for a specific property with any degree of reliability beyond preliminary studies. However, cost models can be very useful for comparison purposes or for acquisition and exploration decisions, particularly when little is known about an ore deposit. They are commonly used to establish cut-off grades for preliminary reserve estimates.

The 10,000-t/d surface mine models in Table 14A.1 were constructed using Sherpa for Surface Mines software, and the block caving underground models in Tables 14B.1 and 14C.1 were constructed using Sherpa for Underground Mines software, both published by Aventurine Engineering in cooperation with CostMine, a division of InfoMine USA (Aventurine Engineering 2014a, 2014b). Following is a list of what is and is not included in the models.

Included

- All labor, material, supply, and equipment operation costs incurred at the mine or mill site, including supervision, administration, and on-site management
- Benefits and employment taxes
- All on-site development
- Mine and mill equipment and facilities purchase and installation or construction
- Limited haul road construction
- Engineering and construction management fees
- Working capital
- Tailings disposal
- Contingencies

Not Included

- Preproduction exploration
- Permitting and environmental analysis costs
- Access roads, power lines, pipelines, or railroads to the mine or mill site
- Home office overhead
- Taxes (except sales taxes)
- Insurance
- Depreciation

* The text and tables in this section are provided by InfoMine USA and Aventurine Engineering in Spokane Valley, Washington.

- Townsite construction or operation
- Off-site transportation of products
- Incentive bonus premiums
- Overtime labor costs
- Sales expenses
- Smelting and refining costs (except doré production at hydrometallurgical mills)
- Interest expenses
- Start-up costs (except working capital)
- Postclosure reclamation

10,000-t/d Surface mine models—four stripping ratios. The 10,000-t/d surface mine models (Table 14A.1) includes all labor, material, supply, and equipment operating costs incurred at the mine site, including supervision, administration, and on-site management. Preproduction development and purchase, installation, or construction of all equipment and facilities necessary to operate the mine at full design capacity are included.

The following facilities and operations are included for each model:

- Drilling, blasting, and excavating of ore, waste, and overburden
- Hauling of ore by truck out of the pit and to a mill site
- Hauling of waste and overburden out of the pit and to a dump site
- Construction, installation, and operation of facilities and equipment necessary for equipment maintenance and repair, electrical systems, fuel distribution, water drainage, sanitation facilities, offices, labs, powder storage, and equipment parts-and-supply storage

The mines are assumed to be in areas of moderate relief with warm summers and moderate, snowy winters. The 10,000-t/d option uses bulk (ANFO [ammonium nitrate and fuel oil]) explosives.

Haul road widths are determined by the size of the haul trucks. Haul road gradients are assumed to be +10% coming out of the pit and 0% to the mill or dump site. Rolling resistance is assumed to be +3%.

Supply prices used in the models include those shown in Table 14.11. It is important to note that while diesel fuel and electricity are categorized as supplies in Table 14.11, they are included as a portion of the equipment operating costs in the models.

The wage and salary scales shown in Table 14.12 are common to all the surface models. These wages were determined using Infomine USA's annual *Mining Cost Service* wage and salary survey for U.S. metal and industrial mineral mines (Infomine USA 2016b). In keeping with the results of the survey, wages for the smaller mine models are less than those for the larger models.

TABLE 14.11 Supply prices in surface mine models (2014 dollars)

Item	Price per Unit, US$
Diesel fuel	0.75/L
Electricity	0.089/kW·h
Bulk-type explosives	1.23/kg
Caps	3.65 each
Primers	5.32 each
Detonation cord	0.733/m

Source: Infomine USA 2016b

TABLE 14.12 Personnel salaries and wages for surface mine model (2014 dollars)

Salaried Personnel	Annual Salary, US$ (includes a 38% burden factor)	Hourly Personnel	Hourly Wages, US$ (includes a 38% burden factor)
Manager	135,900	Driller	31.67
Superintendent	127,100	Blaster	32.43
Foreman	97,300	Excavator operator	35.08
Engineer	131,200	Truck driver	29.37
Geologist	97,600	Equipment operator	30.08
Supervisor	96,600	Utility operator	25.79
Technician	77,700	Mechanic	34.61
Accountant	90,900	Maintenance worker	25.53
Clerk	56,600	Laborer	24.26
Personnel manager	116,600	Electrician	36.57
Secretary	55,600		
Purchasing agent	91,500		

Source: Infomine USA 2016b

Block caving underground mine model. The underground mine models in Appendixes 14B and 14C include all labor, material, supply, and equipment operating costs incurred at the mine sites, including supervision, administration, and on-site management. The costs of purchasing and installing all equipment are included, as are preproduction development costs and the costs of constructing surface facilities.

The models include at least two routes of access, two primary excavations (adits or shafts; Tables 14B.1 and 14C.1, respectively), and at least one secondary excavation (raise). Additional raises are excavated as needed over the life of the operation to provide adequate ventilation pathways and access routes. Preproduction and production development requirements are specified in the detailed descriptions of each model.

Each model includes the costs of purchasing, installing, and operating all equipment required for the following:

- Drilling
- Hauling
- Hoisting
- Compressed air
- Freshwater pumping
- Support installation
- Exploration drilling
- Mucking
- Underground crushing
- Ventilation
- Drainage pumping
- Backfilling
- Maintenance
- Raise boring

All shaft entry models rely on hoisting for their primary method of ore transportation.

Sufficient working capital is provided for two months of operation. Mine facilities, including shops, offices, worker changehouses, warehouses, and mine plants are located on the surface. The costs of the structures containing these facilities are included in the estimate. Wages and salaries shown in Table 14.13 were determined using InfoMine USA's annual *Mining Cost Service* wage and salary survey for U.S. metal and industrial mineral mines (Infomine USA 2016b).

TABLE 14.13 Personnel salaries and wages for block caving underground mine models (2014 dollars)

Salaried Personnel	Annual Salary, US$ (Includes a 38% burden factor)	Hourly Personnel	Hourly Wages, US$ (includes a 38% burden factor)
Manager	165,600	Stope miner	36.57
Superintendent	133,600	Development miner	38.36
Foreman	113,200	Mobile equipment operator	32.73
Engineer	121,400	Hoist operator	33.12
Geologist	97,600	Motorman	28.35
Shift boss	82,800	Support miner	36.36
Technician	71,800	Exploration driller	34.11
Accountant	90,900	Crusher operator	31.71
Purchasing agent	91,500	Backfill plant operator	36.36
Personnel manager	116,600	Electrician	35.88
Secretary	56,600	Mechanic	34.87
Clerk	55,600	Maintenance worker	27.12
		Helper	27.26
		Underground laborer	23.92
		Surface laborer	22.36

Source: Infomine 2016b

Selected supply prices used in the models include those shown in Table 14.14. Other supply prices, including those for steel pipe, ventilation tubing, rock bolts, drill bits and steel, electric cable, and rail are dependent upon the size required for the specific model.

TABLE 14.14 Supply prices in block caving underground mine models (2014 dollars)

Item	Price per Unit, US$
Diesel fuel	0.75/L
Electricity	0.089/kW·h
Explosives	1.69/kg
Caps	2.16 each
Boosters	4.05 each
Fuse	0.942/m
Timber	455.00/m^3
Lagging	338.14/m^3
Steel liner material	3.20/kg
Cement (for backfill)	126.32/t

Source: Infomine USA 2016b

DETAILED COST ESTIMATING METHODS FOR ALL STUDIES BEYOND PRELIMINARY

The basic method of cost buildup by applying a functional analysis of each element of cost is covered in Chapter 15. The method is applied to both the intermediate feasibility (or prefeasibility) and final feasibility studies, as described in Chapter 11. The primary difference between cost estimating for the two stages is the amount of engineering that has been expended on the design of the project facilities. For the intermediate feasibility study, approximately 15%–20% of the useful design engineering will have been applied to the project. For the final feasibility study, approximately 20%–30% of the design engineering will have been applied. Another major difference is that in the intermediate study, several alternatives are investigated to try to optimize the design approach at this level. Thus, there are several alternatives that must have estimated the costs, but only one of each alternative goes forward to the final feasibility study. Consequently, there is always some engineering and cost estimating that is lost in the intermediate feasibility study that is not counted in the useful design engineering.

For the intermediate feasibility study, costs should be based on suppliers' written quotes and bench-scale metallurgical testing. Environmental baseline studies will be initiated, impact assessments will be conducted, and some permit applications may be made. Thus, the environmental work may lead to additional engineering that will require costing. Results of this study will be adequate for determining economic feasibility and additional predevelopment and/or metallurgical testing requirements. The PE of cost estimates should be about ±15%–20% Contingencies of 15%–20% will apply.

Based on the results of the intermediate feasibility study showing a project that still has the potential to achieve the desired company goals, the final feasibility study should be initiated. Test mining with bulk sampling and pilot-plant testing may have been carried out since the intermediate feasibility study; consequently, more engineering will have been completed that will need costing. Final environmental impacts will be determined, which could also lead to additional engineering and cost estimating. Applications for construction and operating permits will usually be made early in this phase of study (subject to later modification). The PE of cost estimates should be about ±10%–15%. Contingencies of 10% may apply to some well-defined engineered structures. Other less well-defined aspects of the project (e.g., mine development) should have contingencies of at least 15%. Evaluation work can be terminated at any time if it is determined that a project is not, or cannot, be designed and built to be profitable or otherwise meet the company objectives. Project approval and appropriation of funds for design and construction will normally occur after the final feasibility phase. The most attractive alternative will be presented in sufficient detail for a design basis report (DBR; sometimes called a design basis memorandum or design basis document) to be produced. The results of the DBR will enable bids to be solicited and final design and construction to begin.

Most textbooks give considerably more information on capital cost estimating methods than they do operating cost methods. Both are equally important.

Types of Capital Cost

There are two types of capital cost: fixed cost and working capital cost. Fixed costs are categorized as follows:

- Land and water acquisition
- Project development
- Preproduction development
- Environment studies and permitting
- Mining equipment, building, and all other facilities
- Milling equipment, building, and all other facilities
- Support facilities (administration and support buildings)
- Infrastructure facilities (railroad, roads, power lines, houses, schools, hospitals, clinics, etc.)
- Design and engineering costs
- Contingency

All funds that are needed to get the operation running beyond the fixed capital are considered working capital, categorized as follows:

- All funds needed for operating expenses
- Inventories
  - Raw material products
  - Spare parts
  - Supplies
  - Materials in process
  - Finished products
- Accounts receivable (+)
- Accounts payable (–)
- Cash on hand (payroll, utilities, etc.)

Working capital is sometimes estimated as 10%–20% of the capital cost, but this is a risky practice. It is really more related to the operating cost.

$$\text{working capital} = \text{operations cost/ton} \times \text{tons/year} \times Y\,\text{months}/12\,\text{months} \quad \textbf{(EQ 14.6)}$$

where Y is the number of months it takes to get paid for the product that you sell to the customer, smelter, refinery, and so on. It is not unusual for it to take three to four months to get paid from a non-company-owned smelter. This author once did a study in Ireland, where it consistently took six months for the company to be paid for their product from a smelter in Spain.

Noakes (1993) presented a valid method of calculating working capital:

First, as was stated above, determine the time (T) for the first revenue return. Then the total capital is the sum of

- *Fixed operating cost × T;*
- *Variable operating cost × T;*
- *Financing and head office administration cost × T;*
- *Raw material inventory (1–3 months) at cost;*
- *Material in process inventory (1–3 months) at cost;*
- *Spares inventory (1–3 months) at cost including major maintenance spares items (e.g., complete set of mill liner, major spare motors, etc.)* [Complete tire inventory added]

General Approach to Detailed Capital Cost Estimates

Some of the activities that must go into the planning that leads to a detailed capital cost estimate beyond the preliminary feasibility study can be found in the discussion that follows and the contents of Chapters 15 and 16.

Mine Planning Leading to Cost Analysis

Once it has been decided whether the mine should be developed as an open pit mine or an underground mine, then detailed mine planning is required in preparation for costing and economic evaluation. However, for any mine plan, the basic parameter of the mining system must have been agreed on between the project management team and the management that will eventually operate the mine. These parameters would have included the following:

- The mine life or daily rate of production
- The work schedule
- The probable physical characteristic of the waste and mineralized material to be excavated
- The ground control that will probably be required
- Some basic preferences on drilling, blasting (or ripping), excavating, and hauling equipment
- Whether or not contract development, mining, or maintenance is preferred
- The equipment replacement policy of the operating group

While all of these design parameters used would have been worked out prior to the engineering that was done during that phase of the study, it helps the cost estimators to review these elements again with the operations people. This ensures there is no misunderstanding that the parameters to be used in the cost and evaluation economics are the same as those that will go into the final design and construction of an approved project. If this is the final feasibility study, then this is developed into the form of the DBR, covered in Chapter 12.

Cost estimating for open pit mines. For preproduction stripping of an open pit mine, the unit cost for drilling, blasting, loading, and hauling must be calculated as part of the capital costs. Capital costs for equipment for detailed cost estimates should be from the final vendors' written quotations. For equipment selection, one could consider using the materials generated from computer simulation programs developed by vendors sizing the equipment and establishing fleet sizes. Other capital costs are built up by doing material quantity and labor units of work take-offs from the drawings, however imprecise they might be for that phase of the project. The mine capital costs will, of course, include the maintenance facility for the mine equipment.

A checklist of items that must be considered in constructing the capital cost varies depending on the type of mine and process plant. For an open pit mine, several of the following should be considered:

- Service equipment and facilities
  - Mine shop buildings
  - Mine shop bridge cranes and initial tools and equipment
  - Power lines and transformation
  - Communications and operations control
  - Lighting plants
  - Pumping equipment
- Rock fracturing equipment
  - Drilling equipment
  - Explosive-charging equipment
  - Dozers for ripping (possibly)

- Excavating, loading, and hauling equipment
  - Electric shovels
  - Hydraulic shovels
  - Front-end loaders
  - Draglines
  - Bucket-wheel excavators
  - Haul trucks
  - Scrapers
- Pit service equipment
  - Dozers
  - Graders
  - Water trucks
  - Service trucks
  - Pickup trucks

Cost estimating for underground mines. For an underground mine, the capital cost for the initial entries into the mine are necessary, whether entry is by adit, decline, shaft, or combinations thereof. Included are the following:

- The ore storage excavations or structure, between the operating production units and the shaft or decline, the crusher station, and skip station
- The shaft stations at each level
- The horizontal development to set up the haulage system for the mine and to the various portions of the ore body
- The vertical development of orepasses, and slot and vent raises related to stoping
- Main ventilation adits or shafts
- Pump excavation rooms, along with the cleanable water sumps
- Underground mine shops
- Supply storage room
- Lunch room/rescue chambers
- Fuel and lubrication storage rooms

To capital cost for the development of these openings and facilities, all material and labor costs will have to be included; and through a functional analysis of time to perform each and every task, a unit cost can be built up for each type of opening. An example of a worksheet for one such opening (crusher station) is shown in Table 14.15, so the reader can better understand the process if he or she is not already familiar with this type of system. This type of detailed cost estimate on hundreds of underground development openings must be completed. All cost estimators have their own type of worksheet, and there is nothing special about this one. Next, it is a matter of documenting and totaling the capital cost of all such worksheets for all of the openings and facilities.

TABLE 14.15 Example of Acme mine construction cost estimate*

WBS Number: 40721 Task Title: Crusher Station Construction Task Duration: 25 days								Acme Mine Construction Cost Estimate Golden Promise Mine							Date: 5/15/01 Pages: 57 of 113 Quantity: 3,700 yd^3
LABOR SHIFT							MATERIALS				EQUIPMENT RENTALS				
Days	Labor Title	1	2	3	Daily Rate ($)	Labor Cost ($)	Qty.	Consumable Description	Unit Price ($)	Material Cost ($)	Days	No. Each	Item	Unit Price ($)	Equipment Cost ($)
25	Development miner	4	4	4	176	52,500	1000	Roof bolts	25.00	25,000	25	1	75-Ton Truck	1,300	32,500
25	Hoist operator	1	1	1	200	15,000	9000	ANFO	0.25	2,250					
25	Electrician	1	1	1	193	14,475	2750	Detonator	2.25	6,189					
25	Maintenance worker	2	2	2	184	27,600	85	Shotcrete	115.00	9,583					
25	Surface tender	1	1	1	144	10,800	10	Drill steel	110.00	1,100					
25	Helper/truck driver	1	1	1	128	9,600	75	Drill bits	15.00	1,125					
25	Shifter	1	1	1	208	15,600									
								CAPITAL ($)							
Total Labor Cost:						145,575	Consumables Total:			45,247	Equipment Total:			32,500	
Applied to Labor:		0.65				94,624	Capital Total (not shown on this page):			0	Gen. & Adm.:		0.16	5,200	
Gen. & Adm.:		0.16				23,292	Freight:		0.11	4,977	Equipment Rental Total:			37,700	
Labor Total:						263,491	Handling:		0.16	7,240					
							Gen. & Adm.:		0.16	7,240					
							Material Totals:			64,704					
Remarks: Average Employees 33							Cost with Contingency:				Capital Cost:			0	
							Labor:			$303,015	Construction Cost:			$365,895	
							Consumables:			$74,409	Subtotal			$365,895	
							Rental Equipment:			$43,355	Total Contingency @ 15%			$54,884	
							Total Cost			$420,779	Total Cost			$420,779	

*For illustration only; not an accurate estimate.

For the capital equipment of an underground mine, you would need some of the following:

- Surface service equipment
  - Hoists and related control equipment
  - Ropes and attachments
  - Conveyances and counterweights
  - Headframe(s), sheave wheels, dump scrolls, and other control equipment
  - Surface ore bin
  - Compressor
  - Communication equipment (for surface and underground)
  - Ventilation equipment
  - Power transformations equipment and power cable
- Underground service equipment
  - Shaft hardware
  - Skip loading measuring and load control equipment
  - Crusher and conveyance to skip loading
  - Pump, piping, and pump control equipment
  - Bridge cranes, jib hoist, and other shop facilities and equipment
  - Initial setup of shop tools
  - Fueling and lubrication equipment
  - Supply house equipment
  - Underground power stations
- Rock-fracturing equipment
  - Development drills and drill jumbos
  - Stoping drills and drill jumbos
  - Explosives loading equipment
- Mechanical excavating equipment
  - Longwall units
  - Shearers
  - Continuous miner
  - Roadheaders
  - Tunnel boring machine
- Loading and hauling equipment
  - Load-haul-dump equipment (diesel, electric, or battery)
  - Front-end loaders
  - Trucks
  - Shuttle cars
  - Rail locomotives, initial rails and ties, and haulage cars

  - Conveyor system
- Stope equipment
  - Slushers and scraper buckets
  - Backfill equipment
  - Feeder breaker/mobile crusher
- Miscellaneous mobile service equipment
  - Rock and/or cable bolting jumbo(s)
  - Maintenance/lube truck(s)
  - Prospecting drill rigs
  - Personnel transportation vehicles
  - Foreman's vehicles

In-situ solution mine, heap or dump leach. The capital cost for an in-situ solution mine is the required drilling of the initial well field, the well completions, casings, pumps, distribution field piping, booster pumps, and storage tanks and/or tank trucks to move the pregnant solution to the process plant. The capital cost of the heap or dump leach is the heap leach pad and liner, the barren pond and pregnant pond liner and construction, any necessary diversion channel or ditches, collection ditches of systems, pumps distribution and heading piping going to the field of emitter lines and emitters, pumps and piping to deliver the pregnant solution to the process plant, safety berms, and monitoring wells.

Process plant planning leading to capital cost analysis. The capital cost for a process plant is variable, depending of the type of mineral product being processed. Most process plants are surrounded by buildings, which need to be costed. Inside the process plant, some of the following equipment would be included:

- Comminution equipment
  - Primary crushers (which may be located at the mine)
  - Secondary and tertiary crushers
  - Roll crushers (for unique applications, roll crushers may be primary, secondary, or tertiary)
  - Screening plant
  - Conveyor system
  - Grinding mills (including initial rod and ball charge)
  - Tower mills
  - Air cleaning equipment
    - Multi-cyclones
    - Electrostatic separators
    - Baghouse
    - Scrubber systems
  - Classifying system and pumping stations
    - Hydrocyclones

    - Spiral or rake classifier
- Concentrating/beneficiation equipment
  - Gravity and classification systems
    - Tables (wet and dry)
    - Mineral jigs (pulsating and centrifugal)
    - Coal jigs
    - Gravity cones, spirals and bowls
    - Dense media systems
    - Dense media cyclones
  - Magnetic separators
    - Low intensity
    - High intensity
    - Superconducting magnets
  - Electrostatic plate separator
  - Ore sorters
    - Color
    - Reflectivity
    - Radioactivity
  - Flotation systems (including associated pumps and piping)
    - Mechanical/pneumatic machine
    - Column
    - Electroflotation
  - Leaching plant systems
    - Mixing tanks
      - Stirred tank
      - Pressure tank
      - Percolation tank
    - Solution purification equipment
      - Ion-exchange columns
      - Elution tanks
      - Precipitation tanks
      - Solvent extraction
    - Metal recovery systems
      - Electrowinning system
      - Precipitation tanks
    - Agglomeration systems
      - Pug mills
      - Pelletizing mills
      - Sintering system

- Thickening, filtering, and drying (including associated pumps, compressors, and piping)
  - Thickeners
    - Rake
    - Lamella
  - Clarifiers
  - Centrifuge system
  - Vacuum filters
    - Drum
    - Disk
    - Belt
  - Pressure filters
    - Plate frame
    - Leaf
  - Dryers
    - Rotary
    - Flash
    - Fluosolids systems
- Materials handling systems
  - Plant conveyors
  - Weigh scales
  - Belt magnet
  - Sampling system
  - Pumps and piping
  - Product load-out system
  - Switch engine
- Plant control system
  - Onstream analyzer system
    - Individual probes and sensors
    - Multiple sample systems
  - Computer control system
  - Control console room
  - Automatic device control system on every piece of automated operating equipment activated by programmable logical control, fuzzy logic, or neural network system
  - Density meters on grinding feed lines
  - Particle size analyzers on mill discharge lines

Operating Cost Estimates

Operating costs are those ongoing, recurring costs incurred during normal functioning of the operations. In general, total operating costs can be divided into three primary classifications: direct, indirect, and general overhead. In the following lists, overhead is included with the indirect cost.

Direct operating costs include the following:

- Labor
  - Direct operating and stope development labor
  - Operating supervision
  - Direct maintenance
  - Maintenance supervision
  - Payroll burden (all of the labors benefit over and above the base rate)
- Materials
  - Maintenance and repair materials
  - Process materials
  - Raw materials
  - Consumables (fuel, power, water, etc.)
- Royalties
- Development (for production area)

Indirect operating costs include the following:

- Labor
  - Administrative
  - Maintenance
  - Technical staff
  - Service (clerical, accounting, and general office)
  - Shop (fabrication and major overhaul)
  - Payroll burden on the foregoing
- Utilities and telephone
- Insurance (property and liability)
- Depreciation
- Interest on investment
- Taxes
- Reclamation (bonding and accruals)
- Travel, meetings, and donations
- Office supplies, lubricants, and oil
- Public relations
- General mine development
- General overhead (sometimes time charged to the operating costs; these may include marketing/sales and a proportionate amount of corporate administrative labor and costs.)

Measured time units are usually satisfactory for operating cost estimates. The development of owning and operator costs, using manufacturers' procedures and historical information, is a good method of estimating that portion of the operating cost. Caterpillar's Fleet Production and Cost Analysis software (Caterpillar 2016) is a good example of a computerized program for estimating surface machinery. For most preliminary feasibility studies, the previously mentioned techniques will suffice.

For labor and materials costs, at the intermediate and final feasibility studies, performing a thorough functional analysis on each unit operation will lead to labor and material breakdowns that can be costed more accurately than historical similar mines. These estimates will develop such unit costs as dollars per foot of drift development, dollars per ton overburden removed, dollars per ton drilled, dollars per ton mined, and so on. On top of these costs, the repair and maintenance, general and administrative, insurance and taxes, and indirect and payroll burden will have to be added. This method of functional analysis is described in more detail in Chapter 15.

The detailed development and operating cost estimate should be built up by function. All costs are incurred for some specific purpose, and it is this purpose that becomes the function which defines how each cost is incurred. Mining productivity of the operation must also be developed by elemental time analysis of each mining function—that is, drilling, charging, loading, hauling, mine development (stripping, drifts, crosscuts, raises, etc.), road building, and maintenance—as well as for every other operational service productivity for the maintenance, ground support, prospecting, and so forth. This will then lead to the building of the labor required for each function. Subsequently, a table of labor productivities for each and every activity in the mining operation can be created. Next, when the dollars per shift are applied to the labor productivities, the unit labor dollars per foot or dollars per ton can be derived for each of the functions.

Likewise, material and supply cost, by function, will need to be developed from the volumes of material that will be excavated and moved. As stated previously, for equipment loading and hauling operating cost, the vendor's computer simulation programs can be used for the direct estimates of cost, but the sometimes-optimistic productivities and unit cost may need adjusting. If the vendor simulation models are used, care must be given so that double-dipping into the labor and material previously added in does not occur for that particular function.

Therefore, when all of the above is put together, one should be able to develop a table of costs per unit (whether it be \$/ft, \$/ton, or \$/specific activity).

More detailed information for the method and description of developing operating costs for processing plants is provided in Chapter 16.

REFERENCES

AACE International. 2005. *Cost Engineering Terminology.* AACE International Recommended Practice No. 10S-90 (revised 2014). Morgantown, WV: AACE International.

Aventurine Engineering. 1998a. Sherpa, Cost Estimating for Surface Mines. Proprietary/commercial software, Version 1.98. Elk, WA: Aventurine Engineering.

Aventurine Engineering. 1998b. Sherpa, Cost Estimating for Underground Mines. Proprietary/commercial software, Version 1.97. Elk, WA: Aventurine Engineering.

Aventurine Engineering. 2014a. Sherpa for Surface Mines. Proprietary/commercial software, Version 2.14. Spokane Valley, WA: InfoMine USA, CostMine Division. http://costs.infomine.com/software/

Aventurine Engineering. 2014b. Sherpa for Underground Mines. Proprietary/commercial software, Version 2.14. Spokane Valley, WA: InfoMine USA, CostMine Division. http://costs.infomine.com/software/

Camm, T.W. 1994a. Simplified cost models for prefeasibility mineral evaluations. *Mining Engineering,* 46(6):559–562.

Camm, T.W. 1994b. The Public Version of CES, The USBM Cost Estimating System, *Branch of Engineering and Economic Analysis, Western Field Operations Center,* U.S. Bureau of Mines, Spokane, Washington. Software issued for beta testing, 1994.

Caterpillar. 2016. Caterpillar Fleet Production and Cost Analysis (CD), Version 3. Peoria, IL: Caterpillar

Clements et al. 1975. *USBM Mineral Availability System, Cost Estimating System.* Denver, CO: USBM Denver Research Center.

Compass International. 2017. *2017 Global Construction Costs Yearbook.* Morrisville, PA: Compass International Consultants. www.compassinternational.net/construction-estimating-yearbooks/physical-books/2016-global-construction-costs-yearbook/

Darling, P., ed. 2011. In *SME Mining Engineering Handbook*, 3rd ed. Englewood, CO: SME.

Grogan, T., et al. 1998. Summary: Currents behind calm costs, first quarterly cost report. *Eng. News Record* (March 30): 29–45.

Henry, J.R., and Wagner, L.A. 1995. Economic Analysis Tools for the Mineral Industry, CES-Software CD, No. SP 17-95. U.S. Bureau of Mines, Denver Federal Center, December 1995.

Hickson, R.J., and Owen, T.L. 2015. *Project Management for Mining: Handbook for Delivering Project Success.* Englewood, CO: SME.

Hustrulid, W.A., and Bullock, R., eds. 2001. *Underground Mining Methods: Engineering Fundamentals and International Case Studies.* Littleton, CO: SME.

Infomine USA 2016a. *Mine and Mill Equipment Costs: An Estimator's Guide.* Spokane Valley, WA: InfoMine USA, CostMine Division.

Infomine USA 2016b. *Mining Cost Service*, a subscription cost data update service. Spokane Valley, WA: InfoMine USA.

JORC (Joint Ore Reserves Committee). 2012. *The Australasian Code for Reporting of Exploration Results, Mineral Resources and Ore Reserves* (The JORC Code). Gosford, New South Wales: JORC. http://jorc.org/docs/jorc_code2012.pdf

Lemmons. 1984. *USBM Mineral Availability System, Cost Estimating System.* Mainframe computer tape for beta testers, Denver, CO: USBM Denver Research Center.

Mular, A.L. 1978. *Mineral Processing Equipment Costs and Preliminary Capital Cost Estimations.* CIM Special Edition 18. Montreal: Canadian Institute of Mining and Metallurgy.

NI 43-101. *Standards of Disclosure for Mineral Projects.* Canadian National Instrument 43-101. Ontario Securities Commission. www.osc.gov.on.ca/en/15019.htm.

Noakes, M. 1993. Using the handbook. In *Cost Estimation Handbook for the Australian Mining Industry.* Edited by M. Noakes and T. Lanz. Victoria, Australia: Australasian Institute of Mining and Metallurgy. pp. 6–7.

Numbeo.com. 2017. Cost of living index by country. https://www.numbeo.com/cost-of-living/gmaps_rankings_country.jsp?indexToShow=getCpiIndex&title=2015-mid&OK=OK (accessed March 2017).

O'Hara, T.A. 1979. Quick Guide to Evaluation of Orebodies, *81st CIM General Meeting*, Apr., 1979, Montreal.

Schumacher, O.L. 1993. Mining Cost Service, a subscription cost data update service. Spokane, WA: Western Mine Engineering.

Scott, J.B. 1988. Computerized cost estimating for underground mines, *CIM* 81(916):44–50.

SEC (U.S. Securities and Exchange Commission). 1992. *Industry Guide 7: Description of Property by Issuers Engaged or to Be Engaged in Significant Mining Operations.* sec.gov/about/forms/industryguides.pdf (accessed Jan. 17, 2017).

SME (Society for Mining, Metallurgy & Exploration). 2017. *The SME Guide for Reporting Exploration Information, Mineral Resources, and Mineral Reserves.* Englewood, CO: SME.

Stebbins, S.A. 2011. Cost estimating for underground mines. In *SME Mining Engineering Handbook*, 3rd ed. Edited by P. Darling. Englewood, CO: SME. pp. 263–279.

Stebbins, S.A., and Leinart, J.B. 2011. Cost estimating for surface mines. In *SME Mining Engineering Handbook*, 3rd ed. Edited by P. Darling. Englewood, CO: SME. pp. 281–293.

Stebbins, S.A., and Schumacher, O.L. 2001. Cost estimating for underground mines. In *Underground Mining Methods: Engineering Fundamentals and International Case Studies.* Edited by W.A. Hustrulid and R.L. Bullock. Littleton, CO: SME. pp. 49–72.

USBLS (U.S. Bureau of Labor Statistics). 2015. BLS database. https://data.bls.gov/cgi-bin/surveymost

USBM (U.S. Bureau of Mines). 1994. Cost Estimating System software CD. Washington, DC: USBM.

Wiggins, T. 2013. International construction intelligence, parity index Q2. New York: Faithful+Gould. https://www.fgould.com/americas/articles/international-construction-intelligence-usbase-13/ (accessed March 2017).

APPENDIX 14A

Cost Models for a Surface Mine*

Equipment purchase and operating costs used in the models shown in Table 14A.1 are current costs drawn from *Mine and Mill Equipment Costs: An Estimator's Guide* by CostMine, a division of InfoMine USA (2016a). A sales tax rate of 7.24% is added to all equipment and nonfuel supply prices. Sufficient working capital is included to cover two months of operations. All units are metric and all costs are in 2014 U.S. dollars.

TABLE 14A.1 Cost models for a surface mine (10,000 metric tons of ore per day)

	Stripping Ratio (waste/ore)			
Cost Parameters	**1:1**	**2:1**	**4:1**	**8:1**
Ore production, t/d	10,000	10,000	10,000	10,000
Waste production, t/d	10,000	20,000	40,000	80,000
Total resource, t	37,440,000	37,440,000	37,440,000	37,440,000
Final Pit Dimension				
Pit depth, m	193	222	264	322
Pit floor length, m	352	404	481	587
Pit floor width, m	176	202	241	293
Pit wall slope, degrees	50	50	50	50
Haul Profile, Ore				
Face to pit ramp				
Distance, m	198	208	225	253
Gradient, %	0.0	0.0	0.0	0.0
Ramp entrance to pit exit				
Distance, m	947	1,290	1,760	2,360
Gradient, %	12.0	12	12.0	12.0
Pit exit to mill				
Distance, m	783	901	1,072	1,308
Gradient, %	2.0	2.0	2.0	2.0
Haul Profile, Waste				
Face to pit ramp				
Distance, m	268	298	345	412
Gradient, %	0.0	0.0	0.0	0.0

(Table continues)

* Appendixes 14A, 14B, and 14C are provided by InfoMine USA and Aventurine Engineering in Spokane Valley, Washington.

TABLE 14A.1 (Continued)

	Stripping Ratio (waste/ore)			
Cost Parameters	**1:1**	**2:1**	**4:1**	**8:1**
Ramp entrance to pit exit				
Distance, m	254	394	572	782
Gradient, %	12.0	12.0	12.0	12.0
Pit exit to waste stockpile				
Distance, m	392	451	536	654
Gradient, %	2.0	2.0	2.0	2.0
Stockpile base to surface				
Distance, m	489	616	776	977
Gradient, %	12.0	12.0	12.0	12.0
Across stockpile to dump				
Distance, m	249	313	395	497
Gradient, %	0.0	0.0	0.0	2.0
Production				
Hours per shift	10	10	10	10
Shifts per day	2	2	2	2
Days per year	312	312	312	312
Bench height, ore, m	4.60	4.60	4.60	4.60
Bench height, waste, m	6.72	6.72	6.72	6.72
Powder factor, ore, kg/t	0.33	0.33	0.33	0.33
Powder factor, waste, kg/t	0.29	0.29	0.29	0.29
Development				
Preproduction stripping, t	300,000	600,000	1,200,000	2,400,000
Haul road construction, m	3,580	4,470	5,681	7,244
Equipment, (Number) Size				
Hydraulic shovels, m^3	(1) 8.4	(1) 8.4	(1) 8.4	(1) 8.4
Front-end loaders, m^3	(1) 12.2	(1) 16.1	(2) 16.1	(3) 19.9
Rear-dump trucks, t	(8) 54.0	(10) 77.0	(20) 91.0	(26) 136.0
Rotary drills, cm	(3) 20.00	(2) 25.08	(3) 27.94	(4) 31.12
Bulldozers, kW	(3) 110	(4) 140	(5) 140	(7) 180
Graders, kW	(1) 115	(1) 140	(2) 140	(2) 160
Water tankers, L	(1) 19,000	(1) 19,000	(1) 26,500	(1) 30,000
Service/tire trucks, kg gvw	(4) 6,800	(5) 11,000	(8) 11,000	(11) 11,000
Bulk trucks, kg/min	(1) 450	(1) 450	(1) 450	(2) 450
Light plants, kW	(4) 8.9	(4) 10.1	(5) 10.1	(6) 10.1
Pumps, kW	(3) 37.3	(3) 74.6	(4) 74.6	(5) 93.2
Pickup trucks, kg	(7) 680	(8) 900	(12) 900	(17) 900

(Table continues)

TABLE 14A.1 (Continued)

Cost Parameters	Stripping Ratio (waste/ore)			
	1:1	2:1	4:1	8:1
Buildings				
Shop, m^2	906	1,063	2,404	4,348
Dry, m^2	349	447	720	1,173
Office, m^2	587	715	1,047	1,533
Warehouse, m^2	466	506	776	1,361
ANFO storage bin, m^3	64	81	132	234
Hourly Personnel Requirements				
Drillers	4	3	4	5
Blasters	2	2	2	4
Excavator operators	4	4	6	8
Truck drivers	13	18	34	47
Equipment operators	7	9	11	15
Utility operators	3	3	4	5
Mechanics/electricians	10	14	24	45
Laborers/maintenance	17	24	39	73
Total hourly personnel	**60**	**77**	**124**	**202**
Salaried Personnel Requirements				
Managers	1	1	1	1
Superintendents	1	1	1	1
Foremans	2	2	4	4
Engineers	2	2	3	5
Geologists	1	2	3	4
Supervisors	3	3	6	9
Technicians	5	6	8	11
Accountants	1	1	2	3
Clerks	2	3	4	7
Personnel managers	1	2	2	4
Secretaries	3	4	5	8
Security	1	1	2	3
Total salaried personnel	**23**	**28**	**41**	**60**
Primary Supply Requirements				
Diesel fuel, L/d	9,527	17,130	33,183	53,514
Powder, kg/d	7,250	9,102	14,903	26,505
Caps, caps/d	72	55	67	92
Primers, primers/d	68	51	63	88
Drill bits, bits/d	1.570	1.253	1.654	2.372
Detonation cord, m/d	861	688	907	1,301

(Table continues)

TABLE 14A.1 (Continued)

Cost Parameters	Stripping Ratio (waste/ore)			
	1:1	2:1	4:1	8:1
COST SUMMARY				
Operating Costs				
Supplies, $/t ore	1.30	1.57	.54	$4.38
Hourly labor, $/t ore	2.22	2.94	4.65	7.66
Equipment operation, $/t ore	1.60	2.72	5.10	10.57
Salaried labor, $/t ore	0.78	0.95	1.38	2.00
Miscellaneous, $/t ore	0.59	0.82	1.37	2.46
Total operating costs	**6.49**	**9.00**	**15.04**	**27.07**
Capital Costs				
Equipment purchase	17,270,800	24,984,100	45,976,600	105,264,500
Haul roads/site work	2,299,900	3,095,700	4,339,600	6,931,200
Stripping	813,500	1,455,100	2,900,200	6,028,600
Buildings	5,272,800	6,256,600	11,195,600	18,940,800
Electrical	211,100	223,200	488,100	512,300
Sustaining capital	3,187,900	6,675,700	8,548,400	17,283,200
Working capital	1,582,800	2,039,200	3,188,900	5,336,600
Engineering	4,791,400	6,702,600	12,149,100	25,983,800
Contingency	3,065,900	4,271,700	7,704,900	16,366,100
Total capital costs	**38,496,100**	**55,703,900**	**96,491,400**	**202,647,100**
Total capital cost per daily metric ton ore and waste	**1,925**	**1,857**	**1,930**	**2,252**

Source: Infomine USA 2016b

APPENDIX 14B

Cost Models for a Block Caving Underground Mine with Adit Access

Equipment purchase and operating costs used in the models shown in Figure 14B.1 and Table 14B.1 are current costs drawn from *Mine and Mill Equipment Costs, An Estimator's Guide* by CostMine, a division of InfoMine USA (2016a). A sales tax rate of 7.24% is added to all equipment and nonfuel supply prices. The underground mine models were constructed using Sherpa for Underground Mines, published by Aventurine Engineering in cooperation with CostMine (Aventurine Engineering 2014b). All units are metric and all costs are in 2014 U.S. dollars.

These three models represent mines on large, bulk deposits, roughly 450, 525, and 600 m^2. Access is through three to five adits, 1,605 m long. Ore is collected using slushers, and haulage from the stopes is by diesel locomotive. Diesel locomotives also haul ore to the surface. Stope development includes driving drifts (haulage, slusher, and undercut) and raises (stope draw, orepass, and boundary weakening). Caving is initiated through minor blasting from the undercut drifts.

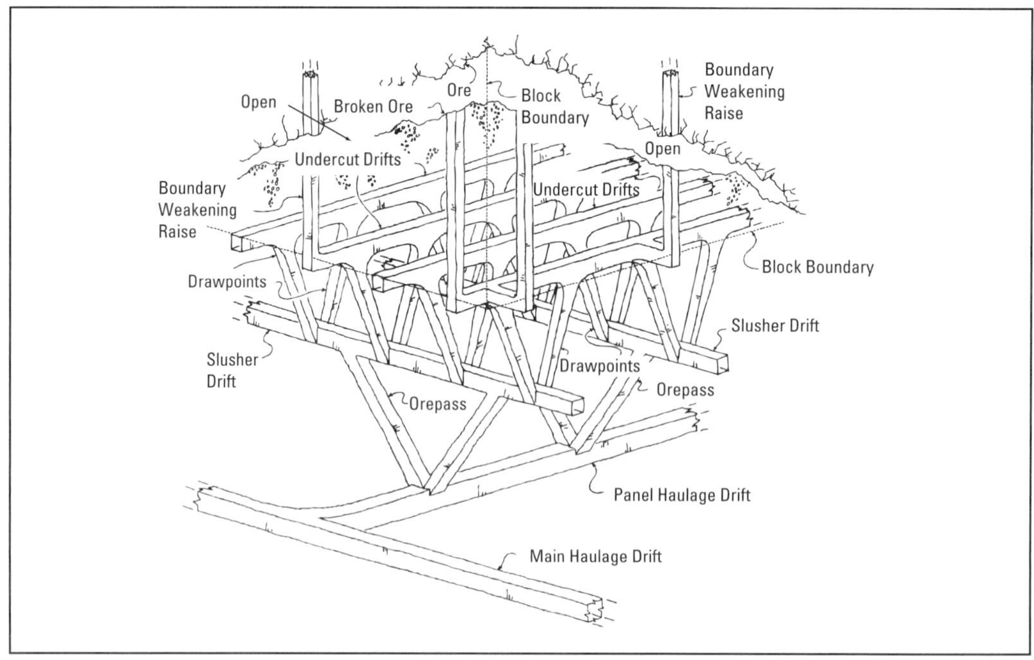

Source: Stebbins 2011

FIGURE 14B.1 Block caving, adit access

TABLE 14B.1 Cost models for block caving, adit access

	Daily Ore Production, metric tons		
Cost Parameters	**20,000**	**30,000**	**45,000**
Production			
Hours per shift	8	8	8
Shifts per day	3	3	3
Days per year	365	365	365
Deposit			
Total minable resource, t	84,000,000	147,000,000	252,000,000
Average maximum horizontal, m	450	525	600
Average minimum horizontal, m	450	525	600
Average vertical, m	150	200	250
Ore			
Density, m^3/t	2.73	2.73	2.73
Swell, %	55	55	55
Compressive strength, kPa	103,420	103,420	103,420
Rock quality designation, %	65	65	65
Footwall			
Density, m^3/t	2.49	2.49	2.49
Swell, %	45	45	45
Compressive strength, kPa	172,370	172,370	172,370
Rock quality designation, %	80	80	80
Hanging Wall			
Density, m^3/t	2.49	2.49	2.49
Swell, %	45	45	45
Compressive strength, kPa	103,420	103,420	103,420
Rock quality designation, %	65	65	65
Blocks			
Block length, m	150	175	200
Block width, m	150	175	200
Block height, m	150	200	250
Development Openings			
Adits	3 ea.	4 ea.	5 ea.
Face area, m^2	26.0	26.0	26.0
Preproduction advance, m	1,605	1,605	1,605
Cost, adit 1, $/m	3,496	4,180	4,194
Cost, adit 2, $/m	3,541	4,225	4,239
Cost, adit 3, $/m	3,541	4,225	4,239
Cost, adit 4, $/m	—	4,225	4.239
Cost, adit 5, $/m	—	—	4,239
Drifts			
Face area, m^2	21.2	35.2	38.7
Daily advance, m	5.1	5.8	7.0
Preproduction advance, m	463	531	637
Cost, $/m	1,796	1,859	2,222

(Table continues)

TABLE 14B.1 (Continued)

	Daily Ore Production, metric tons		
Cost Parameters	**20,000**	**30,000**	**45,000**
Crosscuts			
Face area, m^2	21.2	35.2	38.7
Daily advance, m	3.5	4.0	4.8
Preproduction advance, m	320	366	439
Cost, $/m	1,653	1,712	2,072
Drawpoints			
Face area, m^2	23.2	23.2	23.2
Daily advance, m	14.8	16.8	20.0
Preproduction advance, m	1,352	1,536	1,828
Cost, $/m	994	1,000	1,007
Orepasses			
Face area, m^2	18.8	27.9	41.5
Daily advance, m	0.42	0.65	1.07
Preproduction advance, m	1,219	1,600	2,000
Cost, $/m	1,516	2,100	2,989
Boundary raises			
Face area, m^2	4.1	4.3	4.5
Daily advance, m	3.6	4.1	4.9
Preproduction advance, m	331	376	450
Cost, $/m	795	797	801
Ventilation raises			
Face area, m^2	38.0	56.3	83.7
Daily advance, m	0.18	0.20	0.21
Preproduction advance, m	396	500	600
Cost, $/m	1,525	2,106	2,974
Hourly Labor Requirements			
Undercut miners	24	24	30
Development miners	22	24	28
Motormen	9	12	15
Equipment (loader/truck) operators	41	64	100
Support miners	46	58	80
Diamond drillers	6	10	18
Electricians	6	6	6
Mechanics	15	15	17
Maintenance workers	24	30	38
Helpers	8	9	10
Underground laborers	30	38	48
Surface laborers	24	30	38
Total hourly personnel	**255**	**320**	**428**

(Table continues)

TABLE 14B.1 (Continued)

Cost Parameters	Daily Ore Production, metric tons		
	20,000	30,000	45,000
Salaried Personnel Requirements			
Managers	1	1	1
Superintendents	4	4	4
Foremen	30	42	63
Engineers	8	10	12
Geologists	9	11	14
Shift bosses	6	9	12
Technicians	16	20	24
Accountants	9	11	14
Purchasing agents	14	17	22
Personnel managers	13	16	21
Secretaries	24	30	38
Clerks	30	38	48
Total salaried personnel	**164**	**209**	**273**
Supply Requirements (Daily)			
Explosives, kg	1,863	2,516	3,494
Caps, each	633	707	871
Boosters, each	622	688	837
Fuse, m	3,407	4,923	6,913
Drill bits, each	18.70	26.78	35.87
Drill steel, each	1.042	1.481	1.915
Freshwater pipe, m	12.2	13.9	16.7
Compressed air pipe, m	12.2	13.9	16.7
Electric cable, m	12.2	13.9	16.7
Ventilation tubing, m	12.2	13.9	16.7
Rock bolts, no.	30	37	48
Shotcrete, m^3	1	1	1
Concrete, m^3	55	70	97
Buildings			
Office, m^2	4,190	5,339	6,974
Changehouse, m^2	2,961	3,716	4,970
Warehouse, m^2	511	556	601
Shop, m^2	1.080	1,182	1,283
Mine plant, m^2	150	150	150
Equipment Requirements, (Number) Size			
Undercut drills, cm	(2) 4.76	(2) 5.08	(2) 5.72
Drawpoint scoop trams, m^3	(2) 7.6	(3) 7.6	(4) 7.6
Horizontal development jumbo drills, cm	(2) 3.17	(2) 3.17	(3) 3.49
Raise drills, cm	(4) 4.76	(6) 5.08	(7) 5.72
Development scoop trams, m^3	(3) 7.6	(3) 7.6	(3) 7.6
Locomotives, t	(7) 31.8	(8) 31.8	(10) 31.8

(Table continues)

TABLE 14B.1 (Continued)

Cost Parameters	Daily Ore Production, metric tons		
	20,000	30,000	45,000
Rock-bolt drills, cm	(3) 3.81	(3) 3.81	(3) 3.81
Freshwater pumps, hp	(6) 2.5	(8) 3.5	(10) 5.2
Service vehicles, hp	(5) 210	(5) 210	(6) 210
Compressors, m^3/min	(1) 142	(1) 227	(1) 227
Ventilation fans, cm	(2) 366	(4) 366	(6) 366
Exploration drills, cm	(1) 4.4	(2) 4.4	(3) 4.4
Equipment Costs ($/unit)			
Undercut drills	515,300	515,300	515,300
Drawpoint scoop trams	1,340,500	1,340,500	1,340,500
Horizontal development jumbos	1,145,500	1,145,500	1,149,300
Vertical development stopers	10,500	10,500	10,500
Development scoop trams	1,340,500	1,340,500	1,340,500
Main haulage locomotives w/cars	1,033,100	1,033,100	1,033,100
Rock bolt jacklegs	9,200	9,200	9,200
ANFO loaders	47,200	47,200	47,200
Freshwater pumps	10,700	11,500	11,500
Service vehicles	185,900	185,900	185,900
Compressors	108,900	108,900	108,900
Ventilation fans	521,000	451,200	521,000
Exploration drills	292,000	292,000	292,000
	COST SUMMARY		
Operating Costs ($/t ore)			
Equipment operation	2.12	2.16	2.32
Supplies	0.90	0.72	0.64
Hourly labor	3.65	3.03	2.71
Administration	2.26	1.91	1.67
Sundries	0.89	0.78	0.73
Total operating costs	**9.82**	**8.60**	**8.07**
Unit Operating Cost Distribution ($/t ore)			
Stopes	2.81	2.62	2.61
Drifts	0.45	0.36	0.30
Crosscuts	0.31	0.25	0.21
Drawpoints	0.67	0.54	0.43
Boundary raises	0.22	0.18	0.14
Orepasses	0.03	0.04	0.05
Ventilation raises	0.01	0.01	0.01
Main haulage	0.87	0.88	0.89
Services	1.25	1.00	0.93
Ventilation	0.21	0.21	0.27
Exploration	0.11	0.13	0.16
Maintenance	0.32	0.24	0.21

(Table continues)

TABLE 14B.1 (Continued)

	Daily Ore Production, metric tons		
Cost Parameters	**20,000**	**30,000**	**45,000**
Salaries	1.67	1.36	1.13
Miscellaneous	0.89	0.78	0.73
Total operating costs	**9.82**	**8.60**	**8.07**
Capital Costs			
Equipment purchase	15,649,100	19,105,200	24,457,800
Preproduction underground excavation			
Adit 1	5,611,100	6,708,700	6,731,100
Adit 2	5,683,700	6,781,300	6,803,700
Adit 3	5,683,700	6,781,300	6,803,700
Adit 4	—	6,781,300	6,803,700
Adit 5	—	—	6,803,700
Drifts	832,100	986,400	1,414,800
Crosscuts	528,100	626,700	910,000
Drawpoints	1,343,200	1,536,100	1,841,000
Boundary raises	263,000	300,200	360,400
Orepasses	1,848,200	3,360,000	5,976,900
Ventilation raises	604,400	1,052,700	1,784,400
Surface facilities	15,139,000	17,373,100	21,504,900
Working capital	11,919,700	15,721,800	22,114,300
Engineering and management	6,914,100	9,281,100	11,985,500
Contingency	5,318,600	7,139,300	9,219,600
Total capital costs	**77,338,000**	**103,535,200**	**135,515,500**
Total capital cost per daily metric ton ore	**3,867**	**3,451**	**3,011**

Source: Infomine USA 2016b

APPENDIX 14C

Cost Models for a Block Caving Underground Mine with Shaft Access

Equipment purchase and operating costs used in the models shown in Table 14C.1 are current costs drawn from *Mine and Mill Equipment Costs, An Estimator's Guide* by CostMine, a division of InfoMine USA (2016a). A sales tax rate of 7.24% is added to all equipment and nonfuel supply prices. Sufficient working capital is included to cover two months of operations. The underground mine models were constructed using Sherpa for Underground Mines, published by Aventurine Engineering in cooperation with CostMine (Aventurine Engineering 2014b). All units are metric and all costs are in 2014 U.S. dollars.

These three models represent mines on massive deposits, roughly 450, 525, and 600 m^2. Access is through three to five shafts, 430, 530, and 630 m deep, and by secondary access/ventilation raises. Ore is collected using slushers, and haulage from the stopes is by diesel locomotive. Stope development includes driving drifts (haulage, slusher, and undercut) and raises (stope draw, orepass, and boundary weakening). Caving is initiated through minor blasting from the undercut drifts.

TABLE 14C.1 Cost models for block caving, shaft access

	Daily Ore Production, metric tons		
Cost Parameters	**20,000**	**30,000**	**45,000**
Production			
Hours per shift	8	8	8
Shifts per day	3	3	3
Days per year	365	365	365
Deposit			
Total minable resource, t	84,000,000	147,000,000	252,000,000
Average maximum horizontal, m	450	525	600
Average minimum horizontal, m	450	525	600
Average vertical, m	150	200	250
Ore			
Density, m^3/t	2.73	2.73	2.73
Swell, %	55	55	55
Compressive strength, kPa	103,420	103,420	103,420
Rock quality designation, %	65	65	65
Footwall			
Density, m^3/t	2.49	2.49	2.49
Swell, %	45	45	45

(Table continues)

TABLE 14C.1 (Continued)

	Daily Ore Production, metric tons		
Cost Parameters	**20,000**	**30,000**	**45,000**
Compressive strength, kPa	172,370	172,370	172,370
Rock quality designation, %	80	80	80
Hanging Wall			
Density, m^3/t	2.49	2.49	2.49
Swell, %	45	45	45
Compressive strength, kPa	103,420	103,420	103,420
Rock quality designation, %	65	65	65
Blocks			
Block length, m	150	175	200
Block width, m	150	175	200
Block height, m	150	200	250
Development Openings			
Shafts	3 each	4 each	5 each
Face area, m^2	43.2	48.4	54.2
Preproduction advance, m	1,290	2,120	3,150
Cost, shaft 1, $/m	19,458	20,720	21,825
Cost, shaft 2, $/m	19,850	20,980	22,251
Cost, shaft 3, $/m	19,990	21,124	22,400
Cost, shaft 4, $/m	—	21,188	22,462
Cost, shaft 5, $/m	—	—	22,503
Drifts			
Face area, m^2	21.2	23.0	24.8
Daily advance, m	5.2	5.8	7.0
Preproduction advance, m	472	531	637
Cost, $/m	1,786	1,859	2,212
Crosscuts			
Face area, m^2	21.2	23.0	24.8
Daily advance, m	3.6	4.0	4.8
Preproduction advance, m	325	366	439
Cost, $/m	1,642	1,712	2,062
Drawpoints			
Face area, m^2	23.2	23.2	23.2
Daily advance, m	15.1	16.8	20.0
Preproduction advance, m	1,376	1,536	1,828
Cost, $/m	993	1,000	1,007
Orepasses			
Face area, m^2	18.8	27.9	41.5
Daily advance, m	0.43	0.65	1.07
Preproduction advance, m	1,200	1,600	2,000
Cost, $/m	1,518	2,100	2,989

(Table continues)

TABLE 14C.1 (Continued)

Cost Parameters	Daily Ore Production, metric tons		
	20,000	30,000	45,000
Boundary raises			
Face area, m^2	4.1	4.3	4.5
Daily advance, m	3.6	4.1	4.9
Preproduction advance, m	329	376	450
Cost, $/m	795	797	801
Ventilation raises			
Face area, m^2	38.1	56.3	83.7
Daily advance, m	0.19	0.20	0.21
Preproduction advance, m	400	500	600
Cost, $/m	1,526	2,106	2,974
Hourly Labor Requirements			
Undercut miners	24	24	30
Development miners	22	24	28
Equipment operators	41	64	100
Hoist operators	18	24	30
Support miners	47	58	80
Diamond drillers	6	10	18
Electricians	6	6	6
Mechanics	15	15	17
Maintenance workers	24	30	38
Helpers	8	9	10
Underground laborers	30	38	48
Surface laborers	24	30	38
Total hourly personnel	**265**	**332**	**443**
Salaried Personnel Requirements			
Managers	1	1	1
Superintendents	4	4	4
Foremen	30	42	63
Engineers	8	10	12
Geologists	9	11	14
Shift bosses	6	9	12
Technicians	16	20	24
Accountants	9	11	14
Purchasing	14	17	22
Personnel managers	13	17	22
Secretaries	24	30	38
Clerks	30	38	48
Total salaried personnel	**164**	**210**	**274**

(Table continues)

TABLE 14C.1 (Continued)

	Daily Ore Production, metric tons		
Cost Parameters	**20,000**	**30,000**	**45,000**
Supply Requirements (Daily)			
Explosives, kg	1,880	2,515	3,494
Caps, each	640	707	871
Boosters, each	629	687	837
Fuse, m	3,407	4,923	6,913
Drill bits, each	18.70	26.78	35.87
Drill steel, each	1.042	1.481	1.915
Freshwater pipe, m	12.4	13.9	16.7
Compressed air pipe, m	12.4	13.9	16.7
Electric cable, m	12.4	13.9	16.7
Ventilation tubing, m	12.4	13.9	16.7
Rock bolts, no.	30	37	48
Shotcrete, m^3	1	1	1
Concrete, m^3	56	70	97
Buildings			
Office, m^2	4,190	5,365	7,000
Changehouse, m^2	3,077	3,855	5,144
Warehouse, m^2	481	516	551
Shop, m^2	1,012	1,091	1,170
Mine plant, m^2	150	150	150
Equipment Requirements, (Number) Size			
Undercut drills, cm	(2) 4.76	(2) 5.08	(2) 5.72
Drawpoint scoop trams, m^3	(2) 7.6	(3) 7.6	(4) 7.6
Horizontal development jumbo drills, cm	(2) 3.17	(2) 3.17	(3) 3.49
Raise drills, cm	(4) 4.76	(6) 5.08	(7) 5.72
Development scoop trams, m^3	(3) 7.6	(3) 7.6	(3) 7.6
Hoists, hp	(3) 3,176	(4) 3,979	(5) 5,518
Rock-bolt drills, cm	(3) 3.81	(3) 3.81	(3) 3.81
Freshwater pumps, hp	(6) 0.5	(8) 0.5	(10) 0.5
Drain pumps, hp	(9) 593	(16) 550	(20) 781
Service vehicles, hp	(5) 210	(5) 210	(6) 210
Compressors, m^3/min	(1) 142	(1) 227	(1) 227
Ventilation fans, cm	(2) 366	(2) 366	(2) 366
Exploration drills, cm	(1) 4.4	(2) 4.4	(3) 4.4
Equipment Costs ($/Unit)			
Undercut drills	515,300	515,300	515,300
Drawpoint scoop trams	1,340,500	1,340,500	1,340,500
Horizontal development jumbos	1,145,500	1,145,500	1,149,300
Vertical development stopers	10,500	10,500	10,500
Development scoop trams	1,340,500	1,340,500	1,340,500

(Table continues)

TABLE 14C.1 (Continued)

Cost Parameters	Daily Ore Production, metric tons		
	20,000	30,000	45,000
Hoists	3,898,400	5,070,400	5,817,400
Rock-bolt jacklegs	9,200	9,200	9,200
ANFO loaders	47,200	47,200	47,200
Drain pumps	163,500	163,500	193,600
Freshwater pumps	7,620	7,620	7,620
Service vehicles	185,900	185,900	185,900
Compressors	108,900	108,900	108,900
Ventilation fans	217,100	238,000	308,100
Exploration drills	292,000	292,000	292,000
COST SUMMARY			
Operating Costs ($/t Ore)			
Equipment operation	2.85	3.06	3.36
Supplies	0.91	0.73	0.64
Hourly labor	3.79	3.16	2.81
Administration	2.25	1.92	1.68
Sundries	0.98	0.89	0.85
Total operating costs	**10.78**	**9.76**	**9.34**
Unit Operating Cost Distribution ($/t Ore)			
Stopes	2.81	2.62	2.61
Drifts	0.46	0.36	0.30
Crosscuts	0.32	0.25	0.21
Drawpoints	0.68	0.54	0.43
Boundary raises	0.22	0.18	0.14
Orepasses	0.03	0.04	0.05
Ventilation raises	0.01	0.01	0.01
Main haulage	1.68	1.80	1.94
Services	1.47	1.27	1.25
Ventilation	0.04	0.03	0.04
Exploration	0.11	0.13	0.16
Maintenance	0.31	0.27	0.21
Salaries	1.66	1.37	1.14
Miscellaneous	0.98	0.89	0.85
Total operating costs	**10.78**	**9.76**	**9.34**
Capital Costs			
Equipment purchase	28,931,700	42,694,300	58,188,800
Preproduction underground excavation			
Shaft 1	8,375,900	10,990,700	13,759,300
Shaft 2	8,534,900	10,594,200	14,017,200
Shaft 3	8,595,300	10,667,200	14,111,300
Shaft 4	—	10,699,300	14,150,100
Shaft 5	—	—	14,176,000

(Table continues)

TABLE 14C.1 (Continued)

	Daily Ore Production, metric tons		
Cost Parameters	**20,000**	**30,000**	**45,000**
Drifts	842,500	986,400	1,408,700
Crosscuts	534,300	626,700	905,800
Drawpoints	1,366,600	1,536,100	1,841,000
Boundary raises	261,200	299,800	360,400
Orepasses	1,821,100	3,360,000	5,976,900
Ventilation raises	610,300	1,052,700	1,784,400
Surface facilities	14,563,000	17,499,400	21,561,300
Working capital	13,117,100	17,803,100	25,582,100
Engineering and management	9,676,800	14,430,900	21,091,400
Contingency	7,443,700	11,100,700	16,224,100
Total capital costs	**104,674,400**	**154,341,500**	**225,138,800**
Total capital cost per daily metric ton ore	**5,234**	**5,145**	**5,003**

Source: Infomine USA 2016b

CHAPTER 15

Intermediate and Final Estimates for Mining by Cost Buildup

Rachal H. Lewis Jr.

This chapter defines and broadly describes the scope and purpose of the intermediate feasibility study, and provides procedures and methodologies generally used to build the operating costs suitable for a final feasibility study. The material presented has been taken, for the most part, from actual experience in preparing final feasibility studies and from conducting due diligence reviews of final studies conducted by others. This chapter also incorporates the experience gained from engineering, procurement, and construction management activities and mining operations overseas and in the United States.

INTERMEDIATE FEASIBILITY STUDY

An intermediate feasibility study (or prefeasibility study, as designated in the Canadian NI 43-101 nomenclature) is, in part, an update of the preliminary feasibility study, and it incorporates any new information that has become available. The intermediate feasibility study involves both capital and operating cost changes and a revised economic analysis and project schedule for the entire project. A few examples of new information would include

- Additional development drilling, such as core drilling and/or reverse circulation drilling;
- Results of continued geological mapping;
- Further metallurgical sampling and testing;
- Adjustments in the quality and quantity of the proven and probable ore reserves and in the geologic resource;
- Cost adjustments of labor, parts, and consumables;
- Equipment sizing and performance;
- Refinements in the mining method;
- Update of the capital and operating costs;
- Project schedule update;
- New developments related to permitting and environmental considerations;
- Availability of more geotechnical or hydrological data; and
- Changes in metal price forecasts.

An intermediate feasibility study will also include the results of any trade-off studies, referred to sometimes as alternatives, conducted following completion of the original preliminary or scoping study. Some typical examples of trade-off studies would include

- Contract mining versus owner mining;
- Owner maintenance versus some form of maintenance and repair contracts;
- Leasing versus buying the equipment;
- Types of mining equipment;
- The mining method;
- Process changes or refinements;
- Refining charges;
- Concentrate or final product shipping charges;
- Utility power versus on-site power generation;
- Changes required to meet environmental laws;
- Company town versus a worker camp and company-provided transportation; and
- Contractor blasting, a growing trend in the mining industry.

The refinements of the original preliminary study, through the various trade-off studies conducted during the intermediate feasibility study phase, provide the optimization needed for a sound basis for conducting the final feasibility study, which is very often referred to as the design basis. The final feasibility study involves, therefore, what will be designed and built. The techniques used to perform the cost estimates for the trade-off studies are really no different than those described in the following section for the final feasibility study, except that they must be done repeatedly for each alternative considered in the intermediate feasibility study. The final feasibility study is the focus of this chapter and is the basis for detailed engineering.

FINAL FEASIBILITY OPERATING COSTS

The procedures for estimating quarterly and annual open pit ore and waste tonnage requirements, referred to as the production schedule, were developed and presented in Chapter 4. The buildup of operating costs for a final feasibility study are based on the size and units of equipment, staffing levels, parts and consumables, explosives, and miscellaneous items needed to meet the requirements of the production schedule.

The general methodology and procedures used to develop final feasibility study operating costs for open pit mining are reviewed in the sections that follow. These are procedures generally used and accepted in the industry. This section does not address preparing cost estimates for scraper, dozer, scraper-dozer, dragline, or bucket-wheel operations. However, procedures similar to those described would be followed regardless of the type of equipment.

Equipment Selection

Equipment selection is very important and yet complex. Although a detailed analysis of this subject is beyond the scope of this chapter, a few pertinent factors are worthy of mention and, in general, cover the main features of equipment selection. Hendricks and Dahlstrand (1979) list eight factors that are basic to the selection of equipment, which include the following:

1. Required production (annual ore and waste tonnage)
2. Haul distances
3. Operating room within the mine design
4. Availability and cost of power and fuel
5. Weather conditions
6. Type of material (rock, alluvium, etc.)
7. Mine life versus capital required for a specific mining system
8. Operating characteristics of the equipment

In the preceding list, it is suggested that ramp grades and the number of lifts be included as part of haul distance considerations, and that fragmentation characteristics be included along with classifications of material types.

Couzens (1979) lists similar factors as the previous list but adds the very important consideration of ore selectivity, which is extremely critical in digging at the erratic ore–waste boundaries in precious metal open pit operations. The Couzens list includes the following:

- Ore and waste requirements
- Rock characteristics
- Shape and continuity of the mineralization
- Selectivity needed in mining the ore
- Size and geometry of the mine
- Distances required to move the material
- Efficient size of units
- Geographical distribution within the pit of working areas required for ore blending
- Backup equipment
- Mobility of the equipment

In actuality, all of the previously listed factors should be taken into account in selecting the appropriate equipment. The collective experience of the personnel of the engineering/consulting firm preparing the feasibility study provides important input to the selection process. In addition, the equipment manufacturer can provide worldwide expertise and should be contacted. Practical, hands-on experience of other operators under similar conditions, if available, is always a valuable source of information for equipment selection and performance and should not be overlooked.

The trend in the mining industry is to purchase large and more productive primary equipment units, such as drills, trucks and shovels, or loaders, in an effort to reduce operating costs. There is, however, usually a trade-off between the size of the equipment and such important factors as impact on ore grade, separation of material types, and productivity.

Savings can be realized in many cases if the bench height is made approximately equal to the maximum vertical working height of the loading equipment (Hendricks and Dahlstrom 1979). However, several other considerations need to be taken into account in setting the bench height. A good general list by Hendricks and Dahlstrom (1979) includes the following:

- Vertical distribution of the ore
- Production requirements and equipment size
- Existing equipment and the availability of capital for purchasing new equipment
- Safety
- Weather

Mine Safety and Health Administration regulations, or those regulations of the host country, should be carefully reviewed because the bench height is usually set at slightly lower than the maximum vertical reach of the loading equipment selected. This enables the equipment to dig up cleanly through the crest of the bench to remove overhangs.

Slope stability and dilution constraints are two additional considerations that should be added to any list of factors affecting bench height.

Bench height can be determined by comparing the grade and tonnage of each bench and the total mining reserve at various bench height assumptions. This technique has been used successfully for precious metal deposits.

Work Schedules and Effective Operating Time

Work schedules vary considerably and are generally structured based on management's production goals, local labor laws, national holidays, and such traditional customs as lunch and rest periods. Weather delays need to be taken into account because of the large, negative effect that the weather can have on operating time.

The example shown in Table 15.1 for mine A calls for two 12-hour shifts per day, 365 days per year, and shows the effective operating time per shift.

TABLE 15.1 Effective work time per 12-hour shift, proposed mine A

Item	Minutes	Hours
Scheduled shift hours	720	12
Less:		
Travel, inspection/start-up	30	0.5
Lunch	30	0.5
Two 15-minute breaks	30	0.5
Theoretical time available	630	10.5
Less:		
55-Minute hour allowance	52.5	0.875
Effective operating time per shift	**577.5**	**9.63**

Another example of a work schedule and effective working time is presented in Tables 15.2 and 15.3 and demonstrates the wide variance that is encountered in final feasibility studies. In the case of mine B, additional time is lost because of weather. Mine B is scheduled to operate two 8-hour shifts per day, six days per week.

Both mine A and mine B are proposed open pit precious metal operations. The properties are located overseas, but in different countries, and reflect local conditions, management philosophy, labor laws, and customs.

TABLE 15.2 Scheduled operating days and shifts, proposed mine B

Category	Drills, Loaders, and Trucks
Days per year	365
Sundays off	52
Holidays per year	11
Weather delays (entire two-shift day)	20
Estimated production days per year	282
Scheduled shifts per day	2
Number of expected operating shifts per year	**564**

TABLE 15.3 Effective operating time per shift, proposed mine B

Category	Operating Time per Shift, minutes
Scheduled time per shift	480
Scheduled nonproductive time per shift:	
Trip-in and start-up	15
Shutdown and trip-out	15
Lunch	30
Net scheduled time	420
Additional nonproductive time:	
Operational delays at 10%	42
Weather delays at 5%	21
Effective operating minutes per shift	357
Effective operating hours per shift	5.95

In all cases, the due diligence or cost estimating engineer must make sure that any overtime associated with the selected schedule has been taken into account and verify that the proposed work schedule is in accordance with the law. He or she must also verify that a realistic amount of time has been deducted for all types of delays.

Primary Units of Equipment Required

The terms *mechanical availability* and *use of availability* are used in this chapter to determine the required number of units of equipment and are defined as follows:

- Mechanical availability, which declines with the age of the equipment, is estimated by dividing the total number of hours worked by the sum of the total hours worked, plus the total number of repair hours. The result is expressed in percent.
- Use of availability is calculated by dividing the total hours worked by the sum of the total hours worked, plus the total number of standby hours. The result is expressed in percent and is usually accepted as 90% for estimating purposes.

It is worthy of mention that other methods of measuring mechanical availability and use of availability do exist, although a discussion of the other techniques is beyond the scope of this chapter. It is important, however, to fully understand the method and definitions used in a feasibility study, or at an operating property, to be sure that realistic and accurate operating time, standby time, and repair time are properly taken into account, and that the combination of operating time and hourly output is not overstated.

Drills

The example in Table 15.4 shows the procedure generally used to estimate the number of drills needed for a proposed hard rock, open pit mining operation. In this case, the drill selected is a crawler-mounted, straight-rotary-type drill capable of drilling a 165-mm (6.5-in.) diameter hole to the required depth of 7.35 m (24.1 ft) in one pass. Blasthole drilling in many open pit mines is dominated today by rotary drilling equipment. The most common bit sizes for these rigs are 175, 200, 250, 311, and 381 mm (6⅞, 7⅞, 9⅞, 12¼, and 15 in.). Blasthole drilling applications requiring DTH (down-the-hole) drills vary in bit size from 121 to 229 mm (4¾ to 9 in.).

There can be several sources for the blasthole pattern layout, explosives requirements, and drill productivity design criteria needed for a final feasibility study. These include

- Explosives manufacturers,
- Drill manufacturers,
- Actual experience under similar conditions,
- Experience gained during exploration and predevelopment drilling,
- Handbooks on drilling and blasting, and
- Consulting and engineering firms specializing in open pit drilling and blasting.

All of the sources in Table 15.4 rely on good technical information regarding the geologic, hydrologic, and geotechnical characteristics of the various ore and waste rock types that are present in the deposit to be mined. With adequate technical information, good design criteria can be developed for each ore and rock type.

Under some extreme conditions, such as high water flows, high rock temperatures, and unusual hardness, field drilling tests may have to be conducted to obtain the necessary design criteria. In many cases, however, small adjustments in the drill pattern and powder factor

TABLE 15.4 Drill requirements and design criteria

Crawler-mounted, straight-rotary drill One-pass drilling 9.63 effective hours per shift	
Types of Material	**Ore/Waste**
Basic drilling information	
In-place density, g/cm^3	2.60
Bench height, m	6.00
Subgrade drilling, m	1.35
Hole depth, m	7.35
Burden, m	4.50
Spacing, m	4.95
Hole diameter, mm	165.1
Explosives information	
Explosive density, g/cc	0.85
Powder factor, kg/t	0.23
Column load, kg/m	18.19
Explosive rise, m	4.35
Stemming, m	3.00
Effective drilling time per hole	
Moving and setup time, min	2.50
Penetration rate, m/hr	33
In-hole drilling time, min	13.36
Drill-rod coupling time, min	0.00
Pull out of hole, min	0.50
Sampling, min/hole	1.00
Drilling time per hole, min	17.36
Drilling production	
Metric tons per meter drilled, t	47.3
Metric tons per hole, t	347
Holes per operating hour per drill, no.	3.46
Metric tons drilled per operating hour, t	1,201
Metric tons drilled per drill per shift, t	11,563
Number of drills required	
Metric tons of ore per year, t	5,475,000
Metric tons of waste per year, t	16,425,000
Total metric tons per year, t	21,900,000
Scheduled operating shifts per year, no.	730
Mechanical availability, %	84
Use of availability, %	90
Shifts operated per unit per year, no.	**551.88**
Metric tons drilled per unit per year, t	**381,388**
Number of drills required, no.	**3.43**
Drill fleet requirement, no.	**4**

should provide the desired results. Dowding and Aimone (1992), and many others, point out that blasting is a continuous process of learning and fine tuning.

Dowding and Aimone (1992) point out that burden and borehole diameter are the key factors in obtaining the desired fragmentation and particle size distribution in the blasted material. Borehole diameter is usually set by the drill capacity, and the burden can be determined initially from formulas such as those developed by Ash (1990) and by consulting with the explosives and drill manufacturers who have developed detailed computer programs for determining the many variables in blasting patterns and explosives requirements.

The height of the stemming can be set initially by taking two-thirds of the burden (Ash 1990). The powder factor is then determined from the metric tons per hole, the explosive rise, and the column load, as shown in Table 15.4. Ash (1990) states, "other than obtaining the powder factor for cost purposes, its use as a basis upon which to design blasts is not recommended."

The type, source, and quality of stemming material should be given careful consideration. Drill cuttings, which are still commonly used as stemming, may not be the best cost-effective material. It has been shown, for example, that crushed stone produces better fragmentation at a lower overall cost at some operations.

In many hard-rock cases, subgrade drilling is set at one-third of the burden in the initial analysis (Ash 1990) and is adjusted in the field as needed. Spacing is normally a multiple of the burden and may vary from 1.0 to as high as 2.0 theoretically. Here again, the explosives manufacturer's technical representative or consultant can be of assistance in designing the drill-hole spacing.

Specific gravity for each ore and rock type for a final feasibility study should be based on extensive sampling in the field during exploration and development drilling, followed by laboratory determinations.

The penetration rate is determined, in most cases, from the drill manufacturer, who makes an estimate based on all of the technical parameters involved. The manufacturer also takes into account the experience gathered by the drilling crews and contractors during exploration and development drilling.

Delay blasting between holes and between rows is commonly practiced today. The design of delay blasting is complex and is best left to the explosives manufacturer or explosives consultant. Timing between holes and rows and proper sequencing are critical to blast design. These factors may be the source of a major problem if not correctly estimated. Again, the explosives technical representative can be of great assistance.

Mechanical availability for a new blasthole drill usually starts out in year 1 at 90% and decreases by about 3% per year thereafter until leveling out at around 73%. Some engineering and consulting firms have a detailed database that is based on actual experience, under certain conditions, and equates mechanical availability to a range of operating hours, such as 0 to 5,000, 90%, and 5,000 to 10,000, 87%, and so forth. In any event, the due-diligence or estimating engineer needs to be sure that a reasonable declining mechanical availability is used in the feasibility study and that it is applicable to local conditions. Industry-wide studies have their application, but local conditions can be at large variance from the average. Use of availability is usually accepted at 90%.

Current emphasis and practice for production blasting in most open pit mines is on using somewhat elevated powder factors and closer drill-hole spacing to achieve more effective fragmentation, providing that it does not result in increased dilution or in unacceptable mixing of rock types that may have to be separated because of acid-generating characteristics. In practice, better fragmentation will generally improve loading productivity and, therefore, will lower the costs of material handling and the costs of crushing as well. Nevertheless, some form of controlled blasting is required at many open pit operations to minimize damage to the final pit wall. Drilling equipment, drill-hole patterns, and explosive loading, different from that used in production blasting, may be required and should be designed by the explosives manufacturer or consultant, taking into account the recommendations of the geotechnical firm that designed the pit slopes. Controlled blasting, for example, may take the form of presplitting in conjunction with special loading and delays in the row or rows in front of the presplitting.

By following the guidelines discussed previously, and by referring to any modern text on the subject (Hustrulid 1999; Hopler 1998; Olofsson 1997), one should be able to design and correctly estimate the blasting patterns (including delay blasting), explosives requirements, and accessories needed with sufficient accuracy for the feasibility study.

Front-End Loader

The example in Table 15.5 shows the procedure generally used to estimate the number of wheel loaders (rubber-tired, front-end loaders) needed for a proposed hard rock, open pit mining operation. In this case, the loader selected is equipped with a 10.5-m^3 (13.7-yd^3) bucket and, at five passes, is a good match with an 86-t-capacity (94.8-st) haul truck. Very similar steps would be followed in determining the number of cable or hydraulic shovels.

TABLE 15.5 Wheel loader requirements and design criteria

Estimated Wheel Loader Productivity Information 10.5-m^3 bucket, 86-t-capacity haul truck, single-side loading 12-hour shift, 9.63 effective hours	
Material to Load	**Ore/Waste**
Basic information	
In-situ density of material, sp. gr.*	2.7
Material swell factor	1.30
Loose density of material, sp. gr.	2.08
Bucket size, m^3	10.5
Bucket fill factor, %	0.90
Metric tons per bucket, t	19.63
Truck size, 86 t, rated capacity	86.00
Loading time	
Theoretical buckets to load, no.	4.38
Actual average buckets per load, no.	5.00
Average swing cycle time, sec	45.00
Average spot time between loads, sec	30.00
Truck delays, sec	10.00
Total time per load, min	4.42
Truck loads per shift	
Effective time per shift, min	577.80
Theoretical truck loads per shift, no.	131
Actual truck loads per shift (95%), no.	124
Truck load factor (90%)	0.90
Average truck load, t	77
Wheel loader productivity	
Loader production per shift, t	9,138
Fleet requirements	
Annual ore tonnage, t	5,475,000
Annual waste tonnage, year 3, t	16,425,000
Total annual tonnage, t	21,900,000
Scheduled shifts per year, no.	730
Mechanical availability, year 3, %	85
Use of availability, %	90
Shifts operated per unit per year, no.	**558.45**
Metric tons loaded per unit per year, t	**5,103,350**
Number of loaders required, no.	**4.08**
Loader fleet size, no.	**4**

*Naturally occurring in-place density including moisture.

Generally speaking, wheel loaders at many open pit mining operations vary in capacity from 6.0 m^3 (7.8 yd^3) at small operations to 18 m^3 (23.5 yd^3) at medium-sized properties. The larger mining operations commonly use hydraulic shovels equipped with either backhoe or front shovel attachments and vary in size from 14 to 29 m^3 (18.3 to 37.9 yd^3) in the backhoe option and from 10.5 to 27 m^3 (13.7 to 35.3 yd^3) for front shovels. Electric-powered cable shovels at very large mines generally range from 22 to 44 m^3 (28.8 to 57.5 yd^3).

The size and type of loading equipment depends, in part, on several variables, including

- Tonnage requirements,
- Digging and other operating conditions,
- Mine design,
- Ore-body characteristics, and
- Power source and cost.

Additional criteria for equipment was discussed in the "Equipment Selection" section earlier in this chapter. The design parameters listed in Table 15.5 are within a range known to exist at different open pit mining operations. They will vary, however, depending on local physical operating conditions, the mechanical condition of the equipment, and operator skills. Wherever operating conditions permit, double-side loading is recommended for hydraulic and cable shovels because it can increase both loader and truck output by as much as 15%. Several sources for wheel loader design parameters include

- Various equipment manufacturers' published performance data;
- The *Caterpillar Performance Handbook* (Caterpillar 2015);
- Practical on-the-job experience of consultants (such as the authors of this book);
- Various textbooks and handbooks, including those listed in the reference section of this chapter; and
- Personal contacts at existing surface mining operations.

General practice calls for rounding the theoretical buckets to load (swings) over 0.10 to the next higher number of buckets. The example in Table 15.5 has been rounded from 4.38 buckets to 5.0 for estimating purposes. To achieve maximum loader or shovel productivity in practice, the loading equipment must be adequately covered with the proper number of trucks. The number of trucks is usually rounded to the next higher number starting with 0.2 units, as shown later in Table 15.7.

Truck coverage delays can be expected when there are insufficient numbers of trucks, in addition to the 10 seconds shown in Table 15.5 under the heading "Loading time." Therefore, the theoretical truck loads per shift shown in Table 15.5 under "Truck loads per shift" are usually reduced by an experience factor that may be on the order of 5%.

Planning and estimating haul truck requirements can also be aided by using computer programs such as Caterpillar's Fleet Production and Cost Analysis (FPC) software (2016). These types of programs have proven to be reasonably accurate and acceptable for feasibility studies.

The more common sizes of haul trucks vary from 31.8 t (35 st) at small operations up to 345 t (380 st) at vary large mines.

Mechanical availability for a new wheel loader will probably be about 90%–92% in year 1 and will decrease thereafter by about 5% per year.

The swell of the various rock types in the deposit to be mined can be taken from published tables such as those available in the *Caterpillar Performance Handbook* (Caterpillar 2015). Swell used in feasibility studies is very often based on the actual operating experience of the engineering firm or consultant under similar conditions. It is important to determine the swell for each ore and rock type, such as surface overburden, clay, waste rock, alluvium, oxide ore, sulfide ore, and so on.

The due-diligence or estimating engineer needs to verify that a generally accepted procedure has been used to estimate the number of units needed and that the design parameters are realistic for the specific site under consideration. One method of doing this is by using computer simulation software. Most of the current manufacturers have developed their own software that they use in-house or, at least in one case, make it available commercially to the mining industry. While no endorsement for a particular brand of equipment is implied from this author, the application of Caterpillar's FPC software has been found to be reasonably reliable and accurate for feasibility work (Caterpillar 2016). The latest version of FPC is available on CD and is considered to be relatively easy to use.

Haul Trucks

The haul cycles needed to determine the number of haul trucks required by mining period can be determined from any one of several sources:

- Equipment manufacturer's truck simulation program, considered to be the best source for feasibility work
- Equipment manufacturer's published tables, graphs, and formulas
- In-house computer program

The basic information needed to determine the number of haul trucks includes the following:

- The tonnage of ore and waste to be mined by pit phase and by period, the point of origin (the centroid of the bench), and destination (either the waste dump or the mill).
- The measured up, down, and flat haul distances for each component of the haul road, referred to as the haul profile, from the point of origin to the final destination, less the allowance for acceleration and deceleration.
- The percent grade and the rolling resistance in terms of percent grade equivalent for the loaded and empty runs for each profile.
- An allowance of 60 m (197 ft) for acceleration at the beginning of the haul profile and another 60 m for deceleration at the end, unless an equipment manufacturer's simulation program is used that includes acceleration and deceleration.
- Truck speeds, which can vary over a wide range, should be based on the manufacturer's published curves, but might need to be adjusted for safety considerations, operating conditions, and good judgment, especially when hauling loaded downhill.
- Rolling resistance can be determined from the published information and formulas found in the *Caterpillar Performance Handbook* (2015).
- Truck speeds should be based on the total wet metric-ton load.

- The total cycle time consists of the fixed time, hauling time (loaded), and return time (empty).
- The loaded haul and empty haul times of the haul cycle should be increased by 10%–15%, or separate segments estimated, if sharp curves are present in one or more of the haul segments of the haul profile.

Excessively sharp curves should be avoided because they can cause serious damage to truck final drives and electric-wheel motors. In addition, sharp curves are detrimental to tire wear.

The fixed time is the sum of the loading time, as developed and shown in Table 15.5, which amounts to 4.42 minutes in this example, plus 1.0 minute for turning and dumping at the destination, for a total of 5.42 minutes.

Tables 15.6, and 15.7 are theoretical and are presented here to show the basic procedures generally used to develop the haul profile times and to estimate the number of trucks needed to haul the desired tonnage to its destination. The ore and waste tonnage used in the tables was selected to coincide with that used to estimate the number of drills and loaders required, as shown in the preceding examples, and should be considered for illustrative purposes only. In practice, the actual tonnage to be mined would come from several benches, not one location as shown, as well as from different points along each bench. Each location would have its own haul profile. Haul profiles are developed by period, usually quarterly for the first two years and annually thereafter. The more common sizes of haul trucks vary from 31.8 t (35 st) at small operations to 290 t (320 st) at very large mines.

The same methodology illustrated in Tables 15.6 and 15.7 for ore haul cycles and fleet requirements can be applied for estimating the number of trucks needed for waste haulage. The procedures and basic assumptions are essentially the same.

TABLE 15.6 Theoretical ore haul cycle (86-t mechanical drive haul truck)

Loaded				Empty				Combined
% Grade Including 3% RR*	Actual Distance, m	Speed, km/hr†	Elapsed Minutes	% Grade Including 3% RR*	Actual Distance, m	Speed, km/hr†	Elapsed Minutes	Total Elapsed Time, min
Acc. 3%‡	60	—	0.40	Acc. 3%	60	—	0.25	
Flat 3%	240	43	0.33	Flat 3%	140	55	0.15	
Up 13%	480	8	3.60	Flat 3%	2,000	55	2.18	
Flat 3%	2,000	43	2.79	Down 7%	480	35	0.82	
Flat 3%	140	43	0.20	Flat 3%	240	55	0.26	
Dec. 3%	60	—	0.40	Dec. 3%	60	—	0.25	
						Loaded time		7.72
						Empty time		3.92
						Fixed time§		5.42
						Total haul cycle time		17.06

*RR = Rolling resistance of 3% is assumed, from tables and formulas in the *Caterpillar Performance Handbook* (2015). Total resistance = uphill (+) grade + RR. Effective resistance = downhill (–) grade + RR. Measured grade is 10%.
†Speed is based on a total wet metric-ton load (Caterpillar 2015). Maximum downhill speed is set at 35 km/hr.
‡Allow 60 m (197 ft) for acceleration (Acc.) and deceleration (Dec.). Elapsed time is assumed to be as shown. Actual acceleration and deceleration is included in most haul truck simulation programs.
§ Fixed time equals the loading time of 4.42 minutes from Table 15.5 plus 1.0 minute for turning and dumping at the destination.

TABLE 15.7 Ore haul truck requirements, 5,475,000 metric tons of ore per year

Metric tons per truck load*	77
Cycles per operating hour†	3.71
Effective hours per shift	9.63
Metric tons per shift per truck	2,821
Shifts per year	730
Mechanical availability, %	90
Use of availability, %	90
Metric tons per year per truck, in thousands	1,668,057
Annual metric tons, in thousands	5,475
Number of trucks‡	3.28
Number of fleet trucks	4
Fleet operating hours per year	19,446

*Trucks capacity metric tons, Table 15.5 = 77.
†Haul cycle minutes, Table 15.6 = 17.06.
‡The number of trucks should be rounded to the next higher number, starting with 0.2 units.

Support Equipment Requirements

The following auxiliary equipment checklist should cover the needs of most open pit precious and base metal mines. The size and number of each unit will vary, of course, with different operations and must be adapted for each particular case. The sizes of the units listed here are adequate for a total annual ore and waste tonnage of around 20,000,000 t:

- Track dozers, with ripper (400 fwhp [front-wheel horsepower])
- Wheel dozers (rubber-tired) (450 fwhp)
- Road grader (280 fwhp)
- Secondary breakage rock drill
- Drills for controlled blasting at the final pit wall
- Water truck (50,000 L)
- Hydraulic crane (120 t)
- Dispatch system
- Heavy equipment lowboy trailer and tractor (100 t)
- Explosives truck
- Blasting crew truck
- Stemming truck
- Tire truck
- Lube/fuel truck
- Welding truck
- Mechanic's truck
- Mechanic's flatbed truck with crane
- Cable reel truck for cable shovels
- Aerial maintenance truck (some operations)

- Overhead cable towers (cable-shovel operations)
- Backhoe loader (some operations)
- Large fuel truck (some operations)
- Rough terrain forklift (5 t)
- Mobile light plants
- Pumps for dewatering
- Light vehicles (pickups)
- Bus(es) for mantrips

In general, there should be one track dozer for each waste dump and surroundings, if the dumps are far apart, and one for general pit operations. Two rubber-tired dozers are generally needed for cleanup around the shovels and blasthole drill sites, and for haul-road maintenance and spillage cleanup. One or more graders will be needed, depending on the length and distribution of haul roads to both the waste dumps and the mill. The remaining items usually involve one unit, with the exception of pumps, pickups, light plants, and possibly buses.

The other items on the list are fairly standard but may or may not be needed depending on several conditions including the following:

- The type of loaders selected (wheel loaders, cable shovels, front-loading hydraulic shovels)
- Number and location of waste dumps
- Haul distance to the waste dumps and mill
- The employment of contract blasting
- Pumping requirements for water flows into the pit and for large amounts of rainfall
- Special drilling requirements as mining approaches the final pit wall
- The amount of secondary blasting that may be required
- Fueling operations

Maintenance Organization

The procedures used in this chapter to develop a maintenance organization and maintenance costs are based on the assumption that a formal preventive maintenance program will be conducted in-house. This chapter also assumes that a component exchange system will be in place and that component rebuilding will be done off-site.

Maintenance and repair contracts and other forms of conducting maintenance are available and should be investigated as part of an intermediate feasibility study. They may result in lower maintenance labor and repair costs.

Salary and Labor Requirements and Cost

Basic salary and hourly wage rates are usually determined from known and published rates within a certain district where the proposed operation will be located. In the case of a remote area location, an experienced firm should conduct a wage and salary survey. In either case, benefits as required by law have to be added to the basic rate. Benefits vary widely from country to country, and may be in the range of 50% to 100% in some cases and need to be thoroughly researched. Any built-in overtime resulting from the work schedule must be included.

Mine staffing varies, of course, and depends on the type, complexity, and location of the operation. Generally speaking, the following positions will be needed, although the number of each will vary and, of course, there are exceptions:

- Mine manager
- Maintenance superintendent
- Maintenance supervisor(s)
- Maintenance planner
- Chief engineer
- Senior mining engineer
- Mining engineer
- Senior geologist
- Geologist
- Grade control technicians
- Operations supervisors
- Surveyor
- Surveyor technician(s)
- Draftsperson
- Secretary
- Clerk

A large mining operation, such as found at many porphyry copper deposits, may employ an operations superintendent who would report to the mine manager. In these cases, the operations superintendent is usually responsible for drilling and blasting, shovel and truck operations, training, and mine area environmental control.

The total annual salary plus benefits divided by the annual ore tonnage provides the staff labor cost per metric ton of ore mined.

Operating labor requirements are determined from the average number of all units of equipment required per shift and rounded to the next highest number. An adequate number of relief operators to cover vacations and sick leave is usually added and may range from 5% to 15%. The total annual operating labor requirements for each wage scale times the annual wages for that scale equals the total annual labor cost by wage category or scale. The sum of the totals for all wage categories divided by the total annual ore tonnage yields the labor operating cost per metric ton of ore mined.

The number of hourly operating labor requirements vary according to the size of the operation, both in daily ore production and stripping ratio. Table 15.8 shows the estimated range of size of the hourly staff from a very small mine to a very large mine with stripping ratios from 1:1 to 8:1. These numbers do include the mine maintenance workers. The breakdown of the types of labor is given in the source document available from InfoMine's *Mining Cost Service* (InfoMine USA 2014).

Maintenance labor requirements are more difficult to estimate and are usually based on published maintenance labor costs per equipment operating hour such as found in *Mining Cost Service* (Schumacher 1998; InfoMine USA 2015) and the *Cost Reference Guide for Construction*

TABLE 15.8 Total hourly personnel requirements for various size open pit mines with different stripping ratios

	Stripping Ratios (metric tons waste/metric tons ore)			
	1:1	2:1	4:1	8:1
Metric Tons of Ore Per Day	Estimated Number of Hourly Personnel*			
500	13	20	21	30
1,000	25	31	39	53
2,000	38	47	60	100
5,000	42	52	69	109
1,0000	60	77	124	202
20,000	135	200	271	446
40,000	230	299	479	876
80,000	292	452	800	1,245

Source: InfoMine USA 2014
*Includes maintenance.

Equipment (Primedia Information 1999). Annual equipment operating hours are estimated as shown by these formulas:

$$\text{annual operating hours for each type of equipment} = (\text{effective hours per shift}) \times (\text{required number of shifts per year}) \quad \textbf{(EQ 15.1)}$$

$$\text{required number of shifts per year} = (\text{total metric tons required per year}) \div (\text{metric tons per machine per shift}) \quad \textbf{(EQ 15.2)}$$

In the case of more than one unit of the same type of equipment, the annual operating hours are usually distributed equally among the number of units.

The year for that unit divided by the annual mechanic's salary yields the number of maintenance personnel for a particular type of equipment:

$$\text{number of maintenance personnel} = (\text{annual operating hours per year}) \times (\text{published maintenance cost per hour}) \div (\text{annual mechanics salary}) \quad \textbf{(EQ 15.3)}$$

The preceding steps are repeated for each type of equipment required by year, which yields the total number of maintenance personnel. However, this total does not normally include pickups and other light vehicles such as mechanics trucks, and so on. Therefore, a realistic number of maintenance personnel needs to be added for these vehicles and is usually based on experience. As in the case of equipment operators, an adequate number of relief operators to cover vacations and sick leave should be added and may range from 5% to 15% additional personnel.

Cost of Parts and Consumables

The cost per operating hour is often derived from published tables such as those developed and maintained up-to-date by Primedia Information (1999) and Western Mine Engineering and/or InfoMine USA (Schumacher 1998; InfoMine USA 2015). These tables show the cost per operating hour for maintenance labor, repair parts, fuel/electricity/gas, lube, tires, and ground-engaging components. The *Caterpillar Performance Handbook* (2015) also contains tables and sample forms for estimating operating costs, but without maintenance or operating labor.

The estimator needs to be cautious in using these tables and must carefully review all of the cost components to be sure that nothing has been left out, and that such items as wages, fuel, lubricants, and tires reflect the location and expected working conditions. If properly used, the tables should provide good initial cost-per-hour data for parts and consumables. The cost per hour needs to be adjusted upward annually based on the total operating hours of the equipment and may amount to about 10% per year.

One very important cost item, which is frequently overlooked or underestimated and sometimes not included in the tables, is the cost of component changes. If this cost is not included in the tables, it needs to be developed and incorporated. The frequency and cost of these overhauls for the equipment selected can be obtained from equipment manufacturers and engineering/consulting firms.

As the equipment ages, the hourly maintenance cost will increase, and this should be taken into account in estimating the number of maintenance personnel needed by year. Depending on the type of equipment, the hourly rate may increase in the range of 10% to 15% per year, assuming that that the equipment will run about 5,000 hours per year.

It is important to realize and take into account that the annual operating cost of the equipment is not constant and will vary over the life of the unit. This may seem obvious, but it is frequently underestimated or improperly represented in the annual cost analysis.

As an example, Table 15.9 lists the components that make up the estimate for maintenance and repair (M&R) costs per operating hour for a 10.5-m^3 wheel loader. These costs are based on frequent surveys of manufacturers, suppliers, and operators conducted by *Mining Cost Service* (InfoMine USA 2015).

The estimator needs to adapt each of the components described in Table 15.9 to local conditions, as they will vary considerably from country to country and even within the United States. Overhaul costs (component changes) and normal operating repairs are included in Table 15.9 up to the point that the unit is either replaced or requires a major rebuild of the engine, transmission, and entire drive chain. The trade-off between buying new replacement equipment and the cost of a major rebuild warrants careful study and should take into account the remaining life of the mine and the operating cost reduction resulting from new and more efficient equipment.

TABLE 15.9 Maintenance and repair costs* per operating hour for a 10.5-m^3 wheel loader (2015 dollars)

Item	Cost, $
Parts	21.47
Maintenance labor†	14.64
Diesel fuel‡	43.71
Lube§	23.66
Tires	38.83
Total	142.31

Source: InfoMine USA 2015

*Repair labor: $37.30 per hour.

†This item should be left out if determined using the formula as described in the text.

‡Diesel fuel: $2.09 per gallon.

§ Lube oil: $11.18 per gallon.

The total annual M&R cost is the product of the hourly cost times the annual operating hours as defined previously. The cost per metric ton of ore mined is the annual cost divided by the annual ore tonnage. A reasonable multiplier, as mentioned previously, should be used for the M&R costs once the equipment exceeds 10,000 operating hours, and is usually 10%–15% in each of the succeeding 5,000-hour increments, which is roughly equivalent to one year. This procedure is repeated for each type of equipment by quarter for the first two years and by year thereafter. The total M&R costs and cost per metric ton of ore mined are then computed.

As in the other steps in estimating operating costs, the estimator needs to consider the location, operating conditions, management maintenance philosophy, availability of spare parts, amount of off-site repair work, and the experience levels of maintenance and operating personnel. Equipment manufacturers as well as engineering and consulting firms should be contacted. Another excellent source of maintenance organizations and maintenance cost information is personal contact with the management of known operations who have had experience under similar operating conditions.

An additional method of estimating maintenance labor costs, which is sometimes used, is to determine the ratio of repair hours to equipment operating hours for a particular type of equipment and set of operating conditions, as explained and illustrated by Wescott and Hall (1993) and Zimmer (1990). The various maintenance ratios are selected from handbooks, historical records, and, as suggested by Wescott and Hall (1993), by back-calculating from actual repair costs per equipment operating hour. In this case, the maintenance labor cost for a particular type of equipment is estimated by the following formula:

$$\text{maintenance labor cost per hour} = (\text{unit labor cost}) \times (\text{maintenance ratio}) \qquad \textbf{(EQ 15.4)}$$

The cost of bits and drill steel for blasthole drilling, which are not found in the tables, can be obtained from the supplier and is based on the technical information made available by the operator. The technical information would include all pertinent geologic, hydrologic, and geotechnical data. Any information on the hardness and abrasiveness of the rock as experienced during development drilling is also very helpful. Another excellent source of bit and drill steel cost information, if available, is a mining company or engineering/consulting firm that has a large database and can provide these costs based on actual experience under similar conditions.

Cost of Explosives

The cost of explosives for a final feasibility study, which would include blasting agents, boosters, delays, connectors, primacord, and so on, should be based on written quotations from several manufacturers, including delivery charges, and in accordance with the initial blasting design criteria, an example of which is shown in Table 15.4 under "Explosives information." The total cost per period for explosives for each rock and ore type is divided by the metric tons in each case, which yields the explosives cost per metric ton. The cost of loading the holes and carrying out the blasting procedures is included in the labor costs.

Blasting can also be done by a contractor who provides all of the explosives needed, loads the holes, and conducts the blasting procedures. The contractor also provides, in many cases, storage facilities for the explosives.

Miscellaneous Operating Costs

The cost of assaying the blasthole samples, and other pit samples, is charged at many operations to the mine general and administrative costs or to a technical services category. The important detail is to make sure that these costs are represented in the operating costs. Assay costs should be developed by estimating the number of samples required per day times the cost of the assay per sample. The total annual cost of assaying for the mine samples divided by the metric tons of ore mined per year yields the assaying cost per metric ton. Blastholes in precious metal mines are commonly sampled on 5-ft (1.5-m) centers, which results in a large number of samples per day. The example in Table 15.4 would result in about 87 samples per 24-hour period. Subgrade drilling is not sampled at precious metal mines because of the erratic nature of these deposits. Many base metal mines take only one sample per blasthole as a composite.

It is common practice at many mining operations to conduct additional development drilling and/or fill-in drilling during the production period to support both short- and long-range mine planning. An allowance for these costs should be included in the feasibility study.

The cost of pumping mine water inflows and/or large amounts rainwater runoff can be significant and need to be addressed for each particular operation.

An allowance for office supplies and minor repairs should be included and is generally in the range of $0.03 to $0.06 (current dollars) per metric ton of ore mined.

Operating Cost Summary

Operating cost summaries are usually presented in table form and show unit costs for the operation. Of course, they vary greatly by the size of the operation (metric tons per day) and the stripping ratio. The unit cost summary is very often broken down into functions such as supplies and consumable materials, labor, equipment cost (i.e., ownership and operating cost), administration, and sundries items. Table 15.10 was partly taken from Zimmer (1990) and then extracted from the work of Stebbins and Leinart (2011) but escalated to 2015 dollars using the escalation factors from the U.S. Bureau of Labor Statistics (USBLS 2015), as described in Chapter 14. This type of information is sufficiently accurate for benchmark comparisons, but it is no substitute for detailed cost "estimate buildup" by function analysis.

An illustration of breaking down the unit cost is by operating functions—drilling, blasting, loading, hauling, roads and dump construction and maintenance, equipment maintenance, operations management, and administration—is shown in Table 15.10, which has been partially taken from a paper by Zimmer (1990) and is shown for illustrative purposes as a good example of an operating unit cost by function by period for a large base metal operation. Years 6 and 7 of the 20-year operation were selected as a typical example. For the final feasibility phase, to get accurate cost into the economic analysis, cost should be broken into time periods of the operation. The accuracy of Table 15.10 is questionable since it was escalated 25 years, but it illustrates a good method of cost buildup.

FINAL FEASIBILITY CAPITAL COSTS

This section provides procedures and methodologies generally used to build the capital costs suitable for a final feasibility study. The material presented has been generated from actual experience in preparing final feasibility studies and from conducting due diligence reviews of

TABLE 15.10 Unit operating cost summary of unit cost, dollars per metric ton of material and dollars per metric ton of ore (2015 dollars)

Combined period of year 6 and year 7 Total metric tons of material: 85,200,000 t Total metric tons of ore: 35,000,000 t	
Drill	0.1671
Blast	0.2449
Load	0.3298
Haul	0.7696
Roads and dump	0.2065
Equipment maintenance*	0.2766
Operations management†	0.1153
Administration‡	0.5555
$ Per metric ton of material	2.6564
$ Per total cost of all material	226,240
$ Per metric ton of ore	6.46

Source: Adapted from Zimmer (1990), and escalated with data from Stebbins and Leinart (2011) using data from USBLS 2015.
*The cost buildup for equipment maintenance is part of and included in Table 15.7.
†The cost buildup for operations management is part of and included in the "Salary and Labor Requirements and Cost" section earlier in the chapter.
‡The cost buildup for roads and dumps is also part of and included in Tables 15.6 and 15.7.

final studies conducted by others. This section also incorporates the experienced gained from engineering, procurement, and construction management activities and mining operations overseas and in the United States.

The capital costs for a final feasibility study should be based on having completed 20%–30% of the engineering. This information was given in Chapter 11 and was based on an industry-surveyed recommendation (Bullock 2013). The basic engineering design for the mine maintenance and office buildings, pit slopes, waste dumps, roads, retention dams, drainage ditches, and so on, in turn depends on the completion of several key activities, including

- Geological and structural mapping;
- Field sampling from surface outcrops, test pits, and bore holes of all soil and rock types and laboratory strength testing of these samples;
- Research of all pertinent data and information related to surface and groundwater hydrology;
- Local conditions such as climate and altitude; and
- Water well drilling and pump testing if needed as a basis for a pit dewatering system or for concerns regarding pit slope design.

The cost of all equipment should be based on written quotations from several manufacturers or suppliers. Factoring or the use of broad-based, industry-wide, generalized cost curves is not used. The cost of spare parts, local materials and supplies, construction labor, equipment rental, and so forth, should be based on quotations, area surveys, and direct contact. Specific actual cost from similar other mining operations (especially within the same mining district) should be used as benchmarked data, if available.

Detailed plans for mine closure and reclamation, as required by law, should be developed as a basis for estimating the costs of these activities. Although these expenditures occur at the end of the mine's life, they are significant, even when discounted in the economic analysis, and should be shown as part of the project's cash flow.

A 15% contingency is, with some exceptions, usually recommended for a mine's capital cost estimate for a final feasibility study. In theory, the amount of contingency should reflect the degree of engineering that has been completed. In other words, the more engineering that has been completed, the lower the contingency requirement. If 20%–30% of the engineering has been completed, as recommended in Chapter 11, a 10%–15% contingency can be used, with the 10% contingency only applied to structural facilities and not mining operations. A 15% contingency, however, is strongly recommended for a final feasibility study if the project is located in a remote area.

Cost of Preproduction Stripping

Most open pit mining projects require at least some degree of pioneering work, which usually includes temporary road access to the future mine area, clearing, and topsoil removal and storage. In many cases, pioneering work is done with a different type of equipment or different sized equipment than used for full-scale stripping and is often contracted. Pioneering work generally continues until enough area is open and available for operating the equipment designed to meet the waste and ore mining schedule. Pioneering work may be small in scope or quite extensive and costly. Whatever the circumstances, the front-end pioneering work should be carefully checked for adequacy, cost, and schedule.

The volume and tonnage of preproduction stripping for each waste rock type, soil and alluvium, and each ore type by phase and period are developed in the mine planning stage. This provides the basis for estimating the cost of stripping required to expose enough ore to sustain production at the designed rate.

Preproduction stripping is accomplished either by contractor or with the owner's fleet and personnel. In the case of contracting, formal and detailed bid packages should be sent to at least three qualified contractors. Preproduction stripping by the owner requires purchasing of equipment and extensive personnel training far in advance. This ensures that the number and types of equipment and trained operators are available when needed as dictated by the production schedule. Contract stripping is often selected especially where conditions dictate the use of different equipment, such as scrapers, rather than the drills, loaders, and trucks specified for full-scale ore and waste production.

It is in the owner's best interest to evaluate, at the intermediate feasibility study level, which of the two approaches best suits the owner's economic goals and project execution philosophy. The total cost difference between the two alternatives, when evaluated in detail, frequently represents large sums of money and, therefore, warrants careful analysis.

Capital Cost of Primary and Support Equipment

The capital cost of the equipment should be based on written quotations from several suppliers. The information needed for a final cost evaluation includes

- FOB (free on board) factory price with options selected by the owner;
- Applicable taxes;

- Freight charges—inland and ocean if both apply—to the site, including crating or placement in containers;
- Import duty, taxes (including value-added taxes), handling charges, and so forth, if an overseas shipment is involved; and
- Assembly costs.

Equipment Replacement

The decision to replace a unit of equipment or rebuild it and keep it in operation is difficult to do and requires careful analysis. Theoretically, a unit of equipment is replaced when the actual cost of owning and operating the equipment exceeds the cost of owning and operating a new unit, as discussed and explained by Spark (1993). This type of analysis requires, of course, that accurate maintenance and cost records are kept for each unit of equipment in the fleet.

There are other considerations, however, that should also be taken into account when deciding whether to replace a unit of equipment or to spend the time and money for a major overhaul and keep it in service. These considerations include

- The remaining life of the mine,
- Cost of the overhaul,
- Expected mechanical availability of the unit after the rebuild,
- Cost and productivity of the new unit,
- Availability of spare parts for the aging equipment,
- Sale price of the old unit,
- Whether the old unit can produce the required output,
- Whether adequate and temporary replacement is available for the unit being rebuilt,
- Future production from the mine (increasing or decreasing), and
- Costs and constraints involved in increased maintenance staffing and facilities for the older equipment.

Based on the analysis proposed previously by Spark (1993), if the mine has a long life—say, 15–20 years—most of the fleet will have to be replaced, depending on when the units were placed in service. As a general rule of thumb, and with the possible exception of cable shovels, most of the primary units of production are replaced between 7 and 10 years of service. Support equipment is usually replaced after five years and pickups before three years of service. Both the cost of equipment replacement and that of major overhauls should be shown on the project's cash-flow spreadsheet.

As a matter of reference, typical replacement lives used in some recent feasibility studies include 40,000 hours for a 6½-in. rotary drill, 35,000 hours for a 14-yd^3 wheel loader, and 60,000 hours for an 88-t mechanical-drive haul truck. These numbers are presented to give an idea as to what life might be expected, and they were estimated for a potential operations forecast to have good operating conditions and skilled operators and maintenance personnel.

Additional equipment is sometimes added to the existing fleet during the mine life to meet increased stripping requirements. In these cases, the ongoing capital investment costs should show on the cash-flow spreadsheets.

Maintenance and Office Buildings

The design and cost estimate for a maintenance facility is commonly prepared by an experienced engineering firm, selected by the owner through competitive bidding, to prepare the final feasibility study. The cost estimate should be based on material take-offs from engineering drawings. The basic design and layout of these facilities should cover several considerations, including

- Maintenance philosophy (in-house or contract maintenance);
- Sufficient maintenance bays based on the number of each type of equipment anticipated to be undergoing maintenance in the shop at any one time;
- Bay doors designed for all-weather conditions such as rain, mud, snow, ice, wind, and dust without getting stuck;
- Adequate room for equipment parking;
- General layout and location that includes room for future expansion;
- Wash and steam cleaning facility for the equipment
- Spare parts and tool room;
- Welding and cutting area;
- Area for electrical repairs;
- Adequate space for maintenance and repair of gas-engine vehicles;
- Machine shop area;
- Service area for fueling, lubrication, and coolant fluids and materials;
- Offices for the maintenance superintendent, maintenance planner, maintenance general foreman and shift foreman, clerk, and warehouse personnel;
- Training room and equipment;
- Medical and safety facilities;
- Fire truck building, fire truck, and equipment;
- Building fire protection;
- Ambulance building, ambulance, and equipment;
- Fenced dry storage area;
- Adequate clean area for storage of rebuilt components;
- Adequate HVAC design for local climatic conditions;
- Clearance and capacity of overhead crane; and
- Maintenance equipment and mechanics tools.

Offices for the mine superintendent, mine operations supervisors, engineering and geology, surveying, and so on, are frequently grouped in one building with offices for the general manager, mill superintendent, administrative manager, and other technical, accounting, and administrative personnel. This building is usually a preengineered structure designed for the life of the property and built in such a way that it can easily be removed as part of the reclamation and closure plan. The basic layout would include office space as required, and a staff meeting room, training room, and first-aid facility.

The cost of the office building would be estimated by the same engineering firm selected to conduct the final feasibility study. Some key considerations for the office building design would include the life of the project, number of personnel needed, distance between the mine and the mill, topography, geotechnical criteria, and the climate. Each property is different and the layout and design of all buildings, and the associated access and communications, need to be carefully planned and coordinated.

A noted exception to the previous discussion would be if the operation decided to use contract mining and stripping, and then the contractor would supply much of what would go into the open pit cost estimate. But even in this situation, remember that the owner must still provide maintenance facilities for those pieces of equipment that the mine operates and must also provide offices for those professionals listed in the preceding paragraphs.

An allowance for computers, software, drafting and printing equipment, surveying instruments, and initial office supplies and furniture should be included in the estimate. Training aids and materials should also be included.

Pit Infrastructure

Pit infrastructure is another area that the engineering firm selected to conduct the final feasibility study would design and cost. This area will include such infrastructure as access and haul roads, overland belt conveyor, power transmission and distribution, and communications network. It may also include relocating existing state or county roads, natural drainage patterns, and bridges.

Initial Spare Parts Allowance

The cost of the initial stock of spare parts for a final feasibility study may range from a low of 6% of the total purchase price of the equipment to as high as 12%. The location of the mine, operating conditions, operator experience, and maintenance quality are some of the factors affecting the stock of spares. The stock of spare parts is an issue that warrants careful consideration not only because of the cost involved but also to ensure that an adequate supply is on hand or available on very short notice. Again, the use of contract mining for mining or stripping will affect that which the mine operator will need to stock as inventory.

In all cases, the potential manufacturer/supplier should be contacted and consulted regarding what arrangements can be expected for providing spares and consignment items and what safe minimum will be needed for the location in question. The experience of other operators in the area, if any, can be very helpful in arriving at the requirement for spares at the final feasibility stage.

Waste Dump Design and Construction

Waste dump location, design, and method of construction should be studied and designed by a geotechnical/engineering firm experienced in the design of waste dumps. Extensive geological mapping, rock type sampling, laboratory strength testing and analysis, and hydrological field and research investigations are all a prerequisite for establishing criteria for waste dump design.

The location and design also needs to take into account all applicable environmental laws and regulations for the site under study before making the cost estimate. The cost of waste dump construction may be quite high if drainage channels, retention dams, and extensive diversion ditches are required.

Drainage Control, Dewatering System, Ditches, and Dams

Diversion ditches or canals will be needed at locations that experience heavy rainfall or snowmelt runoff. Retention dams may also be needed to temporarily hold the runoff from a 100- or 500-year event. An experienced geotechnical firm should be contracted to design and estimate the cost of this type of work.

Extensive dewatering systems surrounding the mining operation may be required at some locations to prevent excessive groundwater flows into the pit. These systems require water-well drilling and pump testing in the field, followed by design and cost estimating, and is usually done by an experienced firm as a subcontract to the feasibility study contractor.

Hiring and Training Costs

Recruiting, hiring, and training activities must start far enough in advance so that the required number of equipment operators, maintenance personnel, and supervisors are available when needed and have the necessary skills to achieve design productivity as required in the production schedule. These combined activities are frequently referred to as the hiring schedule, and they require careful consideration and detailed planning as part of the final feasibility study.

The cost of hiring and training should include temporary trainer-operators who are brought in to conduct the hands-on training of the new and inexperienced equipment operators. Even for a small mining operation, these costs can easily exceed $1 million. If the operations choose to use contract stripping and mining, then this cost will be reduced.

Example Capital Cost Estimate

Table 15.11 has been extracted from the cost estimating models of *Mining Cost Service* by InfoMine USA (2015). This service supplies detailed cost elements where a company can build its own cost estimate, or extract cost models. From this information, a small sample is shown in Table 15.11. InfoMine has surface mining models from 250–80,000 t per day, with various stripping ratios, 1:1 to 8:1.

PROBLEM AREAS IN COST ESTIMATING

Problems can crop up anywhere in developing costs for a final feasibility study; therefore, each component of the entire capital and operating cost estimate should be carefully scrutinized for completeness and compliance with accepted standards. Several areas, for some reason, seem not to get the attention that they deserve. Consequently, they can cause large cost underestimates for both capital and operating costs. Some of the more important of these neglected areas include the following.

Capital Costs

- Omission or insufficient funds for replacement of the mine equipment fleet
- Insufficient funds and/or an overly optimistic schedule for recruiting, hiring, and training mine personnel
- Insufficient planning and cost estimating for mine closure and reclamation
- Incorrect densities of the various types of ore and waste, leading to over- or underestimating the number of primary units of production as well as affecting the reserves

TABLE 15.11 Cost estimating summary for surface mines (2015 dollars)

	10,000 metric tons of ore per day				80,000 metric tons of ore per day			
Stripping Ratio	1:1	2:1	4:1	8:1	1:1	2:1	4:1	8:1
Operating Cost, $/t of Ore								
Supplies	1.35	1.63	2.64	4.54	0.88	1.67	2.07	3.65
Hourly labor	2.34	3.10	4.92	8.10	1.13	1.74	3.07	4.83
Equipment operations	1.35	2.29	4.47	9.20	1.71	3.14	7.02	13.78
Salaried labor	0.85	1.02	1.49	2.16	0.35	0.44	0.63	1.00
Miscellaneous items	0.59	0.80	1.35	2.40	0.41	0.77	1.28	2.33
Total operating cost	6.48	8.84	14.87	26.40	4.48	7.69	14.07	25.59
Capital Cost, $ × 1,000								
Equipment	16,343.0	23,688.2	50,177.6	98,937.6	122,824.6	254,012.2	503,413.3	965,635.7
Haul roads/site work	2,507.8	3,337.8	4,847.4	7,935.2	9,610.5	13,282.8	22,609.0	28,184.7
Stripping	799.5	1,410.3	2,830.6	5,806.9	4,615.2	11,520.8	23,449.5	50,099.8
Buildings	5,400.8	6,408.6	11,467.5	19,401.0	21,985.0	35,057.9	58,899.1	112,755.2
Electrical system	208.6	220.8	480.3	500.7	1,786.3	2,217.6	3,262.3	4,190.8
Sustaining capital	3,211.9	6,195.3	8,504.5	17,359.5	12,553.4	22,915.3	28,698.8	51,897.7
Working capital	3,369.1	4,598.9	7,733.4	13,728.0	18,624.3	31,986.2	58,527.0	106,437.8
Engineering and management	5,010.0	6,941.3	13,568.9	25,604.8	31,403.0	60,913.6	117,278,.0	221,778.2
Contingency	3,149.2	4,363.1	8,529.0	16,094.5	19,739.0	38,258.5	73,717.6	139,403.4
Total capital cost	39,999.9	57,164.3	108,139.2	205,372.2	243,141.3	470,194.9	889,854.6	1,680,383.3
Total capital cost/ day/t, ore and waste	2,000	1,905	2,163	2,282	1,520	1,959	2,225	2,334

Source: InfoMine USA 2015

- Failure to properly develop a detailed project schedule, including procurement time frames, and a means of monitoring and controlling the schedule, leading to schedule overruns

Operating Costs

- Incorrect estimate of the total cost of fringe benefits
- Failure to completely develop the maintenance organization, leading to a shortage of maintenance personnel and an understatement of operating costs
- Underestimating dilution, resulting in a lower head grade
- An inadequate grade control organization, resulting in higher dilution and loss of ore
- Inadequate short- and long-term mine planning
- Pit slopes that are too steep as a result of incomplete or inadequate pit slope design and which result in the need to flatten the slopes at the cost of increased waste removal
- Failure to discount operating time where climatic conditions are extreme for prolonged periods of time

In recent years, overruns of capital cost over the final cost estimate amounted to 60% in South America, 51% in North America, 40% in Australia, and 30% in South Africa (Deloitte

2012). To keep such massive overruns from occurring, it is obviously a serious problem that exists and it is global. It is imperative that after the final feasibility study, a design basis report (Chapter 12) should be written, and this is what the final cost estimate will be based on. If it is a major project (say, more than a billion dollars) or it is in an isolated area for which the owners have no experience, has multiple construction sites, or multiple ownerships, then a project risk appraisal and adjustment should be completed (see Chapter 17).

REFERENCES

Ash, R.L. 1990. Chapter 6.2.2, Design of blasting rounds. In *Surface Mining*, 2nd ed. Littleton, CO: SME. pp. 565–583.

Bullock, R.L. 2013. Mine Feasibility Studies Still Need Improvement! Presented at PDAC Annual Conference, May 3–6, Toronto, Canada.

Caterpillar. 2015. *Caterpillar Performance Handbook*. Edition 41. Peoria, IL: Caterpillar.

Caterpillar. 2016. Caterpillar Fleet Production and Cost (CD). Version 3, Peoria, IL: Caterpillar.

Couzens, T.R. 1979. Chapter 16, Aspects of production planning: Operating layout and phase plan. In *Open Pit Mine Planning and Design*. New York: American Institute of Mining, Metallurgical and Petroleum Engineers.

Deloitte. 2012. *Tracking the Trends 2013*. https://www2.deloitte.com/content/dam/Deloitte/ca/Documents/international-business/ca-en-ib-tracking-the-trends-2013.pdf. Deloitte Global Services.

Dowding, C.H., and Aimone, C.T. 1992. Chapter 9.2, Rock breakage: Explosives. In *SME Mining Engineering Handbook*, 2nd ed., Vol. 1. Edited by H.L. Hartman. Littleton, CO: SME.

Hendricks, D., and Dahlstrand, A. 1979. Chapter 15, Selected aspects of production planning—Berkeley Pit. In *Open Pit Mine Planning and Design*. New York: American Institute of Mining, Metallurgical and Petroleum Engineers.

Hopler, R.B., ed. 1998. Chapter 23. In *Blasters' Handbook*, 17th ed. Cleveland: International Society of Explosives Engineers. pp. 313–340.

Hustrulid, W. 1999. Chapters 4 and 5. *Blasting Principles for Open Pit Mining*, Vol. 1. Brookfield, VT: A.A. Balkema.

InfoMine USA. 2014. *Mining Cost Service*, a subscription cost data update service. Spokane Valley: WA: InfoMine USA. pp. CM 15, CM 21.

InfoMine USA. 2015. *Mining Cost Service*, a subscription cost data update service. Spokane Valley: WA: InfoMine USA.

NI 43-101. *Standards of Disclosure for Mineral Projects*. Including 43-101F1. www.osc.gov.on.ca/en/SecuritiesLaw_ni_20110624_43-101_mineral-projects.htm (accessed June 21, 2016).

Olofsson, S.O. 1997. Chapter 5, Bench blasting. In *Applied Explosives Technology for Construction and Mining*. Arla, Sweden: Applex. pp. 62–121.

Primedia Information. 1999. *Cost Reference Guide for Construction Equipment*. Primedia Information Inc., Machinery Information Division.

Schumacher, O.L. 1998. *Mining Cost Service*, a subscription cost data update service. Spokane, WA: Western Mine Engineering.

Spark, H. 1993. Chapter 2, Surface Mining Capital. In *Cost Estimation Handbook for the Australian Mining Industry*. Parkville, VIC: The Australasian Institute of Mining and Metallurgy.

Stebbins, S.A., and Leinart, J.B. 2011. Cost estimating for surface mines. In *SME Mining Engineering Handbook*, 3rd ed. Edited by P. Darling. Englewood, CO: SME. pp. 281–293.

USBLS (U.S. Bureau of Labor Statistics). 2015. BLS database. https://data.bls.gov/cgi-bin/surveymost.

Wescott, P., and Hall, R. 1993. Chapter 12, Surface mining equipment and operating costs. In *Cost Estimation Handbook for the Australian Mining Industry*. Parkville, VIC: The Australasian Institute of Mining and Metallurgy.

Zimmer, G.S. 1990. Chapter 7.2.5, Base metal open pit mining. In *Surface Mining*, 2nd ed. Edited by B.A. Kennedy. Littleton, CO: SME. pp. 1008–1020.

CHAPTER 16

Cost Estimating for Processing Plants and Infrastructure

Mark A. Anderson

The estimation of operating costs can be a daunting task to a professional with little or no experience in the operation of concentrators. Operating personnel have, readily available, the current operating history and peer operations of their present company as well as past operations they have served. A wide variety of references are available to an estimator when beginning to assign the proper operating costs to a milling operation. Among these are estimate factors, published generic estimates, and historical information for all of the accounts requiring attention in the concentrator. Operating costs, in the feasibility context, are usually reported both on a "whole dollar" and unit cost basis. Unit costs can be assigned to total tons, tons milled, ounces produced, pounds produced, or any other convenient reference. Contemporary operating costs are usually done in U.S. dollars, metric tons, troy ounces, and pounds.

PRELIMINARY FEASIBILITY STUDIES FOR PROCESSING FACILITIES

At the preliminary (or scoping) level, the estimator has several alternative techniques available. They are, in no particular order or preference, as follows:

- Experience plus factors
- Published factoring equations
- Examples for typical mills

Experience

Experience will enable the estimator to lay out a chart of accounts similar to the one shown in Table 16.1 and to assign values as per company accounting protocol. Table 16.1 gives a summary example of an estimate based on intracompany comparisons. An estimate of this type can be very reliable if the plants under comparison have similar ores, flow sheet, and expected daily tonnage.

The table of accounts will differ between operating companies. It is only important that all costs be identified and assigned to their proper account. An example is mill liners. Some operations carry liners as a discrete account, others lodge the charges under maintenance supplies, and still others place them under operating supplies.

The accounts, when taken from existing accounting formats, will serve the project from the initial feasibility studies through to final operation. This may seem like a trivial comment,

TABLE 16.1 Mill operating costs

Functional Account	Annual Cost, \$	Dollar per Ton Milled	\$/lb Cu
Operating labor	?	?	?
Maintenance labor			
Salaries			
Electric power			
Water			
Maintenance supplies			
Grinding media			
Reagents			
Operating supplies			
Distributed overheads (service)			

Adapted from Mular 1982

TABLE 16.2 Cost estimate for 5,000-stpd mill

Process Function	Mular's Empirical Equation
Process labor, US\$ =	C_{31} = (0.67)(index factor)(tons)$^{0.5}$
Process supplies, US\$ =	C_{32} = (0.67)(index factor)(tons)$^{0.7}$
Processing plant costs, US\$ =	$C_2 = (C_{31} + C_{32})$ US\$/d
Subprocess Function	
Mine/mill electric power	C_4 = US\$/d
Diesel, US\$/d	C_4 = (0.67)(index factor)(tons)$^{0.5}$
Utility, US\$/d	C_4 = (0.67)(index factor)(tons)$^{0.5}$
Coal, US\$/d	C_4 = (0.67)(index factor)(tons)$^{0.5}$
Mill electric power, US\$/d	= 75% (C_4)
General Plant Services	
Supplies, US\$/d=	C_5 = (0.67)(index factor)(tons)$^{0.5}$
Electrical labor, US\$/d =	C_6 = (0.67)(index factor)(108.4)(number of personnel)
Personnel	
Mill personnel (gold mill)	= 1.14 (tons)$^{0.5}$
Mill personnel (simple base metal mill)	= 1.10 (tons)$^{0.5}$
Mill personnel (complex base metal mill)	= 1.06 (tons)$^{0.5}$
Mill electrical services	= 0.03 (number of operations employees)

Adapted from Mular 1982

but it is surprising how many projects start up with no idea of how costs are going to be assigned or reported.

Published Factoring Equations

As estimate of possible mill operating costs can be developed from historical factors following the general steps outlined in Table 16.2 and summarized in Table 16.3. The factors are taken from literature that was published by Mular in 1982. The results must be adjusted to the current year by using the Marshall & Swift index for 1982 and the current index (M&S 1982, 2016). The current relationships are definitely better approximations of costs and should replace the 1982 document. In similar fashion, the number of required personnel can be estimated using

TABLE 16.3 Operating cost estimate, dollar per metric ton of feed, 5,000-t/d flotation plant

Operating Costs	Single Product, $	Two Products, $	Three Products, $
Supplies and materials	7.37	8.39	8.94
Labor	2.55	2.72	2.68
Administration	1.05	1.05	1.05
Sundry items	1.10	1.22	1.29
Total	12.07	13.38	14.16

Source: Infomine USA 2013

the relationships shown in Table 16.2. Again, it must be stressed that the results should be tempered by experience and common sense.

Examples for Typical Mills

InfoMine USA publishes examples of mill operating costs based on comparisons of similarly sized operations through its *Mining Cost Service* (Infomine USA 2014, or latest version). The comparisons provide an excellent basis for estimating both mill capital and operating costs and also offer good data for confirming the composition of prospective operating staff personnel. Table 16.2 is illustrative of the style of data included in the current edition.

However, the *Mining Cost Service* information should be used with a degree of caution. Ongoing operations have often established contracts for heavy supplies and chemicals that are considerably lower than published price data. When available, these costs should be used to support the InfoMine USA data.

Administration

Total administrative costs for a mine/mill operation can be estimated using existing information and/or with representative factoring equations. The number of administrative employees (N_A) can be estimated by using the following equation:

$$N_A = 0.07 \text{ (total number of site employees)} \quad \textbf{(EQ 16.1)}$$

Total administration operating costs can be estimated as follows:

$$\text{total administration wages, \$US} = (0.67)(\text{index factor})(1.35)(9.49 \times \text{total site employees}) \quad \textbf{(EQ 16.2)}$$

$$\text{total administration general expense, \$US} = (0.67)(\text{index factor})(6.38 \times \text{total site employees}) \quad \textbf{(EQ 16.3)}$$

INTERMEDIATE FEASIBILITY STUDIES FOR PROCESSING FACILITIES

Intermediate feasibility (or prefeasibility) level studies usually require a more detailed analysis of operating costs. These costs will be conservative because of the lack of both long-term bulk contract prices and competitive bidding. In addition, a more specific costing of salaries and wages will be used and an accompanying organization chart submitted.

Organization

The concentrator organization can be relatively simple to serve a small 500- to 2,000-t/d plant or rather large in support of daily tonnage that can contemporarily reach 200,000 t/d. The following work centers are usually recognized in the mill organization chart and are represented in Figure 16.1:

- Process operations
- Metallurgical engineering
- Assaying and quality control
- Maintenance

For very-well-automated concentrators in the United States that are milling very easily processed ores, the total complement of employees can be approximately one-fourth that shown in Figure 16.1.

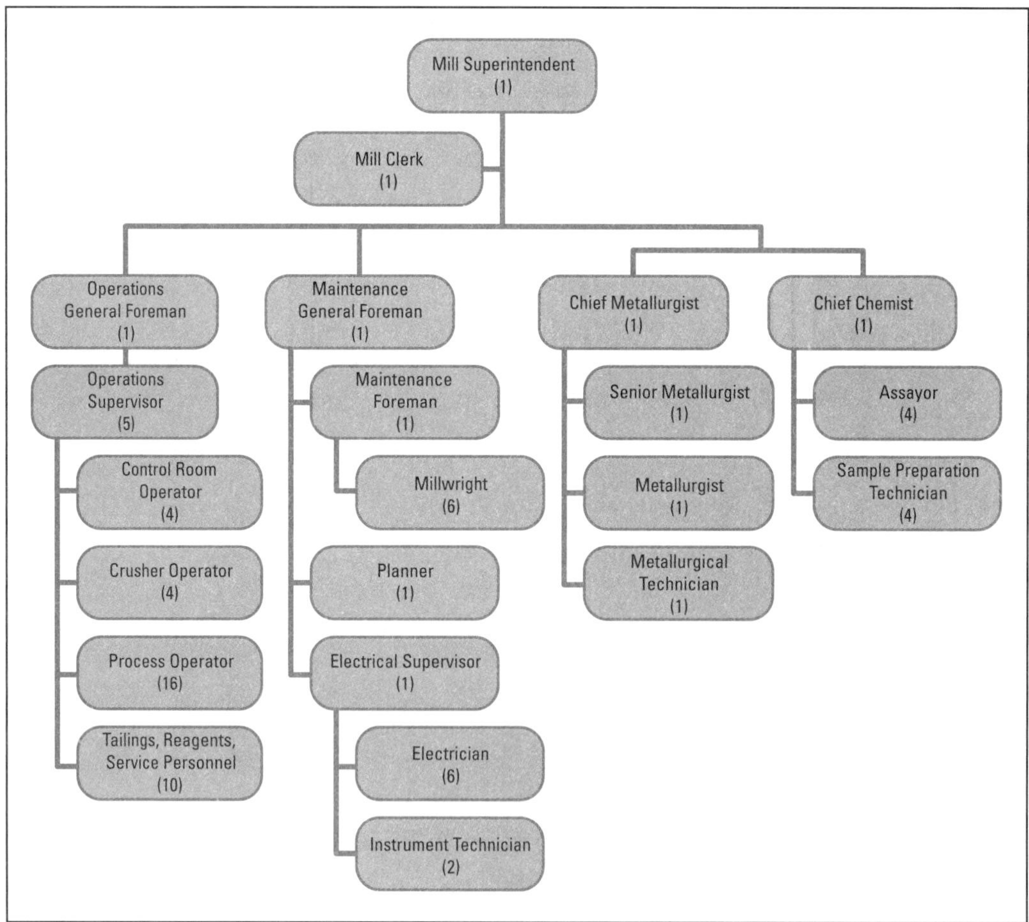

FIGURE 16.1 Mill organization chart

Direct Operating Costs

All operating expenditures must be divided into direct or indirect costs. This section discusses cost of functions, which can be identified as directly chargeable to performing the process, such as labor, materials, power, and overhead directly related to the mill process.

Labor

Labor costs can be estimated from Figure 16.1 by using average rates for the geographical area of the project location. Excellent wage and salary information is available for the minerals industry through the Mountain States Employers Council (www.msec.org) and from the U.S. Department of Labor, Bureau of Labor Statistics (USBLS; www.bls.gov). Important values to be included in labor statistics are estimates of overtime and fringe benefits. Typically, U.S. fringe benefit rates are estimated at approximately 35% in contrast to that of certain Latin American countries, which have fringe benefit loading that can be 200% or more.

Salaries

Salary data can be estimated in the same manner as labor. Existing operations have access to current rates and can also be predicted by the Mountain States Employers Council or from USBLS. It is also important to remember that both wages and salaries can differ widely between competing industries (e.g., coal versus hard rock) and by relative size of operation. In all cases, the estimator must exercise judgment and experience to avoid over- or underestimation of wage and salary data. In existing operations, the logic of salaried data is usually well understood and can be easily forecasted for positions up to and including mill manager or superintendent. Frequently, existing operations use a system of salary management such as the Hay system, and positions can be translated in cost both laterally and vertically. However, the logic and equity of such systems frequently breaks down for positions higher than the superintendent level and therefore are of little use.

Electric Power

Base electric power rates can be determined from existing published information in the most recent InfoMine USA data and by calling the public utilities directly. Similarly, the existing (non-negotiated) rates may be obtained from public information in most of the free world.

Operations that will be required to self-generate must estimate the cost of power generation based on the capital cost of generating equipment and fuel costs at the site.

It is important to recognize that utility electric rates are inclusive of connection, use rate, and demand charges and are usually negotiated on a site-by-site basis. Large concentrators with the extensive application of synchronous engines can be expected to gain favorable rates in that they will correct the dismal power factor demonstrated by the high inductive loads of community and light industrial use.

Reagents, Grinding Media, and Liners

At the intermediate feasibility level, metallurgical data are developed for the usage rates for mill reagents and the consumption rates for grinding media and liners. Sources such as InfoMine

USA can give an appropriate approximation of expected spending levels. If the estimator wishes a higher level of accuracy, a series of simple emails, faxes, or direct mail communications can usually obtain reliable cost data for consumables. Direct contact with suppliers often includes bulk commodity rates that are much improved over the rates charged for small or jobber quantities of chemicals and heavy supplies.

Operating Supplies

Operating supplies are usually factored based on the daily tonnage of the operation or past experience.

Water

One of the most frequently overlooked operating costs at any level of feasibility analysis is the cost for water. The operating costs for water, including labor, maintenance, electric power, and supplies, are collected for the water account and then distributed to the departments that use it, which include the mine, mill, and/or smelter. Importantly, water is not free and must be adequately provided for.

Maintenance Supplies

Maintenance supplies are also factored with relationships to mill tonnage or maintenance labor costs.

Maintenance Labor

Maintenance labor is detailed from the suggested organization chart, and hourly rates are estimated from published labor statistics. It is typical for maintenance labor to be maintained at a higher level during start-up and initial operations with gradual reductions as the operations mature and gain efficiency.

Fuel

As in the case of water, fuel costs are frequently overlooked or minimized in feasibility documents. In cold weather climates, the mine, mill, and maintenance facilities must be heated to an acceptable level. Mills can be heated by using waste heat from the grinding circuits, recuperated heat from electrical generation, and heat from supplemental heating systems that are fired by oil or natural gas.

Indirect Operating Costs

Indirect operating costs are those expenditures that cannot be identified as solely applied to a particular function of the operation but rather are applicable to more than one operating function. A good example of an indirect operating cost would be all the administration costs related to all functions of the operation.

Administration

An initial step in developing administrative costs is to construct an initial organization chart. The chart should include provisions for the management of warehousing, purchasing, accounting, personnel, health and safety, environmental, and general management of the property. Figure 16.2 is illustrative of a typical administrative organization chart.

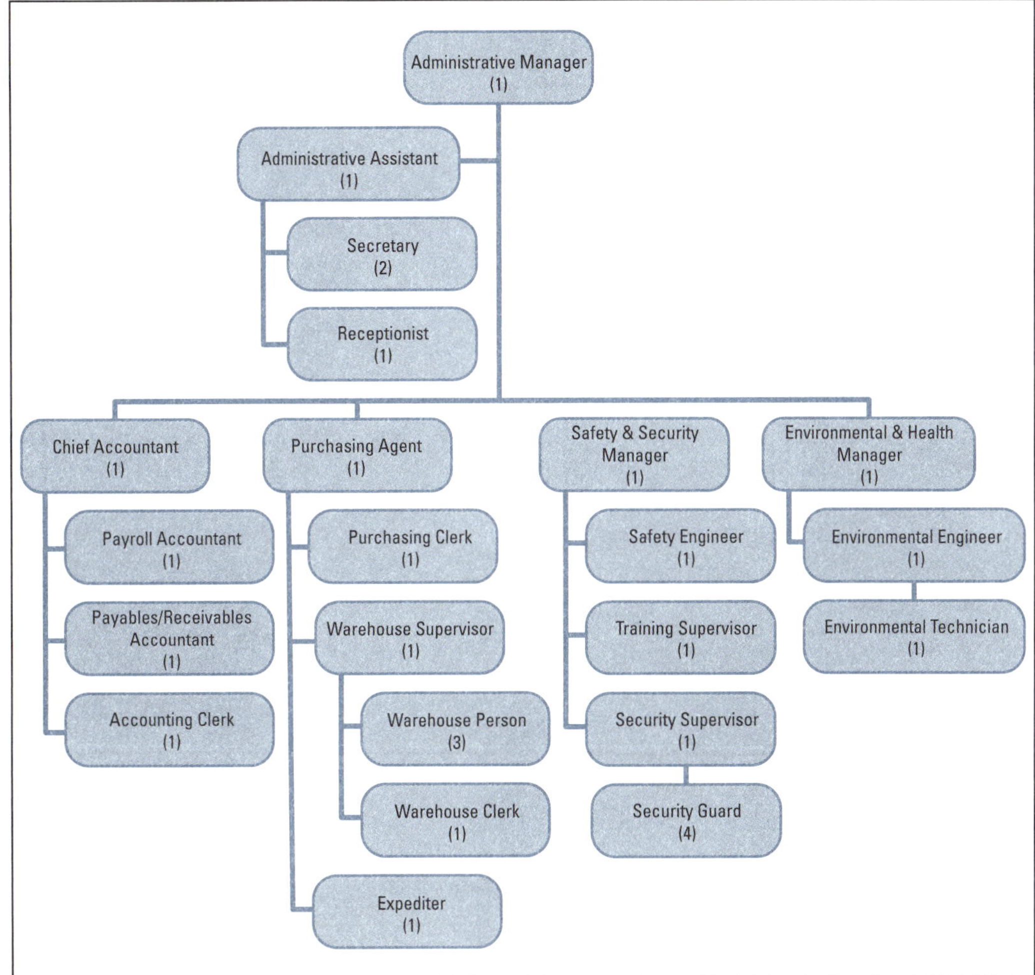

FIGURE 16.2 Administrative organization

Salaries/Clerical

Constructing an organization chart similar to the one in Figure 16.2 allows the estimator to approximate the burden for administrative salaries and the nonexempt salaried personnel in the operation. It is not critical for the organization to be in its final form at the intermediate feasibility level, only that all activities are accounted for.

The relative costs for salaried and nonexempt salaried personnel can be obtained from the Mountain States Employers Council for metal and coal mining, and for large and small organizations. Actual salaried data from within an existing organization is perhaps the most reliable.

Professional Services

The professional services accounts are given an allowance based on experience. Typically, these charges include the costs of training consultants, independent engineers, and vendor start-up personnel who are not included in the purchase price of equipment or systems.

Operating and Office Supplies

Operating and office supplies budgets are estimated by the number of personnel assigned to the overhead and administrative departments.

Small Vehicles

The small vehicles account includes the cost of fuel, maintenance, licensing, and maintenance and repairs for pickup trucks and automobiles assigned throughout the operations. These costs can be estimated by taking into account the number of vehicles assigned, the service into which they will be placed, and normal industrial multipliers.

Travel Expenses

Travel expenses include travel to and from the home office for company officials, expenses incurred in training, and travel to and from technical society meetings. These costs are frequently overlooked and can swell considerably in new organizations.

Employment Expenses

A new operation typically incurs expenses for employee moving costs for both salaried and hourly personnel, employment bonuses, and the costs of management recruiters ("headhunters").

Postage and Freight

Costs for postage and freight are usually estimated from the number of total salaried and non-exempt salaried personnel.

License Fees

License fees are estimated as an allowance based on the size of the operation and a certain knowledge of the administrative requirements of the local and federal government requirements. They frequently include environmental permit costs, communications fees, and so forth.

Equipment Rentals

Equipment rentals, in the initial stages of a project, are usually high because they compensate for overlooked equipment, such as large construction cranes, unique equipment for specialized tasks, and so on. The costs are usually handled by an allowance that is historical or resident with the estimator's experience.

Communications

Historically, communication was limited to telephone and perhaps telegraph. Contemporary communication systems include telephone (voice, fax, and data transmission), email, inter-company networking, teleconferencing, and other aspects of the rapidly expanding plant systems, such as pagers, intercoms, cellular phones, and so forth. The costs can be prohibitive if not closely budgeted and accounted for.

Insurance

Insurance charges, including production interruption, bonding, and so on, can be estimated based on the size of the operation, the production rate, and the location of the project. At the intermediate feasibility level, these will almost certainly be allowances based on experience.

Property Taxes

Property taxes can be estimated by using InforMine USA's latest published data for their *Mining Cost Service* (Infomine USA 2016) and by having conversations with local and state tax assessors. Property taxes vary widely in their scope and application, and the estimator usually consults a tax expert to plan for these costs.

Reclamation Costs

Reclamation costs, including the cost of bonding, are usually booked in a separate account. The costs are accrued against a future estimated cost and will probably not be a direct cash charge, with the exception of the reclamation bond cost.

FINAL FEASIBILITY STUDIES FOR PROCESSING FACILITIES

The final feasibility study differs from the intermediate feasibility (or prefeasibility) study only in the level of detail and certainty of specific costs. This often includes additional metallurgical data not yet available in the previous studies. All costs for labor, supplies, energy, water, maintenance, and reagents are supported by firm cost quotations and are negotiated to the lowest possible level.

Organization

At the final feasibility level, the organization should be completely delineated with key personnel identified by name in the four general areas of work. Costs to recruit and train personnel are also fully developed. A quality feasibility study anticipates the recruitment of all operations and maintenance personnel well before the initiation of detailed engineering so that field input to construction can be realized. As in the intermediate feasibility study, the following operational and operating units are assigned appropriate costs:

- Process
- Metallurgical engineering
- Assaying and quality control
- Maintenance

Direct Operating Costs

All operating expenditures must be divided into direct or indirect costs. This section discusses cost of functions, which can be identified as directly chargeable to performing the process, such as labor, materials, power, and overhead directly related to the mill process.

Labor

All hourly labor are identified for both operations and maintenance. Nonexempt salaried positions are also assigned schedules and payment. The work schedules for each department are finalized and training programs for all employee levels established. The costs for training are developed.

Salaries

Following completion and approval of the final feasibility study, the management team will be in place and active recruitment of supervisory personnel is initiated. It cannot be overstated that management must be on the project early to provide continuity to operations and training for all levels of personnel.

Consumables

Detailed bids based on agreed consumption rates are negotiated for all major plant consumables, including

- Electric power,
- Reagents,
- Grinding media,
- Liners, and
- Water.

Allowances

Allowances from the intermediate feasibility study are reviewed and will control the anticipated spending levels for the following areas:

- Operating supplies
- Maintenance supplies
- Fuel

Indirect Operating Costs

Indirect operating costs are those expenditures that cannot be identified as solely applied to a particular function of the operation but rather are applicable to more than one operating function. A good example of an indirect operating cost would be all the administration costs related to all functions of the operation.

Administration

The administrative organization is finalized and the key accounting, purchasing, environmental, and safety personnel hired.

Clerical

The clerical functions within the administrative function are identified and project costs are tracked in parallel with construction costs. The transition of project to operations accounting and purchasing is always clumsy and often results in significant control problems during start-up.

Allowances

Allowances for the following accounts are reviewed and finalized for inclusion in the final project operating budget:

- Professional services
- Operating and office supplies
- Small vehicles
- Travel expenses
- Employment expenses
- Postage and freight
- License fees
- Equipment rentals
- Communications
- Insurance
- Property taxes

Reclamation Costs

The approval of the final feasibility study and the commencement of construction reflects the completion of all environmental permitting requirements, including closure. Budget costs for the accrual of reclamation costs can now be inserted in the final budget figures.

REFERENCES

Infomine USA. 2013. *Mining Cost Service*, a subscription cost data update service. Spokane Valley, WA: InfoMine USA.

Infomine USA. 2014. *Mining Cost Service*, a subscription cost data update service. Spokane Valley, WA: InfoMine USA. p. CM 140.

Infomine USA. 2016. Taxes section. In *Mining Cost Service*, a subscription cost data update service. Spokane Valley, WA: InfoMine USA. pp. TA-Canada, 1-21; TA USA, 1-71, TA Appendix 1-16.

M&S (Marshall & Swift). 1982. *Equipment Cost Index*. Los Angeles: M&S.

M&S (Marshall & Swift). 2016. *Equipment Cost Index*. Available from www.equipment-cost-index.com/eci/UserContentStart.aspx (accessed April 5, 2017). Irvine, CA: CoreLogic.

Mular, A.L. 1982. *Mining and Mineral Processing Equipment Costs and Preliminary Capital Cost Estimations.* Montreal: Canadian Institute of Mining and Metallurgy.

CHAPTER 17

Investment Risk in Mining

Richard L. Bullock

What makes investing in mining any more of a risk than investing in any other industrial development? This is the multitudinous topic that is covered in this chapter. In fact, when one considers how few of the important aspects of a mineral property are known with 100% certainty during the feasibility phase of development versus how many of the important aspects of the property are still at least somewhat uncertain, then one begins to realize why it is so important to quantify the associated risk in every aspect of the mineral property feasibility study.

Unlike almost all other industries (except the petroleum industry), the location of the mining project has to be developed wherever creation placed the mineral resource. It may be north of the arctic circle, in the Amazon jungle, at a 4,000-m (13,123-ft) elevation in the high Andes, in the Mojave Desert, or where it can be more challenging, such as in a pristine area close to a national park. Many risks are associated with this lack of choice for where to locate a mine. Mineral development sometimes involves areas or countries where mining has a bad reputation. In those cases, it will be difficult to get the opportunity to show that your mine can and will be environmentally and socially in harmony with the existing conditions, and there might never be an opportunity to demonstrate that this so. The mining opportunity may exist where mining has never been practiced, and therefore there are few legal provisions for these activities as well as a strong resistance to changing anything in the area. It may be that the government is weak and unstable so that the concessions that have been granted may become worthless pieces of paper, after many millions of dollars have been invested in land and water acquisitions. These are just a few of the problems related to the location that must be considered.

What about the geologic risk that the mineral resource will not become the minable ore body that the geologic experts have projected? What about the technical risk involved with the assumption that all of the ore will be recoverable in the metallurgical process as identified in the laboratory samples that were taken? What about the risk in mining? Can the ore be recovered in the cost-effective, safe manner that is being projected by the mining engineers? Once the product is produced, will there be an economically, safe method to move it to the market and will there be a market? These are just some of the technical issues that must be considered.

Then there are the project execution issues to deal with. Will it be possible to construct the mine, plant, and infrastructure in the area at a reasonable, predictable cost and in a reasonable, predictable time schedule? Is there overall labor stability in the area, or are labor strikes likely? Are there some dependable sources of construction material available in the area, or is there a risk in relying on anything produced locally even while the local government is insisting on local procurement of certain items? Has the company worked in the region before or is it totally unfamiliar with labor practices and the work ethic of the area? What risks are imposed by the remoteness or the severe weather conditions in the area? Will an avalanche, tidal wave, or earthquake destroy the plant that has just been built? Has the project been timed to be caught in an upswing of construction activity that will greatly increase construction costs? All of the preceding issues and many other such items have contributed to severe cost overruns in previous project executions.

All mining projects have risk components to a varying degree. Major components of risk include technical issues, environmental aspects, market and financial concerns, and political factors. Tolerance for risk is dependent on the size of the parent company and the financial and business approach to the project. A partnership (joint venture) approach may be fraught with induced risk pertaining to the smooth development and operation of the project. Adding a third-party lending agency or government financial institution, or even a vendor financing organization, can sometimes cause problems not recognized during a feasibility study. Such complications can cause project management to delay making timely decisions or to make poor decisions, leading to less-than-optimal financial returns for the project. With all of the related risks, it is a wonder that any mine ever gets the financial backing to be developed. Proper quantification and management of the risks are essential for mineral properties to progress to development.

Risk at mining projects can be accounted for in numerous ways: raising the rate-of-return threshold on specific projects, using higher discount rates for cash flows performed on projects at different development points, or performing an iterative Monte Carlo type of analysis. Many projects warrant a detailed engineered risk assessment. The author is not endorsing any given method, as different methods may be more suitable than others, depending on each individual project.

PART I: USING DISCOUNT RATES TO ACCOUNT FOR PROJECT RISK*

Every mining company needs to establish a risk policy so that all decision-making employees will be "working on the same page" and will know that top management supports all decisions following this policy. In the author's opinion, a company cannot survive in the mineral/mining industry without taking some measure of calculated risk. In taking the risk, however, one must be accountable and prepared to manage the risk. By following this practice, one will achieve greater returns on the investment than those companies that will not tolerate any risk. Using the discount rates to account for project risk at the feasibility evaluation level is one way to manage risk.

At the feasibility stages of the mineral project, all the expended effort from the years of work comes down to the financial analysis of what the profitability of the projected operation will yield when built and operating, as is described, engineered, and estimated in the feasibility reports. That yield is determined from the project's cash flow; this is discounted by an interest rate that needs to reflect the risks associated with the project. These include the risks that are involved with this particular project, at this particular time and place, with this particular company.

---

* This section is primarily reporting the work of others who are the experts. The author makes no personal claim as being an authority of calculating discount rates, but those who are authorities are cited.

DETERMINING THE RISK-ADJUSTED DISCOUNT RATE

The components of the discount rate are made up of basic opportunity cost, transaction cost for acquiring the capital for the investment (weighted average cost of capital, or WACC), and appropriate additions for the added risk of this particular project versus a non-risk investment. The risk-adjusted discount rate (RADR) is well described in the literature (Smith 2013). It is also described by Guarnera and Martin (2011) when applying valuations to mineral properties. How to determine the "added risk" is somewhat controversial. Nevertheless, it is an important topic in this chapter. For other approaches to determining the best interest rate, the reader is referred to literature by Runge (1998) and Torries (1998). A preferred method is described in "Discount Rates and Risk Assessment in Mineral Project Evaluations." Lawrence D. Smith (1995) cites this specific approach as an applicable way of determining discount rates for projects in the minerals industry. His assumptions for this situation are that

- Constant dollar analysis is used,
- Financing is 100% equity (which removes the owner's financial condition from the equations), and
- The analysis is after tax (since it is a cost).

Smith (1995) presents three principal factors to be considered in the determination of the project-specific discount rate (D):

1. Risk-free return (R_f);
2. Risk related to this particular project because of product, mineral, metallurgical, geotechnical, or mining characteristics (R_p); and
3. Risk for location in a country or particular state or province within a country (R_c).

If the inflation rate (I) is considered, then the equation becomes

$$D = (R_f - I) + R_p + R_c \qquad \textbf{(EQ 17.1)}$$

Risk-Free Interest Rate

One way to develop the risk-free interest rate is (Guarnera and Martin 2011)

$$Rfr = \frac{(1 + Rfn)}{(1 + Ie)} - 1 \qquad \textbf{(EQ 17.2)}$$

where
R_{fr} = *real risk-free rate of return*
R_{fn} = *nominal risk-free rate offered by U.S. Treasury notes*
I_e = *expected inflation rate*

Accordingly, assuming a 10-year mine life and 10-year U.S. Treasury notes yielding 4% with inflation at 1.5%, the real risk-free rate of return is

$$\frac{(1 + 0.04)}{(1 + 0.015)} - 1 = 0.025 \text{ or } 2.5\%$$

Mineral Project Risk at the Evaluation Phase (Smith 1995)

The following areas of risk should be considered:

- Reserves (sampling, grade, density, tonnage, and mine life)
- Mining (rock quality, mining method, recovery, dilution, layout, mechanization, and automation)
- Process (sampling, lab testing, labor factors,* plant availability, metallurgical recoveries, material balances, reagent consumption, water balance, waste disposal, and automation)
- Construction (labor availability, equipment and material delivery delays, costs, schedules, group interference, and weather delays)
- Environmental permitting and compliance
- New technology (if used)
- Cost estimation (capital and operating)
- Prices and markets (particularly for industrial minerals)

It is apparent that the best way to reduce risk during the evaluation period is to do more engineering. Given that more engineering and geological work will have been completed during each phase of the feasibility period, the project risk will obviously be reduced. The amount of engineering/study recommended is discussed in Chapter 11.

Country Risk

Some companies believe that a degree of country risk should be added to all projects, including those in the United States (primarily depending on which state the project is in). As previously discussed, the degree of risk would be higher for new, proposed projects as compared to those that are well established and are continuing to meet changing regulations regarding operations and environmental compliance. Because of difficulties in permitting and establishing new projects anywhere in the United States, this author recommends a 1.0% to 2.0% (averaging 1.5%) risk factor for operations in the United States (assuming it is not accounted for in other risk-adjustment categories). In some countries, such as the United States, Argentina, and even Australia and Canada, the permitting time and potential failure is quite varied, which must also be taken into account. As an example, permitting in Nevada or Wyoming may take many months, but it is probable that a reasonable project will eventually obtain its permits. In contrast, for permitting a large underground mine in northern Wisconsin (which is generally easier to permit than an open pit mine), one would have to consider that this may be an extremely lengthy process (years), if not impossible. The risk of investing money early in the Wisconsin project could well be equivalent to investing in much riskier, problematic countries where the risk is more apparent. The risk associated with projects in other countries may be higher or lower (than in the United States), based on the specific site/country involved, and, for these, the risk rate selected must be based on detailed study of all the relevant factors.

Sources for determining country risk can be found in the following references:

- *Behre Dolbear Newsletter:* Since 1999, Behre Dolbear has annually compiled political risk assessments in the global mining industry. Over time, their efforts have revealed a positive correlation between the growth of a nation's wealth and the prosperity of its mining industry. The company has observed that when most countries recognize

* Labor factors should also be considered in the risk of "mining, construction, environmental compliance, and cost estimation. See Tables 14.3 and 14.4 for mine and construction labor cost indexes.

their critical need to adapt and restructure burdensome policy, it begins to optimize its economic potential. This "Where to Invest?" analysis, published once a year, presents a report discussing world political and security risk, both the positive as well as the negative sides of the various issues. It ranks countries, but not individual states or provinces within countries. The current analysis is available at www.dolbear.com, but the list is limited to the *25 least riskier countries* where mining is ongoing or contemplated (Behre Dolbear Group 2015).

- The Fraser Institute, a Canadian group, produces a report that ranks areas in North and South America and other major mining areas in terms of their investment attractiveness (Jackson and Green 2017). Recent survey information may be downloaded by visiting their website at www.fraserinstitute.org and typing "Annual Survey of Mining Companies" into their search bar. This biannual report is a very comprehensive risk analysis published on the worldwide industry. One particular value of this survey is that it ranks individual states, provinces, or territories within countries. However, although it does list all of the provinces, states, and territories of Canada, Argentina, and Australia, it does not list all of the U.S. states. Even in states like Missouri, which is a major mining state, it can be impossible to permit a mine in certain areas.
- The *SME Mining Engineering Handbook* (Darling 2011) includes four chapters in Part 17 that discuss community and social issues and can be quite useful. Although the subject is not directed at the comparative risk of various countries, it does cover circumstances that need to be considered when mitigating social and community risk.
- The U.S. State Department generates considerable information in the form of Country Security Alerts on personnel security issues in countries around the world; much of this can be found on the Internet (https://travel.state.gov, then click on the "U.S. Passports & International Travel" tab, then the "Country Information" tab). The website includes an interactive map as well as travel alerts and warnings where one can obtain country information. The author has used this service many times to evaluate where it was safe to accept or reject assignments. This same information helps to verify what other sources are publishing on country risk.

In considering the country risk ranking, Behre Dolbear (2000) uses seven criteria:

1. The country's economic system.
2. The country's political system, which affects
   - Political parties,
   - Constitutional risk,
   - Quality of government,
   - Foreign ownership policy (risk of nationalization),
   - Foreign policy,
   - Government crises, and
   - Taxation instability.
3. The degree of social issues affecting mining in the country, including
   - Distribution of wealth,
   - Ethnic or religious differences within indigenous population,
   - Literacy rate,

  - Corruption, and
  - Labor relations.
4. Delays in receiving permits due to bureaucratic and other issues, such as
- Environmental policy and environmental protectionism, and
- Land claims and protected areas.
5. The degree of corruption prevalent in the country.
6. The stability of the country's economy and currency, such as
  - Currency stability, and
  - Foreign exchange restrictions.
7. The competitiveness of the country's tax policy.

Each country has its own set of risks, be they social, economic, environmental policy, government crisis, and so on. As previously stated, country risks vary with area, state, or province, so the evaluator must determine how to apply the overall risk.

Adjusting the Corporate WACC for Risk

One can account for all of the known risks by a composite buildup of the various components of risk by adding to the WACC. Smith (2013) illustrates this in Figure 17.1. Starting with the WACC, he adds the project stage (feasibility study level), plus any other risk that is recognized, and then adds the country risk to reach a total RADR of about 11.5%. Figure 17.1 also illustrates how this same result can be obtained starting with the risk-free interest rate given in Equation 17.1, but adding a market premium risk. Other likely risks are technology risk, remoteness risk, extreme-weather risk, market risk, or any other condition identified for that unique site that can be quantified.

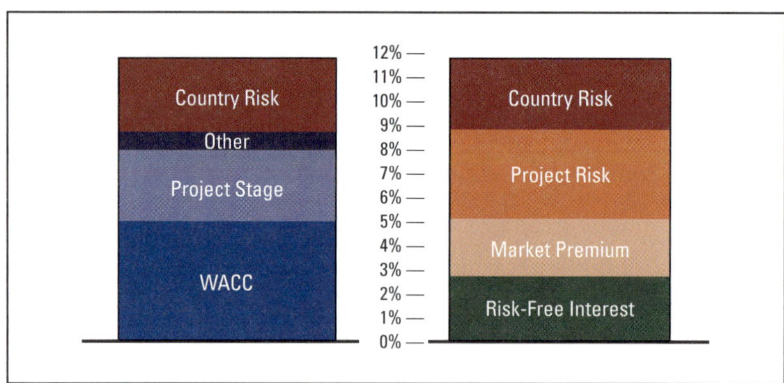

Source: Smith 2013

FIGURE 17.1 Risk-adjusted discount rate buildup

Level of Study and Commodity-Specific Risk

Risks can also vary with the level of study and commodity under consideration. Smith (2013) recognized that certain commodities had lower risk than others. Applying his observations during the period from 1996 to 2005, he developed Table 17.1.

Smith observed that gold companies, when developing gold properties, typically develop a WACC of 4% to 5%, whereas this usually is increased by 2% if they are developing a copper property. In contrast, base metal companies usually develop a WACC of 7% to 8% for base metal properties, which is typically increased by 3% when they are developing a gold property (Smith 2013).

Company- and Property-Specific Risk

Many variables must be considered when assessing the risk at each property and company—not only those risks listed in the "Country Risk" section but also the geologic/reserve issues and the mining, processing, environmental, operating, and closure issues as well as those of costs, pricing, and marketing.

Behre Dolbear developed a list of such risks to be considered, which has been extensively modified and is shown in Table 17.2. Each of these risks found to be applicable must then be

TABLE 17.1 Risk-adjusted discount rates at different project stages for gold and base metal projects

	Discount Rate		RADR	
Level of Feasibility Study	Gold Project, %	Base Metal Project, %	Gold, %	Base Metals, %
Preliminary	12.4	14.0	6.9	5.5
Intermediate (or prefeasibility)	11.2	13.0	5.7	4.5
Final	8.8	11.0	3.3	2.5
Operating mine	5.5	8.0	0.0	0.0

Source: Smith 2013 (data from CIM-MES [Management and Economics Society of Canadian Institute of Mining, Metallurgy and Petroleum] Industry Surveys: 1996, 1999, 2005)

TABLE 17.2 Company and property feasibility study risk identification list

Risk Category	Risk Description
Level of project and operating status	The level or phase that project is in: exploration; preliminary, intermediate (or prefeasibility), or final feasibility; project execution/construction/mine development; start-up or operational; shutting down; and the reclamation phase.
State or country risk status	The risk and stability associated with the political system that governs the rules and laws under which the mining operation will be built and operated. Although federal level laws must be followed, state laws in some places are more restrictive but nevertheless must be followed. This is not only true of health, safety, and environmental laws and regulations, but it is also true of taxation and royalty laws. This also includes the country nationalization of property risk, import/export, and currency restrictions; over- or under-enforcement of police or military forces; and the restrictions of removing your product from the country by the most economical method.
Social and civil risk (these two are closely related)	This social and civil risk may be countrywide, or very local, but even if it is local, it can stop the project. Of course, if there is political instability in that local area, the risk is higher. The risk could be as high and widespread as a civil war or civil terrorism, or as local as individual roadblocks by the resident population. It may result in employees being harmed or kidnapped, or simple harassment of employees coming to and leaving from the project. But it all illustrates why a social license to operate (SLO) is so important. The risk seems to be higher in emerging nations where there is (1) larger disparity in wealth; (2) more government corruption; and/or (3) a division between ethnic, indigenous, or religious groups.

(Table continues)

TABLE 17.2 (Continued)

Risk Category	Risk Description
Climate and natural disaster risk related to geography and location	These risks are related to the specific site location of the project. Although companies have no control regarding where the mineral resource is located, they do have control over where they locate many of the facilities that must be constructed. Such events as snow avalanches, earthquakes, floods, tidal waves, droughts, and tornadoes must be considered in certain areas, and companies need to build their facilities accordingly. Such facilities as power lines, buildings, ore and product transport features, waste impoundments, and all dwellings must be protected or considered at risk in some areas. There is also risk when building mining facilities close to restricted areas, such as national parks or game refuges.
Land, water, and property risk	The risk of ownership and control of land and water rights, and the ability of the mining company to control its property-related assets and all facilities, both surface and subsurface, is of primary importance. However the land and water were acquired—whether by purchase/deed, land grant, claim/patent system, or similar—the resulting documents must be in perfect, unchallengeable order or the entire project will be at risk. This risk will involve the surface, mill site, tailings site, underground area, and, if patented claims of vein outcrop, the extra-lateral rights beyond the surface boundaries.
Geologic and reserve risk	The risk of the identified "reserve" depends on the understanding of the geology, mineralogy, and mineral geologic controls—those that led to the perception of the grade, tonnage, mineralogy, and ore resource type—that led to geologic models correctly interpreted and the tonnage developed using the correct density of the ore. This geologic and reserve risk is one of the key failures when projects become exploited.
Mining risk	The mining risks are very much dependent on the correct geological/mineralogical and geotechnical interpretations and projections from the work of the project studies. If there are serious questions or doubts, then a test mine may need to be developed. These risks involve ore and rock hardness, abrasiveness, ore and wall/roof (back) strength such that the correct mining method and equipment are chosen that will yield the productivity, dilutions, and recovery as predicted during the feasibility studies.
Metallurgical/ Product recovery risk	The metallurgical risk is related to being able to recover the product at a cost and with the quality as predicted by the feasibility study and the resulting positive economic forecast. This involves the correct selection of processing method, equipment selection and utilization, extraction and recovery volume, productivity, technology, and operational coordination. Also included in the risk is transporting the product to the point of sales with minimum loss. The reality of the metallurgical risk is one of the key failures of many projects from becoming successful.
Infrastructure risk (both industry and community)	The biggest risk in this category is for major items to be left out entirely or grossly undersized. In most mining communities, the infrastructure serves both the company facilities and the community, so it must be sized for both. It must accommodate all health and safety requirements, including fire and ambulatory services as a minimum, but may also include hospital and educational facilities, in which case they must be coordinated with local authorities in the field. In some communities, adequate housing must be provided. Other social facilities, such as stores, recreational facilities, electrical and communications networks, freshwater and sewage facilities, must be provided. All of these facilities must, at a minimum, meet state and local standards of approval.
Product quality risk	There is a risk that the quality of the product that was predicted in the feasibility study may not be produced. This of course will result in a penalty to the profitability of the enterprise. This is especially critical in industrial mineral products, where quality specifications are very important and extremely vital to the price received for the product. There is also the risk of competition delivering a better-quality product if this operation and its product are not meeting specifications.

(Table continues)

TABLE 17.2 (Continued)

Risk Category	Risk Description
Marketing risk	Marketing risks include all risks that occur in getting the product to market. This includes the risk of losses in handling and packaging, storage, and shipping. It involves local and foreign competition as well as product substitution. Also included is the risk of the market's normal volatility; to minimize this risk, the operation needs to be in the lower quartile of the cost seriatim of this product to survive the worst of the market recessions.
Environmental risk	Environmental risks are those that may occur during and after construction, which may be temporary, or those of a more permanent environmental impact, which may or may not have been predicted to be caused by the operation. These risks include those associated with physical, human, and ecologic factors such as soil, water, air, noise, flora, fauna, changes to the biodiversity index, and archaeology sites showing the presence of former human habitat. The real risks are related to environmental policy, environmental restrictions, costs, technology, legislation, regulations, and implementation, which are constantly changing. Where there is not a SLO, opponents will seek to promote such changes to try to stop the project either with fines or infinite delays.
Labor risk	The labor risks start at project execution. Lack of skilled labor sufficient to construct the plant or develop the mine (especially during "up" periods of the mineral industry), or labor strife, can stop or slow projects at extreme cost to the owner. When the operation begins, labor risks may occur from union versus nonunion confrontations, and from lack of skilled or trained plant operators and miners. Or there may be the special requirement to use aboriginals or a given percentage of indigenous peoples versus expatriate employees, which will require much more training and a steeper learning curve and needs to be defined in the feasibility study.
Management risk	From the feasibility level, the first management risk is that of not having a management team for the execution phase that has successful experience (1) in this type of operation, (2) for a development in this country, (3) with this type of management venture (single company versus joint venture); and that understands project management governance and controls. For the operations phase, the management risks are the same: not having successful experience in managing operations such as the one being considered, in the country being considered, and under the same ownership arrangement.
Financial/Economic risk	Feasibility study risk is primarily related to the insufficient engineering expended to properly design and correctly estimate the capital cost and schedule to construct the plant that will be built; and to estimate the correct operating cost, working capital, contingency, royalties, taxes, insurance, financing requirements, the state and federal accounting and repatriation restrictions.
Other site risk	Each project will have unique conditions that need to be flushed out between the end of the final feasibility study and the awarding of the contract to construct. Part II of this chapter deals precisely with the mechanisms needed to assess these conditions through a project risk appraisal and adjustment procedure. There may be other associated risks that are unique to the particular project being evaluated. For example, the contractor may have a history of underestimating projects, which can be documented, or the project will be executed in a very active economic time or place and costs are very likely to be much higher than estimated. In such cases, some adjustment needs to be made. It may be that the data from the geologic or metallurgical sampling have not had the security or quality assurance normally attended to such information, and again, an adjustment may need to be made.

Information in this table is from work with Behre Dolbear associates 2000–2015.

quantified and used in the overall discount rate used. A quantifying method is described in Chapter 19.

Tax Considerations

The "pretax versus after-tax" topic as well as a general discussion on taxes are covered in Chapter 20.

Conclusion

Because of the inherent nature of mining projects, it is critically important to investors and shareholders that all the foreseeable risks are accounted for in the discount factor used in the financial evaluation of the proposed project. It would be a relatively simple exercise to exaggerate the risks and thus stop an otherwise worthwhile project. Similarly, it is all too easy to underestimate the risks of a not-very-robust project to move forward and hope for a good outcome. The former course of action (overestimation) results in postponement or cancellation and at least temporary loss of money already invested. The latter course results in, at best, a less-than-expected return on investment or, at worst, financial disaster. It is believed that the discussion in this chapter will enable the reader to land in the middle ground.

PART II: AN ENGINEERING APPROACH TO RISK APPRAISAL AND ADJUSTMENT

The primary purpose of a risk appraisal is to determine, and thereby be able to mitigate, the risk of engineering, procurement, and construction of the project that is being advanced and thus control the potential of major cost overruns during the execution phase of the project. This risk appraisal is usually completed after the final feasibility study and after the design basis report has been completed. However, it can be done at the end of each phase of the feasibility study to help assess what the real cost of the project is likely to be and thus adjust the cost estimate and related net present value and internal rate of return accordingly.

This portion of the chapter takes a different approach to risk assessment than that previously described. The multitude of risks are organized and evaluated by breaking project elements into the various components and then examining areas of concern to determine if the project being evaluated supports development. To complete the method described in this section is a very tedious and labor-intensive task. But for a very large project, which is potentially at high risk, it can be a worthwhile exercise.

The approach entails making lists of the potential risks and then quantifying those risks for which there is no economical mitigation. The process is called project risk appraisal and adjustment, or PRAA. The team that will perform the risk appraisal for projects is called the *appraisal team*. The same approach should be taken by individuals knowledgeable in both the technical and business aspects of the company. The timing for this appraisal for all projects could be at the end of each evaluation phase, after each cost estimate for that phase of the project. What will be adjusted is the *standard cost estimate* from that study, with the project risk adjustment (PRA) applied to that estimate.

Potential PRAA Guidelines

A PRAA is a structured analysis of a project situation and conditions developed for uniform application to all projects. Appraisals are made to identify potential occurrences that can influence project costs and schedules, both favorably and unfavorably. The appraisal results in a list of such occurrences and a judgment of their likelihood of the event happening. Estimates are prepared to quantify the effect of the events on the project cost and schedule, and the PRA is developed.

The PRAA is a probability-weighted cost estimate of *potential cost growth* for a specific project. It is developed from a combination of potential occurrences and the cost/schedule impact of those occurrences on the project. Although it is intended to develop a range with upper and lower limits, a single number showing the most likely cost resulting from the combination of potential occurrences should be used for project evaluation. This probability-weighted total relates the high risk of a minor problem with the low risk of a catastrophe. This limitation is recognized and addressed by testing the PRAA for full cost (non-weighted) as well as probability-weighted totals.

The PRAA is added to the standard cost estimate to provide the total project estimate. It is not an alternative to traditional contingency allowances and does not reduce them. A *cost estimate contingency* is an allowance for items of cost not thought of or considered in the cost estimate. It is intended that issues relevant to a PRAA not overlap traditional contingencies. For rare cases, and to the extent that factors or issues were recognized by the cost estimators but they chose not to apply the cost directly and chose to cover them with a cost contingency, those recognized contingencies may be eliminated from quantification of the PRAA analysis. PRAAs will likely add to total project cost.

An appraisal does not underscore operating issues that may develop after project completion, except insofar as they have a direct impact on project construction costs or schedules. Examples of the issues not emphasized are: potential operating inefficiencies, operating mechanical reliability problems, and the inability to meet productivity goals. Such concerns are handled as traditional project sensitivities in the evaluation process. However, management may want the PRAA teams to also identify and evaluate those operating items as a separate and simultaneous exercise with the PRAA. Such operating items could be quantified as to their impact on the project discounted cash flow, just as the PRAA items are quantified as to their impact on the project cost. These items could then be used by project management in presenting project sensitivities to the company's management.

The total project estimate (with the standard cost estimate and PRAAs identified) should be included in project economics and budget appropriations, which are subject to review by company management in accordance with established budgetary procedures. This chapter does not attempt to define how PRAA adjustments should be handled in project execution budgets. Such control steps must be jointly worked between the company's accounting department and project management when the project is appropriated.

Estimates Requiring an Appraisal and PRAAs

PRAAs should be generated for all projects as part of the estimate development for all classified project estimates (from preliminary estimate to construction estimate). Appraisals and adjustments should be generated for all projects depending on their complexity or special characteristics. Table 17.3 shows factors that may dictate the need for an appraisal if several factors are relevant or if one or more factors may have significant project impacts beyond the assumptions recognized in the standard cost estimate.

Sometimes it is even appropriate for unclassified study estimates to require appraisals and PRAAs if the unclassified estimates involve potentially large investments and the estimates will

TABLE 17.3 Relevant factors that may affect projects, requiring the need for a PRAA

Complexity Factor	Project Characteristics
Unclear or ambiguous basic project definition	Rapidly changing or uncertain market conditions Poor geologic or metallurgical resource definition Long duration of project (more than three years) Estimating/Scheduling tools unproven on this type of project First time for this company in this country Property ownership or partnership uncertain
Novel or unfamiliar technology	Basic design schedule and procurement when novel technology is adapted to commercial scale Cost estimate when novel technology is employed
Remote, isolated site or hostile weather	Far from center of human population Historically harsh weather Complicated labor facilities Extensive infrastructure required Off-site preassembly and/or modularization of facilities Parts of project spread over large distances
Complicated project construction resources	Multiple prime contracting beyond normal mine/surface facilities split Labor scarcity in the area or country Scarcity of material supply sources Overextension of engineering and design resources
Partnership, financier, or stakeholder complications	Joint ownership, thus joint decision making Financing procedures causing on-site delays Complicated project organization and chain of command Complicated permitting difficulties Complicated social issues at the project site
National political sensitivities	Complicated or especially sensitive political (or other) issues Risk of nationalization Risk of confiscatory taxation on property Changing of important restrictive rules that alter project character or economics

be used for making major decisions (e.g., resource acquisition, business commitments) regarding engineering and construction after the investment. Appraisals and PRAAs will not usually be required for unclassified estimates used for project screening, process selection, or where the financial risk of the resulting decision is small.

Timing of the PRAA

Appraisal and generation of the PRAA can proceed essentially in parallel with the development of the standard cost estimate; however, there must be documentation of three fundamental project elements that are required before making an appraisal and generating the PRAA:

1. Project basis, which defines the facilities and location
2. Execution basis, which defines how engineering, labor, and materials resources will be obtained to design and build the project, and the schedule for implementing these resources
3. Cost estimate basis, which defines the database that is used for the estimate

When these elements are available and have been reviewed by the project executive, the appraisal team can be selected and scheduled.

TABLE 17.4 Approximate labor and time estimates to perform a PRAA

Description	Preliminary Phase	Intermediate Phase	Final Phase
Total team size (11%–17%)	4.5%–11%	9%–11%	12%–17%
Approximate estimated worker-hours (11%–17%)			
Organizational meeting	400	700	1,150
Analysis and quantifying meeting	600	800	1,500
Final meeting*	200	450	850
Consolidation of information and reporting	400	600	600
Total worker-hours	**1,600**	**2,550**	**4,100**
Cumulative total	**1,600**	**4,150**	**8,250**

*It is assumed that only 75% of the appraisal team will take part in the final meeting.

Initiating an Appraisal

Appraisals will normally be sponsored by a group formed as the Technical Evaluation Board (TEB), which is headed by the senior company officer in charge of project development. The need for a PRAA, the level of resources to be assigned, and the critical personnel to be involved will be decided by the TEB in consultation with the project executive or manager or, where a company has many projects in a foreign country, the country manager. The objective should be to identify and evaluate project risk exposure in a cost-effective manner. Early in a project's development (preliminary feasibility stage), the appraisal should address broad project exposures to risks. Senior members of the company organization should participate with project team members to conduct the appraisal. By the final feasibility stage, prior to budgeting, the appraisal should concentrate on completeness with broadest practical inputs from company management and other affected company affiliates, if applicable.

The amount of labor devoted to making an appraisal will depend on the size of the project, its complexity, its locations, and the feasibility level that is being assessed. Table 17.4 is an approximate guide to the estimated amount of time that may be required to perform an appraisal for a very large project. (This estimate was taken from an actual complex mining project in an isolated area in the Andes of South America. The capital cost estimate in 2000 dollars would have been approximately $3.1 billion.) For simpler, smaller projects or those requiring less infrastructure and logistical problems, the labor and time estimates required will be much less than what is shown in Table 17.4. Normally, the PRAA would not be performed until the completion of the final feasibility study, but the two lower phases are shown in case the owner would want to know the real constructed cost at the lower phase.

Appraisal Tasks

The appraisal team coordinator gathers data pertaining to the project and will provide this information to all appraisal team members. The information that describes the scope and basis of the project, as well as the documents that describe the entire appraisal process, should also be provided. At the organizational meeting, the appraisal scope and goals, team organization, individual team members, and the scheduled plan for the PRAA process should be developed. At this stage of the appraisal, the group might wish to obtain input from management groups not represented on the appraisal team, as this is necessary for operating groups that may not have representation but need to have input to the project. Survey forms, which will bring in

TABLE 17.5 Cost risk appraisal information form for a mining or concentrating project*

Name of Project: ______ Person's Name: ______ Date: ______

Item	Estimated Weight of Project Cost <0.5% LOW	0.5–5.0% MED	>5.0% HIGH	Level of Risk in the Subject Area LOW	MED	HIGH	1. Reason for Risk 2. Identification Data That Could Mitigate Risk
Geology Items							
Ore Reserve Accuracy							
Geostatistical Analysis							
Grade Accuracy							
Mineralogy Accuracy							
Other: ______							
Mining Items							
Mine Development Plan							
Mine Plan/Design							
Mine Recovery							
Rock Hardness/Abrasiveness							
Ground Control/Support							
Hydrology							
Mine Equipment Selection							
Other: ______							
Processing Items							
Flow Sheet							
Process Design							
Ore Complexity							
Water Supply							
Process Recovery							
Product Quality							
Tailings Disposal							
Downstream Processing							
Other: ______							
Infrastructure Items							
Land and Water Ownership							
Transportation							
Electricity/Utilities							
Other: ______							

(Table continues)

TABLE 17.5 (Continued)

Name of Project: ______	Person's Name: ______					Date: ______
	Estimated Weight of Project Cost			Level of Risk in the Subject Area			1. Reason for Risk 2. Identification Data That Could Mitigate Risk
	<0.5%	0.5–5.0%	>5.0%				
Item	LOW	MED	HIGH	LOW	MED	HIGH	
Operating Plan							
Maintenance/Plan Organization	____	____	____	____	____	____	________
Plant Reliability	____	____	____	____	____	____	________
Design Redundancy	____	____	____	____	____	____	________
Other: ________	____	____	____	____	____	____	________
Project Execution Plan							
Remoteness	____	____	____	____	____	____	________
Climate Hostility	____	____	____	____	____	____	________
Labor Relations/Productivity	____	____	____	____	____	____	________
Government Relations	____	____	____	____	____	____	________
Partner Relations	____	____	____	____	____	____	________
Contractor Adequacy	____	____	____	____	____	____	________
Public Relations	____	____	____	____	____	____	________
Construction Deviation	____	____	____	____	____	____	________
Permits	____	____	____	____	____	____	________
Government Stability	____	____	____	____	____	____	________
Other: ________	____	____	____	____	____	____	________
General Business Issues							
Financing	____	____	____	____	____	____	________
Economic Climate	____	____	____	____	____	____	________
Marketing Plan	____	____	____	____	____	____	________
Downstream Transportation	____	____	____	____	____	____	________
Royalty Issues	____	____	____	____	____	____	________
Tax Issues	____	____	____	____	____	____	________
Other: ________	____	____	____	____	____	____	________
Environmental Issues							
Air Problems	____	____	____	____	____	____	________
Surface Water Problems	____	____	____	____	____	____	________
Groundwater Problems	____	____	____	____	____	____	________
Waste Disposal Problems	____	____	____	____	____	____	________
Species Problems	____	____	____	____	____	____	________
Other: ________	____	____	____	____	____	____	________

*Add to the form any other items that may be needed. Add more pages for additional comments.________

the opinions of those not represented, can be used. A sample of one type of questionnaire is presented in Table 17.5, though it should be more readable and have more space than provided in this text version. It would work very well as a multiple-page spreadsheet.

Occurrences Leading to Project Adjustments

The appraisal should consider occurrences that will cause divergence from the standard cost estimate in three areas:

1. **Technical:** Design, development, operation, and equipment considerations, as well as estimating data and definition of facilities
2. **Project execution:** Contracting and construction plan, fabrication and procurement plan, and worker productivity
3. **Business:** Impact of governments and third parties, including loan groups and partners, market, and competition

Any potential occurrences related to these areas are reasonable components for consideration in developing an occurrences list. Issues that are normally excluded from the appraisal and occurrences list are the following business areas:

- Open-ended scope and basis items of the companies' business should normally be excluded given that their adoption will likely be justified on their own economics. However, non-economic scope and basis changes may be mandated by the third parties or partners. These should then be considered in the appraisal.
- The appraisal will not normally address changes in the parent companies' business plans, because such changes will likely alter the project basis.
- The appraisal will not normally address out-of-the-ordinary indiscriminate occurrences (war, embargos, etc.), because these occurrences are usually not predictable.

Tables 17.6, 17.7, and 17.8 provide lists of technical, execution, and business occurrences that have influenced former mineral property projects in the cost and schedule areas. The occurrence descriptions are shown in the left-hand column and typical concerns related to the descriptions are given in the right-hand column. These descriptions and concerns are intended to stimulate discussion and set the starting point for generating the project-specific occurrences list for each individual project. It is left to the ingenuity of the PRA team to build the occurrences list for their specific project.

Determining Project Risk Adjustment

Determining the effect of separate occurrences is normally a cost/schedule engineering task, using methods and data from normal cost engineering practices. The cost estimation methods are not covered in this chapter. Further, for the purposes of simplifying this text, discussion material will express comments as if the occurrence increases both cost and schedule. The same concepts and approaches apply should the specific occurrence reduce cost or schedule.

Any given occurrence will affect the project in cost, schedule, or both. For each occurrence on the occurrences list, an appraisal should include the following:

TABLE 17.6 Sample technical occurrences list for a mineral property project

Potential Occurrences List	Typical Concerns
T-1.0 Project design information for performing engineering is incomplete or subject to change as a result of ongoing work.	**T-1.0 Given that both the cost and schedule estimates are based on the design basis, any change creates risk.**
T-1.1 Resource mineralogy or concentrate characteristics definition is uncertain.	T-1.1 Further evaluation could show ▪ Wider variability, ▪ Unanticipated impurities, and/or ▪ Lower/Higher recoveries.
T-1.2 Site characteristics not fully identified and/or engineering analysis lacks complete details with which to make sound engineering designs or judgments.	T-1.2 Data may be insufficiently accurate. For example, a contour map of 15-m (50-ft) contours for railroad layout may have spacing contours that are too wide to provide accurate design.
T-1.3 The design database, used for developing engineering, changes as a result of outside actions by others.	T-1.3 Changes in environmental regulations may impose any number of more-stringent requirements on the design process.
T-2.0 Project is located in frontier or area of hostile weather, creating additional technical problems.	
T-2.1 Project site is remote. Modularization of facilities is employed to minimize site activity.	T-2.1 Designs may not recognize shipping problems or extraordinary bracing requirements for shipping.
T-2.2 Site weather is hostile or subject to periods of extremely bad conditions.	T-2.2 Designs may not recognize potential for long shutdowns in bad weather or design for these severe conditions.
T-2.3 The frontier site has unusual characteristics and unfriendly indigenous personnel.	T-2.3 Designs may not reflect the need for extraordinary security measures.
T-3.0 Design is based on new or emerging technology for which the company has no project management experience to rely on.	
T-3.1 Emerging technology is being employed at commercial scales for the first time.	T-3.1 Design may not recognize soft areas subject to major modifications.
T-3.2 Technology new to the company is being employed in contractor design.	T-3.2 Design may not incorporate all of the company's desired features.

TABLE 17.7 Sample execution occurrences list for a mineral property project

Potential Occurrences List	Typical Concerns
E-1.0 Business and technical occurrences create execution difficulties to be addressed by the Execution Subcommittee.	**Note: An accelerated mineral industry demand may put extreme pressure on contractor and labor availability and thus cause project delays.**
E-2.0 Project is being developed in a location where your company has not worked before or for which your company has no recent project experience. Lack of knowledge of the site and lack of experience can introduce risks into design basis for the estimate and its execution plan.	**Note: This risk may apply to most companies for foreign minerals projects. It should be checked, but the appraisal should maintain an acknowledged recognition of favorable mining industry experience in the area, even if the area is new to the company.**
E-2.1 Estimate of material cost levels and availability could be optimistic.	E-2.1 Basis may be reflecting optimistic local procurement.
E-2.2 Estimate of labor productivity and availability is optimistic.	E-2.2 Optimistic estimate will minimize the adjustment in estimate for training.
E-2.3 Estimate of tax effect on foreign personnel is not firm (project management team personnel, contract and vendor representative personnel).	E-2.3 Estimated funds incorporated to cover tax impact are too low.

(Table continues)

TABLE 17.7 (Continued)

Potential Occurrences List	Typical Concerns
E-2.4 Import procedures, restrictions, or customs clearance are not fully understood; may require more time to process.	E-2.4 Overconfidence in ability to secure prompt clearance may lead to unrealistic schedules, large project delays, and cost overruns.
E-2.5 System of obtaining labor is not finalized (i.e., direct hire or subcontract).	E-2.5 Contractor capability to operate in preferred mode may not be properly evaluated when contract is awarded.
E-2.6 Work location's climate, culture, and political/social unrest create unfavorable working/living conditions (may also include site where engineering work is planned).	E-2.6 Added incentives will be necessary to induce personnel into area and to keep them there. This will affect projected cost if it was not planned.
E-3.0 Project site is in frontier area and/or with hostile weather.	**E-3.0 This could extend construction schedules and impact labor productivity.**
E-3.1 Transportation and coordination problem is more complex than assessed in the standard cost estimate.	E-3.1 Increased cost may be incurred to improve local transportation and communications network.
E-3.2 The effects of weather conditions are more severe than used in the standard cost estimate (i.e., hurricanes, earthquakes, avalanches).	E-3.2 This will require adaptive plans and safety precautions beyond those costed in the standard cost estimate.
E-4.0 Project development requires employing contractors for early development work. The company lacks experience that lends direction to the development and provides a database for evaluating the work product and contractors' performance (work includes technical, planning, estimating, and scheduling).	**Note: This element applies to most new companies in the industry and often results in extensive engineering design work at early stages of a project. Treatment of this factor should be recognized in the contractor selection for the level of design work, demonstrated expertise of contractor, and similar effects that could mitigate the potential impact.**
E-4.1 Construction planning, by contractor, does not optimize facilities or does not fully reflect owner requirements.	E-4.1 Construction planning may require recycling of plans several times to achieve desired results.
E-4.2 Schedule for completion of planning is optimistic primarily based on contractors' appraisal. Contractor has not worked with your company before on a similar project.	E-4.2 The schedule is probably optimistic and this will promote unattainable expectations.
E-4.3 Accurate information required by contractor to carry out facilities planning is not available in a timely fashion.	E-4.3 The operation will not meet completion date.
E-4.4 Contractor estimating and scheduling work is completed on a casual, undocumented foundation, with no historical basis.	E-4.4 Their estimate and schedule will require a more detailed check by the project management team to feel comfortable with results.

TABLE 17.8 Sample business occurrences list for a mineral property project

Potential Occurrences List	Typical Concerns
B-1.0 Financial plan involves other parties (i.e., a partner organization, a third-party lending agency, or a lending agency such as a government financial institution or vendor financing).	
B-1.1 Financial plan is identified but not completed.	B-1.1 Until plan is completed, schedule implications are at risk.
B-1.2 Financing partner/lending agency applies restrictions on various material and supply sources.	B-1.2 Free-market forces of competition will not exist on the job as anticipated in the standard cost estimate and may cause overruns.
B-1.3 Financing partner/lending agency insists on longer bidders list.	B-1.3 Procurement and/or contract problems will be compounded.
B-1.4 Financing partner/lending agency requires independent/more extensive review of technical product.	B-1.4 This will introduce extra work on project management team, which could cause schedule delays.

(Table continues)

TABLE 17.8 (Continued)

Potential Occurrences List	Typical Concerns
B-1.5 Financing partner/lending agency requires review/ approval authority over expenditures, even with relatively low values.	B-1.5 This will inhibit the project management team from promptly moving with unexpected but low-value procurement, delaying project.
B-1.6 Financing partner/leading agency requires more extensive reporting or documentation of expenditures. May also require more frequent forecasts of commitments.	B-1.6 This increases the burden on the cost reporting/ control system. It will increase administrative cost.
B-2.0 Project has a high profile and consequently a high level of government interest at local, regional, and national levels. The government's indirect involvement filters through all aspects of the project, and government cooperation is vital to timely achievement of specific objectives.	
B-2.1 Government imposes (directly or by inference) maximum use of indigenous resources.	B-2.1 Non-optimum functioning personnel structure may be required due to resource limitation; thus more personnel are needed.
B-2.2 Government permits required for construction, rights-of-way, project approval, etc., or other government approvals at various levels of government are required.	B-2.2 Timing of approval cycle may be optimistic. System may also be conducive to and breed corruption through unwarranted compensation, if not illegal bribes.
B-2.3 Stability of existing government regime is tenuous. Elections are in the offing.	B-2.3 New regime may change project ownership structure, objectives, and/or approvals.
B-2.4 Various activist groups are pushing government toward actions unfavorable to the project or future operation.	B-2.4 Activist groups may delay approvals or may even cause financial backing to withdraw support.
B-2.5 Government imposes unusual requirements of the project for "political" reasons.	B-2.5 Requirements are likely to impose much more social infrastructure than actually required for the project.
B-2.6 Government entities are expected to provide utilities or other services that are not now in place and which must be constructed prior to, within, or overlapping the execution period. This is particularly critical if the government has little money or experience for such construction items.	B-2.6 Government schedules are likely to be grossly optimistic.
B-2.7 Government's tax policies are not fixed.	B-2.7 There is strong political pressure to push the confiscatory tax well beyond that which is applied to other industries, and well over 50%.
3.0 Economic conditions, either nationally or worldwide, lack stability, which impacts forecasts of escalation, currency reevaluation, and economic activity. (This may be a factor when government policies are perceived to have a major impact on economic growth, and the project is perceived as capable of stimulating or even carrying this growth.)	
B-3.1 Economic activity forecasts indicate low or normal levels based on current plans, but pending actions in government or private sectors could heat up the economy. (Note: The converse is true when the standard cost estimate is predicted on normal or high activity.)	B-3.1 Business activity level changes can alter a resource acquisition plan.
B-3.2 Project duration extends beyond three years, increasing potential that occurrences will take place which are not contemplated at time of estimating.	B-3.2 This marketing plan may change the project's economic viability. This is true for most mineral projects.
B-4.0 Project involves one or more partners undertaking their first major construction project.	

(Table continues)

TABLE 17.8 (Continued)

Potential Occurrences List	Typical Concerns
B-4.1 Partnership management imposes a combination of unusually restrictive approval procedures.	B-4.1 This may create ineffective management of the job.
B-4.2 Partnership management requires unusual or duplicative reporting or documentation procedures.	B-4.2 This creates an extra burden of all cost control personnel involved.
B-4.3 Partnership management seeks to over-control expenditures and release funds in relatively small amounts (i.e., multiplicity of authorization for expenditure against which job is to be controlled).	B-4.3 Delays in the project are caused by management that inhibits prompt action due to multiple administrative reviews.
B-4.4 Personnel assigned to the project by parties involved have limited large-project experience.	B-4.4 This requires additional training to attain proficiency and close supervision by those who do have the needed experience.
B-4.5 A partner and your company have some dissimilar project objectives, and a mutually compatible, firm project basis is not documented.	B-4.5 This results in excessive confusion on what to do and instigates confrontation between the partners. This can result in incompatible operation but is less likely to affect the project.
B-4.6 Partnership management creates a parallel or an overlapping monitoring organization.	B-4.6 This causes animosity, confusion, duplication of effort, and nonproductive second-guessing.
B-4.7 Partnership management frequently rotates their personnel due the duration of the project.	B-4.7 This adds to training problems.
B-5.0 Business plans and project schedule are dictated by external forces requiring a force fit to achieve target and dates irrespective of schedule restraints inherent in the project.	
B-5.1 Early in the project development phases, a firm commitment is made on project completion date or product delivery date to third parties (government, product purchaser). Project team may also establish unrealistic milestones for internal reviews.	B-5.1 This creates an atmosphere that emphasizes schedule rather than cost and quality.

- Pinpoint the relevant cost and schedule basis in the standard cost estimate.
- Describe the change to the basis should the event occur.
- Describe the impact on the cost and schedule differences stemming from the occurrences.

The cost impact of any occurrence is made up of two elements. One is the *explicit cost* related to the physical changes resulting from the occurrence. Direct costs will arise if the event occurs, regardless of the happening of any other occurrences. The second element is the *implicit cost*, primarily resulting in schedule effects. Whether a schedule effect impacts the project cost or schedule depends on whether it is on the critical path and whether other occurrences are affecting the same time frame. For example, if there are two independent occurrences and one occurrence creates a three-month delay while the other one creates a one-month delay, the net effect on the project may total only three months, because the implicit schedule effect of the one-month occurrence may not influence the project cost or schedule.

The independent, or stand-alone, cost of an occurrence is the sum of the explicit and implicit costs. This represents the cost of the occurrence if it were the only event that occurred. The independent cost for each occurrence, after weighting the likelihood of occurrence, provides information for ranking the occurrences in priority order. That is, the probability-weighted, stand-alone cost of the event provides a measure of its importance to the other occurrences. In developing action steps to control project costs, the ranking lists the occurrences in order of importance to the project.

The range of the PRAA is an indication of the variability of the total project estimate from the standard cost estimate. It shows the maximum cost increase if all unfavorable events occur, as well as the maximum cost reduction if all favorable events occur.

The upper level is developed from the sum of all unweighted explicit cost increases, plus the net unweighted cost increases resulting from the implicit schedule effects. The lower level is the sum of all unweighted explicit cost reductions, plus the net unweighted cost reductions resulting from the implicit schedule effects. Figure 17.2 shows an occurrence evaluation example. This same type of analysis must be done for every major element of the project.

Schedule Adjustments

Three separate schedule evaluations are made covering the design, management, and construction activities. Within an activity, schedule impacts should be evaluated as a series of individual occurrences and tested for overlap and mutual exclusivity. They are adjusted for probability and combined in a probability-weighted impact. The activity analyses are combined in an overall schedule. This overall sensitivity becomes the calculation basis for the implicit occurrence cost adjustment discussed previously.

The PRAA or PRA is the probability-weighted sum of explicit occurrence costs plus the net probability-weighted implicit schedule cost. The PRAA plus the standard cost estimate is the total project estimate used for an economic analysis. An example summary of PRA impacts is included as Table 17.9 showing PRA and PRA ranges. To illustrate the effectiveness of this method of a project's capital cost expenditures in the execution of the project, Table 17.10 illustrates how the PRAA results should be applied to the standard capital cost estimate. The probably adjustment was applied to the standard estimated capital cost in 2000 dollars. This is what would then be used to rerun the economic analysis and then using the lower and upper adjustment as sensitivities to the base case. When applied in 2015 dollars, this looks very much like many of the actual overruns that occurred in 2012 to 2014.

Operating Cost Sensitivity Evaluation

An operating cost sensitivity evaluation may be completed simultaneously with the PRAA. Structure methodology would be similar to the PRAA, except the operating cost issues would cover the following:

- Areas not addressed in the PRAA, particularly those having longer-term impacts on the project, for example,
  - Higher (or lower) taxes,
  - Governmental changes,
  - Labor conditions/productivity changes,
  - Operational reliability problems,
  - Earnings repatriation issues,
  - Operational flexibility restraints,
  - Tightened environmental standards, and
  - Mine plan changes as the ore body is developed;

MINERAL PROPERTY PROJECT OCCURRENCE
COST AND SCHEDULE APPRAISAL
(Example of only one element of analysis: B-2.1)

OCCURRENCE DESCRIPTION—Execution Occurrence No. B-2.1

Country Requires Use of Local Procurement
This occurrence describes the likely impact of local material procurement on the project schedule and cost. The effect will be in two areas:

Equipment and materials cost expected to be 20%–30% higher than world open-market prices for comparable items with competitive bidding.

Home office cost due to a requirement for additional resources attributable to

- Delivery delays—more expediting staff.
- Lower quality—more inspection and contractor procurement activity.

Probability of the Occurrence Taking Place
Equipment and materials cost: High (75%)
Delivery delays/lower quality: Medium (50%)

Timing of Occurrence
Equipment and materials cost—Year 2015
Delivery delays/lower quality—Year 2015

Schedule Appraisal
While it is anticipated there will be no schedule impact associated with higher cost of materials, there will be a 4-month schedule impact due to late delivery compounded by low quality, which requires repair time. This 4-month delay could be masked by concurrent customs clearance delays, leaving only the cost impact of additional inspection/expediting for this occurrence.

Summary
Occurrence schedule impact: 4 months.

Cost Appraisal
The 25% premium for locally purchased materials (over world open-market procurement) has been applied to 10% of the material costs. Additional Procurement Services personnel for inspection and quality control, combined with a 10% growth of contractor procurement staff to reflect the joint impacts of delivery delays and lower quality.

SUMMARY
Stand-alone occurrence cost (unweighted): $103 million.
Stand-alone occurrence cost (weighted): $78 million.

PREPARERS
Joe Miner and I.M. Copperas

FIGURE 17.2 Example occurrence evaluation

- A basis for developing traditional project sensitivities for reporting to your parent company; and
- Discounted-cash-flow impacts rather than cost/schedule impacts.

PART III: SOCIAL AND POLITICAL RISKS

The mining industry has made great strides in creating social improvements in many Third World countries. Like the improvements made in the environment, this did not start yesterday, but the perception of the general public universally is that the industry has yet to begin. Most of the public perceives the mining industry as moving in, stealing the mineral wealth, raping the land, and then leaving the ravaged country penniless, jobless, hungry, and destitute. Certainly this industry has inherited the reputation from the ways of the distant past. But

TABLE 17.9 Sample project PRA cost and schedule summary

	Percent of Base Standard Project Case		
	PRA Cost, %	Range of PRA Cost, %	
Specific Occurrence Cost	Probability (weighted)	Lower	Upper
Explicit Costs			
Technical (12 items)	2.0	(1.0)	6.0
Execution (16 items)	5.0	(0.5)	8.0
Business (4 items)	0.5	—	0.5
Implicit Costs			
Support	8.5	(1.0)	16.0
Escalation	4.0	(2.0)	8.0
Total PRA cost impact	20.0%	(4.5%)	38.5%
Initial product shipment date	February 2015	January 2015	November 2015

TABLE 17.10 Example of the adjustment application of the PRAA of an actual project estimate (in billion dollars)

Standard Estimate	Probable Adjustment +20%	Lower Adjustment –4.5%	Upper Adjustment +38.5
2000: $3.233	$3.880	$3.705	$5.374
2015: $4.717	$5.660	$4.504	$6.533

Note: This table illustrates how the PRAA results should be applied to the standard capital cost estimate. The actual standard estimate of a project in 2000 dollars was used and applied to the probable adjustment. This is what would then be used to rerun the economic analysis and then using the lower and upper adjustment as sensitivities to the base case. Then for illustration, 2000 was simply inflated to 2015 dollars and the same percentages applied. This example illustrates how a nearly 1-billion-dollar to 1.8-billion-dollar blowout could occur if it were not adjusted.

today's forward-thinking minerals companies are trying to do whatever is economically feasible to improve the destiny of their workers through education and educational facilities, medical facilities, charities, and, in some cases, actually establishing businesses in the region where the locals can earn a living not related to mining (Attenborough 1999). In general, the industry has not done well in "tooting their own horn" on these and similar issues.

Social Concerns

As pointed out by Attenborough (1999), most if not all major mining companies have documented their company-wide policy statements that attest to their intent to build and operate mineral facilities that are in harmony with the environment and acceptable social standards, from exploration through to closure. This is now what is expected and it must be done. But many believe that the social concerns of today are where the environmental movement was back in the 1970s. The mining industry must, over the long term, accept this as standard industry practice and require that it make durable contributions to social, environmental, and economic progress. Without taking this approach to mine-building in developing nations, the industry runs the risk of a disastrous human relations problem before they even get started.

Such was the plight of one junior mining company in 1996 in the impoverished Department of Potosí in southwestern Bolivia (Anonymous 1996). After a local revolution of sorts and with people killed and injured, the government forces restored order. An investigation into the incident by the Inter-American Commission on Human Rights of the Organization of American

States concluded that although the police presence was justified, poor communication and misunderstandings between security forces and the villagers were to blame for the deaths. The examination of events leading up to the occupation of the mine by workers concluded that the initial dispute between the mining company and the miners "was not handled appropriately" and, moreover, that general labor unrest was the result of "extreme poverty that has predominated in an area that has known a past of great wealth, which did not come to benefit the population as a whole." This situation is common throughout Latin America's 400 years of mining history (i.e., this mining company was caught in the trap of being blamed for transgressions of the past). The mining company signed an agreement with communities near the Amayapampa and Capa Circa mines to provide funds for educating and job training. These concessions went a long way toward reestablishing peace in the region.

Other such cases involving environmental incidents are cited by Evans and Kemp (2011), including the Bougainville copper mine crisis in 1989, the Ok Tedi incident in 1999 in Papua New Guinea, the 1996 Marcopper disaster and uprising in the Philippines, and the Baia Mare cyanide spill in 2000 in Romania. The most recent dam failure and disaster in Brazil's iron mine, with a pending lawsuit of $40 billion involving Vale, BHP, and Samarco (Els 2016), is a strong reminder that no matter where the mine is located, and even when thorough and complete geotechnical engineering must be applied, accidents that affect the local population will occur.

There are other such stories, but the reader should also be aware of many companies that take the initiative from the start to communicate to the local "stakeholders" of what positive actions the company will do for their community. In a presentation by Patricia Bennett (1998), representing the Institute of the Americas, she outlined what problems are to be expected and what is really required of a company coming into the area of a developing nation in Latin America. Although it is a bit dated, it is an excellent summary:

- *Prejudices in mining enterprise-community relations have historical roots;*
- *A good relationship with the community is crucial to the smooth operations of a mining project;*
- *The company must set forth a Declaration of Principles:*
  - *Recognize the land's original ownership;*
  - *Respect the culture, values, customs, and environment;*
  - *Recognize the status of communities as stakeholders and get involved with them in an effective communication process;*
  - *Contribute and participate in their social, economic, and institutional development;*
  - *Integrate mining activities to the regional, state and national objectives;*
- *The real challenge is to create activities which are economically sustainable, by training and building capacity;*
- *It is important to avoid paternalistic patterns while supporting a community:*
  - *There should be participatory involvement of the community in designing processes for equitable sharing of benefits; more equitable sharing of benefits;*
  - *A reduction of social costs at the local community level;*

- *During the community discussion meetings:*
  - *Recognize the need to reach consensus;*
  - *Include every stakeholder;*
  - *Allow equitable roles;*
  - *Work with transparency and respect;*
- *Develop comprehensive strategies to ensure that all parties gain from the mines near established communities;*
- *Find ways to increase "added-value" to the community: Increased emphasis of the mining sector as a source of growth; create methods to capture the economic benefits of the mining operations by taking advantage of the related activities;*
- *The governments and the private sector have a joint responsibility for sustainable development; and*
- *The mining investor could lead modern social change by taking the responsibility for social and economic development of the operations region.*

Bennett went on to describe the success of two properties that included many of the above-listed actions: the Antamina zinc property in Peru and the Inti Raymi property in the Kori Kollo gold mines of Bolivia.

The Social License to Operate

In a paper titled "Earning a Social License to Operate: Social Acceptability and Resource Development in Latin America," Susan Joyce and Ian Thomson (2000) discuss what it will take for the mining communities to overcome 400 years of mining abuses in Latin America. The key word in the title is *earning*.

> *The point is that most of the rural population of most Latin American countries have no more respect and allegiance for the current central government regime that granted the foreign mining entity the right to come in their community and disrupt their entire way of life, then* [sic] *they have for these foreigners that are there.*

Joyce and Thomson (2000) identify four basic problems that pose significant risk to the investor:

1. The legacy of conflict
2. Struggles over the distribution of benefits of mining
3. Legislative inconsistencies between reform processes
4. A perceived lack of legitimacy in the laws and regulations on which foreign companies rely

But more important to this discussion are the solutions that they raise (i.e., "the mining industry is now faced with the fact that in many countries the legal, government-awarded right to explore or mine does not bestow universal approbation on a project" [Joyce and Thomson 2000]). The authors propose that

...a Social License to Operate (SLO) exists when a mineral exploration or mining project is seen [by the locals] *as having the approval and the broad acceptance of society to conduct its activities. It is a license which cannot be provided by civil authorities, by political structures, or even by the legal system, and must begin with, and be firmly grounded in the social acceptance of the resource development by local communities.*

Earning a SLO starts in the exploration phase with the arrival of personnel at the project site. First impressions are long lasting....Conflict can arise very quickly if there is a failure to respect local customs of land use and religious sites, give notice of actions, pay fair market compensation and so on.

During exploration, and later during mine development, when a project is most exposed financially, the greatest risk of social conflict is probably created by the mismatch of expectations that may exist between the company and the community. These expectations must be managed through respect and communication and are a challenge to the community relations during this period of project development.

During the operational phase, the company needs to ensure that the promised benefits of mining really do accrue to the local communities. Some level of direct corporate investment will be needed, and projects should be integrated with the existing infrastructure, meet community needs and be capable of continuing successfully without the presence of the company. Failure to observe these simple guidelines risks establishing "patronism," engendering a dependency and indifference to the benefits being received, and leading, eventually to resentment from the community and a rising risk of social conflict. Properly applied, a programme of community relations and investment in community development can create true sustainability.

The similarity of what was said by Patricia Bennett and then later independently by authors Joyce and Thomson seems amazing.

In summary, Patrick James, president and CEO of Rio Algom, says it best, "If our industry cannot effectively combine social, economic and environmental goals, then we will gradually find ourselves unable to operate wherever we turn" (Bennett 1998). The social concerns of what could be required in a community may be considered a risk if we do not recognize them and plan how they should be addressed with the communities affected. If we do recognize and plan for these things that have been outlined above, then they are not a risk, but planned, scheduled, and budgeted activities that the property must support to be developed.

More information on the SLO and sustainability is included in Chapters 9 and 10.

Other Political and Security Risks

There are many other forms of risk that mining investors must be aware of when selecting areas of investment for mineral exploration or property acquisition. Some of these risks manifest as financial investment risk, while other appear as plant security or human safety risk.

Many sources of information appear in the literature, where rating of risks is compared by various surveys. In addition to the sources listed in the "Country Risk" section earlier in this chapter, another source used by some is the Gini coefficient (also known as Gini index) (Anonymous 2015), as published by such organizations as the U.S. Central Intelligence Agency and World Bank. This is a measure of statistical dispersion intended to represent the income distribution of a nation's residents and is the most commonly used measure of inequality that exists within the country. Zero means perfect equality, while 100 would be perfect inequality. It measures the extent to which the distribution of income (or, in some cases, consumption expenditure) among individuals or households within an economy deviates from a perfectly equal distribution. The spread of the index is believed by some to be an indicator of likely unrest, and therefore instability. However, when the index for a few familiar countries is considered, it can be difficult to be used as a risk indicator for investments. For instance,

- Australia with a 30.2 index compared to Pakistan's 29.6 or Egypt's 30,8; or
- Canada with 32.1 versus Bangladesh with a 32.1 and Croatia with a 32.0; or
- The United States with 45.0 versus Peru's 45.3 or Iran's 44.5.

Using the index in these types of situations would be troublesome because there are so many other more significant factors than wealth inequality. Nevertheless, some like to use it.

Other than the previously mentioned surveys that appear in the literature, simply keeping up with all the major mining, exploration, and financial publications is a must for potential mineral property investors or industry suppliers. When one consistently follows worldwide activities, it becomes obvious where the risk areas are, but one also begins to realize that there are still a few good places for mineral investment, even for mid-cost ore bodies. Furthermore, there are still a lot of good places for mineral investment for those with ore bodies that will produce their marketable product in the lower quartile of the industries cost seriatim for that commodity. Such properties can afford to do the needed infrastructure, environmental, and socioeconomic work that will be required of newly developed properties.

The most difficult problem is trying to predict the political stability of the state, province, territory, or country over the life of the project exploration, evaluation, development/construction, and operation to receive the return on the investment that is expected. In some countries, it is not the country policies that are a risk to mining but the local/state/provincial policies that must be predicted. The United States must be considered in this category. This usually means trying to consider the risk potential 10–20 years in advance, which would be no small challenge in any country, state, or community.

REFERENCES

Anonymous. 1996. Conclusions. In *Report of the Situation of Human Right in Amayapampa, Llallagua and Capasirca, Northern Potosi, Bolivia.* Inter-American Commission on Human Rights. https://cidh.oas.org/countryrep/bolivia-eng/v.conclusions.htm (accessed April 17, 2017).

Anonymous. 2015. Country comparison: Distribution of family income, Gini index. The World Factbook. https://www.cia.gov/library/publications/the-world-factbook/rankorder/2172rank.html.

Attenborough, M. 1999. Social problems in developing countries pose challenge. *The Northern Miner* 85(4):1, 15.

Behre Dolbear. 2000. Company and property specific risk identification list. Intercompany communication. Denver, CO: Behre Dolbear.

Behre Dolbear Group. 2015. Where to invest 2015. In *High-Grade Resource.* Denver, CO: Behre Dolbear Group.

Bennett, P. 1998. *Mining and the Community in Latin America.* La Jolla, CA: Institute of the Americas.

Darling, P. (ed.). 2011. Part 17, Community and social issues. In *SME Mining Engineering Handbook*, 3rd ed. Englewood, CO: SME. pp. 1767–1825.

Els, F. 2016. BHP, Vale hit with $44 billion lawsuit over deadly spill. May 3, 2016. Mining.com. www.mining.com/bhp-vale-hit-44-billion-brazil-lawsuit/?utm_source=digest-en-mining-160503&utm_medium=email&utm_campaign=digest (accessed March 28, 2017).

Evans, R., and Kemp, D. 2011. Chapter 17.1, Community issues. In *SME Mining Engineering Handbook*, 3rd ed. Edited by P. Darling. Englewood, CO: SME. p. 1769.

Guarnera, B.J., and Martin, M.D. 2011. Valuations of Mineral Properties. In *SME Mining Engineering Handbook*, 3rd ed., Vol. 1. Englewood, CO: SME. pp. 219–226.

Jackson, T., and Green, K.P. 2017. *Fraser Institute Annual Survey of Mining Companies, 2016.* Fraser Institute. www.fraserinstitute.org.

Joyce, S., and Thomson, I. 2000. Earning a social license to operate: Social acceptability and resource development in Latin America (synopsis). *CIM Bulletin* 93(1037):49–53.

Runge, I.C. 1998. *Mining Economics and Strategy.* Littleton, CO: SME. pp. 199–201.

Smith, L.D. 1995. Discount rates and risk assessment in mineral project evaluations. *CIM Bulletin* 88(989):34–43.

Smith, L.D. 2013. Risk adjusted cash flows: Discount rates, risk and long life projects. www.wimmongolia.org/uploads/2/0/9/4/20941268/smith_racf_discount_rates_risk_long_life_projects_cim_May_2013_r13.pdf (accessed March 28, 2017).

Torries, T.F. 1998. *Evaluating Mineral Projects: Applications and Misconceptions.* Littleton, CO: SME. pp 38–39, 106–107.

CHAPTER 18

Professional Ethics in Mineral Property, Mineral Resource, and Mineral Reserve Estimates, and Feasibility and Evaluation Studies

David M. Abbott Jr.

When conducting mineral property, mineral resource, and mineral reserve estimates as well as feasibility and evaluation studies, one is expected to work competently and ethically. Failure to do so can subject one to legal and other liabilities resulting from the harm done. This chapter focuses on those aspects of generally recognized professional ethics directly applicable to mineral property feasibility and evaluation studies. The chapter begins with a brief review of some basic moral and ethical concepts that must underlie any discussion of professional ethics. The chapter then reviews those professional ethics concepts that most commonly arise in connection with mineral property, mineral resource, and mineral reserve estimates, and feasibility and evaluation studies.

Herbert Hoover (1909) noted,

> *Every year the mining investors of the new order are coming more and more to the engineer for advice, and they should be encouraged, because such counsel can be given within limits, and these limits tend to place the industry upon a sounder footing of ownership. As was said before, the lamb can be in a measure protected. The engineer's interest is to protect him, so that the industry which concerns his own life-work may be in honorable repute, and that capital may be readily forthcoming for its expansion.*

To illustrate Hoover's point, look what happened after the 1997 Bre-X gold mining fraud (Francis 1997; Goold and Willis 1997; Danielson and Whyte 1997): seed money that was desperately needed by junior mining companies, which find most of the ore reserves in the world, dried up. No one wanted to invest their money in an industry that could allow such a scam and drain the public of their millions for investment. It became a matter of who could be believed. The mining industry had lost its "honorable repute." Although various regulatory reforms followed the Bre-X scandal, particularly the adoption of National Instrument (NI) 43-101 in Canada, such reforms cannot prevent fraud. The accounting profession has shelf-feet of documents designed to prevent financial fraud, but accounting frauds continue to occur, perpetrated by those who know how to defeat the accounting system. Only honest, competent, and ethical practice truly prevents fraud.

MORAL FUNDAMENTALS

Too many works on professional ethics assume that everyone understands basic ethical principles. This is an unwarranted assumption. Few people have taken the time to carefully study accepted moral and professional ethical principles and their application. In *Common Morality: Deciding What to Do*, Gert (2004) provides an excellent and readable summary of basic moral principles. Gert's most important insights are the following:

1. There is an informal but universally recognized set of basic moral rules with which everyone must comply.
2. There is a difference between these moral rules (e.g., do not injure) and moral ideals (such as feed the hungry, work to end a particular disease, etc.) in that while everyone must obey the moral rules, we can pick and choose among the moral ideals we support.
3. There are recognized exceptions to the moral rules and a logical set of procedures for recognizing these exceptions. For example, the general moral rule, do not injure, means that most of us cannot cut into someone's abdomen with a knife. However, appropriately skilled doctors are allowed to do so to mitigate disease or injury.
4. Clear moral or ethical reasoning cannot answer all moral or ethical questions because different people rank moral or ethical principles differently and so, with firm ethical bases, people can reach differing conclusions on a moral or ethical question. Debates over capital punishment and abortion are examples of such unresolvable moral questions.

Anyone wishing to seriously study professional ethics should begin by reading Gert's *Common Morality* (2004).

The words *ethics* and *morals* are interchangeable in general usage. The distinction made between the two by this author is that ethics are formally written down while moral principles need not be. The subject of ethics (or morals) concerns itself with distinguishing right from wrong and the character, volitions, and actions of responsible persons.* In addition to general moral principles, there are specific ethical statements that may apply to particular groups, such as religions or professions.

General morals or ethics cover universal notions of what is right and wrong. *Professional morals or ethics* pertain to the practice of a particular profession. Professionals are recognized as having a recognized set of knowledge, skills, and experience that provides expertise in the professional subject. Society grants professionals certain rights and privileges in recognition of this expertise, for example, the right to appear as expert witnesses rather than lay witnesses in court. Professional ethics statements express the responsibilities professionals assume in practicing their profession. Professional ethics are generally based on written guidelines or codes.† Because these guidelines and codes have been written down and formally adopted by professional organizations, they constitute ethical rather than moral principles given the preceding distinction between ethics and morality.

* The restriction of ethics or morals to rational persons or beings is important in general application. We do not hold children or the mentally disturbed or disabled to the same ethical standards and responsibilities for their actions as we do "normal adults," who are presumed to be rational persons. Gert (1998) addresses this topic in detail. Because professionals are presumed to be rational persons, this distinction will not be pursued further in this chapter.

† *Professional Ethics and Insignia* (Stierman et al. 2000) contains many professional ethics codes. *Eighty Exemplary Ethics Statements* (Murphy 1998) provides examples of corporate ethics statements.

Because professions differ, so will their ethics statements. The health-care and biological professional ethics statements contain sections dealing with the ethical treatment of living research subjects. Such sections are lacking in geoscience ethics statements because rocks and fossils are inanimate.

Because of differences between professions and their associated professional ethics statements, the question arises, Can those who are not members of the profession in question adequately judge the ethical practices of that profession? In other words, can a lawyer or layperson judge the ethical practices of a doctor, or engineer, or hydrologist? In fact, they can, and they do. Many state licensing boards specifically require that some of the members be public members, that is, not members of the licensed profession. This practice is believed to serve as a check against a professional group becoming a mutual admiration society. On the other hand, we also have a legal concept that we should be judged by our peers. In the professional context, that would indicate our professional peers. Because of the basic moral underpinnings of professional ethics statements, these two positions, (1) being judged only by professional peers, or (2) being judged by those who are not members of the profession, are not mutually exclusive.

RELEVANT CODES OF PROFESSIONAL ETHICS AND PROFESSIONAL PRACTICE

The relevant professional ethics guidelines, codes, and legal requirements for professional practice depend on

- One's professional specialty (geologist, mining engineer, metallurgist, etc.), including one's particular expertise within the specialty;
- The laws of one's home jurisdiction and the jurisdiction in which one is working regarding professional licensing;
- The ethical obligations of the professional associations to which one belongs; and
- The general expectations of the profession and of society.

Society has a set of moral or ethical standards that all members of society are expected to follow. Likewise, legal jurisdictions pass laws that must be obeyed. Chief among the general ethical and legal obligations are those against killing and injuring. However, death and injury are not normally major concerns in a property evaluation except that the property operator will be expected to comply with appropriate local and national health, safety, and environmental laws, regulations, and standards for its operations; social licensing issues; and international standards, such as the Equator Principles if international bank financing is being used.

The coverage of professional ethics extends to areas outside the mining industry. Geohazards is an example. Peppoloni and Di Capua's *Geoethics: The Role and Responsibility of Geoscientists* (2015) contains papers addressing issues that could affect the mining industry. The April 6, 2009, earthquake (moment magnitude 6.3) in L'Aquila, Italy, led to the indictment and initial conviction of six seismologists and a seismic engineer for multiple manslaughter and serious injury counts for their negligent conduct in failing to warn the public of the imminence of earthquake. They were sentenced to six years in jail, perpetual interdiction from public office, and a fine of several million euros. In November 2014, the convictions of the six seismologists were overturned and the six-year jail sentence of the seismic engineer was reduced to two years. Cocco et al. (2015) review the technical aspects of the L'Aquila trial. Albarello's (2015) paper in the same book, "Communicating Uncertainty: Managing the Inherent Probabilistic

Character of Hazard Estimates," makes an important ethics point regarding geoscience predictions. As geoscientists, we should effectively communicate the probabilities of a geohazard event occurring. Thus, we are effectively the bookmakers. But we are not the policy makers deciding what the public should do. We should not "place the bet" by specifying a particular outcome (Abbott 2016). Abbott (2016) also points out that just because some author or other person believes that a particular issue is a matter of professional ethics does not make it so by citing the example of a paper on the development of solar salt ponds along the Pilbara Coast of Western Australia.

Protection of the public's health, safety, and welfare is the primary basis for regulation of professions by licensing and is included in the ethics statements of most self-regulating professional bodies, including all those that are recognized for "competent person" or "qualified person" status in Canada, Australia, and many other countries. In the wake of the Bre-X fraud of 1997, one of the recommendations of the Mining Standards Task Force (1999) was that a professional association's ethics statements explicitly include protection of the public's financial welfare as part of protecting the public's welfare. Recognition that protection of the public's financial welfare is part of the professional's ethical responsibility is among the key ethical principles affecting those professionals conducting feasibility and evaluation studies.

More traditional areas coming under public health, safety, and welfare are also of concern. These include miner safety, which is an area receiving enhanced public attention in the United States after the Sago mine disaster (West Virginia) and others resulted in 34 fatalities in 2006. Other mining work-related environment standards continue to evolve or be enacted—for example in the United States, the 2008 diesel particulate standards for underground operations and the environmental regulations covering mining operations.

Professional ethics are obligations imposed on professionals that are in addition to the general obligations imposed on all members of society. Professional ethics concepts are generally developed within the individual professions, although professions dealing with similar problems tend to have comparable ethics provisions. Thus, the various professional organizations involved in mining have similar ethics provisions. Professional groups such as the National Society of Professional Engineers, the American Institute of Minerals Appraisers, the American Institute of Professional Geologists, the Australasian Institute of Mining and Metallurgy, and others have specific, published professional codes of ethics or codes of conduct (Stierman et al. 2000). The Society for Mining, Metallurgy & Exploration (SME) has a code of ethics for its registered members (SME 2016). Likewise, the bodies licensing engineers, geologists, and other professions in a state or province often have codes of ethics or professional conduct.

The difference between ethical guidelines and codes is that guidelines are aspirational—that is, they encourage particular practices whereas codes are backed up with some manner of disciplinary function.* This is particularly true for codes of conduct that are explicitly set up for disciplinary purposes by licensing and similar bodies. The codes of ethics adopted by many professional organizations frequently blend aspirational and disciplinary provisions. An example of a common aspirational statement is one recommending that professionals engage in continuing professional development activities. Such an ethics statement becomes a rule subject to disciplinary action when the professional is required to document participation in at

* The American Geological Institute's "Guidelines for Ethical Professional Conduct" was prepared specifically as a guideline and not as a basis for disciplinary proceedings (AGI 2015).

least a minimum specified amount of continuing professional development (CPD) and education activities within a specified time period. An increasing number of professional organizations are requiring some amount of CPD to be recorded and reported on a regular basis; see the "Technical Competence" section later in this chapter.

Professionals conducting a mineral property feasibility or evaluation study are expected to comply with any relevant legal or regulatory requirements pertinent to the job, with the general legal requirements pertaining to their profession for working on the job (licensing, etc.), the general professional ethical requirements of their profession, the specific ethical requirements of any organizations to which they belong, and the legal and regulatory requirements of the jurisdiction (e.g., a stock exchange) with whom the study is to be filed. Within the international mining community, the only material differences between jobs will be those relating to the specific legal requirements of the countries and other jurisdictions involved. There is not major disagreement internationally among professional organizations regarding general professional ethical principles. However, there obviously are international cultural differences that do affect the application of moral principles, even among the developed western countries. But these differences are less significant than the areas of agreement. And there are significant differences in the application of some principles in different countries. For example, miner safety concerns and environmental concerns differ among countries, with some third-world countries currently having dismal records. But there is also recognition, even within those countries, that currently accepted methods of operation require change.

The principal professional ethical requirements involved in mineral property feasibility and evaluation studies are those involving professional technical competence, scientific honesty, confidentiality of client information, and conflicts of interest. In each of these areas, professionals have ethical obligations that are additional to those normally required of all members of society and that are therefore appropriately covered under the heading of professional ethics.

PRACTICE STANDARDS AND PROFESSIONAL JUDGMENT

Formal and informal standards of practice exist for all varieties of work. These standards range from rules of thumb to detailed specifications, such as those promulgated by the American National Standards Institute and ASTM International or the International Organization for Standardization. Various regulatory agencies and professional organizations also publish relevant guidelines, definitions, and standards, such as NI 43-101, the U.S. Securities and Exchange Commission's Industry Guide 7[*] (SEC 1992)[†], the JORC Code (2012), the VALMIN Code (2015), and similar codes and guides published by related organizations. A professional is expected to know which standards and regulations apply to his or her scope of work. These standards should be followed where applicable. The scope of work may identify the specific standards to be used. Even where a specific standard, guide, or code is not stipulated by the client, the professional should state which one was used or followed.

* On June 16, 2016, the SEC announced proposed mining disclosure rules that would delete Industry Guide 7 and replace it with definitions and rules fairly closely aligned with international standards such as NI 43-101 and the JORC Code (2012). The process of accepting and reviewing comments on the proposed rules, the issuance of proposed rule revisions, and the following comment period, and so forth, can be expected to take some time.

† As noted in Abbott (2014a), the text that is now in Industry Guide 7 was actually adopted in 1981.

The standards, guides, and codes vary from outlines of the types of inquiries that should be made to very prescriptive standards about how a particular test should be done. Prescriptive standards can create problems when the assumptions on which the standard is based are not applicable to the task at hand. This can be the result of the unique characteristics of the project or new technology that was not available when the standard was written. In these situations, professional judgment is required to determine the appropriate course of action (Abbott 2004, 2014b).

For example, as a result of a series of frauds involving the use of unconventional methods of quantitatively determining the amount of precious metals in the 1990s, the Alberta Stock Exchange promulgated a rule requiring the use of fire assays for precious metal assays. A problem with this rule occurs in the examination of gold placers. Wells' well-known book on placer examination contains the heading "Fire Assay of Placer Samples—Misleading Results" (Wells 1973). The reason for this heading is that fire assays, because of their procedures, report the total precious metal content in the sample assayed. The reported quantity frequently is materially higher than the quantity of precious metal that can be recovered using the gravity concentration techniques employed in placer mining. The value of a precious metal deposit does not depend on the total quantity of precious metal within a specified volume of rock (the in-situ content) but rather on the quantity of precious metal that can be recovered and sold. Placer examination values report the amount of precious metal recovered by particular concentration techniques. In hard-rock mines where fire assaying is the accepted methodology for quantitative analysis, the average assay values must be reduced by various mining and processing losses to determine the recoverable quantity of precious metal.

APPLYING GENERAL ETHICAL CONCEPTS TO ESTIMATES AND STUDIES

This section discusses the major professional ethics issues arising in connection with the estimation of mineral resources and mineral reserves and with feasibility and evaluation studies. These issues are technical competence, scientific honesty and transparency, the confidentiality of client information, conflicts of interest, and determining who is the client. Abbott (2017) contains discussions of many professional ethics issues and should be consulted for a more detailed study of professional ethics.

Technical Competence

The first ethical requirement for the professionals involved in a mineral property valuation is that they possess the required technical knowledge, skills, and experience required to undertake that part of the study for which they are responsible. Evaluating the potential of a mineral property requires knowledge of the relevant geology, mining engineering, processing techniques, environmental characteristics and regulations, mineral title law, economics and financing, social licensing, and potentially other areas. No one professional is knowledgeable in all of these areas; a team effort is required for technical reports that cover these topics. The project manager should be aware of the contributions required from the various professions and assign appropriate parts of the assignment to appropriate professionals. The various members of the evaluation team should also be aware of this necessity and respect the contributions of other team members.

Technical competence comes not only from having had the relevant background training and required years of experience in the relevant area,* it requires professionals to keep improving their knowledge of how the industry is evolving and becoming familiar with new techniques applicable to their field of expertise. Then the professional can select the appropriate techniques, new or old, to apply to the particular project as required. This is the application of professional judgment (Abbott 2014b). The process of improving professional knowledge is through CPD. As mentioned earlier, many professional organizations require specified amounts of CPD to be completed and formally reported to the organization. The specific CPD requirements vary among organizations, although there is general agreement on the types of activities that may be counted. Some study of professional ethics is increasingly being required. This is a currently evolving area of professional practice.

Technical competence also includes knowledge of the various laws, codes, and regulations governing mineral property valuations. Public mining entities are required to follow specific rules relating to the securities markets from which their money has been obtained. Familiarity with the relevant securities regulations is required. The most commonly encountered securities regulations are those of the United States (Industry Guide 7† [SEC 1992]), Canada (NI 43-101 and related policies and forms), and Australia (the JORC [2012] and VALMIN [2015] codes). In the case of bank financing, there may be more flexibility in terms of which mineral resource and mineral reserve definitions will be employed, but usually one of the preceding guidelines or codes is used.

The Australian JORC and VALMIN codes were the first to introduce the "competent person" concept into mineral property evaluation practice. The competent person concept acknowledges the multidisciplinary character of mineral evaluation. It also recognizes that minimum amounts of knowledge and experience are required to adequately perform the parts of the project assigned to particular professionals. Although the project manager is responsible for the entire project, each contributing professional is also responsible for his or her portion of the project. The competent person concept has expanded beyond Australia and is being incorporated into the mineral resource and mineral reserve guidelines of other countries in one form or another. In Canada, NI 43-101 requires that a "qualified person" be involved in reporting estimates of mineral resources and mineral reserves. Although differing from the Australian competent person in detail, the Canadian qualified person is similar in overall concept.

Responsibility brings with it liability if things go wrong. Professional ethics require that a person undertaking an assignment be technically competent and able to complete the assignment. If the assistance of other professionals is required, professional ethics require that the need for this assistance be made known to the client and obtained. Failure to perform competently can result in professional ethics sanctions. Because mineral property evaluation usually involves very large sums of money, the professional faces legal liabilities as well. The securities laws recognize the importance of professional or expert advice in projects. They also assign liability to professionals and experts who do not perform up to expected professional standards. The financial liability can equal the total amount invested in a particular venture, amounts far in excess of most professionals' total net worth. Merely having to defend one's professional

---

* Canadian NI 43-101, for example, requires "at least five years of experience" in its definition of *qualified person*.

† As previously discussed, the SEC is proposing to delete Industry Guide 7 and replace it with new mining disclosure rules.

actions can represent a significant financial liability for attorneys' fees and time spent on the defense rather that on billable project hours.

Avoiding liability requires that all weaknesses in and alternatives to the mineral property evaluation be clearly identified and their impact discussed. In the United States, disclosure about "forward-looking" statements requires that they be "identified as a forward-looking statement, and [be] accompanied by meaningful cautionary statements identifying important factors that could cause actual results to differ materially from those in the forward-looking statement" (Securities Act of 1933, as amended, §27A(c)(1)(A)(i)).

Scientific Honesty and Transparency

Although professional ethics guidelines and codes cover a variety of topics organized by the relationships between the geoscientist and various groups, honesty or avoiding deception is the principal geoscience ethical principle (Abbott 2000, 2001). The greatest transgression a scientist can commit is faking data, lying about it, or otherwise deceiving others about scientific results. The 1997 Bre-X fraud is a notable but by no means unique example.* The honesty required of those acting in their professional capacities is greater than that demanded of them as private individuals. We must make a special effort to ensure that our work is free from undisclosed biases or deceptions. Identifying these biases and unintended deceptions can be extremely difficult (Feynman 1999). But doing so is extremely important when presenting data that the public can or will use.

We must also acknowledge the uncertainties inherent in our work, particularly our projections and estimates (Abbott 2000; de Freitas 2000). The limits of what is known and what is estimated are too often obscured. A common example is the use of too many significant figures in our calculations. Spreadsheets and other programs calculate results with a precision far greater than the accuracy of the input data. We must remember to limit our conclusions to the appropriate number of significant digits, and explain why we are doing so. Uncritical reliance on computer programs that generate excellent graphics is another problem, as pointed out in Martin Geach's "3D Models: Stepping Back" (2016). Barnes and Gossage (2014) also focus on the issues that must be addressed for proper modeling. Stephenson et al. (2014) address similar problems in their paper on the "spotted dog." In summary, achieving the degree of honesty required of us as professionals is difficult to achieve but must be pursued with diligence. It is not enough to avoid conscious lies; we must strive to avoid subtle deceptions and acknowledge the uncertainties that exist.

Although related to *honesty*, *transparency* differs from it. Transparency is achieved by using clear and explicit descriptions written in language readily understood by all potential readers. Transparency's opposite is *obfuscation*: the frequent use of obscure technical terms; the use of long, convoluted, passive-voice sentences; euphemisms; frequent use of numerous unfamiliar acronyms; and other means of cloaking the truth with "smoke and mirrors."

Confidentiality of Client Information

Evaluation of the estimated mineral resources and minerals reserves at a property or held by a firm involves examination of confidential data. Professional ethics codes generally contain

* Francis (1997), Goold and Willis (1997), and Danielson and Whyte (1997) provide three perspectives of the Bre-X scandal written soon after the fraud was discovered, and are the first of several books about Bre-X.

provisions relating to the protection of the confidentiality of this data. In some cases, the confidentiality provisions apply with respect to other divisions within the same company in addition to extra-company confidentiality. This is likely to be true for those companies producing competing product lines, which is not uncommon in the industrial minerals business, although it can also apply elsewhere. It is important to ask the client about the confidentiality rules when starting the job to avoid problems later.

There are two exceptions to the confidentiality provisions. First, there is no "mining industry professional–client privilege" as there is with attorney–client privilege. If a consulting mining industry professional or firm receives a valid subpoena for a client's confidential information, the mining industry professional or firm has no grounds for not complying with it. However, the client may have grounds for opposing the subpoena. Therefore, the first thing to do when a subpoena is received is to let the client know about it so that the client can exercise whatever rights it wishes to assert. Another method of protecting confidential information in appropriate cases is for the consulting professional or firm to be retained by a law firm so that the attorney–client privilege can be asserted.

The second exception to the confidentiality provision stems from the professional's ethical obligation to protect the public's health, safety, and welfare, including financial welfare. Among the consequences of the Bre-X fraud was the requirement that the ethics codes of those organizations whose members are deemed competent or qualified persons provide that, where a client's actions are endangering the public, the professional has the obligation to bring the matter to the attention of the appropriate authorities. Initially, these individuals will be the client's officers and/or board of directors. But if the situation is not corrected following these internal reports, external disclosure to the appropriate authorities may be required by both the law and professional ethics codes. For example, SME's Registered Member Code of Ethics' clause 1 states, "The first responsibility and the highest duty of members shall at all times be the welfare, health and safety of the community." And the first sentence of the interpretation of this clause states, "The principle here is that the interests of the community have priority over the interests of others" (SME 2016). The detailed examination of such cases is beyond the scope of this chapter. When faced with such a situation, one should seek legal counsel about how to proceed.

Conflicts of Interest

Conflicts of interest are the most common type of ethics issue encountered in professional practice. Conflicts of interest occur in a wide variety of situations within firms as well as between individuals and/or firms. Much of the material in *Geologic Ethics and Professional Practices* (Abbott 2017) deals with conflict-of-interest issues. Conflicts of interest are at the center of such topics as improper disclosure of confidential information, insider trading of securities, perceived biases held by an individual or firm, conflicting financial interests, and so on.* The general resolution of conflicts of interest begins with recognition of the conflict followed by disclosure of the actual or potential conflict to the affected parties. In some cases,

* The original insider trading case in the United States involved the discovery of the Kidd Creek copper/zinc massive sulfide deposit near Timmins, Ontario. The District Court case is reported as *Securities and Exchange Commission v. Texas Gulf Sulphur Co.*, 258 F. Supp. 262 (1966) and the Court of Appeals case as *Securities and Exchange Commission v. Texas Gulf Sulphur Co.*, 401 F.2d. 833 (1968). Both reported opinions are worth reviewing.

the existence of the conflict of interest will terminate a proposed relationship. For example, those engaged as independent estimators and valuators of a mineral property may not have any financial interest in the property or the property's owners or operators.

There are also cases where a conflict of interest may arise through no fault or action on the part of the individual professional or consulting firm. For example, a consultant (professional or firm) has two clients for whom he does independent mineral resource and reserve reviews. Suppose that both clients make a bid for the same property, which is held by a third firm. The consultant will probably not be able to advise both firms about their intended acquisition. Likewise, if one client makes an unfriendly takeover bid for the other, the consultant's ability to continue working for both clients may cease. Disclosure of the conflict to both clients is required, and the resolution of the conflict will depend on the circumstances but is likely to result in the loss of at least one client.

Who Is the Client?

The issue of client identity can arise when working for a corporation. The question of client identity does not generally arise unless problems occur. But when a problem does happen, there are three groups within a corporation whose interests may differ. They are the officers, the board of directors, and the shareholders. Suppose that a professional learns in the course of a mineral reserve audit that the reserve estimates have been manipulated in an inappropriate manner. In uncovering the manipulation, the professional learns that the corporation's officers are aware of and are supporting the manipulation to increase the value of their stock options and bonuses. Here is a situation in which the interests of the corporation's officers clearly conflict with the interests of the shareholders. Whether the corporation's board of directors will side with the officers or shareholders is unknown in the situation described. However, if the professional recognizes that conflicts of interest between a corporation's officers, directors, and shareholders can arise, and if the professional decides ahead of time that if a conflict arises, the shareholders will be considered the client whose interests must be protected, then an important piece in determining how to proceed in such a situation has been avoided.

LIABILITIES ASSOCIATED WITH ESTIMATES AND STUDIES

The liability associated with mineral resource and mineral reserve estimates, and feasibility and evaluation studies is that the individual professional and/or the firm for whom he or she works can be sued for incompetent or unethical practice. Such legal actions are most commonly civil cases, but in exceptional cases, criminal charges can be brought. Independent consulting professionals and their firms may be charged in legal actions aimed at the officers, directors, and insurance and financial firms employed by a client company because the independent consultants are alleged to have assisted in the alleged violations of the securities or other applicable laws. In addition to legal actions, the professional organizations whose members are deemed to be competent or qualified persons are expected to bring disciplinary proceedings against their members who have allegedly violated the organization's code of ethics.

The consequences of such proceedings can be very costly regardless of the outcome. Legal counsel usually charge much more per hour than consulting individuals or firms. Aside from the real threat of bankruptcy, the individual professional, if found to have violated the code of ethics of a professional organization, may find that every professional organization to which he or she belongs and that has an enforceable code of ethics can piggyback on the first organization's charges, as can state licensing bodies. In short, the individual may lose the right to practice his or her profession in addition to the financial costs of the action.

To some extent, protecting assets by forming a limited liability company or incorporating a company through which you conduct your professional practice can assist in limiting the financial consequences. Competent legal advice should be sought regarding these options. But these methods of separating personal from corporate assets may not prevent the loss of one's professional standing, or the cost of defending that standing. Professional liability insurance can protect against these costs, but it is very expensive. Furthermore, having such insurance may induce lawsuits because the suing party sees that there is some money, the insurance amount, that can be obtained through the suit.

The best protection is to avoid being sued in the first place. The following sections address some ways of avoiding or limiting your liability.

Scope of Work: Defining Your Areas of Responsibility

Every report should describe your scope of work. This scope of work defines your area(s) of professional responsibility, and thus liability, if a subsequent problem arises. For example, if your responsibility is mine design, you rely on someone else's geologic work, even if the geologist is someone on your own team. Your description of the scope of work should also identify the level of accuracy or reliability your work is expected to meet. A preliminary scoping study is not expected to have the level of data reliability and accuracy that is associated with a final feasibility study.

Your reliance on another's work should not prevent you from asking questions about issues that affect your work. Have you been provided with the data you need? Is that data accurate, reliable, and consistent? If not, then you should ask questions to get any issues resolved, or if the issues cannot be resolved, you can report on your concerns, the manner in which the unresolved issues affect your ability to fulfill your scope of work, and any resulting project risks.

Disclaimers

Disclaimers are statements made to limit liability in some fashion, for example, stating the professional's right to change his or her opinion should additional relevant information come to light. Although disclaimers are not uncommon parts of professional reports, their effectiveness may be subject to legal challenge. The use of disclaimers may be restricted by a disclosure regulation. For example, Canadian NI 43-101 prohibits the use of disclaimers except in limited situations (Section 6.4 of NI 43-101 and Item 3 of NI 43-101F1). Similar international mining codes, including the proposed new SEC rules, contain similar provisions. It is important to consult with competent legal counsel about disclaimers and their effectiveness.

Indemnity Clauses in Contracts

Professional service contracts frequently contain indemnity clauses designed to shift the financial burden of legal and related actions to another party. The wording and effectiveness of such clauses should be discussed with competent legal counsel.

REFERENCES

Abbott, D.M., Jr. 2000. Honesty: The principal geoscience ethical principle (abs). *Geological Society of America Abstracts with Programs,* 32(7): A293.

Abbott, D.M., Jr. 2001. Scientific honesty. *European Geologist,* no. 11, pp. 18–19.

Abbott, D.M., Jr. 2004. Are scientific honesty and "best practices" in conflict? *European Geologist* no. 18 (December): 34–38 (reprinted in *The Professional Geologist,* 42(4):46–51).

Abbott, D.M., Jr. 2014a. A historical review of recommendations for reporting exploration results, mineral resources and mineral reserves. *Mining Engineering,* 66(2):38-40.

Abbott, D.M., Jr. 2014b. Use professional judgment: Know the assumptions underlying a best practice. *Transactions SME,* 366:421–425.

Abbott, D.M., Jr. 2016. Professional Ethics & Practices—Column 157: Geoethics. *The Professional Geologist,* 53(1):32–33.

Abbott, D.M., Jr. 2017. Geologic ethics and professional practices. Reprint Series No. 2. CD. Arvada, CO: American Institute of Professional Geologists.

AGI (American Geosciences Institute). 2015. Guidelines for ethical professional conduct. www.americangeosciences.org/sites/default/files/2015%20AGI%20Guidelines%20for%20Ethical%20Professional%20Conduct.pdf (accessed June 20, 2016).

Albarello, D. 2015. Communicating uncertainty: Managing the inherent probabilistic character of hazard estimates. In *Geoethics: The Role and Responsibility of Geoscientists.* Edited by S. Peppoloni and G. Di Capua. Special Publication 419. London: Geological Society. pp. 111–116.

Barnes, J.F.H., and Gossage, B.L. 2014. The do's and don'ts of geological and grade boundary models and what you can do about it. In *Mineral Resource and Ore Reserve Estimation—The AusIMM Guide to Good Practice,* 2nd ed. Monograph 30. Carlton, VIC: Australasian Institute of Mining and Metallurgy. pp. 175–188.

Cocco, M., Cultrera, G., Amato, A., Braun, T. Cerase, A., Margheriti, L. Bonaccorso, A., Demartin, M. de Martini, P.M., Galadeni, F., Meletti, C., Nostro, C., Pacor, F., Pantosti, D., Pondrelli, S., Quareni, F., and Todesco, M. 2015. The L'Aquila trial. In *Geoethics: The Role and Responsibility of Geoscientists.* Edited by S. Peppoloni and G. Di Capua. Special Publication 419. London: Geological Society. pp. 43–56.

Danielson, V., and Whyte, J. 1997. *Bre-X: Gold Today, Gone Tomorrow.* Toronto: Northern Miner Press.

de Freitas. C. 2000. Uncertainty in science: What should we believe? *The Professional Geologist,* 37(1):5–7.

Feynman, R.P. (1974) 1999. Cargo cult science: Some remarks on science, pseudoscience, and learning how not to fool yourself in engineering and science (Caltech's 1974 commencement address). Reprinted in *The Pleasure of Finding Things Out: The Best Short Works of Richard P. Feynman.* Edited by J. Robbins. Cambridge, MA: Helix Books, Perseus Books. pp. 205–216.

Francis, D. 1997. *Bre-X: The Inside Story.* Toronto: Key Porter.

Geach, M. 2016. 3D models: Stepping back. *The Geoscientist,* 26(5):9. http://geolsoc.org.uk/Geoscientist/June-2016/3D-models-stepping-back (accessed June 13, 2016).

Gert, B. 1998. *Morality: Its Nature and Justification.* New York: Oxford University Press.

Gert, B. 2004. *Common Morality: Deciding What to Do.* New York: Oxford University Press.

Goold, D., and Willis, A. 1997. *The Bre-X Fraud.* Toronto: McClelland and Stewart.

Hoover, H.C. 1909. *Principles of Mining.* New York: McGraw-Hill.

JORC (Joint Ore Reserves Committee). 2012. *The Australasian Code for Reporting of Exploration Results, Mineral Resources and Ore Reserves* (The JORC Code). Gosford, New South Wales: JORC. http://jorc.org/docs/jorc_code2012.pdf

Mining Standards Task Force. 1999. *Setting New Standards, Recommendations for Public Mineral Exploration and Mining Companies.* Final report. Toronto: Toronto Stock Exchange, Ontario Securities Commission.

Murphy, P.E. 1998. *Eighty Exemplary Ethics Statements.* Notre Dame, IN: University of Notre Dame Press.

NI 43-101. *Standards of Disclosure for Mineral Projects.* Including 43-101F1. www.osc.gov.on.ca/en/SecuritiesLaw_ni_20110624_43-101_mineral-projects.htm (accessed June 21, 2016).

Peppoloni, S., and Di Capua, G. (eds.). 2015. *Geoethics: The Role and Responsibility of Geoscientists.* Special Publication 419. London: Geological Society.

SEC (U.S. Securities and Exchange Commission). 1992. *Industry Guide 7: Description of Property by Issuers Engaged or to Be Engaged in Significant Mining Operations.* sec.gov/about/forms/industryguides.pdf (accessed Jan. 17, 2017).

SEC (U.S. Securities and Exchange Commission). 2016. 17 CFR Parts 229, 239, and 249. Modernization of property disclosure rules for mining registrants; proposed rule. Securities Act Release 33-10098. www.sec.gov/rules/proposed/2016/33-10098.pdf (accessed June 17, 2016).

Securities and Exchange Commission v. Texas Gulf Sulphur Co., 258 F. Supp. 262 (1966) (the District Court case) and the Court of Appeals case, *Securities and Exchange Commission v. Texas Gulf Sulphur Co.*, 401 F.2d. 833 (1968).

SME (Society for Mining, Metallurgy & Exploration). 2016. SME Registered Member Code of Ethics. Englewood, CO: SME. www.smenet.org/getmedia/d2e293a7-460f-4fde-90a5-a80571359560/SME_RegisteredMember_CodeofEthics_2016.pdf.aspx (accessed June 21, 2016).

Stephenson, P.R., Allman, A., Carville, D.P., Stoker, P.T., Mokos, P., Tyrrell, J., and Burrows, T. 2014. Mineral resource classification—It's time to shoot the "spotted dog." In *Mineral Resource and Ore Reserve Estimation—The AusIMM Guide to Good Practice*, 2nd ed. Monograph 30. Carlton, VIC: Australasian Institute of Mining and Metallurgy. pp. 799–804.

Stierman, J., Joswick, K.E., Stierman, J.K., and Sharpe, R.L. 2000. *Professional Ethics and Insignia.* Lanham, MD; London: Scarecrow Press.

VALMIN Committee. 2015. *Australasian Code for Public Reporting of Technical Assessments and Valuations of Mineral Assets (The VALMIN Code).* Carlton, VIC: Australasian Institute of Mining and Metallurgy.

Wells, J.H. 1973. *Placer Examination: Principles and Practice.* Technical Bulletin 4. Washington, DC: U.S. Government Printing Office, U.S. Bureau of Land Management.

CHAPTER 19

Due Diligence Reports

Richard L. Bullock

Every industry has an obligation to its investors to ensure that their investment is secure against harm from fraudulent, misleading, or faulty decision-making regarding the transaction entered into by the company receiving the investment. The terms *requisite effort* or *ordinary care* covered this action in the 15th century and had the intention of protecting the investor's property. *Merriam-Webster's Unabridged Dictionary* (n.d.) still defines *ordinary care* the same way:

> *The care that an average reasonable man exercises to prevent harm to the person or property of others and failure to exercise which when under a duty to do so constitutes actionable negligence on the part of one causing such harm.*

Variants of ordinary care are *ordinary diligence* or *ordinary prudence*, which have evolved to *due diligence* for the mineral industry. Very early on, this notion of ordinary prudence was cited in the case of a mineral property by Georgius Agricola (1556) in his *De Re Metallica* as

> *Moreover, a prudent owner before he buys shares, ought to go to the mine and carefully examine the nature of the vein, for it is very important that he should be on his guard lest fraudulent sellers of shares should deceive him.*

And 350 years later, Hoover (1909) and his advice is still needed for the investor to guard against false claims and misleading information:

> *The old terms, "ore in sight" or "profit in sight" have been as of late years subject to much malediction on the part of engineers because the terms have been so badly abused by the charlatans of mining in attempts to cover the flights of their imagination.*

But while both Agricola and Hoover have offered adequate warnings, there are those in the industry who, either from just sloppy work or deliberately misleading information, fail to deliver from the projected return on investment that which was implied. It is useful to cite a few examples of these types of mining investments over the years:

- Naxos Resources (Carey 2010)
- International Precious Metals (Moukheiber 1997)
- Delgratia Mining Corporation (*Northern Miner* 1997)
- Crystallex International Corporation (Asenio and Company 1998)
- Cartaway Resources (Pelkey 2000)

- The Poseidon bubble (Wikipedia n.d.)
- Bre-X, the most famous of all (Ro 2012)

Many other problems are uncovered by due diligence study other than fraud, but these are the ones that the public remembers. Ever since the Bre-X fraud, however, government regulations have been directed toward restricting what the companies can and cannot announce to the public concerning a public mineral company. But the extent of these regulations has focused primarily on the definition of the mineral resource as it is described. Other than the reserve issues in the past, there have been very few regulations setting minimum standards for the feasibility and evaluation studies that are presented to the public and investors. These studies may or may not depict what the true return on an investment would be if it is indeed put into production. Sometimes these are investigative engineering errors in sampling, testing, design, or judgment; or they are caused by inadvertently overlooking some costly item. Yet the same effect occurs when these errors are intentional. Flyvbjerg et al. (2002) calls them *strategic misrepresentations*, Lumley (2012) dubs them *deliberate deceptions*, and this author simply calls them *scams* if they are shown to be deliberate.

These types of failures usually lead to massive overruns in the capital expenditure for the project's construction and/or a failing to reach the projected return on investment. This problem is also discussed in Chapter 17 (see "Part III: Social and Political Risks").

The preceding examples illustrate why, at least several times a year, project consultants are requested by individuals or companies who are looking to or have spent several millions of dollars on a mining property, or have been approached to lend money for a project, to show them why their investment in a mining property or company is going bad, or verify that it will yield that which has been projected by the mineral company. Thus, prior to completing an acquisition of a mining company or mining property, or providing financing (either debt or equity) for a property's development, *a due diligence study is a necessity.*

TYPES, CATEGORIES, AND SUBCATEGORIES OF DUE DILIGENCE STUDIES AND REPORTS

The type of due diligence study to be performed varies in scope and rigor according to the size of investment and the end purpose of the study. These factors, of course, will affect the qualification and number of members of the project team and the length of time and cost for the due diligence project and report. As an example, this author completed a very limited-scope due diligence audit (including a report) on the reserve status of a small underground mine in a very isolated area of southern Peru in one week, including the report. In contrast, this author took eight months to perform an acquisition due diligence for a detailed audit on 40 coal mines: to address reserves, machinery, people, production capability, productivity, and labor relations stability, as a part of a larger team analyzing the entire coal company being acquired by a diversified metal company.

But the normal time for a due diligence audit is a week or two on-site, and another three to six weeks of research, study of information acquired from the company, performance of risk assessments on all aspects, and production of the due diligence report. Various kinds of due diligence studies are discussed in the following subsections. The functions studied during acquisition due diligence versus the due diligence audit are very similar but vary mostly in depth of the study.

Acquisition Due Diligence for Individual Mining Operations or the Entire Mining Company

This is the most extensive type of due diligence study. Every detail of the operation and organization must be thoroughly examined. This study will be accomplished using experienced professionals in every discipline involved in running the operation. It is an exhaustive study and may take several months. The cost can be very high, depending on the size of the acquisition and the size of the due diligence crew. This is true because the acquiring company must understand every aspect of the company to be acquired and the synergistic relationship, or lack thereof, for the newly acquired company with the existing company, which is a critical point. Sometimes it makes sense to use an all-consultant team to perform this study. But at other times, when one operating company acquires another operating company, it makes sense to use a mixture of the acquiring company personnel as part of the team with consultants in order to have a successful acquisition.

It is emphasized that every aspect of the company being considered for acquisition must be a part of the due diligence audit. And every aspect of the target company, which can affect profitability, must especially be included in sufficient depth to be totally understood in the due diligence study. Besides the reserve base that must be verified, all operation costs should be verified with benchmark data. Particular areas of the targeted mineral company need a specific scrutiny study, and these are discussed in a later section.

Due Diligence for a Property or Project Audit

A *due diligence audit* is an in-depth audit but is not exhaustive when compared to the acquisition due diligence audit. This type of audit is normally favored by financial institutions prior to lending on a minerals project that is being studied or is about to be developed. It is usually completed within 60 days. This author was project manager for many of these types of audits. One small, high-grade gold mine in northern British Columbia, Canada, took four days on-site and two weeks for reporting. Another fairly large open pit mine project in Bolivia took five days on-site and five weeks to complete the report. The extra time needed to complete the report was due to errors in the density factor used for many of the drill cores and mistakes in recording data from the assay sheets to the computer model, both of which contributed to having to recalculate the reserve estimate. A mineral property due diligence audit usually has four major purposes:

1. To provide assurance to the lenders (with the qualified risks identified) that the information presented by the company, looking forward to when the project is completed and put into operation, will produce the projected return and allow their loan to be collateralized
2. To determine where there are deficiencies in the operation or operation planning that may cause the operation to fail to produce the return expected in the future
3. To identify risk to the operations performance and quantify those risks (Where mitigations have been suggested or planned, quantify the cost of those mitigations and their likely success.)
4. To develop the basis for converting to a nonrecourse loan in the future

As in the acquisition due diligence, particular areas of the targeted mineral company need a specific scrutiny study. These are usually the same areas of the facility but are typically not studied to the same depth as the audit.

Due Diligence Failure Risks (Santini 1999) or Fatal Flaw Studies

Failure risks or *fatal flaw studies* are not in-depth studies. Failure risks involve any item that could result in complete failure of the operation at some predictable point. The second type of risk, "impact risk," will not cause a complete failure but will severely impair the projected return on investment or create a serious employee safety risk for the operation. This type of study gives a cursory investigation of all project functions by very experienced, qualified professionals (i.e., every function of the operation must be quickly studied by a vastly skilled professional(s) by visiting the site and studying the critical reports). If a problem is suspected, then to protect the investor, of course, a follow-up must be completed to a more in-depth analysis of that particular function. The study should be conducted by only a few professionals working on-site for a few days, unless a serious problem is found. The report could be completed very quickly, possibly by a "letter report." This author and two others, a geologist and a metallurgist associate, completed such a study after spending two days on-site and providing a letter report the following week. It was, however, a fairly small projected placer operation in the northwestern United States.

Typical Checklist for Due Diligence Studies and Reports

Figure 19.1 is a simple outline of items to be reviewed during the development of a due diligence report (Behre Dolbear 2008).

This outline demonstrates the typical requirements for conduct of such studies. Each due diligence effort requires intensive, detailed investigation of the issues unique to the particular project. In most cases, due diligence studies require on-site visits and investigations in addition to the review of documentation. The on-site investigations should include the review of exploration results, operational idiosyncrasies, infrastructure limitations, regulatory influences, and socioeconomic aspects of a project.

1.0 Executive Summary

2.0 Project Background

- 2.1 Location and Access
- 2.2 Climate and Topography
- 2.3 Existing Site Conditions, Available Services and Facilities
- 2.4 Nature and Type of Project
- 2.5 Overview/Summary Description of Operator

3.0 Study Overview

- 3.1 Purpose and Scope
- 3.2 Work Undertaken to Date
- 3.3 Status of Project
- 3.4 Role and Responsibility of [Examiner]
- 3.5 List of Definitions, Abbreviations and Conventions

4.0 Land Status

- 4.1 Constitution of Land "Package"
- 4.2 Issue Affecting Control (if any)
- 4.3 Survey Control

FIGURE 19.1 Typical outline for due diligence studies and reports (Figure continues)

5.0 Project(s)
- 5.1 Summary Description
- 5.2 Historical Synopsis
- 5.3 Historical Production
- 5.4 Current Operation and Facilities

6.0 Mineral Resource, Geologic
- 6.1 Geological Setting
- 6.2 Data Base/Status of Exploration
  - 6.2.1 Surface Programs
  - 6.2.2 Drill Programs
    - Location and Attitude Control
    - Logging and Sampling
    - Sample Preparation and Assaying [chain of custody—if applicable]
    - Check Assay and Quality Control Program
    - Specific Gravity/Density Determination
    - Bulk Sampling Programs
- 6.3 Mineral Resource Inventory
  - 6.3.1 Methodologies Applied
  - 6.3.2 Summarized Estimates
  - 6.3.3 Established Mineral Resource
  - 6.3.4 Deposit
    - Background
    - Statistics and Variograms
    - Model and Model Checks
    - Mineral Resource Estimate and Classification
    - Confidence Limits

7.0 Ore Reserve, Mineable
- 7.1 Applied Methodologies—Surface
  - 7.1.1 Pit Design Parameters/Criteria
  - 7.1.2 Cut-off Grade
  - 7.1.3 Dilution Estimates
- 7.2 Applied Methodologies—Underground
  - 7.2.1 Underground Parameters/Criteria
  - 7.2.2 Thickness/Width
  - 7.2.3 Dip/Slope
  - 7.2.4 Cut-off Grades
  - 7.2.5 Dilution Estimates
- 7.3 Summarized Mineable Ore Reserve Estimate

8.0 Reserve Estimate Support
- 8.1 Special Studies That Support Reserve Tonnage and Quality
- 8.2 Other Supporting Documentation

9.0 Mine Engineering and Planning
- 9.1 Mining Method
  - 9.1.1 Current Operations
  - 9.1.2 Surface
  - 9.1.3 Underground
  - 9.1.4 Geotechnical Conditions
- 9.2 Hydrology and Dewatering
  - 9.2.1 Project Hydrology
  - 9.2.2 Methods of Hydrological Control
  - 9.2.3 Continued Monitoring and Control Program
  - 9.2.4 Water Treatment and Disposal

FIGURE 19.1 (Continued) (Figure continues)

9.3 Mining Operations and Mine Plans
- 9.3.1 Operations—Surface
  - Design Adequacy
  - Roads and Haulages
  - Benches
  - Pit Slopes and Slope Stabilities
  - Detailed Development Plan
  - Cost Estimates
  - Labor Productivity
- 9.3.2 Operations—Underground
  - Design Adequacy
  - Ventilation Adequacy and Design
  - Escapeway and Evacuation Plans
  - Hoisting, Haulage, and Conveying Equipment
  - Detailed Development Plan
  - Cost Estimates
  - Labor and Productivity

9.4 Equipment
- 9.4.1 Existing Equipment
- 9.4.2 Proposed Equipment
  - Mining Equipment
  - Equipment Availability
  - Support
  - Manning
  - Maintenance

10.0 Development (New Project Implantation—if any)
- 10.1 Pre-Development Preparation
- 10.2 Final Engineering Design
- 10.3 Procurement
- 10.4 Site Development and Construction Management
- 10.5 Project Schedule

11.0 Mine Operations
- 11.1 Annual Ore and Waste Production
- 11.2 Design of Stopes
- 11.3 Waste Volumes and Waste Handling
- 11.4 Operations and Quality Control

12.0 Processing
- 12.1 Background
- 12.2 Metallurgical Testwork
  - 12.2.1 Integrity of Samples
  - 12.2.2 Crushing
  - 12.2.3 Processing
  - 12.2.4 Differences Among Deposit Test Results
  - 12.2.5 Products
- 12.3 Process Facilities
  - 12.3.1 Crushing System
  - 12.3.2 Concentrator
  - 12.3.3 Handling and Bagging
  - 12.3.4 Operating Schedule
  - 12.3.5 Personnel Requirements

FIGURE 19.1 (Continued)

(Figure continues)

13.0 Operations and Support Services
- 13.1 General and Administration
  - 13.1.1 Accounting and Purchasing
  - 13.1.2 Secretarial and Legal
  - 13.1.3 Environmental and Safety
  - 13.1.4 Security and Inventory Control
- 13.2 Management and Technical/Engineering
- 13.3 Warehouse/Parts/Supplies
- 13.4 Maintenance Facilities
- 13.5 Assay Laboratory
- 13.6 Personnel Qualifications and Requirements

14.0 Infrastructure
- 14.1 Power Supply and Distribution
- 14.2 Water Supply and Distribution (process, potable, fire)
- 14.3 Fuel Supply, Storage and Distribution
- 14.4 Communications
- 14.5 Labor Source and Housing
- 14.6 Access Roads, Fences, Guardhouse, etc.
- 14.7 Sewage Treatment
- 14.8 Social Infrastructure (schools, churches, hospital, etc.)
- 14.9 Transportation
- 14.10 Concentrate Loading Facilities

15.0 Environmental Controls and Compliance
- 15.1 Plan of Operations
- 15.2 Required Pollution Controls
- 15.3 Reclamation Plan and Bonding
- 15.4 Reclamation and Mine Closure Liabilities

16.0 Permitting
- 16.1 Baseline Studies and Data Collection
- 16.2 Permit Adequacy for Future Operations
- 16.3 Compliance Record—Outstanding Violation

17.0 Marketing and Sales
- 17.1 Marketing
  - 17.1.1 Products
  - 17.1.2 Distribution Channels
  - 17.1.3 Markets
- 17.2 Sales Mechanisms and Sales Revenue Projections

18.0 Capital Cost Estimates
- 18.1 Pre-Production Capital Costs—if applicable
- 18.2 On-Going Capital Costs During Project Operating Life
- 18.3 Working Capital Requirement

19.0 Operating Costs
- 19.1 Royalty Burdens
- 19.2 Summary of Personnel Requirements
- 19.3 Salary and Wage Rates, and Payroll Burdens
- 19.4 Indirect (Overhead) Burdens
- 19.5 Summary of Materials and Consumable Supplies Requirements
- 19.6 Startup Production Schedule
- 19.7 Operating Costs, by Function

FIGURE 19.1 (Continued)

(Figure continues)

20.0	**Economic Analysis**		
	20.1	Critical Parameters and Assumptions Underlying Cash Flows	
	20.2	Tax Regulations and Considerations	
	20.3	Product Price Determinations	
	20.4	Base Case (Most Likely Case) Cash Flows	
	20.5	Possible Worst and Possible Best Case Cash Flows	
	20.6	Sensitivity Analysis	
		20.6.1	Break-Even Points
		20.6.2	Multiplier Effect of Selected Parameter Pairs or Groups
		20.6.3	Monte Carlo Simulation
21.0	**Political and Deal Analysis**		
	21.1	Current Political Structure	
	21.2	Credit Standing	
	21.3	Political Risk Level	
	21.4	Deal Analysis and Structure	
22.0	**Risk Analysis**		
	22.1	Geological Risks	
	22.2	Engineering or Project Risks	
	22.3	Economic and Market Risks	
	22.4	Political Risks	

Source: Behre Dolbear 2008.

FIGURE 19.1 (Continued)

An itemized risks list must be developed for each element of the targeted company operation being studied. Table 17.2 in Chapter 17 contains a suggested list to be considered. The risks should be descriptive (e.g., the mine ventilations will not meet MSHA [Mine Safety and Health Administration] diesel particulate standards) and should be categorized as low, moderate, or high. The following standard definitions can be used for these risk categories:

- ***High***—*The risk factor poses the prospect of an immediate, major failure, which if uncorrected, will have a material (+15%) impact on cash flow and/or operating performance and could potentially lead to project failure.*
- ***Moderate***—*The risk factor* [if uncorrected] *could have a significant (±10%) effect on the project cash flow and performance, if not mitigated by some corrective action.*
- ***Low***—*Even if uncorrected, the risk factor will have little or no effect on project cash flow and performance* (Behre Dolbear 2008).

Steps to mitigate each of the risks must also be described and, where practical, an estimated range of costs of mitigation should be established.

ENGINEERED RISK ASSESSMENT OF FACILITIES

An *engineered risk assessment* (ERA) is an evaluation of engineered systems to determine the likelihood of system failure. The ERA is a formal risk analysis procedure (Behre Dolbear 1999). This is very important, particularly for new operations that often fail in the first few months of operation. Such failures may completely destroy the opportunity to meet the needed return on investment. These failures may occur in the operational, environmental, or marketing parts of the new operation. (This is not to be confused with the project risk appraisal and adjustment described in Chapter 17, which is also an engineering risk appraisal, but specific to the project execution phase.)

The objective of the ERA is to identify and quantify risk of every vulnerable area of each operating system (such as mine, mill, waste disposal, product shipping, and marketing problems), evaluate mitigation alternatives, and prioritize resource allocation to optimize use of limited resources.

A team of qualified engineers (and in the case of industrial minerals, a qualified geologist to evaluate product quality) and environmental specialists, in concert with the operation's professionals, then analyzes each component (e.g., the correct ore hardness and work index on the grinding of the ore) for its risk of failure. The team then quantifies, in terms of financial losses caused by downtime and environmental damages, the consequences resulting in failures of high-risk components. The consistent global tailings dam failures are classic examples of an engineered area that needs very close scrutiny. These failure types are exemplified by the Brazilian Samarco iron mine dam failure, for which the direct damage cost will likely be in the hundreds of millions of dollars, followed by the multiple billion-dollar lawsuits, which are ongoing.

The benefits of an ERA to the operations' technical issues, such as its geomechanical, geotechnical, and environmental damage mitigation systems, are

- Clear definition of environmental liabilities,
- Quantification of uncertainties,
- Prioritization of problems to be tackled, and
- Reduction in failures.

Once these are determined, requirements to mitigate risks can be determined and should be described and quantified.

SOCIAL AND LOCAL COMMUNITY ISSUES

Twenty years ago, the subject of social and community issues would not have been an item considered in a due diligence report. But those days are long past. It is not that many mining companies have not been providing many of the social needs of the communities for scores of years, but that the industry has done a very poor job of publicizing their good works, especially in the early planning of the new mines or mining district. For the past hundred years, some mining companies did an excellent job of providing many of the social needs of the communities in the original plans of development. This author observed many such developments by St. Joe Minerals Corporation, New Jersey Zinc Company (now called Horsehead Holding Corporation), and Cominco Ltd. But the difference now is that these community plans must be highlighted and featured in the feasibility study, and they must be created from the meeting results and agreements of the local stakeholders to receive a *social license to operate*. Otherwise, there is severe risk that the operation will, at some point, receive such strong opposition that the project will fail or, if allowed to operate, will not be as profitable as the feasibility study forecast. Thus it is a necessary item to be considered in the due diligence report.

Chapters 10 and 17 describe many of such problems that may occur, the results of the problems, and the potential mitigation of the problems, which should be audited in the due diligence report. They need not be repeated here, except to summarize some of the salient points that must be apparent in the feasibility study report and/or in actions of the company being audited:

- Every company performing exploration or developing mineral properties should document its company-wide policy statements, which attest to its intent to build and operate mineral facilities that are in harmony with the environment and the existing social system, from exploration through to closure and beyond.
- This stated policy, or a summary thereof, needs to be present in the feasibility for all stakeholders to see.
- The company must also be proactive in organizing and conducting open meetings and providing good faith communication with the principal stakeholders in establishing the needs of the community as they relate to the development of the project.
- Comprehensive strategies need to be developed to ensure that all parties have the opportunity to gain from the mining operation near established communities.
- Where it is mutually advisable, the company must work with various government agencies to meet the needs of the community as they relate to the development of the project.

Along with the preceding actions, there are other things companies need to be aware of:

- Political systems of some countries demand more than can ever be provided by the proposed project. This author was involved with one such company. A very large copper project was proposed near the Ngäbe-Buglé comarca (a *comarca* is an administrative division) in northwestern Panama. It would have been a joint venture of two very large multinational companies (one company buying into the other company) that had invested millions of dollars in exploration. It looked very feasible from the initial study. The company representatives then sat with various government dignitaries, including someone from what would be the equivalent of the U.S. Department of Health and Human Services. The regulator explained that for this very large project (approximately US$1 billion in 1985), they were very happy to see jobs developed in this area, but since there were thousands (probably close to 200,000) poverty-stricken indigenous people in the area (the Ngäbe-Buglé indigenous groups), the government could not allow the same disparity of "haves and have-nots" to take place as it had along the U.S.-operated Panama Canal. Therefore, the future mining operation would be responsible for all social support of the entire area (although what was meant by the "entire area" was not defined, but it was clearly at least the comarca of the Ngäbe-Buglé indigenous groups). Obviously, the return from the project could not at that time support such an effort and leave anything for the investors, which killed the joint venture project. Since the buy-in company had nothing invested, it disengaged from the project.
- Sometimes the project happens to be in an area where those responsible for issuing environmental permits have a different agenda than those of the company and even those of the local community. An example is the Crandon project in northern Wisconsin. In the project area, hundreds of feet below the surface, lies a truly world-class copper/zinc resource—probably one of the best unmined resources in the world. The mine would have been an underground mine, with all waste material going back in the mine beginning about two to three years after start of mining, and all of it disposed underground by the end of mining, scores of years later. Every environmental issue had a potentially

successful mitigation plan presented to the Wisconsin Department of Natural Resources (WDNR). The large company worked in good faith with the local stakeholders and the WDNR for many years and completed the intermediate feasibility study. Every issue was resolved except two: (1) After several years of exploration and land acquisition, a local Native American tribe announced that this was sacred ancestral ground and could not be disturbed; and (2) there were many summer cottages in the area owned by residents of Madison, Milwaukee, and Chicago. These residents did not want their pristine north woods vacation areas developed. Both groups had tremendous influence on the Wisconsin legislature. The permitting agencies continued to find reasons to add more conditions to the permitting and never granted the final permits. After about US$100 million was spent on the property, and three different companies tried to get it permitted and developed, the last company practically gave it away to the Native American tribe, and the land sits unmined.

- Other such projects in recent years that have run into serious trouble because of escalated social problems are operations at Bougainville in Papua, New Guinea; Grasberg in Irian Jaya, Indonesia; Amayapampa in Bolivia; and Diavik and Voisey's Bay in Canada. Failure to consider these needs can be costly and result in delays in obtaining permits.

A review of Chapter 10 will remind the reader of what must be included in the sustainable development plan that needs to be in the feasibility study.

REASONS WHY MINING PROJECTS FAIL TO PRODUCE PROJECTED RETURNS

When performing a due diligence audit, one must constantly be aware of the bottom line and what return is projected, which will allow the return of the investors' capital. If one considers the measure of *achieving a projected return on investment as the criteria of success*, then probably 80% of developed mining projects will fail. This is one of the primary purposes of the due diligence study: to ferret out mistakes, poor practices, and omissions, both inadvertent and intentional, that may decrease profitability. The three major categories of failures are gross overruns in capital cost investments, serious technical errors (both problems occurring during the feasibility studies), and poor management recognition of potential problems occurring during construction and operations. All of the items listed earlier in the acquisition due diligence study are the same functions that must be again tested in the due diligence audit.

Technical and Economic Factors

Although some of the technical and economic failures were discussed earlier, here is a summary of the factors that must be correctly accomplished:

- Geologic audit
  - Proper construction of geologic models to produce a verifiable reserve
  - For precious metal resources, verifiable chain of custody for all samples handling and processing
  - Proper sample drilling
  - Correct sample splitting
  - Verification of correct assaying

- Correct interpretation of structural geology and mineral continuity
- Correct dilutions and recoveries included in the block model
- Correct calculation of the resource/reserve categories, according to the applicable country codes

- Mining and geologic cut-off grade audit
- Mining problems audit
  - Mining recovery, dilutions, and losses
  - Mining density factor
  - Geomechanical review
  - Detailed mine plan
  - Mining plans meeting all MSHA and Occupational Safety and Health Administration codes
  - Realistic (not overoptimistic) mine design and productivity estimates
  - Realistic (not overoptimistic) mine development schedule and start-up (learning curve) time
- Metallurgy and processing
  - Verification of locked cycle or pilot-plant testing
  - Verification of processes sampling and controls
  - Verification of the processing method based on testing
  - Inclusion of proper processing of metallurgical recoveries and losses
  - Verification of material balance
  - Verification of water balance
  - Feasible process plant designed correctly to meet design production expectations
  - Mine, plant, and other waste disposal methods and procedures in compliance with government standards and regulations
- Cost estimating and marketing
  - Using the commodity trend price (see Chapter 20) rather than some speculated higher price
  - Identification of where the project's costs lie in the seriatim of other operations of the same commodity
  - Capital and operating cost that is not understated
  - All major capital cost items considered and nothing omitted
- The correct working capital and contingencies used
  - Correct estimation of marketability of the commodity
  - Recognition of differential price inflation on specific consumables versus product
  - Recognition of differential exchange rates between home country and development country

In addition to correctly accomplishing the preceding factors, environmental audits must examine all of the planned environmental actions. Care must be taken that those who are securing the permits are allowing the project to move forward. And all aspects of the environmental plan must be properly evaluated and the correct capital and operating costs must be applied. Chapter 9 is the best guide as to what must be audited.

- Environmental audit mistakes
  - Underestimated cost of discharge water recirculation and/or cleanup
  - Underestimated extreme-weather conditions
  - High-altitude conditions for personnel and equipment ignored for productivity projections
  - Unidentified environmental problems missed by the feasibility study
- Social license to operate issues
  - Unpredicted variations of social and/or business attitudes of the community, state, or national government's reaction to the project
  - Verification that stakeholder meetings are required to be conducted, and that the needs of the community and nongovernmental organizations have been heard and addressed to the extent economically possible
- Management mistakes
  - Lack of experience of company and/or contractor in developing projects, especially in a country where the company or contractor have no project experience
  - New joint venture management

Project Cost Overruns

As recorded recently in the financial community, 60% of projects were overrun in South America, 51% in North America, 40% in Australia, and 31% in South Africa (Deloitte 2012). Mine project overrun failures are not uncommon occurrences. The reasons for these failures can usually (but not always) be identified from the feasibility studies due diligence review. The weighted average amount of the overrun for 221 projects between 1965 and 2006 was 26%. This author identified four megaprojects in 2013 that had total capital cost overruns that averaged $US5.5 billion and had a weighted average of 352% overrun (Bullock 2013a, 2013b). This trend is totally unacceptable to the mineral industry and to the financial community that supports mineral investments.

In examining projects over the past 20 years, the following factors should have been identified and often were not:

- Construction inflation between final feasibility and construction time is not identified during the feasibility studies.
- There is a lack of available experienced project management talent.
- The activities of special groups who are obstructing projects are not mitigated during feasibility studies.
- There is poor system analysis or engineering and/or under-engineered feasibility studies.

- Major facilities items are completely missed or all items are not updated during each phase of the feasibility studies.
- The metallurgical results are incorrect either because of poor testing or incorrect sampling (geological problem).
- Information and engineering trade-off studies are insufficiently developed during feasibility studies.
- Change orders are issued for problems that were not identified in feasibility studies (may or may not be justified).
- Unidentified environmental problems are discovered after construction is begun.
- The potential for inclement weather was not identified during the feasibility study, so there is no allowed downtime for usual bad weather conditions.
- Poor estimating techniques are used during feasibility studies, pointing to a *lack of benchmarking and reference class forecasting*.
- The contingency allowance is inadequate because it was calculated incorrectly during feasibility studies.
- The owner's cost is left out or the cost was underestimated during the feasibility studies.
- The working capital is incorrectly calculated or was simply underestimated during the feasibility studies.
- Construction productivity is lower than was projected during the feasibility studies (no benchmarking).
- Because of improper planning during the feasibility studies, there are time schedule overruns (*averaging about 1% of increased capital cost per month*). According to a survey by Golder Associates of several hundred mining companies in 10 countries (Kuestermeyer 2016), the principle causes of project delays identified were
  - Aggressive and unrealistic schedules,
  - Inadequate construction plans,
  - Permitting,
  - Inexperienced/inadequate project management, and
  - Late and/or inadequate data.

The following are events that are not likely to be identified in due diligence feasibility studies, but they still need to be cited if there are any indications of future occurrence:

- There is an unexpected delay of equipment deliveries because of customs or manufacturers.
- There are accidental events that occur during construction.
- There are unpredicted changes in environmental regulations after the feasibility study.
- An engineering contractor may submit a low bid during feasibility studies and lowball the capex estimates to get the next contract.

Great works by various countries in mineral resource/reserve reporting code requirements have all contributed to the much improved standards of defining the resources and reserves (see Chapter 2). These have served to minimize fraud in these areas, and this author believes it has been successful.

All of the preceding information is meant to outline the broad scope of reasons why the project may fail to meet the return on investment as projected by the feasibility study. Many of these problems should have been identified during a due diligence study. The project, when it becomes an operation, may or may not fail completely and end up in Chapter 11 bankruptcy (in U.S. financial terms), however, such a project may still be able to provide a break-even economic position, jobs, and a mineral product that is needed or can be exported. *It remains that the job of the due diligence auditor of the feasibility study is to identify flaws in the study that will result in not producing the return on investment projected*, and this handbook addresses the in-depth technical information required for each area.

To its credit, the Society for Mining, Metallurgy & Exploration has set some minimum standards with its new recommendations in *The SME Guide for Reporting Exploration Information, Mineral Resources, and Mineral Reserves* (SME 2017) as to the amount of engineering that should go into engineering feasibility studies at the three levels as well as the expected accuracy of the cost estimates and the amount of contingency that should be used. This is a great step in the right direction that hopefully will start a trend in the codification of these standards in the codes of the various countries.

CONCLUSION

It has been demonstrated many times—by the past experiences of many consulting organizations—that due diligence audits or reviews, while considered costly and time-consuming, should be a requirement for an investor in ensuring the success of an acquisition, joint venture, or financing investment. A proper due diligence analysis will help protect the financial loss and personal costs inherent in a mine failure or encourage a promising investment with minimal risks.

REFERENCES

Agricola, G. 1556. *De Re Metallica* (Translation by H.C. Hoover and L.H. Hoover, 1912), 1950 edition. New York: Dover Publications. p. 29.

Asenio and Company. 1998. Crystallex fails to deny new fraud charges. Press release. March 19. www.prnewswire.com/news-releases/Asenio--company-crystallex-fails-to-deny-new-fraud-charges-77167782.html (accessed November 2016).

Behre Dolbear. 1999. Engineering risk assessment. PowerPoint presentation for management and associates. Denver, CO: Behre Dolbear.

Behre Dolbear. 2008. Typical checklist and guidelines for due diligence studies and technical review. Intercompany communication. Denver, CO: Behre Dolbear.

Bullock, R.L. 2013a. Mine feasibility studies still need improving. Presented at the PDAC (Prospectors and Developers Association of Canada) Annual Meeting, Toronto, ON, March 3–6.

Bullock, R.L. 2013b. Mine feasibility studies still need improving. Presented at the Innovations in Mining Engineering: Exploring Global Mining Frontiers Conference, Missouri University of Science and Technology, St. Louis, Sept. 9–12.

Carey, C. 2010. Small Texas company promotes big South American oil venture. Share Sleuth. June 28. http://sharesleuth.com/investigations/2010/06/both_of_the_oil_companies (accessed November 2016).

Deloitte. 2012. *Tracking the Trends 2013: The Top 10 Issues Mining Companies May Face in the Coming Year.* https://cmo.deloitte.com/content/dam/Deloitte/ca/Documents/international-business/ca-en-ib-tracking-the-trends-2013.pdf. p. 13 (accessed January 2017).

Flyvbjerg, B., Holm, M.S., and Buhl, S. 2002. Understanding costs in public works projects: Error or lie? *Journal of the American Planning Association,* 68(3):279–295.

Hoover, H.C. 1909. *Principles of Mining.* New York: McGraw-Hill. p. 16.

Kuestermeyer, A. 2016. The whys and wherefores for project delays. Presented at the SME Financial Conference, New York, April 26.

Lumley, G. 2012. Reducing project risk through fact-based mine planning. Presented at the AusIMM (Australasian Institute of Mining and Metallurgy) Project Evaluation Conference, Melbourne, Australia, May 24–25.

Merriam-Webster's Unabridged Dictionary. n.d. Ordinary care. http://unabridged.merriam-webster.com/unabridged/ordinary%20care (accessed November 2016).

Moukheiber, Z. 1997. Glitter, but not gold. *Forbes.* June 16. www.forbes.com/forbes/1997/0616/5912046a.html (accessed November 2016).

Northern Miner. 1997. Little light shed on Delgratia salting scandal. *Northern Miner,* 83(18). http://www2.northernminer.com/news/little-light-shed-on-delgratia-salting-scandal/1000161719/ (accessed November 2016).

Pelkey, D. 2000. Two Vancouver brokers found to be in conflict of interest in Cartaway Resources stock promotion. News release 00/42. October 2, British Columbia Securities Commission. www.bcsc.bc.ca/News/News_Releases/Two_Vancouver_brokers_found_to_be_in_conflict_of_interest_in_Cartaway_Resources_stock promotion/ (accessed January 2017).

Ro, S. 2012. Bre-X: Inside the $6 billion gold fraud that shocked the mining industry. *Business Insider.* July 1. www.businessinsider.com/bre-x-6-billion-gold-fraud-indonesia-2012-7 (accessed November 2016).

Santini, K. 1999. Due diligence review and valuation of industrial minerals acquisitions. SME Short Course, Denver, CO. pp. 6–7.

SME (Society for Mining, Metallurgy & Exploration). 2017. *The SME Guide for Reporting Exploration Information, Mineral Resources, and Mineral Reserves.* Englewood, CO: SME.

Wikipedia. n.d. Poseidon bubble. https://en.wikipedia.org/wiki/Poseidon_bubble (accessed November 2016).

CHAPTER 20

Using Technical Economic Evaluations and Cash-Flow Analyses in Feasibility Studies

Amy E. Jacobsen

The cash-flow model is a tool for identifying the strengths and weaknesses of a project through the analysis of key engineering and physical inputs in combination with the economic factors that impact the feasibility of a project. The cash-flow model incorporates a project's central engineered attributes, such as the mine plan and processing rates, mineral or metal recoveries, and final product production and combines them with revenue and cost factors, including both capital and operating costs, to determine the project's underlying economic viability. The cash-flow model is where the technical and economic aspects of the project are combined, allowing one to have an overall view of the project.

Cash-flow models can be used for a number of purposes, such as determining economic viability, analyzing development and operating options, investigating financing and investment considerations, and establishing project value. In the case of a *technical cash-flow model*, such as those used for feasibility studies, one of its key purposes is to assess the economic viability of a project and to test the impact of the engineering inputs on the project economics through a series of sensitivity analyses. In addition, the technical cash-flow model may be used to evaluate and compare development and operating alternatives.

In comparison, a *financing cash-flow model* is designed for a specific audience looking at the financing and investing options for the project. This type of model specifically incorporates the effect of debt and other financing options into the cash-flow model.

Similarly, the *valuation cash-flow model* is designed to be used in determining a project's fair market value or to assist in determining a company's value. It should be noted that the valuation cash-flow model usually only comprises part of the valuation process and should not be used as the sole method for determining fair market value.

TECHNICAL CASH-FLOW MODEL INPUT DEFINITIONS AND CONSIDERATIONS

Several inputs need specific consideration and would be expected to be included in the cash-flow model, regardless of its purpose. In particular, the treatment of reserves and resources, mine plan, processing approach, capital and operating costs, and final product output are generally considered to be the key engineered inputs. The nonengineered inputs are product marketability and expected sales, product sales price, and income tax calculations.

Every mining project is unique. Trying to apply standard "out-of-the-box" cash-flow modeling procedures and methodologies to every project is not appropriate; however, there are numerous standard considerations and practices that need to be considered when developing a technical cash-flow model. Although this list will vary with each project, generally the following considerations will need to be addressed by the modeler and will provide the basis for the cash-flow model:

- The desired level of accuracy
- Justification of the cash-flow inputs and the level of acceptable subjectivity of those inputs
- Effects of inflation
- Pretax versus after-tax presentation
- Inclusion of debt and financing components
- Cash versus noncash provisions
- The key cash-flow components, both engineered and economic based

Accuracy and Subjectivity

A cash-flow model can only be as good as the information included in the model. To state the obvious, the higher the level of accuracy of the inputs, the more accurate the results of the cash-flow model. In other words, there is less chance of variation between the projected values and what happens in reality. In some cases, such as a preliminary economic analysis, a lower level of accuracy is to be expected (estimates may be within ±30%); however, in the case of a feasibility study, a higher level of accuracy is expected. In a feasibility study, the inputs tend to have a greater confidence level because of more extensive engineering and confirmation of the inputs (i.e., equipment cost quoted directly from vendors versus estimates based on engineering tables).

Because a cash-flow model is based on future events and conditions, there is a need to make certain assumptions. In actuality, the project performance can vary significantly from the cash-flow model assumptions and forecasts due to unanticipated events and circumstances. Some of the inputs to a cash-flow model are inherently more subjective than others. For example, the production plan should be based on an engineered mine plan, which is less subjective than, say, the projected commodity prices.

For others to understand the level of subjectivity and the resulting accuracy of a cash-flow model, it is critical that the modeler make clear statements of and justification for all the input and variable assumptions used in developing the cash-flow model. In many cases, the modeler will not be the person who has generated and determined the inputs. The modeler will need to rely on information and data provided by others. The assumptions behind the inputs will need to be clearly communicated to ensure consistency between the feasibility study and the cash-flow analysis.

Effect of Inflation

Cash-flow models are generally prepared using constant currency terms without including the effect of inflation or deflation. When *real* currency terms are used, the cash flows begin with the actual dollars of the period and incorporate the effects of inflation through the term of analysis. The use of real cash flows are appropriate for budgets and short-term forecasts; however, in longer-term projections, as those usually associated with feasibility study cash-flow models, costs and revenues are reported in constant dollars without inflation or escalation. A constant currency approach is recommended as estimating the change in the model inputs that affects costs and revenues becomes complicated and more subjective.

- Mineral or metal prices are generally not escalated to reflect the effect of inflation. As a result, it has not been appropriate to annually escalate production costs.
- Errors or omissions in estimating production costs are further compounded in later years as the operating costs are escalated to reflect the effect of inflation.
- The cash-flow model tends to be more easily understood and audited. The cash model in constant dollars of the day avoids adding variations every year to input costs and revenues.

Pretax Versus After Tax

Generally, feasibility study cash-flow models are prepared on an after-tax basis. These cash-flow models include the effects of depletion, depreciation, and income tax considerations (including tax loss carried forward). A typical feasibility study cash-flow model is unleveraged, and the income tax calculations do not take into account either debt- or equity-related cash flows.

In some cases, a pretax cash flow is justifiable. This is a common practice when comparing development and operating alternatives. It is prudent to ensure that one or more of the alternatives does not disproportionally affect the taxes.

Debt and Financing Considerations

As the purpose of the feasibility cash-flow model is to analyze the underlying economics of a project given a set of physical inputs, most cash-flow models are unleveraged and do not include debt or other financing considerations such as dividend payments. This allows the evaluation of the project to be done on the basis of the technical inputs and economic assumptions rather than the financing terms. Financing terms can vary by company over time and jurisdiction, making direct comparisons difficult.

Cash Flows Versus Noncash Accounting Provisions

It is important to realize that when used in a feasibility study, the cash-flow model reflects the anticipated *cash flows* of a project and should be prepared using the cash method of accounting where expenses and revenues are recorded when the transaction occurs. In comparison, the accrual method of accounting matches the revenues with the expenses or assets that are used to generate that revenue. In other words, the revenues and expenses are accrued or distributed over the period of time that they were earned or used and the recording of the revenues and expenses does not necessarily reflect the actual timing of the cash flow associated with the revenues or expenses. It is necessary to distinguish between the noncash accounting provisions and actual cash flows in the feasibility study cash-flow model. For example, the timing of an environmental bond may involve noncash accruals with specific cash-flow payments. The actual cash payments and deductions need to be reflected in the technical cash-flow model, which are not the same as the accrual totals.

In a feasibility study cash-flow model, noncash items, such as depletion and depreciation, are used only in the determination of income taxes. These noncash items are deducted in the final cash-flow analysis for the determination of the net present value (NPV).

Incorporation of the Key Cash-Flow Components

Generally, certain key components and inputs are expected in a technical cash-flow model. The following discussion outlines the generally accepted inputs for the technical cash-flow model as well as the general considerations that need to be made to appropriately incorporate these key inputs. The following list is provided as a guideline. It is by no means exhaustive and should be modified as appropriate for specific projects and commodities.

Cash-Flow Timeline

- The cash-flow model should reflect the life of a project from present day through the end of the project, with the inclusion of closure and long-term care requirements ("first to last dollar spent"). It should be noted, however, that the NPV can be determined for any period of time, not just the project life. Although this is not a standard approach in a feasibility study, this approach may be useful in determining the economic viability of an event in the future, such as a plant expansion or in the case of a shutdown and mothballing of a mine because of market or technical reasons.
- The cash-flow model should not include sunk costs; in other words, costs that have occurred in the past, unless service cash payments of these costs are ongoing and occur during the period of time modeled in the cash-flow model.
- A common practice is that the cash flows be determined on an annual basis; however, it is sometimes necessary and prudent to include in the cash-flow model either monthly or quarterly cash flows. This is especially true during the construction period where the cash flows can vary greatly from month to month or if a project has significant annual seasonality in either production or price.
- End-of-the-year cash-flow convention is typically used in calculating NPV. In this methodology, all of the annual cash flows are assumed to occur at the end of the year. This is appropriate for long-term projects where the annual cash flows do not have much variability from year to year. In projects with seasonality or variability from year to year, it may be more appropriate to use a midyear cash-flow convention where it is assumed that all of the cash movements occur in the middle of the year. As mentioned earlier, during the construction period, it may be more appropriate to determine the NPV on a month-to-month or quarterly basis. It should be clearly stated which practice is being used.

Reserves and Resources

- The cash-flow model should be based on the reported reserves and the mining and processing rates; however, with some caveats, the measured and indicated resources can be included if these resources are an integral part of the mine plan. Inferred resources are typically not included because of the high level of uncertainty. If measured and indicated resources are incorporated, it must be clearly indicated since resources have not been classified as reserves as they have not demonstrated economic viability. In some cases, the cash-flow analysis and inclusion of resources will be limited to a minimum standard as dictated by a particular entity, such as a stock exchange or regulatory agency.
- The tonnage and grade of the projected mined material that is incorporated into the model must reconcile with the reported reserves and resources of the project.

- It is useful to calculate the remaining reserves and resources on an annual basis as they are depleted.

Mine Plan and Mine Production Schedule

- The projected mine plan should be based and correlated to the reserve and resource model.
- The projected mine plan is used to generate the costs and revenues within the model and should include both ore and waste tonnages, strip ratios (as appropriate), dilution factors, mine recovery, and operating days. Equipment availability, capabilities, performance, and utilization should be reflected in the mine plan and forecasted production.
- The cash-flow model must include forecasted tonnage and grade of the mined material in appropriate time increments as described in the cash-flow timeline (i.e., monthly, quarterly, or annually).
- The mine plan should include an appropriate production ramp-up period. In other words, the cash-flow model and the mine production plan should reflect a reasonable anticipated time required to achieve the designed production output. This is an area that is commonly overlooked and minimized, yet it is critically important because NPV is most significantly affected by events in the near future rather than those in the distant future. It is unusual, and almost unheard of, for a mining project to "hit the start button" and to be operating at full production capacity immediately.

Stockpile Management

- As appropriate, stockpile management needs to be incorporated into the cash-flow model to account for any inventories that may be present in stockpiles between the mine and the mill. This is especially important during start-up periods because of the greater effect on NPV. The greatest changes in stockpile inventory typically occur near the beginning of the project or when production rates in either the mine or the processing facility change, but they can also occur in the future if changes are made to the production plan. Stockpile inventory changes indicate a change in the production costs and revenues.
- Stockpile inventories and accounting must also reflect any blending processes between the mine and the mill.

Processing and Product Production

- Processing feed tonnages and grades must be based directly on the mine plan and resource and reserve model.
- An appropriate ramp-up period should be incorporated into the cash-flow model that reflects the complexity of the process and equipment. A longer ramp-up period is necessary with complicated or untried processes and equipment.
- Milling and processing technical parameters (such as feed grade, metal recoveries, and concentrate grades) need to be confirmed through appropriate test work or historical operating data and should be appropriately based on representative samples.

- The processing technical parameters incorporated into the cash-flow model must reflect the processing flow sheet and equipment capabilities, as well as the mineralogical and metallurgical factors that affect metal or mineral recoveries.
- Equipment availability, capabilities, performance, and utilization should be reflected in the forecasted metallurgical and processing performance and production.
- Projected variability in the ore-body grade and other physical characteristics that may affect ore processing should be included in the cash-flow analysis and should be reflected in the timeline of the cash-flow model as well as the projected performance of the processing facility.
- Feed grades should not differ from what is available in the stockpiles or what is available during a period of time.
- The cash-flow model must include forecasted processing feed and product tonnage and grade in appropriate time increments (i.e., monthly, quarterly, or annually).

Product Marketability and Revenues

- The sales volume and quality of product available for sale must reflect the production capacities, outputs, and variations of the proposed processing facility. In the case of direct sales of run-of-mine ore, the quantity and quality of the produce must reflect the mine plan and resource and reserve model.
- Forecasted sales prices should be based on market studies, forecasted market conditions, sales or offtake contracts, or long-term historical price trends. Whatever the source, it must be clearly indicated.
  - The sales price assumptions must be clearly stated.
  - As previously stated, sales prices can be the most subjective input into the cash-flow model. The more definitive sales prices are those based on actual sales or offtake contracts while forecasted market conditions tend to be more variable from the prices that are realized in actuality.
  - Using averaged historical price trends is a common practice for projecting average sales prices going forward; however, it is important to use a sufficiently long enough time horizon to determine a reasonable average historical price. In the case of steady long-term prices, two or three years is sufficient. In the case of volatile markets, a much longer time horizon of 5–10 years may be more appropriate.
  - Although in reality, commodity prices vary over time, the cash-flow model should incorporate a steady or constant long-term commodity price. Nevertheless, it is an acceptable practice to incorporate consensus pricing in the first years of a model (typically, no more than three years). Consensus pricing is the average of the forecasted commodity prices compiled from many forecasters, such as banks and various analysts. Consensus pricing is available from several groups, including Consensus Economics (see www.consensuseconomics.com).
- Revenues should take into consideration key factors such as the size, nature, and location of markets.

- Revenue calculations must reflect the terms of any offtake agreements, such as smelter treatment and refinery charges, transportation costs (i.e., shipping terms of sale), and penalty components of price.
- Exchange rates, like forecasted sales prices, can be a subjective input that can have a significant effect on revenues as well as costs. The general practice is to use an averaged historical exchange rate as the basis for the feasibility study cash-flow model. Once again, the more volatile the exchange rate, the longer the time horizon needed to determine an average value.

Royalties

- Royalties are a payment to the owner of the mineral rights as an exchange for being able to extract the minerals. Generally, royalty payments are based on the earnings from production. In the cash flow, royalties are generally shown as a direct deduction from the revenues.
- Royalties can be paid to private owners, other companies who hold the mineral rights, or to governments.
- Royalties can vary by type of mineral and commodity prices. In some countries, royalties are based on a sliding scale, which changes according to variations in commodity prices. If this is the case, then the cash-flow model needs to appropriately address this requirement for the purposes of the sensitivity analysis.

Operating Costs

- The operating costs include direct cash production costs.
- The forecasted operating cost estimates should be based on supporting data such as engineering studies, historical operating costs, or comparable operations. The operating costs need to be delineated to within a target level of accuracy, depending on how much engineering has be expended on the study (i.e., within ±50% for a scoping-level study; within ±20%–30% for a preliminary-level study; ±15%–20% for a intermediate feasibility (or prefeasibility)–level study; and within ±10%–15% for a final feasibility–level study). As explained in Chapter 11, there should have been 6%–8%, 15%–20%, and 20%–30% of the engineering completed for the preliminary, intermediate, and final feasibility studies, respectively.
- Both fixed and variable or unit operating costs should be incorporated into the cash-flow model.
- Typically, operating costs are presented either under broad functional headings, such as mining costs, processing costs, transportation, environmental costs, and administrative costs, or under specific cost element headings, such as labor, consumables, fuel, utilities, and maintenance.
- Other factors that affect operating costs should be taken into account in developing the operating costs and their incorporation into the cash-flow model. These factors could include labor productivity, specific infrastructure costs (i.e., maintenance and operation of power generation facilities), and the effects of hydrological and climatic considerations. Equipment leasing terms should be included in the operating costs.

General and Administrative Costs

- General and administrative costs refer to the indirect costs associated with the operation of the project. These costs include head office costs, local office costs, accounting, management, and engineering.

Capital Costs

- Capital costs should include development costs (i.e., exploration and development studies), initial capital costs (i.e., construction and equipment), and long-term sustaining capital costs.
- Equipment capital costs should include all the associated costs, such as value-added taxes (VATs), sales taxes, and shipping.
- The forecasted capital cost estimates must be based on supporting data such as engineering studies or historical sustaining capital costs. The capital costs need to be delineated to within a target level of accuracy, depending on how much engineering has be expended on the study (i.e., using the same percentages listed in the second bullet of the preceding "Operating Costs" list). As explained in Chapter 11, there should have been 6%–8%, 15%–20%, and 20%–30% of the engineering completed for the preliminary, intermediate, and final feasibility studies, respectively.
- Both total and unit capital costs should be incorporated into the cash-flow model.
- Typically, capital costs are presented under broad functional headings, such as mining costs, processing costs, transportation, environmental costs, and infrastructure and services costs; however, capital costs can also be incorporated under specific cost element headings such as construction and equipment costs.
- The payment timing for the capital costs should be accurately reflected within the cash-flow model. In some cases, the actual cash payments for the capital costs may not be the same as the installation timing. This is especially true for preorders on long lead-time equipment.
- If the cash-flow model is constructed as a pretax cash-flow model, the classification of a cost as an operating cost or as a capital cost is not as critical as the classification of costs in an after-tax cash-flow model. In an after-tax cash-flow model, the cost classification is essential in determining the taxable income and the taxes. If a cost is to be expensed, it is classified as an operating cost. If a cost is to be amortized or depreciated over time, it is to be classified as a capital cost.

Closure Costs and Salvage Value

- Closure costs incurred throughout the life of the project should be modeled. This includes cash costs associated with reclamation bonds and the final closure and long-term maintenance costs. As regulations and requirements continue to become more stringent, the cost of closure is becoming an integral part of the development and construction costs. As a result, cash closure costs (i.e., reclamation bonds) are being incurred more and more toward the beginning of a project, which can have a more significant effect on the NPV than if the cash closure costs were incurred solely at the end of the project.
- Salvage value is the recoverable salvage value of the mining and processing equipment, spares, and structural and systems equipment. In longer-term projects, it is generally

assumed that the salvage value is zero; however, in short-term projects where the equipment has remaining, useful value, the salvage value can have an effect on the cash-flow analysis. This value can be difficult to assess and may require quotes from equipment manufacturers or salvage experts.

Depreciation, Amortization, and Depletion

- Depreciation, amortization, and depletion are noncash costs that only affect the payable income taxes.
- Depreciation, amortization, and depletion are determined according to the rules and regulations of the particular country where the project is located and taxed.
- Depreciation is the tax deduction that allows the recovery of tangible asset investment costs over the tax life of the asset.
  - *Tax life* is defined as the time that an asset can remain in service and is dictated by the particular tax rules of a country. In the United States, some assets can have a tax life as long as 39 years, although land improvements associated with the mining project (i.e., docks, plant roads, and rail spurs) have a tax life of 15 years and mining equipment has a tax life of 7 years.
  - In determining which expenditures can be depreciated, the rule of thumb is that depreciation is generally applied to equipment and other assets that could be sold or salvaged at the end of the project life (regardless of their actual condition or usefulness).
  - In the cash-flow model, depreciation begins when the asset is placed into service, not when it is purchased.
  - At the end of the project, the cumulative remaining depreciation is added to the income as if the asset had been sold.
  - Depreciation rules vary from country to country. In the United States, most assets used in mining operations are depreciated over a seven-year period using the modified accelerated cost recovery system. This method has higher deduction amounts in the early years of the tax life of the asset.
  - Other general depreciation methods include the straight line method and percent business use method. The straight line method allows the same annual deduction over the tax life of the asset. In the mining industry, the percent business use method is based on units of production and is one of the simplest methods for determining depreciation. This method spreads the depreciation over the life of the project and is one of the more conservative approaches.
- Amortization is similar to depreciation in that amortization spreads the cost of an asset over the period of its useful life. The difference is that amortization is used for intangible assets such as development costs. These can be costs associated with activities such as prestripping, underground development, feasibility studies, and other assets that do not have a separable life from the project (i.e., assets that cannot be salvaged at the conclusion of the project).
  - Typically, amortization has a shorter time frame than depreciation. For example, in the United States, assets can be amortized over a period of only five years.

- In some countries, the governments recognize the value of nonrenewable resources and that mining depletes the resource, which can only be replaced by purchasing another deposit. Depletion is a nontaxable recovery of the value of the resource.
  - In the United States, there are typically two methods for determining depletion: cost depletion and percentage depletion.
  - *Cost depletion* is based on the previously expensed exploration costs and is deducted after the cumulative amount of depletion exceeds the exploration costs.
  - *Percentage depletion* is based on a commodity-driven percentage of the revenue (less any royalties) and is capped according to a specified percentage of the taxable net income. In the United States, the percentage is based on the commodity type (e.g., copper, gold, and silver is 15% of the gross revenue, while uranium is 23%.

Taxes

- When referring to taxes in a cash-flow model, the reference is to income taxes. Other taxes, such as employment taxes, sales taxes, and VATs, should be incorporated as appropriate into the operating and capital costs. Property taxes and provincial or state income taxes may require additional handling in the cash-flow model, especially if these taxes result in significant tax costs to the project.
- Different countries have differing rules regarding the handling of tax losses and tax losses carried forward.
- Income taxes could account for a significant portion of the cash flow and could have a profound effect on the economic viability of the project. In reality, the taxes modeled in the cash-flow model will likely differ from the taxes actually paid because of accounting methods of accrual and corporate tax structures. A company will often use the cash-flow model as a basis for its decision to invest. This is appropriate for individual companies, as different companies have a different tax basis.
- In the cash-flow model, the purpose of the tax determination is to show the effect of the taxes generated *by the project* on a stand-alone basis. The taxes should be calculated in as straightforward of a method as possible to reflect the general tax laws of a country.

Changes in Working Capital

- Changes in working capital take into account the current assets (accounts receivables and inventory, i.e., raw materials, supplies, and materials in process) and current liabilities (accounts payable) associated with a project. The purpose of applying changes in working capital to the cash-flow model is to more accurately reflect the liquidity and timing of the cash flows. For example, when a product is produced and sold to a customer, it is rare that the transaction would be immediate because there is generally a billing period between the time of delivery and payment.
- Often in classifying initial capital costs, there is a cost category referred to as *working capital.* These costs include items such as first fills and the purchase of initial spares. Although this requires working capital, these costs are generally not treated as the working capital (current assets and current liabilities) incorporated into the cash-flow model

to adjust the cash-flow timing. These costs are included in either the capital costs or operating costs, depending on whether these costs are capitalized or expensed.

- Current assets and current liabilities are often determined on the basis of a delay in the payment of revenues from a customer and the delay in paying costs to creditors. These delays can be as little as 10 to 15 days or as great as three to six months.
- Changes in working capital can have a significant effect on the cash flow of a project, especially if there are significant changes in production quantities and sales, commodity prices, or operating costs over time. In a typical project, changes in working capital are most noticeable during the start-up stages of a project. Once steady-state operations are achieved, changes in the working capital tend to be minimized.
- Changes in working capital is the difference between the current assets (accounts receivables and inventory, i.e., raw materials, supplies, and materials-in-process) and current liabilities (accounts payable). For ease of analysis, it is often easier to treat the current assets and current liabilities separately in the cash-flow model rather than combining them under a heading of "changes in working capital."
  - Typically, changes from one period to another in the current assets results in an opposite-direction operating cash-flow adjustment. Conversely, changes from one period to another in the current liabilities results in the same-direction operating cash-flow adjustment.
  - Accounts receivable represent sales that have been made, but the cash has not yet been collected. Since the cash method of accounting assumes that there is no delay between production and recognition of revenues, adjustments are necessary to the cash flow to reflect the delay in cash collection. Increases (decreases) in accounts receivable from one period to the next require a downward (upward) adjustment to the cash flows from operations. In other words, less cash has been received in the later period than the former period, so the revenues reflected in the income statement need to be adjusted downward to reflect the delay in cash collection.
  - Increases (decreases) in the inventory of supplies or raw materials used in the operations is an increase in the actual cash spent on the cost of goods sold (especially if the operating costs are determined on a unit production basis) and as a result, a downward (upward) adjustment in the operating cash flow is necessary.
  - Prepayments and accounts payable (such as employee wages that have been earned but are unpaid) result in higher actual cash flows since the cash has not been paid out. As a result, increases (decreases) in the accounts payable result in upward (downward) adjustments to the operating cash flow.

Net Income and Cash-Flow Determination

- The cash flow is based on the results of the net income where the operating costs and expenses are deducted from the revenues. The *net income after taxes* (if determined on an after-tax basis) is reconciled by adjusting for three items: noncash items such as depreciation, working capital adjustments to reflect the actual cash-flow timing, and nonoperating transactions such as capital costs.

- The typical components and calculation methodology for net income and cash-flow determination are shown in Table 20.1. The treatment of some of the components may be different for different operations, depending on the location of the operation and the rules and regulations of particular countries.

TABLE 20.1 Example calculation of net income and annual cash flow

Calculation	Component
	Revenue
Less	Royalties (if applicable)
Equal	Gross revenue
Less	Operating costs
Equal	Net operating income
Less	Other costs
Less	General and administrative costs
Equal	Net income (EBITDA*)
Less	Depreciation and amortization
Less	Depletion
Equal	Net taxable income
Less	Local and property taxes
Less	Income tax
Equal	Net income after taxes
Add noncash components	Depreciation, amortization, depletion
Equal	Operating cash flow
Less	Nonoperating transactions†
Less (plus)	Increases (decreases) in accounts receivable‡
Less (plus)	Increases (decreases) in inventory‡
Plus (less)	Increases (decreases) in accounts payable‡
Equal	Net annual cash flow (unleveraged)

*EBITDA = earnings before interest, taxes, depreciation, and amortization.
†Typically capital and development costs.
‡Collectively, changes in accounts receivable, inventory, and accounts payable are considered to be changes in working capital.

NET PRESENT VALUE, INTERNAL RATE OF RETURN, AND DISCOUNT RATE

The use of NPV and internal rate of return (IRR) is a common method within the minerals industry for evaluating project economics and for comparing and selecting viable projects. Both NPV and IRR recognize the time value of money at a chosen discount rate and translate the future annual cash flows into today's dollars. The NPV reflects the potential value of the project* and the IRR represents the potential economic rate of return. These values are consid-

* The evaluation of a project or a company should not be based solely on the NPV. An evaluation should be based on the results from several valuation methodologies, such as market comparisons and market multiples. [Editor's Note: The author is absolutely correct about the valuation of a property or project. But there is a distinct difference between *valuation* and *evaluation*, which was discussed earlier in this handbook. Normally for an evaluation, NPV and the rate of return, plus sometimes the payback period, are sufficient.]

ered to be far superior in evaluating the economics of a project than more simplified values of payback periods or cost and revenue ratios.

There are numerous textbooks dedicated to the detailed financial theory and the mathematics and calculation of NPV and IRR (Stermole and Stermole 1987). As such, these calculations are not discussed in detail in this text.

Key to the determination of the NPV and IRR is the discount rate applied to the annual cash flows. In the simplest of terms, the discount rate represents the risk or uncertainty of future cash flows. The higher the discount rate, the higher the uncertainty of the cash flows. This results in a dollar tomorrow being worth less than a dollar today. Discount rates can be calculated using several methodologies that are based on financial theory. The discount rate must be consistent with the methodologies used in the cash-flow model itself and are adjusted accordingly to the use of real and nominal dollars whether the cash flow is based on a pretax or after-tax basis. Financial theory methods tend to rely mostly on the use of the corporate cost of capital, also referred to as the weighted average cost of capital (WACC). This method is based on the proportional cost of equity capital, debt, and preferred stock.

According to Smith (2011), it is a more general practice of mining companies to use a base discount rate of 10% for projects at the feasibility study level* rather than applying a discount rate calculated using WACC or some other financial theory. Smith has based this conclusion on surveys and published evaluations. Smith states that "since this rate is used by major mining investors to make decisions that involve millions of dollars, it must be felt to have validity." It should be noted that a higher discount rate would be expected for projects at less than a feasibility study level or those with greater variability and risk.

Regardless of the base discount rate chosen to evaluate a project, it is prudent to use a range of discount rates to evaluate the economic viability and variability of a project.

SENSITIVITY ANALYSIS

Sensitivity analysis is used to evaluate the robustness of a project's economics to changes in critical input variables. This analysis allows for the introduction of what-if scenarios to various parameters such as production rates, commodity prices, price cycles, operating costs, and capital costs. The sensitivity analysis contributes to the understanding of the effects of uncertainty and the risk associated with the project.

Basic Sensitivity Analysis

The most basic sensitivity analysis is the single-variable analysis where one input value is varied from the base case value. This input value can be varied using a set value, such as moving the gold price from $1,000/oz to $1,200/oz, or by changing the variable on a percentage basis, such as a ±20% change in commodity price. The simplicity of this analysis is the assumption that all the variable parameters are independent of one another. In reality, this is rarely true. If commodity prices drop, there is a likely chance that the mine plan will change and operating costs will likely be reduced in an effort to maintain profitability. As such, the objective of this type of analysis is to determine the parameters that have the greatest effect on the NPV and IRR of a project.

* This is applied to project evaluations conducted in constant dollars, at 100% equity with no debt, on an after-tax basis.

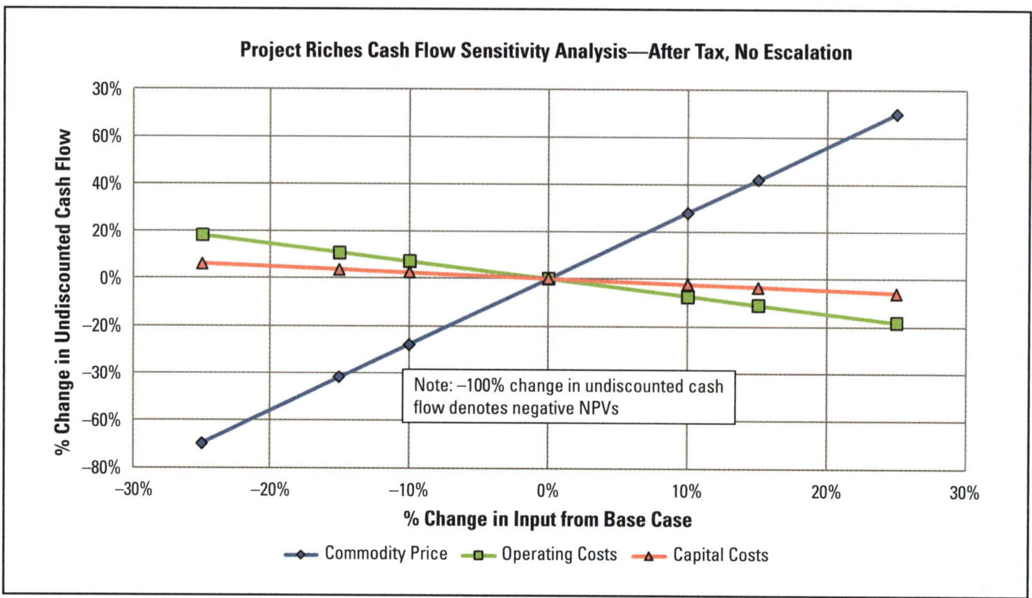

FIGURE 20.1 Spider diagram showing sensitivity to changes of commodity price, operating costs, and capital costs

A variation on the single-variable analysis is to incorporate a likely range of inputs rather than changing a single input value. The range should be estimated on the basis of the best and worst expected values. Once again, this can be accomplished by varying specific values on an annual basis or by using a percentage basis with the input being varied, for example, by ±5%, ±10%, and ±25%. The results from this type of range analysis can be plotted using a spider diagram (Figure 20.1). The spider diagram allows for a visual analysis to determine the most important variable of those analyzed.

It should be noted that in most mining project cash-flow models, the input parameter that has the greatest effect on the NPV and IRR is commodity price followed by operating costs and then capital costs. Varying production output will generally have the same effect as varying the commodity price.

Scenario Analysis

The scenario analysis is based on a specific set of inputs for different scenarios, such as the analysis of two different mine plans or the analysis of two or more processing approaches. In this instance, this analysis allows for the comparison of varying operating scenarios. Other possible comparisons could include delaying or speeding up the construction and start-up of a project or developing a staged start-up plan or the review of future production expansions. Typically, this approach is used for feasibility study alternatives and is based on additional engineering studies where the effect of changing one input variable is taken into account in the other inputs to the cash-flow model.

Monte Carlo Simulation

In each of the techniques previously outlined, the probability of the occurrence of the change in the input variable is not taken into account. For example, let us assume for a particular project that a 20% change in the operating costs results in a negative NPV value, but what is the probability that the operating costs will change by 20%?

Monte Carlo simulation determines a range of possible outcomes by using a computerized iterative process that incorporates probability distribution curves for input parameters that have a level of uncertainty. Monte Carlo simulations allow for multiple variables to be changed simultaneously while a specific calculation is mathematically performed literally thousands of times. The outcome is a distribution of possible results as determined from a range of probabilities assigned to each variable in the analysis (i.e., capital and operating costs and commodity prices). The probabilistic distribution can be used to arrive at a most likely value, or range of values, as based on iterations of cases that test the limits of each input variable.

The use of probability distributions for the input variables allows for a more realistic approach in describing the uncertainty associated with the variable. The distributions of values for a particular value are plotted with the values on the horizontal axis and the probability of the occurrence of the value on the vertical axis. The shape of the resulting probability curve is determined by the nature of the variable described. One of the most common probability distribution curve shapes is the normal or the bell curve, which shows a mean value with standard deviation values used to describe the variation from the mean. It is a symmetric distribution with the values in the middle most likely to occur. Depending on the nature of the variable, it is possible for the curve to be skewed where the distribution of values is not evenly distributed around a mean. Another example of a distribution curve is a triangle curve where three points are plotted: a mean, a minimum value, and a maximum value.

The type of distribution curve is dependent on the variable. The challenge in a Monte Carlo simulation is determining the shape of the probability curves and the degree of deviation for the inputs. Much study has been done regarding the shape of the probability curves for various inputs. There are several software programs currently available that can be used as tools in determining both the shape and variability of the input variables chosen for the analysis (e.g., the Oracle Crystal Ball is one such software application).

In the Monte Carlo simulation, values are randomly selected from each of the distribution curves for each variable that is being analyzed. The selected values are input to the cash-flow model to determine a single NPV for that combination of variable values. This process is repeated thousands of times and a probability distribution curve is generated for the NPV. From the NPV probability distribution curve, it is possible to determine the most likely median NPV and the potential for deviation from that value.

KEY MODELING PRINCIPLES

The construction and architecture of the feasibility study cash-flow spreadsheet is as important as the technical inputs into the model. The spreadsheet model is the mechanism by which the technical inputs are combined with the economic factors to present the overall "view" of the mineral project being addressed in the feasibility study.

Several key principles are to be followed when constructing a cash-flow spreadsheet model:

1. Simplicity
2. Materiality
3. Transparency
4. Consistency

5. Ease of auditing

These five key principles are the basic, universal tenets that apply to all spreadsheet models, regardless of the nature of the project being studied. The appropriate application of these five principles to a spreadsheet model determines how effectively and how accurately the model reflects the project being studied.

Simplistic and Easy to Follow

The feasibility study cash-flow model will be used by many audiences as a project moves toward development and, ultimately, operations. The audience will likely include both technical and nontechnical personnel. Some of these people will have extensive mining backgrounds, while others may have limited backgrounds in the nuances of mining industry terminology and technology. For these reasons, a successful cash-flow spreadsheet is simplistic and easy to follow by all audiences, regardless of their background and expertise.

An effective spreadsheet model is straightforward, intuitive, and visually easy to follow. The spreadsheet and overall model (if multiple worksheets are incorporated into a single workbook) should have a clear visual flow, both across and down the individual worksheets and between the worksheets within the workbook. Headings and subheadings within the spreadsheet should flow similar to a document outline. In a way, a cash-flow spreadsheet is a bit like a book where the "narrative" moves in a logical direction from beginning to end.

The spreadsheet model should be uncluttered. Bright colors, different colored fonts or highlighted rows, and different fonts and text sizes are confusing and distracting. Nevertheless, colored rows or different colored text can be useful in breaking the subsections of a worksheet and highlighting input data (typically done with blue text) or questionable data (typically done with pink text); however, less is always more when it comes to creating a simple, elegant spreadsheet model.

The calculations within a cash-flow spreadsheet should be presented in smaller, "bite-sized" pieces. Complex calculations are better presented as a progression of easy-to-follow steps that lead to a final result rather than a single, one-line, complex calculation. This step-by-step approach accomplishes two things: less opportunity for calculation omissions and errors and a logical flow in achieving the final results.

Materiality

By virtue of the engineering process, cash-flow models used in evaluating feasibility studies are based on potentially large amounts of data and potentially complex background computations. An example of this would be the exploration and drill-core data and calculation process used to generate the resource and reserve model. Another example would be the information used to generate the unit operating costs; unit operating costs are determined using several inputs, including mine rates, fuel costs, equipment sizes, haulage distances, and reagent costs, as well as other data. Although critical in determining the input parameters for developing the cash-flow model, the background data and computations are generally not included in the cash-flow spreadsheet workbook unless specific inputs could materially affect the results or if they are key to analyzing alternative scenarios. In most cases, the input data should be material to the cash-flow model itself and should present the larger picture rather than the minutiae required to determine many of the inputs.

Materiality is also important in the nature of the computations incorporated into the cash-flow spreadsheet. Overly detailed calculations can be distracting. The computations should focus on the key technical and economic drivers, and the attention given to items with a lesser impact should be minimized. As shown in most sensitivity analyses, revenues, followed by operating costs and then capital costs, tend to have the greatest effect on the NPV. As such, greater focus should be given in the cash-flow spreadsheet to the items that affect revenues (production and commodity prices) with a lesser focus to items that affect costs. A more simplistic approach to incorporating operating and capital costs, as well as taxes, may be more appropriate in some cases if these parameters have minimal impact to the project's economic viability.

The purpose of the model also affects the degree of materiality of the spreadsheet model. The size and complexity of the model should be driven by level of study. In other words, a more complex model would be expected for a feasibility study level of detail and a simpler model would be expected for a preliminary economic assessment (PEA). An overly extensive cash-flow model for a PEA could imply a level of accuracy and input materiality that probably does not exist.

Transparency

It is critical that the source of the cash-flow inputs be transparent and based on relevant data and background determinations. The sources of information should include, at a minimum, a document source and the date of that source. Other information might include the person accountable for the information. If assumptions have been made because of lack of solid information, it is important to include a description and justification for those assumptions.

It is normal to include extra rows to record the source (simpler and smaller models, such as a scoping study) or to include and reference a separate worksheet within the workbook that contains the input sources and justification of any assumptions (more complex models for more developed projects). As updates to the model are completed, it is important to update the references, and it is useful to maintain a log that records the updates and changes.

Consistency

Consistency within a particular spreadsheet and throughout an entire workbook contributes to the ease of use and understanding of a model. Consistency is especially important for the following items:

- Calculation methodologies must have continuity.
- For decimal and significant figure conventions, beware that the number of decimal places appropriately reflects the accuracy of the inputs and calculation results (e.g., 0.07 oz per ton of gold is quite different than 0.1 oz per ton of gold).
- Abbreviations and nomenclatures must be used consistently.
- Spreadsheet layouts between worksheets, including column headings, should be cohesive.
- Formatting throughout a spreadsheet and between worksheets in a workbook should be consistent, including fonts, row heights, and column widths.

In addition to the need for consistency within the spreadsheet itself, it is necessary to confirm that the inputs to the spreadsheet model are consistent with the information and findings presented in the rest of the feasibility study. During the preparation of the feasibility study, it is not uncommon for many of the inputs to undergo several iterations before the final number is determined. Often these iterations become more fast and furious toward the end of the feasibility study preparation. As a result, it can be difficult for the cash-flow modeler to keep up with the iterations, and the cash-flow model can become out of sync with the rest of the report, thus creating discrepancies between the cash flow and the feasibility study. This obviously drives part of the need for the final key principle—the auditing process.

Ease of Auditing

If the other four key principles are adhered to, most spreadsheet models will be easy to follow and to audit. Spreadsheet auditing is an essential part of the process of building a cash-flow model that is an integral part of a feasibility study. The model should be audited by the other members of the feasibility team as well as an independent reviewer. The audit should include calculation methodologies, data input, and formatting.

SPREADSHEET AND WORKBOOK ARCHITECTURE

How the workbook and individual worksheets (spreadsheets) are designed and laid out and how the information and computations flow from one place to another is considered to be the *architecture* of the model. The architecture should reflect the key principles just outlined. Following are some generalized concepts and guidelines for the mechanics and architecture of a standard cash-flow spreadsheet model. Obviously, because of the unique nature of every mining project, some of these precepts may not apply or may need to be modified to fit the situation and project. In all cases, however, the key principles just described should be the driver in determining the architecture of the model.

Workbook Layout

The basic structure of the spreadsheet model starts with the overall layout of the model, which can include a single worksheet or multiple worksheets. No matter the number of worksheets, it is essential that the flow between the worksheets is apparent. In many cases, the workbook will evolve as the level of accuracy of the underlying engineering increases.

Single Worksheet Versus Multiple Worksheets

Single worksheet models are most appropriate for PEAs and scoping studies. Not much information is available for the engineered inputs, and the computations tend to be more simplistic. For instance, a PEA would not be expected to have much information regarding the management of stockpiles between the mine and the processing facility. Conversely, in a feasibility study, a more detailed production plan would be expected to have been developed, which could include stockpile management. In this case, it might be more advantageous and easier to understand the model if a separate worksheet is included in the workbook for the computations associated with the stockpile management.

Modular Design and Logical Sequence of Multiple Worksheets

The use of modular worksheets allows for greater simplicity, flexibility, and auditing. In a modular workbook, each worksheet is used to represent the various *unit operations* of the cash-flow model. Each modular worksheet has a clear purpose. Using a modular design allows for several generally simplistic, smaller spreadsheets to be constructed rather than one large, potentially complicated spreadsheet. This design allows for easier modifications as the overall model evolves.

For example, a copper mine cash-flow model prepared at the feasibility study level might have the following modular worksheets:

- Introduction
- Summary of results and sensitivity analysis
- Assumptions and inputs (includes the source and description of the inputs and justifications for the assumptions)
- Net income and cash flow
- Reserves and mine plan
- Stockpile management
- Processing and production
- Net smelter returns
- Operating costs
- Capital costs, depreciation, and amortization
- Depletion
- Taxes
- Working capital

Note how the flow of the modular worksheets follows both the flow of the mining process (reserves → mine plan → processing facility → final production) and the flow of the cash-flow determination (revenue generation → operating costs → capital costs and depreciation → taxes → working capital). Each worksheet builds on the results from previous worksheets, and the results feed into the net income and cash-flow and summary worksheets.

The Introductory Worksheet

All cash-flow models should include an introductory worksheet. This worksheet should include

- The purpose of the model;
- The name and contact information of the modeler;
- An explanation of the worksheets;
- A diagram of the flow of information and computations between the worksheets;
- Dates of modifications and audits;
- Any limitations;
- Description of formatting conventions used throughout the model, such as font colors for input data or the number formatting and rounding practices; and
- Statement of standards used throughout the model, including the use of real versus nominal dollars and whether the model is a pretax or after-tax model.

The Summary Worksheet

The purpose of the summary worksheet is to provide the user with a single source of the key findings. This is particularly important when the model is constructed using modular worksheets or for more complex single-worksheet models. The key findings can be presented on an annual or per-period basis as done throughout the rest of the model or as the total and average values of all the years or several different periods of time. For example, the totals could be presented for the construction period, the first five years of operations, the remaining years of production until the reserves are depleted, and, finally, the closure period. Typically, the summary should include, at a minimum, the following findings in tabular form:

- Available resources and reserves
- Grade and tonnage of mined material
- Totals of product produced
- Commodity price(s)
- Revenues
- Average unit operating costs
- Earnings before interest, taxes, depreciation, and amortization, or EBIDTA
- Net income
- Capital costs
- NPV (at varying discount rates, including a 0% discount rate)
- IRR

Graphical outputs can also be included to provide clarification of the timing of production and cash flows.

The summary worksheet can also include the results of the sensitivity analysis. These results could include a summary table and graphical presentations, such as a spider diagram.

Worksheet Layout

Like the worksheets in a multiple-worksheet workbook, the flow within each worksheet should follow a logical sequence across and down the worksheet.

Title

Each worksheet should have a clear title in the upper left-hand corner containing the project name, purpose or title of the worksheet, the version number of the model, and the date.

Rows and Columns

Normally most cash-flow models are constructed with the time periods across the worksheet with a single year or time period per column. The time periods should be consistent between each worksheet within a workbook and should be located in the same column and row in each worksheet.

The headings or data ranges are listed down the worksheet with a single parameter per row. These headings will vary between the different worksheets as they are specific to the input or result of the particular unit operation of the cash-flow model.

Row and columns should not be hidden within a worksheet. Hidden rows and columns can result in calculation errors if formulas are copied across a row or down a column. Additionally, the use of hidden rows and columns compromises the transparency of the model.

Work Blocks with Subheadings

Work blocks are used in a spreadsheet to separate different functions or unit operations. The objective is to present complicated processes in simple bite-sized pieces. This allows for straightforward modifications, quicker audits, and easier-to-understand computational processes. For example, the work blocks in the processing worksheet for a copper mine could be divided as follows:

- Leaching
- Solvent extraction and electrowinning (cathode production)
- Flotation and concentrate production

The work blocks should be clearly identified and labeled and should follow the flow of the physical processes.

Totals and Averages

As a general rule, totals and average values should be immediately apparent within the spreadsheet. Rather than placing the totals on the right side of the spreadsheet, it is more common to use a column on the left side of the spreadsheet for the totals and averages. In this way, the user does not have to scroll to the far side of a spreadsheet to see the totals. In most cases, nearly every data range should be totaled or averaged. The total column should be highlighted using bold fonts or borders.

An example worksheet is presented in Figure 20.2 showing the worksheet title, column and row titles, work blocks, and totals.

General Calculation Guidelines

Consistent calculation methodologies help to ensure that key modeling principles are adhered to.

Calculate Only Once

Each computation should be done only once. If a calculated value is used in several worksheets, it should be calculated in only one worksheet, preferably in the appropriate worksheet specifically designated for that type of value. The result from that worksheet should be linked to other worksheets as necessary. This avoids the potential for miscalculations. For example, the annual cathode production should be computed in the processing and production worksheet and the result linked to the net smelter return and summary worksheets.

Consistent Formulas

The formulas within a row should be the same for each column. If the formula is different in the first couple of columns, there is a possibility that the wrong formula could be copied across a row. Additionally, the use of different formulas can make the auditing process more difficult. In some cases, it may be necessary to use different formulas for the first couple of years. If this is the case, the cells should be clearly highlighted using a different colored font. It

Row	Item	Unit	Total/Avg	Year -2	Year -1	Year 1	Year 2	Year 3	Year 4	Year 5	Year 6	Year 7	Year 8	Year 9	Year 10
1	Project XYZ														
2	Feasibility Study														
3	Mine and Mill Production Worksheet														
4	Verson 5.1, October 1, 2016														
6	US$														
7–8		Year	Total/Avg	Year -2	Year -1	Year 1	Year 2	Year 3	Year 4	Year 5	Year 6	Year 7	Year 8	Year 9	Year 10
9	**Total Mine Movement**														
10	Ore Mined														
11	Tonnage	kilotons	10,680	-	-	780	1,100	1,100	1,100	1,100	1,100	1,100	1,100	1,100	1,100
12	Grades														
13	Silver	g/t	110.1	-	-	104.8	101.7	111.8	105.5	98.7	123.2	124.7	110.9	105.4	112.3
14	Zinc	%	2.570	-	-	2.542	2.840	2.417	2.200	2.528	2.938	3.109	2.019	2.504	2.594
15	Copper	%	0.660	-	-	0.575	0.598	0.758	0.534	0.590	0.790	0.779	0.839	0.504	0.609
16	Lead	%	0.366	-	-	0.317	0.382	0.358	0.322	0.396	0.481	0.404	0.271	0.333	0.381
18	Metal Content														
19	Silver	kg	1,175,386	-	-	81,737	111,900	123,030	116,000	108,580	135,496	137,164	122,006	115,942	123,531
20		koz	37,790	-	-	2,628	3,598	3,956	3,729	3,491	4,356	4,410	3,923	3,728	3,972
21	Zinc	kilotons	274.5	-	-	19.8	31.2	26.6	24.2	27.8	32.3	34.2	22.2	27.5	28.5
22	Copper	metric tons	70,493	-	-	4,486	6,574	8,340	5,878	6,494	8,685	8,566	9,232	5,539	6,698
23	Lead	metric tons	39,090	-	-	2,469	4,205	3,936	3,545	4,354	5,293	4,447	2,986	3,662	4,193
25	**Processing Summary**														
26	Silver-Lead-Copper Concentrate to Lead Smelter														
27	Tonnage	tonnes	53,895	-	-	3,870	5,300	5,590	5,270	4,935	6,200	6,255	5,550	5,300	5,625
28	Grades														
29	Silver	g/t	12,202	-	-	12,000	12,000	12,000	12,000	12,000	12,000	12,000	12,000	12,000	12,000
30	Lead	%	15.55	-	-	15.55	15.55	15.55	15.55	15.55	15.55	15.55	15.55	15.55	15.55
31	Copper	%	9.00	-	-	9.00	9.00	9.00	9.00	9.00	9.00	9.00	9.00	9.00	9.00
33	Metal Content														
34	Silver	kg	646,740	-	-	46,440	63,600	67,080	63,240	59,220	74,400	75,060	66,600	63,600	67,500
35		koz	21,143	-	-	1,641	2,247	2,157	2,033	1,904	2,392	2,413.2	2,141.2	2,044.8	2,170.2
36	Lead	metric tons	8,379	-	-	601	823	869	819	767	964	973	863	824	875
37	Copper	metric tons	4,851	-	-	349	477	503	474	444	558	563	500	477	506
39	Recovery														
40	Silver	%	56.0	-	-	62.5	62.5	54.5	54.5	54.5	54.9	54.7	54.6	54.9	54.6
41	Lead	%	21.4	-	-	24.4	19.6	22.1	23.1	17.6	18.2	21.9	28.9	22.5	20.9
42	Copper	%	6.9	-	-	7.8	7.3	6.0	8.1	6.8	6.4	6.6	5.4	8.6	7.6
44	Copper Sulfide Concentrate to Copper Smelter														
45	Tonnage	metric tons	150,885	-	-	9,900	15,000	17,650	12,500	13,800	18,400	18,135	19,550	11,850	14,100
46	Grades														
47	Silver	g/t	2,700	-	-	2,700	2,700	2,700	2,700	2,700	2,700	2,700	2,700	2,700	2,700
48	Copper	%	25.00	-	-	25.00	25.00	25.00	25.00	25.00	25.00	25.00	25.00	25.00	25.00
50	Metal Content														
52	Silver	kg	407,389	-	-	26,730	40,500	47,655	33,750	37,260	49,680	48,964	52,785	31,995	38,070
53		koz	13,098	-	-	859	1,302	1,532	1,085	1,198	1,597	1,574	1,697	1,029	1,224
54	Copper	metric tons	37,721	-	-	2,475	3,750	4,413	3,125	3,450	4,600	4,534	4,888	2,963	3,525
56	Recovery														
57	Silver	%	34.7	-	-	32.7	36.2	38.7	29.1	34.3	36.7	35.7	43.3	27.6	30.8
58	Copper	%	53.5	-	-	55	57	53	53	53	53	53	53	53	53

FIGURE 20.2 Example worksheet showing worksheet title, column and row titles, work blocks, and totals

is recommended to insert a comment for those cells, providing an explanation for the deviation from the formulas used in the rest of the row.

Step-by-Step Approach

Simpler is better. It is more desirable to have each data range (or row) present a single computation than to have a complex algorithm that includes multiple functions. By using a step-by-step computational approach, the overall spreadsheet is more transparent, what might be thought of as obvious steps are not skipped, there is less chance for errors or omissions to be built into the calculation itself, and the model becomes more conducive for all users. This approach should also be applied to changes in units.

Conversion Factors and Decimals

Conversion factors should be found in one place within the workbook and linked as necessary. It is not uncommon to use names for the conversion factors.

Multiplying by 1,000 or other factors to adjust decimals should be avoided. There is too much chance for error. It is better to take advantage of the number formatting features of the spreadsheet program. The number formats should be consistent throughout the entire workbook. For instance, if the revenues are reported in thousands ($000s), then the operating costs, capital costs, net income, cash flows, and so forth, should also be reported in thousands ($000s).

Importing Data from External Workbooks

As a general rule of thumb, importing or linking data from external workbooks is not recommended and should be used with caution, for several reasons:

- If the user does not have the external workbook, the link may not work and the data will not import properly.
- Changes to the external workbook may not get captured in the current version of the cash-flow model.
- Auditing is difficult.

It is better to manually enter or copy the results from another workbook and to reference that workbook and not link to another workbook. The reference should include the workbook name, date, and version number. The reference could also include the cell, row, or column number for the input data.

Sensitivity Analysis Calculations

A completed model should include built-in calculation mechanisms for the sensitivity analysis. This may be as simple as multiplying the input that is being analyzed by a sensitivity factor that is entered into a single reference cell. The reference cell value can be manually changed to generate a single result. This is typically done for basic sensitivity analyses that are varied by certain percentages (e.g., a 10% increase in commodity price) or for scenario sensitivity analyses where the actual input value is changed (e.g., gold price is changed from $1,000/oz to $1,200/oz). The model should be built so the input that is being analyzed is always adjusted to take into account the reference cell value (see Figure 20.3). Figure 20.3 is also an example of an

XYZ Gold Corporation
Riches Gold Project - Somewhere in the World
PEA Cash Flow Model - Assumptions and Inputs
Version 3.1 - August 1, 2016

General Assumptions

Item	Value	Unit	Source
Updated	1-Aug-16		
Conversion factors	troy oz » g	31.1034768	
Start of model	Year 0		
Start of oxide production	Year 1		
Transition to sulfide	Year 2		
Start of full sulfide production	Year 3		
Life of mine	10	years	*PEA study prepared by AAA Engineering Inc., dated July 15, 2016*
Ore tonnage	22,000,000	metric tons	*PEA study prepared by AAA Engineering Inc., dated July 15, 2016*
Waste tonnage	100,000,000	metric tons	*PEA study prepared by AAA Engineering Inc., dated July 15, 2016*

Processing Parameters

Item	Value	Unit	Source
Oxide ore	3,500,000	metric tons	*PEA study prepared by AAA Engineering Inc., dated July 15, 2016*
Initial year feed	1,250,000	metric tons	*PEA study prepared by AAA Engineering Inc., dated July 15, 2016*
Steady state feed rate	2,500,000	metric tons per year	*PEA study prepared by AAA Engineering Inc., dated July 15, 2016*
Gold recovery - oxide	88.0%		*Metallurgical study from TESTING, Inc. laboratory dated May 31, 2016*
Gold recovery - sulfide	93.7%		*Metallurgical study from TESTING, Inc. laboratory dated May 31, 2016*

Sensitivity reference cell. This cell is combined with the base case metal price in the rest of the model to allow for the sensitivity analysis. For the base case, the percentage will be set to 0%.

Sales and Revenues

Item	Value	Unit	Source
Base Case metal price	1,000	US$/oz Au	*Average historical gold price*
Sensitivity factor	0%		
Refining, transportation, insurance and sales	1.20%	of revenues	*Preliminary quote from refiner, May 1, 2016*
Land owners royalty	1.50%	of revenues	*Contracted royalty with land owner, dated January 15, 2015*

Operating Costs

Sensitivity F\factor 0%

Item	Oxide	Sulfide	Unit	Source
Mine operating costs - ore	2.00	3.25	US$/metric ton	*PEA study prepared by AAA Engineering Inc., dated July 15, 2016*
Mine operating costs - waste	1.00	2.00	US$/metric ton	*PEA study prepared by AAA Engineering Inc., dated July 15, 2016*
Processing operating costs	4.97	7.40	US$/metric ton	*PEA study prepared by AAA Engineering Inc., dated July 15, 2016*
Labor - metallurgy and production	0.70	0.70	US$/metric ton	*PEA study prepared by AAA Engineering Inc., dated July 15, 2016*
Labor - maintenance	0.25	0.25	US$/metric ton	*PEA study prepared by AAA Engineering Inc., dated July 15, 2016*
Power	1.30	1.90	US$/metric ton	*PEA study prepared by AAA Engineering Inc., dated July 15, 2016*
Maintenance materials	0.35	0.42	US$/metric ton	*PEA study prepared by AAA Engineering Inc., dated July 15, 2016*
Reagents and consumables	2.12	3.88	US$/metric ton	*PEA study prepared by AAA Engineering Inc., dated July 15, 2016*
Miscellaneous	0.25	0.25	US$/metric ton	*PEA study prepared by AAA Engineering Inc., dated July 15, 2016*
General and administration costs	0.89	0.89	US$/metric ton	*PEA study prepared by AAA Engineering Inc., dated July 15, 2016*

Base Case Capital Expenditures and Sustaining Capital (US$)

Sensitivity factor 0% — *PEA study prepared by AAA Engineering Inc., dated July 15, 2016*

Mining Equipment

		Year 0	Year 1	Year 2	Year 3	Year 4	Year 5	Year 6
Primary		-	10,500,000	5,000,000	9,600,000	1,499,000	-	940,000
Secondary		-	4,400,000	2,110,000	900,000	45,000	-	1,900,000
Auxiliary		-	1,800,000	1,180,000	1,180,000	-	-	14,000
Contingency	15%	-	2,500,000	1,250,000	1,800,000	230,000	-	440,000
Total		-	19,200,000	9,540,000	13,480,000	1,774,000	-	3,294,000

Plant and Infrastructure

	Oxide - Incremental	Sulfide - Incremental
	Year 0: 60%	Year 1: 50%
	Year 1: 40%	Year 2: 50%

		Oxide - Incremental	Sulfide - Incremental	Total Plant and Infrastructure
Process plant equipment		17,900,000	7,100,000	25,000,000
Factored construction commodities		20,100,000	7,900,000	28,000,000
Total plant costs		38,000,000	15,000,000	53,000,000
Infrastructure		25,000,000	-	25,000,000
Subtotal direct costs		63,000,000	15,000,000	78,000,000
Contingency	25%	15,750,000	3,750,000	19,500,000
Total direct costs		78,750,000	18,750,000	97,500,000
Indirect costs		19,200,000	4,800,000	24,000,000
Owners costs		4,200,000	2,100,000	6,300,000
Total plant and infrastructure		102,150,000	25,650,000	127,800,000

Depreciation Assumptions - Annual Depreciation Rates

Category	Description	Annual Rate	Lifetime
1	Amortized development costs	10%	10
2	Buildings	20%	5
3	Machinery and equipment	10%	10
4	General vehicles	20%	5

Published regulations - www.example.com

Tax Assumptions

Item	Value	Source
Income tax rate	25.0%	*Published regulations - www.example.com*
Social Contribution Tax	9.0%	
Tax loss carry forward	30.0%	

Working Capital Assumptions

Item	Value	Source
Accounts receivable - months	3	*Assumed values, based on company's experience*
Accounts payable (reagents, consumables) - months	2	
Accounts payable (labor) - months	1	

FIGURE 20.3 Example of an input and assumptions page

input and assumptions page. For this particular model, the input page included the sensitivity analysis reference cells.

Another method of completing the sensitivity analysis would be to take advantage of the data table functionality within Microsoft Excel. Creating a data table in Excel allows for viewing multiple outcomes in a single table. This can be used to view a range of sensitivities for several variables. The use of data tables makes it simpler to generate a sensitivity spider diagram. As with the single-value sensitivity analysis, a reference cell is necessary and the spreadsheet calculations must take into account the reference cell.

Worksheet Functions

Spreadsheet programs, such as Excel, have powerful built-in functions that allow for a wide variety of functionality and calculation capacity. The use of these functions and macros should be used within the limitations of the key principles of spreadsheet modeling discussed earlier. Although the use of Excel functions can make a spreadsheet appear sophisticated, often these functions can make the spreadsheet unnecessarily complicated and can result in calculation errors. They should be used with caution and, in general, their use should be minimized.

It is also necessary to carefully research how the function works. For example, in Excel, the NPV function is useful for calculating NPV; however, the Excel NPV function is based on cash flows occurring at the end of the time period. If the first cash flow occurs at the beginning of the first year, it must be added to the NPV value and not included in the value arguments included in the NPV formula.

Error Checks

The cash-flow model can be built to include error-checking formulas to make the model self-auditing. This might include checking summations from different data ranges or comparing totals and averages.

REFERENCES

Smith, L.D. 2011. Discounted cash flow analysis methodology and discount rates. http://web.cim.org/mes/pdf/VALDAYLarrySmith.pdf.

Stermole, F.J., and Stermole, J.M. 1987. *Economic Evaluation and Investment Decision Methods*, 6th ed. Golden, CO: Investment Evaluations Corporation.

Index

Note: *t.* indicates table; *f.* indicates figure; n indicates note

Q

R

S

V

W

X